# Mastering MATLAB® 5

*A Comprehensive Tutorial and Reference*

## The MATLAB® Curriculum Series

Dabney/Harman, *Mastering Simulink 2*
0-13-243767-8 © 1998

Dabney/Harman, *The Student Edition of Simulink: Dynamic System Simulation for MATLAB User's Guide* 0-13-659699-1 © 1998

Etter, *Engineering Problem Solving with MATLAB, 2/e*
0-13-397688-2 © 1997

Garcia, *Numerical Methods for Physics*
0-13-151986-7 © 1994

Hanselman/Littlefield, *Mastering MATLAB 5: A Comprehensive Tutorial and Reference*
0-13-858366-8 © 1998

Hanselmann/Littlefield, *The Student Edition of MATLAB, Version 5 User's Guide*
0-13-272550-9 © 1997

Marcus, *Matrices and MATLAB: A Tutorial*
0-13-562901-2 © 1993

McClellan, et. al., *Computer-Based Exercises for Signal Processing Using MATLAB 5*
0-13-789009-5 © 1998

McClellan/Schafer/Yoder, *DSP First: A Multimedia Approach*
0-13-243171-8 © 1998

Ogata, *Solving Control Engineering Problems with MATLAB*
0-13-182213-6 © 1994

Polking, *Ordinary Differential Equations Using MATLAB*
0-13-133944-3 © 1995

Roughgarden, *The Primer of Ecological Theory*
0-13-442062-4 © 1997

Van Loan, *Introduction to Scientific Computing*
0-13-125444-8 © 1997

# Mastering MATLAB® 5

## *A Comprehensive Tutorial and Reference*

**Duane Hanselman**
**Bruce Littlefield**
Department of Electrical and Computer Engineering,
University of Maine

The MATLAB® Curriculum Series

Prentice Hall
Upper Saddle River, New Jersey 07458

Library of Congress Cataloging-in-Publication Data

Hanselman, Duane C.
Mastering MATLAB 5: a comprehensive tutorial and reference / Duane Hanselman, Bruce Littlefield.
p. cm.
Includes bibliographical references and index.
ISBN 0-13-858366-8
1. MATLAB 2. Numerical analysis—Data processing.
I. Littlefield, Bruce. II. Title.

QA297.H296 1998
519.4'0285'53042--dc21 97-47330
CIP

Acquisitions editor: **TOM ROBBINS**
Editor-in-chief: **MARCIA HORTON**
Production editor: **IRWIN ZUCKER**
Managing editor: **BAYANI MENDOZA DE LEON**
Director of production and manufacturing: **DAVID W. RICCARDI**
Copy editor: **PATRICIA DALY**
Cover director: **JAYNE CONTE**
Manufacturing buyer: **JULIA MEEHAN**
Editorial assistant: **NANCY GARCIA**

Simon & Schuster / A Viacom Company
Upper Saddle River, New Jersey 07458



Printed in the United States of America

10 9 8 7

ISBN 0-13-858366-8

Prentice-Hall International (UK) Limited, *London*
Prentice-Hall of Australia Pty. Limited, *Sydney*
Prentice-Hall Canada Inc., *Toronto*
Prentice-Hall Hispanoamericana, S.A., *Mexico*
Prentice-Hall of India Private Limited, *New Delhi*
Prentice-Hall of Japan, Inc., *Tokyo*
Simon & Schuster Asia Pte. Ltd., *Singapore*
Editora Prentice-Hall do Brasil, Ltda., *Rio de Janeiro*

# Contents

**PREFACE** **xvii**

**1 GETTING STARTED** **1**

1.1 Introduction 1

1.2 Typographical Conventions 2

1.3 What's New in MATLAB 5 2

**2 BASIC FEATURES** **5**

2.1 Simple Math 5

2.2 The MATLAB Workspace 7

2.3 About Variables 8

2.4 Comments, Punctuation, and Aborting Execution 10

2.5 Complex Numbers 12

2.6 Mathematical Functions 14

3 THE COMMAND WINDOW 19

3.1 Managing the MATLAB Workspace 19

3.2 Number Display Formats 22

3.3 Command Window Control 22

3.4 System Information 23

4 SCRIPT M-FILES 25

5 FILE AND DIRECTORY MANAGEMENT 31

5.1 The MATLAB Workspace 31

5.2 Saving, Loading, and Deleting Variables 32

5.3 Special-Purpose File I/O 34

5.4 Low-Level File I/O 35

5.5 Disk File Manipulation 36

5.6 The MATLAB Search Path 37

5.7 MATLAB at Startup 37

6 ARRAYS AND ARRAY OPERATIONS 39

6.1 Simple Arrays 39

6.2 Array Addressing or Indexing 40

6.3 Array Construction 42

6.4 Array Orientation 43

6.5 Scalar-Array Mathematics 44

6.6 Array-Array Mathematics 47

6.7 Standard Arrays 51

6.8 Array Manipulation 55

6.9 Subarray Searching 64

6.10 Array Manipulation Functions 65

6.11 Array Size 68

7 MULTIDIMENSIONAL ARRAYS 71

7.1 Array Construction 72

7.2 Array Mathematics and Manipulation 73

7.3 Array Size 78

8 RELATIONAL AND LOGICAL OPERATIONS 81

8.1 Relational Operators 82

8.2 Logical Operators 83

8.3 Relational and Logical Functions 85

8.4 NaNs and Empty Arrays 87

9 SET, BIT, AND BASE FUNCTIONS 91

9.1 Set Functions 91

9.2 Bit Functions 94

9.3 Base Conversions 95

10 CHARACTER STRINGS 97

10.1 String Construction 97

10.2 Numbers to Strings to Numbers 102

10.3 String Functions 107

10.4 Cell Arrays of Strings 109

11 TIME FUNCTIONS 113

11.1 Current Date and Time 113

11.2 Date Format Conversions 114

11.3 Date Functions 116

11.4 Timing Functions 117

11.5 Plot Labels 118

**12 CELL ARRAYS AND STRUCTURES 121**

12.1 Creating and Displaying Cell Arrays 122
12.2 Combining and Reshaping Cell Arrays 125
12.3 Retrieving Cell Array Contents 126
12.4 Comma-Separated Lists 129
12.5 Cell Arrays of Character Strings 129
12.6 Creating and Displaying Structures 129
12.7 Retrieving Structure Field Contents 131
12.8 Conversion and Test Functions 133

**13 CONTROL FLOW 135**

13.1 For Loops 135
13.2 While Loops 138
13.3 If-Else-End Constructions 139
13.4 Switch-Case Constructions 141

**14 FUNCTION M-FILES 143**

14.1 M-file Construction Rules 144
14.2 Input and Output Arguments 146
14.3 Function Workspaces 148
14.4 Function M-files and the MATLAB Search Path 151
14.5 Creating Your Own Toolboxes 153
14.6 Command–Function Duality 154
14.7 In-line Functions and `feval` 155

**15 DEBUGGING AND PROFILING TOOLS 159**

15.1 Debugging Tools 159
15.2 Profiling M-files 162

16 NUMERICAL LINEAR ALGEBRA 165

16.1 Sets of Linear Equations 165
16.2 Matrix Functions 169
16.3 Special Matrices 171
16.4 Sparse Matrices 171
16.5 Sparse Matrix Functions 174

17 DATA ANALYSIS 177

17.1 Basic Statistical Analysis 177
17.2 Basic Data Analysis 184
17.3 Data Analysis and Statistical Functions 187

18 POLYNOMIALS 189

18.1 Roots 189
18.2 Multiplication 190
18.3 Addition 191
18.4 Division 192
18.5 Derivatives 192
18.6 Evaluation 192
18.7 Rational Polynomials 193
18.8 Curve Fitting 195

19 INTERPOLATION 199

19.1 One-dimensional Interpolation 199
19.2 Two-dimensional Interpolation 204
19.3 Triangulation and Scattered Data 208

20 CUBIC SPLINES 213

20.1 Basic Features 213
20.2 Piecewise Polynomials 214

20.3 Integration 218

20.4 Differentiation 220

20.5 Spline Interpolation on a Plane 223

**21 FOURIER ANALYSIS 227**

21.1 Discrete Fourier Transform 228

21.2 Fourier Series 231

**22 OPTIMIZATION 237**

22.1 Zero Finding 238

22.2 Minimization in One Dimension 240

22.3 Minimization in Higher Dimensions 242

22.4 Practical Issues 245

**23 INTEGRATION AND DIFFERENTIATION 247**

23.1 Integration 247

23.2 Differentiation 253

**24 ORDINARY DIFFERENTIAL EQUATIONS 261**

24.1 Initial Value Problem Format 261

24.2 ODE Suite Solvers 262

24.3 Basic Use 263

24.4 ODE File Options 265

24.5 Solver Options 268

24.6 Finding Events 273

**25 OBJECT-ORIENTED PROGRAMMING 277**

25.1 Object Identification 278

25.2 Creating a Class 278

25.3 The Constructor 279

25.4 Object Precedence 280

25.5 Displaying Objects 281
25.6 Overloading Functions 281
25.7 Adding Stack Elements 284
25.8 Communicating Between Workspaces 285
25.9 Removing Stack Elements 286
25.10 Examining Stack Contents 288
25.11 Overloading Operators 290
25.12 Converter Functions 293
25.13 Inheritance 294

**26 2-D GRAPHICS 297**

26.1 The `plot` Function 297
26.2 Linestyles, Markers, and Colors 300
26.3 Plotting Styles 301
26.4 Plot Grids, Axes Box, and Labels 302
26.5 Customizing Plot Axes 305
26.6 Multiple Plots 308
26.7 Multiple Figures 310
26.8 Subplots 311
26.9 Interactive Plotting Tools 312
26.10 Screen Updates 314
26.11 Specialized 2-D Plots 316
26.12 Quick Plots 325
26.13 Text Formating 328

**27 3-D GRAPHICS 331**

27.1 Line Plots 331
27.2 Scalar Functions of Two Variables 334
27.3 Mesh Plots 337
27.4 Surface Plots 339

27.5 Mesh and Surface Plots of Irregular Data 346

27.6 Changing Viewpoints 348

27.7 Contour Plots 351

27.8 Specialized 3-D Plots 356

**28 USING COLOR AND LIGHT 365**

28.1 Plotting Styles 366

28.2 Understanding Colormaps 366

28.3 Using Colormaps 368

28.4 Displaying Colormaps 368

28.5 Creating and Altering Colormaps 371

28.6 Using More Than One Colormap 374

28.7 Using Color to Describe a Fourth Dimension 375

28.8 Lighting Models 379

**29 IMAGES, MOVIES, AND SOUND 387**

29.1 Images 387

29.2 Image Formats 389

29.3 Image Files 390

29.4 Movies 394

29.5 Imaging Utilities 396

29.6 Sound 396

**30 PRINTING AND EXPORTING GRAPHICS 399**

30.1 Printing from the Menu 399

30.2 Positioning and Resizing Graphics 400

30.3 Printing from the Command Line 400

30.4 Selecting a Device Driver 401

30.5 Additional Device Drivers 402

30.6 Other Printing Options 405

30.7 Changing Defaults 406

30.8 Exporting Images 407

30.9 Application Notes 407

**31 HANDLE GRAPHICS 409**

31.1 Who Needs Handle Graphics? 410

31.2 Objects 410

31.3 Object Handles 411

31.4 Object Properties 412

31.5 The Universal Functions `get` and `set` 413

31.6 Finding Objects 420

31.7 Selecting Objects with the Mouse 425

31.8 Position and Units 427

31.9 Printing Figures 428

31.10 Default Properties 429

31.11 Common Properties 432

31.12 New Plots 434

31.13 M-File Examples 435

31.14 Callbacks 439

31.15 Summary 442

**32 GRAPHICAL USER INTERFACES 445**

32.1 GUI?: What Is a GUI? 445

32.2 Who Should Create GUIs—And Why? 446

32.3 GUI Object Hierarchy 447

32.4 Menus 448

32.5 Controls 459

32.6 Programming and Callback Considerations 470

32.7 Pointer and Mouse Button Events 480

32.8 Rules for Interrupting Callbacks 482

32.9 M-file Examples 483

32.10 Utility Functions 494

32.11 GUIDE 497

32.12 User-Contributed GUI M-files 498

32.13 Summary 498

**33 DIALOG BOXES AND REQUESTERS 499**

33.1 Dialog Boxes 499

33.2 Requesters 503

33.3 Utility Functions 509

33.4 Maintaining Focus 510

**34 HELP 511**

34.1 Command Window Help 511

34.2 The Help Window 515

34.3 The Help Desk 519

**35 INTERNET RESOURCES 521**

35.1 MathWorks Web Site 521

35.2 Other MathWorks Resources 522

35.3 Other Network Resources 523

35.4 Internet E-mail and Network Addresses 524

**36 THE MASTERING MATLAB TOOLBOX 527**

**APPENDICES**

**A: MATLAB FUNCTION LISTING 537**

**B: *AXES* OBJECT PROPERTIES 571**

**C: *FIGURE* OBJECT PROPERTIES 581**

**D: *IMAGE* OBJECT PROPERTIES 589**

E: *LIGHT* OBJECT PROPERTIES 593

F: *LINE* OBJECT PROPERTIES 597

G: *PATCH* OBJECT PROPERTIES 601

H: *ROOT* OBJECT PROPERTIES 607

I: *SURFACE* OBJECT PROPERTIES 611

J: *TEXT* OBJECT PROPERTIES 617

K: *UICONTROL* OBJECT PROPERTIES 623

L: *UIMENU* OBJECT PROPERTIES 629

INDEX 633

# Preface

This text is about MATLAB. If you use MATLAB or are considering using MATLAB, this text is for you. It represents an alternative to learning MATLAB on your own with or without the help of the documentation that comes with MATLAB. The informal style of this text makes it easy to read. As the title suggests, this text provides the tools you need to master MATLAB. As a programming language and data visualization tool, MATLAB offers a rich set of capabilities to solve problems in engineering, scientific, computing, and mathematical disciplines. The fundamental goal of this text is to help you increase your productivity by showing you how to use these capabilities efficiently. Because of the interactive nature of MATLAB, the material is generally presented in the form of examples that you can duplicate by running MATLAB as you read this text.

This text covers only those topics that are of use to a general audience. The material presented applies equally to all computer platforms, including Unix workstations, Macintoshes, and PCs. None of the toolboxes that are available at additional cost are discussed in depth, although some are referred to in appropriate places. Moreover, aspects of MATLAB that are machine dependent are not discussed, such as the writing of MEX-files.

Since MATLAB continues to evolve as a software tool, this text focuses on MATLAB version 5. For the most part, the material applies to 4.X versions of MATLAB as well. When appropriate, distinctions between versions are made.

We encourage you to give us feedback on this text. What are the best features of the text? What areas need more work? What topics should be left out? What topics should be added? We can be reached at the following e-mail address:

`mm@eece.maine.edu.`

## ACKNOWLEDGMENTS

We would like to thank the folks at The Mathworks Inc., including Liz, Peter, Jim, Naomi, and Cleve, for their cooperation and assistance in this project as well as with our earlier project, *The Student Edition of* MATLAB: *Version 5, User's Guide.*

We also acknowledge the support and input provided by the faculty, staff, and students of Department of Electrical and Computer Engineering at the University of Maine.

I, Bruce Littlefield, would like to dedicate this text to my parents Howard and Minerva Littlefield for showing me the way, and to my wife Hazel for companionship and support on the journey.

I, Duane Hanselman, would like to acknowledge the love and dedication of my wife Pamela and that of our children, Ruth, Sarah, and Kevin. MATLAB may be a great tool, and writing can be very rewarding, but nothing beats the love a family shares.

# 1

# Getting Started

## 1.1 INTRODUCTION

This text assumes that you have some familiarity with matrices and computer programming. Matrices and arrays in general are at the heart of MATLAB since all data in MATLAB are stored as arrays. Besides common matrix algebra operations, MATLAB offers array operations that allow one to manipulate sets of data in a wide variety of ways quickly. In addition to its matrix orientation, MATLAB offers programming features similar to those of other computer programming languages. Finally, MATLAB offers graphical user interface (GUI) tools that allow one to use MATLAB as an application development tool. This combination of array data structures, programming features, and GUI tools makes MATLAB an extremely powerful tool for solving problems in many fields. In this text, each of these aspects of MATLAB will be discussed in detail. To facilitate learning, detailed examples are presented. Many of these examples illustrate construction of M-file functions in the *Mastering MATLAB Toolbox* that accompanies this text.

## 1.2 TYPOGRAPHICAL CONVENTIONS

The following conventions are used throughout this text:

| | |
|---|---|
| ***Bold Italics*** | New terms or important facts |
| Boxed Text | Important terms and facts |
| **Bold Initial Caps** | Keyboard key names, menu names, and menu items |
| `Constant Width` | User input, function and file names, commands, and screen displays |
| `Boxed Constant Width` | Contents of a script, function, or data file |
| *`Constant Width Italics`* | User input that is to be replaced and not taken literally such as » `help` *`functionname`* |
| *Italics* | Window names, book titles, toolbox names, company names, example text, and mathematical notation. |

## 1.3 WHAT'S NEW IN MATLAB 5

This text covers MATLAB version 5. While version 5 has numerous new features, the progression from version 4 to version 5 is more evolutionary than revolutionary. For the most part, M-files created in version 4 should work in version 5 with little or no changes. The biggest changes in version 5 that will require modifying M-files are in the area of Handle Graphics. Warning messages, as opposed to error messages, may be displayed in the *Command* window when you run version 4 M-files under version 5, but they should still work. Warning messages are a new feature in MATLAB. They are used to point out poor programming practices that were not flagged in earlier versions and to point out changes that you should make, but do not necessarily have to make.

If you are wondering what's new in MATLAB 5, here is a brief introduction to new features that are discussed in this text:

- Data types other than double-precision arrays are now supported. Character strings now use only 2 bytes per character. Cell arrays and structures can now be created to group related data together for better data organization and manipulation.
- MATLAB 5 introduces object-oriented programming features whereby you can create your own objects that respond to their own methods. Support is provided to overload most of the syntax of MATLAB and to override any function.

- Arrays can now extend into an arbitrary number of dimensions. You are no longer limited to two-dimensional matrices.
- Function M-files can now have an arbitrary number of input and output arguments.
- The performance of M-files can now be assessed with an integrated M-file profiler.
- The profiler allows you to determine quickly and accurately which lines in your M-files take the most time.
- It is no longer necessary to compute the last element in an array to access it. That is, for vectors `x(length(x))` is the same as `x(end)` and for two-dimensional arrays `A(size(A,1),size(A,2))` is the same as `A(end,end)`.
- Switch-Case constructions have been added to simplify parsing.
- If statements now terminate as soon as possible, rather than performing all tests. For example, `if (isempty(x) | y==0)` does not evaluate `y==0` if `isempty(x)` is True. Likewise, `if (isempty(x) & y==0)` does not evaluate `y==0` if `isempty(x)` is False.
- Scalar expansion has been added so that `A(:,1:5) = pi*ones(size(A,1),5)` can be replaced by `A(:,1:5) = pi`.
- The MATLAB parser has been improved to warn you about the existence of uninitialized variables, noninteger subscripts, multiple `end` statements, and other common programming mistakes.
- There is now a `warning` function that you can use to display warning messages that are not fatal, like calls to the `error` function. In addition, warning messages can be turned off and on.
- M-files can now be stored in P-Code format, which is MATLAB's internal compiled representation. In doing so, initial calls to functions execute much faster since the compile step is avoided.
- Function M-files can now contain more than one function. Appended subfunctions can only be called by other functions in the M-file.
- Directories of M-files can now have subdirectories named `private`, where additional M-files can be placed that are only accessible by functions in the immediate parent directory.
- Empty arrays can now have one or more nonzero dimensions. For example, `zeros(5,0)` creates a 5-by-0 empty column vector.
- The functions `min` and `max` now ignore `NaNs`.
- Default *Figure* window colors and plot attributes have changed to improve visualization.
- The `set` and `get` commands have been vectorized.
- Graphics rendering can be done with either the new Z-buffer renderer or the original Painter's algorithm renderer.
- *Light* is now a child object of *Figures* with its own set of property names and values.

- Multiline text and limited support of LaTeX is provided for graphics annotation.
- Graphics viewing can now be done based on the manipulation of Camera properties, which allow fly-bys and other animation features.
- Listboxes have been added to the set of *uicontrols,* making multiple selections from a list possible.
- Dialog boxes can now be modal. That is, they can force the user to make a choice before any other action is allowed.
- The font characteristics of *uicontrols* can now be specified.
- Twenty-four-bit color is now supported in addition to the older 8-bit RGB model.
- Delaunay triangulation functions have been added to support analysis of scattered data.
- Support is now provided for bitwise manipulation of integers.
- A complete suite of ordinary differential equation solvers now exists to solve a wide range of initial value problems.
- Computations with dates and time are now supported.
- Set arithmetic on arrays can now be performed.
- Numerous new plotting functions have been introduced.

# 2

# Basic Features

Running MATLAB creates one or more windows on your computer monitor. Of these the ***Command*** window is the primary place where you interact with MATLAB. The prompt » is displayed in the *Command* window, and when the *Command* window is active, a blinking cursor should appear to the right of the prompt. This cursor and the MATLAB prompt signify that MATLAB is waiting to perform a mathematical operation.

## 2.1 SIMPLE MATH

Just like a calculator, MATLAB can do simple math. Consider the following simple example: Mary goes to the office-supply store and buys 4 erasers at 25 cents each, 6 memo pads at 52 cents each, and 2 rolls of tape at 99 cents each. How many items did Mary buy, and how much did they cost? To solve this using your calculator, you enter

$$4 + 6 + 2 = 12 \text{ items}$$

$$4 \cdot 25 + 6 \cdot 52 + 2 \cdot 99 = 610 \text{ cents}$$

In MATLAB, this can be solved in a number of different ways. First, the aforementioned calculator approach can be taken:

```
» 4+6+2
ans =
    12
» 4*25 + 6*52 + 2*99
ans =
   610
```

Note that MATLAB does not care about spaces for the most part, and that multiplication takes precedence over addition. Note also that MATLAB calls the result `ans`, which is short for *answer* for both computations.

As an alternative, the preceding problem can be solved by storing information in ***MATLAB variables:***

```
» erasers = 4
erasers =
     4
» pads = 6
pads =
     6
» tape = 2;
» items = erasers + pads + tape
items =
    12
» cost = erasers*25 + pads*52 + tape*99
cost =
   610
```

Here we created three MATLAB variables—`erasers`, `pads`, and `tape`—to store the number of each item. After entering each statement, MATLAB displayed the results, except in the case of `tape`. The semicolon at the end of the line tells MATLAB to evaluate the line but not to display the answer. Finally, rather than calling the results `ans`, we told MATLAB to call the number of items purchased `items` and the total price paid `cost`. At each step, MATLAB remembered past information. Because MATLAB remembers things, let's ask what the average cost per item was:

```
» average_cost = cost/items
average_cost =
    50.833
```

Because *average cost* is two words and MATLAB variable names must be one word, an underscore was used to create the single MATLAB variable `average_cost`.

In addition to addition and multiplication, MATLAB offers the following basic arithmetic operations:

| Operation | Symbol | Example |
|---|---|---|
| addition, a + b | + | 3 + 22 |
| subtraction, a − b | − | 90 − 54 |
| multiplication, a · b | * | 3.14 * 0.85 |
| division, a ÷ b | / or \ | 56/8 = 8\56 |
| exponentiation, $a^b$ | ^ | 2 ^ 8 |

The order in which these operations are evaluated in a given expression is given by the usual rules of precedence, which can be summarized as follows:

> Expressions are evaluated from left to right with the exponentiation operation having the highest precedence, followed by multiplication and division having equal precedence, followed by addition and subtraction having equal precedence.

Parentheses can be used to alter this ordering, in which case these rules of precedence are applied within each set of parentheses starting with the innermost set and proceeding outward.

## 2.2 THE MATLAB WORKSPACE

As you work in the *Command* window, MATLAB remembers the commands you enter as well as the values of any variables you create. These commands and variables are said to reside in the ***MATLAB workspace*** or ***Base workspace*** and can be recalled whenever you wish. For example, to check the value of `tape` all you have to do is ask MATLAB for it by entering its name at the prompt:

```
» tape
tape =
      2
```

If you cannot remember the name of a variable, you can ask MATLAB for a list of the variables it knows by using the MATLAB command `who`:

```
» who
Your variables are:

ans             cost       items    tape
average_cost    erasers    pads
```

Note that MATLAB does not tell you the value of all the variables; it merely gives you their names. To find their values, you must enter their names at the MATLAB prompt.

To recall previous commands, MATLAB uses the **Cursor** keys, ←, ↑, →, ↓, on your keyboard. For example, pressing the ↑ key once recalls the most recent command to the MATLAB prompt. Repeated pressing scrolls back through prior commands one at a time. In a similar manner, pressing the ↓ key scrolls forward through commands. Pressing the ← or → keys moves one within a given command at the MATLAB prompt, thereby allowing it to be edited in much the same way you would edit text in a word processing program. Other standard editing keys, such as **Delete** or **Backspace, Home,** and **End,** perform their commonly assigned tasks. Once a scrolled command is acceptable, pressing the **Return** key with the cursor *anywhere* in the command tells MATLAB to process it. Finally, and perhaps most useful, the **Escape** key erases the current command at the prompt. For those of you familiar with the EMACS editor, MATLAB also accepts common EMACS editing control character sequences such as **Ctrl-U** to erase the current command.

## 2.3 ABOUT VARIABLES

Like any other computer language, MATLAB has rules about variable names. Earlier it was noted that variable names must be a single word containing no spaces. More specifically, MATLAB variable-naming rules are as follows:

| Variable Naming Rules | Comments/Examples |
|---|---|
| Variable names are case sensitive. | `Cost`, `cost`, `CoSt`, and `COST` are all different MATLAB variables. |
| Variable names can contain up to 31 characters. Any characters beyond the thirty-first are ignored. | `howaboutthisvariablename` |
| Variable names must start with a letter, followed by any number of letters, digits, or underscores. Punctuation characters are not allowed, since many of them have special meaning to MATLAB. | `how_about_this_variable_name`<br>`X51483`<br>`a_b_c_d_e` |

In addition to these naming rules, MATLAB has several special variables. They are as follows:

| Special Variables | Description |
|---|---|
| `ans` | Default variable name used for results. |
| `pi` | Ratio of the circumference of a circle to its diameter. |
| `eps` | Smallest number such that, when added to one, creates a number greater than one on the computer. |

| Special Variables | Description |
| --- | --- |
| `flops` | Count of floating-point operations. |
| `inf` | Stands for infinity (e.g., 1/0). |
| `NaN` (or) `nan` | Stands for Not-a-Number (e.g., 0/0). |
| `i` (and) `j` | `i = j =` $\sqrt{-1}$. |
| `nargin` | Number of function input arguments used. |
| `nargout` | Number of functions output arguments used. |
| `realmin` | Smallest usable positive real number. |
| `realmax` | Largest usable positive real number. |

If you reuse a variable such as `tape` in the preceding example, or assign a value to one of the special variables, its prior value is overwritten and lost. However, any other expressions computed using the prior values do not change. For example,

```
» erasers = 4;
» pads = 6;
» tape = 2;
» items = erasers + pads + tape
items =
    12
» erasers = 6
erasers =
     6
» items
items =
    12
```

Here, using the first example again, we found the number of items Mary purchased. Afterward, we changed the number of erasers to 6, overwriting its prior value of 4. In doing so, the value of `items` has not changed. Unlike a common spreadsheet program, MATLAB does not recalculate the number of items based on the new value of `erasers`. ***When MATLAB performs a calculation, it does so using the values it knows at the time the requested command is evaluated.*** In the preceding example, if you wish to recalculate the number of items, the total cost, and the average cost, it is necessary to recall the appropriate MATLAB commands and ask MATLAB to evaluate them again.

The special variables given previously follow this guideline also, with the exception that the special values can be restored. When you start MATLAB, they have the values given previously; if you change their values, the original special values are lost. To restore the special value, all you have to do is ***clear*** the overwritten value. For example,

```
» pi
ans =
        3.1416
» pi = 1.23e-4
pi =
        0.000123
» clear pi
» pi
ans =
        3.1416
```

shows that `pi` has the special value 3.1416 to five significant digits, is overwritten with the value 1.23e-4, and then, after being cleared using the `clear` function, has its special value once again.

## 2.4 COMMENTS, PUNCTUATION, AND ABORTING EXECUTION

As we saw earlier, placing a semicolon at the end of a command suppresses printing of the computed results. This fact is especially useful for suppressing the results of intermediate calculations. For example,

```
» erasers
erasers =
     6
» items = erasers + pads + tape;
» cost = erasers*25 + pads*52 + tape*99;
» average_cost = cost/items
average_cost =
        47.143
```

displays the average cost of the items Mary purchased if she purchased six erasers rather than the original four. The intermediate results `items` and `cost` were not printed because semicolons appear at the end of the commands defining them.

In addition to semicolons, MATLAB uses other punctuation symbols as well. All text after a percent sign (%) is taken as a comment statement. For example,

```
» tape = 2 % number of roles of tape purchased
```

The variable `tape` is given the value 2, and MATLAB simply ignores the percent sign and all text following it.

Multiple commands can be placed on one line if they are separated by commas or semicolons. For example,

```
» erasers = 6, pads = 6; tape = 2
erasers =
     6
tape =
     2
```

Commas tell MATLAB to display results; semicolons suppress printing.

Sometimes expressions or commands get so long that is convenient to continue them onto additional lines. In MATLAB, statement continuation is denoted by three periods in succession, such as

```
» average_cost = cost/items % command as done earlier
average_cost =
       47.143
» average_cost = cost/... % command with valid continuation
items
average_cost =
       47.143
»average_cost = cost... % command with valid continuation
/items
average_cost =
       47.143
» average_cost = cost/it... % command with INvalid continuation
ems
??? age_cost=cost/items
                      |
Missing operator, comma, or semi-colon.
```

As shown, statement continuation works if the three periods appear between variable names and mathematical operators, but not in the middle of a variable name. That is, variable names cannot be split between two lines. In addition, since comment lines are ignored, they cannot be continued either. For example,

```
» % Comments cannot be continued...
» either
??? Undefined function or variable 'either'.
```

In this case the . . . in the comment line is part of the comment and not processed by MATLAB.

Finally, MATLAB processing can be interrupted at any time by pressing **Ctrl-C** (pressing the **Ctrl** and **C** keys simultaneously) on a PC or workstation. Pressing ⌘–.(pressing the ⌘ and period keys simultaneously) on a Macintosh does the same thing.

## 2.5 COMPLEX NUMBERS

One of the most powerful features of MATLAB is that it does not require any special handling for complex numbers. Complex numbers are formed in MATLAB in several ways. Examples of complex numbers include the following:

```
» c1 = 1-2i  % the appended i signifies the imaginary part
c1 =
   1.0000 - 2.0000i
» c1 = 1-2j % j also works
c1 =
   1.0000 - 2.0000i
» c2 = 3*(2-sqrt(-1)*3)
c2 =
   6.0000 - 9.0000i
» c3 = sqrt(-2)
c3 =
        0 + 1.4142i
» c4 = 6+sin(.5)*i
c4 =
   6.0000 + 0.4794i
» c5 = 6+sin(.5)*j
c5 =
   6.0000 + 0.4794i
```

In the last two examples, the MATLAB default values of `i=j=`$\sqrt{-1}$ are used to form the imaginary part. Multiplication by `i` or `j` is required in these cases since `sin(.5)i` and `sin(.5)j` have no meaning to MATLAB. Termination with the characters `i` and `j`, as shown in the first two examples, only works with numbers, not expressions.

Some programming languages require special handling for complex numbers wherever they appear. In MATLAB no special handling is required. Mathematical operations on complex numbers are written the same as those for real numbers:

```
» c6 = (c1+c2)/c3  % from the above data
c6 =
  -7.7782 - 4.9497i
» c6r = real(c6)
c6r =
     -7.7782
» c6i = imag(c6)
c6i =
     -4.9497
» check_it_out = i^2  % sqrt(-1) squared must be -1!
check_it_out =
  -1
```

In general, operations on complex numbers lead to complex numbers. However, in the last case in the preceding code, MATLAB is smart enough to drop the zero imaginary part of the result. In addition, the preceding shows that the functions `real` and `imag` extract the real and imaginary parts of a complex number, respectively.

As a final example of complex arithmetic, consider the Euler (sounds like Oiler) identity, which relates the polar form of a complex number to its rectangular form:

$$M \angle \theta \equiv M \cdot \mathrm{e}^{\mathrm{j}\theta} = a + b\mathrm{j}$$

where the polar form is given by a magnitude $M$ and an angle $\theta$, and the rectangular form is given by $a + b\mathrm{j}$. The relationships among these forms are as follows:

$$M = \sqrt{a^2 + b^2}$$

$$\theta = \tan^{-1}(b/a)$$

$$a = M \cdot \cos(\theta)$$

$$b = M \cdot \sin(\theta)$$

In MATLAB, the conversion between polar and rectangular forms makes use of the functions `real`, `imag`, `abs`, and `angle`:

```
» c1
c1 =
   1.0000 - 2.0000i
» mag_c1 = abs(c1) % magnitude
mag_c1 =
    2.2361
» angle_c1 = angle(c1) % angle in radians
angle_c1 =
   -1.1071
» deg_c1 = angle_c1*180/pi % angle in degrees
deg_c1 =
  -63.4349
» real_c1 = real(c1) % real part
real_c1 =
     1
» imag_c1 = imag(c1) % imaginary part
imag_c1 =
    -2
```

The MATLAB function `abs` computes the magnitude of complex numbers or the absolute value of real numbers, depending on which one you give it. Likewise, the MATLAB function `angle` computes the angle of a complex number in radians.

## 2.6 MATHEMATICAL FUNCTIONS

Lists of the functions that MATLAB supports are shown in the following tables. Most of these functions are used the same way you would write them mathematically:

```
» x = sqrt(2)/2
x =
    0.7071
» y = asin(x)
y =
    0.7854
» y_deg = y*180/pi
y_deg =
   45.0000
```

These commands find the angle where the sine function has a value of $\sqrt{2}/2$. Note that MATLAB only works in radians. Other examples include the following:

```
» y = sqrt(3^2 + 4^2) % show 3-4-5 right triangle relationship
y =
    5
» y = rem(23,4)  % remainder function, 23/4 has a remainder of 3
y =
    3
» x = 2.6, y1 = fix(x), y2 = floor(x), y3 = ceil(x), y4 = round(x)
x =
    2.6000
y1 =
    2
y2 =
    2
y3 =
    3
y4 =
    3
```

| Trigonometric Function | Description |
|---|---|
| acos | Inverse cosine. |
| acosh | Inverse hyperbolic cosine. |
| acot | Inverse cotangent. |
| acoth | Inverse hyperbolic cotangent. |
| acsc | Inverse cosecant. |

| Trigonometric Function | Description |
|---|---|
| acsch | Inverse hyperbolic cosecant. |
| asec | Inverse secant. |
| asech | Inverse hyperbolic secant |
| asin | Inverse sine. |
| asinh | Inverse hyperbolic sine. |
| atan | Inverse tangent. |
| atan2 | Four quadrant inverse tangent. |
| atanh | Inverse hyperbolic tangent. |
| cos | Cosine. |
| cosh | Hyperbolic cosine. |
| cot | Cotangent. |
| coth | Hyperbolic cotangent. |
| csc | Cosecant. |
| csch | Hyperbolic cosecant. |
| sec | Secant. |
| sech | Hyperbolic secant. |
| sin | Sine. |
| sinh | Hyperbolic sine. |
| tan | Tangent. |
| tanh | Hyperbolic tangent. |

| Exponential Function | Description |
|---|---|
| exp | Exponential. |
| log | Natural logarithm. |
| log10 | Base 10 logarithm. |
| log2 | Base 2 logarithm and floating-point dissection. |
| pow2 | Base 2 power and floating-point scaling. |

| Exponential Function | Description |
|---|---|
| `sqrt` | Square root. |
| `nextpow2` | Next higher power of 2. |

| Complex Function | Description |
|---|---|
| `abs` | Absolute value or magnitude. |
| `angle` | Phase angle in radians. |
| `conj` | Complex conjugate. |
| `imag` | Complex imaginary part. |
| `real` | Complex real part. |
| `unwrap` | Unwrap phase angle. |
| `isreal` | True for real arrays. |
| `cplxpair` | Sort vector into complex conjugate pairs. |

| Rounding and Remainder Function | Description |
|---|---|
| `fix` | Round toward zero. |
| `floor` | Round toward minus infinity. |
| `ceil` | Round toward plus infinity. |
| `round` | Round toward nearest integer. |
| `mod` | Modulus (signed remainder). |
| `rem` | Remainder after division. |
| `sign` | Signum function. |

| Coordinate Transformation Function | Description |
|---|---|
| `cart2sph` | Cartesian to spherical. |
| `cart2pol` | Cartesian to polar or cylindrical. |

| Coordinate Transformation Function | Description |
|---|---|
| pol2cart | Polar to cartesian. |
| sph2cart | Spherical to cartesian. |

| Number Theoretic Function | Description |
|---|---|
| factor | Prime factors. |
| isprime | True for prime numbers. |
| primes | Generate list of prime numbers. |
| gcd | Greatest common divisor. |
| lcm | Least common multiple. |
| rat | Rational approximation. |
| rats | Rational output. |
| perms | All possible combinations. |
| nchoosek | All combinations of *N* elements taken *K* at a time. |

| Specialized Mathematical Function | Description |
|---|---|
| airy | Airy functions. |
| besselj | Bessel function of the first kind. |
| bessely | Bessel function of the second kind. |
| besselh | Bessel function of the third kind. |
| besseli | Modified Bessel function of the first kind. |
| besselk | Modified Bessel function of the second kind. |
| beta | Beta function. |
| betainc | Incomplete beta function. |
| betaln | Logarithm of beta function. |

| Specialized Mathematical Function | Description |
|---|---|
| `ellipj` | Jacobi elliptic functions. |
| `ellipke` | Complete elliptic integral. |
| `erf` | Error function. |
| `erfc` | Complementary error function. |
| `erfcx` | Scaled complementary error function. |
| `erfinv` | Inverse error function. |
| `expint` | Exponential error function. |
| `gamma` | Gamma function. |
| `gammainc` | Incomplete gamma function. |
| `gammaln` | Logarithm of gamma function. |
| `legendre` | Associated Legendre function. |
| `cross` | Vector cross product. |
| `dot` | Vector dot product. |

# 3

# The Command Window

As described in the last chapter, the *Command* window is the primary place where one interacts with MATLAB. In this chapter more detailed information about the *Command* window is provided.

## 3.1 MANAGING THE MATLAB WORKSPACE

The data and variables created in the *Command* window reside in what is called the ***MATLAB workspace*** or ***Base workspace.*** To see what variable names exist in the MATLAB workspace, issue the command `who`:

```
» who
Your variables are:

angle_c1       c4            cost          pads
ans            c5            deg_c1        real_c1
average_cost   c6            erasers       tape
```

```
c1          c6i            imag_c1
c2          c6r            items
c3          check_it_out   mag_c1
```

The variables you see may differ from these depending on what you have asked MATLAB to do since starting the program. For more detailed information, use the comand `whos`:

```
» whos
Name              Size          Bytes Class

angle_c1          1x1               8 double array
ans               1x1               8 double array
average_cost      1x1               8 double array
c1                1x1              16 double array (complex)
c2                1x1              16 double array (complex)
c3                1x1              16 double array (complex)
c4                1x1              16 double array (complex)
c5                1x1              16 double array (complex)
c6                1x1              16 double array (complex)
c6i               1x1               8 double array
c6r               1x1               8 double array
check_it_out      1x1               8 double array
cost              1x1               8 double array
deg_c1            1x1               8 double array
erasers           1x1               8 double array
imag_c1           1x1               8 double array
items             1x1               8 double array
mag_c1            1x1               8 double array
pads              1x1               8 double array
real_c1           1x1               8 double array
tape              1x1               8 double array

Grand total is 21 elements using 216 bytes
leaving 47121232 bytes of memory free.
```

Here each variable is listed along with its size, the number of bytes used, and its class. Since MATLAB is array oriented, all the variables belong to the class of double precision arrays, even though all the preceding variables are scalars. Later, as other data types or classes are introduced, this last column will provide more useful information. In addition, depending on your computer platform the statement of free memory available may or may not appear.

In addition to the `who` and `whos` functions, the PC and Macintosh versions of MATLAB have a **Show Workspace** item in the **File** menu that opens a GUI called the *Workspace Browser,* which displays the same information as the `whos` command. This GUI is also displayed by pressing the **Workspace Browser** icon on button on the *Command* window toolbar.

The command `clear` deletes variables from the MATLAB workspace. For example,

```
» clear real_c1 imag_c1 c*
» who
Your variables are:

angle_c1        deg_c1        mag_c1
ans             erasers       pads
average_cost    items         tape
```

deletes the variables `real_c1`, `imag_c1`, and all variables starting with the letter `c`. Other options for the `clear` function can be found by asking for help:

```
» help clear
CLEAR Clear variables and functions from memory.
   CLEAR removes all variables from the workspace.
   CLEAR VARIABLES does the same thing.
   CLEAR GLOBAL removes all global variables.
   CLEAR FUNCTIONS removes all compiled M-functions.
   CLEAR MEX removes all links to MEX-files.
   CLEAR ALL removes all variables, globals, functions, and MEX links.

   CLEAR VAR1 VAR2...clears the variables specified. The wildcard
   character '*' can be used to clear variables that match a pattern.
   For instance, CLEAR X* clears all the variables in the current
   workspace that start with X.

   If X is global, CLEAR X removes X from the current workspace,
   but leaves it accessible to any functions declaring it global.
   CLEAR GLOBAL X completely removes the global variable X.

   CLEAR ALL also has the side effect of removing all debugging
   breakpoints since the breakpoints for a file are cleared whenever
   the m-file changes or is cleared.

   Use the functional form of CLEAR, such as CLEAR('name'),
   when the variable name is stored in a string.
   See also WHO, WHOS.
```

Clearly, the `clear` command deletes more than just variables. Its other uses will become apparent as you become more familiar with MATLAB's other features.

Sometimes the computer memory set aside to store MATLAB variables gets fragmented, making it difficult or impossible to continue work. To alleviate this problem, the `pack` command performs memory garbage collection. This command saves all MATLAB workspace variables to disk, clears all variables from the workspace, and then reloads the variables. Depending on how much memory is allocated to MATLAB on your computer, how long you have been running a particular MATLAB session, and how many variables you have created, you may or may not ever need to use the `pack` command.

## 3.2 NUMBER DISPLAY FORMATS

When MATLAB displays numerical results, it follows several rules. By default, if a result is an integer, MATLAB displays it as an integer. Likewise, when a result is a real number, MATLAB displays it with approximately four digits to the right of the decimal point. If the significant digits in the result are outside of this range, MATLAB displays the result in scientific notation similar to scientific calculators. You can override this default behavior by specifying a different numerical format within the **Preferences** menu item in the **File** menu, if available, or by typing the appropriate MATLAB `format` command at the prompt. Using the special variable `pi`, the numerical display formats produced by the `format` command are as follows:

| MATLAB COMMAND | `pi` | Comments |
|---|---|---|
| `format short` | `3.1416` | 5 digits |
| `format long` | `3.14159265358979` | 16 digits |
| `format short e` | `3.1416e+00` | 5 digits plus exponent |
| `format long e` | `3.141592653589793e+00` | 16 digits plus exponent |
| `format short g` | `3.1416` | best of `format short` or `format short e` |
| `format long g` | `3.14159265358979` | best of `format long` or `format long e` |
| `format hex` | `400921fb54442d18` | hexadecimal, floating point |
| `format bank` | `3.14` | 2 decimal digits |
| `format +` | `+` | positive, negative, or zero |
| `format rat` | `355/113` | rational approximation |

It is important to note that MATLAB does not change the internal representation of a number when different display formats are chosen; only the display changes. All calculations with common numbers are performed in double precision arithmetic.

## 3.3 COMMAND WINDOW CONTROL

Depending on your hardware setup, you may not have a menu bar or scroll bars on the *Command* window. For this reason, MATLAB provides several commands for managing the *Command* window. These commands include the following:

| *Command* Window Control Commands | Description |
|---|---|
| `clc` | Clear the *Command* window and move the cursor to the upper left corner. |
| `home` | Move cursor to upper left corner. |
| `more` | Pages the *Command* window. |

These commands provide rudimentary cursor control within the *Command* window. The command `more` keeps the *Command* window from scrolling text beyond view, as described by its help text:

```
» help more
MORE  Control paged output in command window.
  MORE OFF disables paging of the output in the MATLAB command window.
  MORE ON enables paging of the output in the MATLAB command window.
  MORE(N) specifies the size of the page to be N lines.
  When MORE is enabled and output is being paged, advance to the next
  line of output by hitting the RETURN key; get the next page of
  output by hitting the spacebar. Press the “q” key to exit out
  of displaying the current item.
```

## 3.4 SYSTEM INFORMATION

From the *Command* window it is possible to determine information about the computer platform that MATLAB is running on as well as information about the MATLAB application itself.

The command `computer` returns a character string describing the computer that MATLAB is running on:

```
» computer
ans =
MAC2
```

In this case, MATLAB is running on a Macintosh computer. `PCWIN` signifies a PC running Windows.

Many computers utilize IEEE arithmetic, while some do not. The command `isieee` returns 1 if the computer platform running MATLAB utilizes IEEE arithmetic and 0 if it does not:

```
» isieee
ans =
     1
```

The Macintosh performs IEEE arithmetic.

The version of MATLAB running at the current time is found by typing the command `version`:

```
» version
ans =
5.0.0.4064
```

This character string signifies version 5.0 of MATLAB.

More detailed information regarding the version of MATLAB running can be found by issuing the command `ver`:

```
» ver
---------------------------------------------------
MATLAB Version 5.0.0.4064 on MAC2
MATLAB License Identification Number: 51483
---------------------------------------------------
MATLAB Toolbox                        Version 5.0       05-Nov-1996
Control System Toolbox.               Version 4.0       15-Nov-1996
Mastering MATLAB Toolbox              Version 5.0       31-Dec-1997
Optimization Toolbox.                 Version 1.5.0     31-Oct-1996
Signal Processing Toolbox.            Version 4.0       15-Nov-1996
SIMULINK                              Version 2.0       15-Nov-1996
```

Your computer will undoubtedly return a different result. In addition to the MATLAB application, `ver` returns information about MATLAB toolboxes you have installed. In the preceding example, the *Mastering MATLAB Toolbox* is a set of MATLAB functions developed by the authors of this text.

Finally, the commands `hostid` and `license` return licensing information for the version of MATLAB you are running.

# 4

# Script M-files

For simple problems, entering your requests at the MATLAB prompt in the *Command* window is fast and efficient. However, as the number of commands increases or in the case when you wish to change the value of one or more variables and reevaluate a number of commands, typing at the MATLAB prompt quickly becomes tedious. MATLAB provides a logical solution to this problem. It allows you to place MATLAB commands in a simple text file and then tell MATLAB to open the file and evaluate commands exactly as it would if you had typed them at the MATLAB prompt. These files are called ***script files*** or simply ***M-files.*** The term "script" symbolizes the fact that MATLAB simply reads from the script found in the file. The term "M-file" recognizes the fact that script file names must end with the extension `'.m'` (e.g., `example1.m`).

To create a script M-file on a Macintosh or PC, choose **New** from the **File** menu and select **M-file.** This procedure brings up a text editor window where you enter MATLAB commands. On workstation platforms, it is common to open another *terminal* window and run your favorite text editor for creating and modifying M-files in that window. The following script M-file shows the commands from an example considered earlier.

```
% script M-file example1.m

erasers = 4; % number of each item
pads = 6;
tape = 2;

items = erasers + pads + tape
cost = erasers*25 + pads*52 + tape*99
average_cost = cost/items
```

On the Macintosh, this file can be saved on disk and executed immediately by choosing **Save and Execute** from the **File** menu when the *Editor* window is active or by using the keyboard shortcut ⌘-E. The PC version of MATLAB does not have this feature. On the PC, this file can be saved as the M-file `example1.m` on your disk by choosing **Save** from the **File** menu. On workstation platforms, save the file using the appropriate save command in your text editor.

To execute the script M-file choose **Run Script...** from the **File** menu on a PC or Macintosh, or simply type the name of the script file without the `.m` extension at the MATLAB prompt:

```
» example1
items =
    12
cost =
   610
average_cost =
       50.833
```

When MATLAB interprets the preceding `example1` statement, it follows the hierarchy described in the next chapter. In brief, MATLAB prioritizes current MATLAB variables and built-in MATLAB commands ahead of M-file names. Thus if `example1` is not a current MATLAB variable or a built-in MATLAB command (it is not), MATLAB opens the file `example1.m` (if it can find it) and evaluates the commands found there just as if they were entered directly at the *Command* window prompt. As a result, commands within the M-file have access to all variables in the MATLAB workspace and all variables created in the M-file become part of the workspace. Normally, the commands read in from the M-file are not displayed as they are evaluated. The `echo on` command tells MATLAB to display or echo commands to the *Command* window as they are read and evaluated. You can probably guess what the `echo off` command does. Similarly, the command `echo` by itself toggles the echo state.

With this M-file feature it is simple to answer "what if?" questions. For example, one can repeatedly open the `example1.m` M-file, change the number of `erasers`, `pads`, or `tape`, save the file, and then ask MATLAB to reevaluate the commands in the file. The power of this capability cannot be overstated. Moreover, by creating M-files your commands are saved on disk for future MATLAB sessions.

The utility of MATLAB comments is readily apparent when using script files, as shown in `example1.m`. Comments allow you to document the commands found in a script file so that they are not forgotten when viewed in the future. In addition, the use of semicolons at the end of lines to suppress the display of results allows you to control script file output so that only important results are shown.

Because of the utility of script files, MATLAB provides several functions that are particularly useful when used in M-files. They are as follows:

| Function | Description |
|---|---|
| `disp(`*variable*`)` | Display results without identifying variable names. |
| `echo` | Control the *Command* window echoing of script file commands. |
| `input` | Prompt user for input. |
| `keyboard` | Give control to keyboard temporarily. Type `return` to return control to the executing script M-file. |
| `pause` | Pause until user presses any keyboard key. |
| `pause(`*n*`)` | Pause for *n* seconds, then continue. |
| `waitforbuttonpress` | Pause until user presses mouse button or keyboard key. |

When a MATLAB command is not terminated in a semicolon, the results of the command are displayed in the *Command* window with the variable name identified. For a prettier display it is sometimes convenient to suppress the variable name. In MATLAB this is accomplished with the command `disp`:

```
» items
items =
    12
» disp(items)
   12
```

Rather than repeatedly edit a script file when computations for a variety of cases are desired, the `input` command allows one to prompt for input as a script file is executed. For example, reconsider the `example1.m` script file with modifications:

```
% script M-file example1.m

erasers = 4;  % Number of each item
pads = 6;
tape = input('Enter the number of rolls of tape purchased > ');

items = erasers + pads + tape
cost = erasers*25 + pads*52 + tape*99
average_cost = cost/items
```

Running this script M-file produces

```
» example1
Enter the number of rolls of tape purchased > 3
items =
    13
cost =
   709
average_cost =
       54.538
```

In response to the prompt, the number 3 was entered and the **Return** or **Enter** key was pressed. The remaining commands were evaluated as before. The function `input` accepts any valid MATLAB expression for input. For example, running the script file again and providing different input gives

```
» example1
Enter the number of rolls of tape purchased > round(sqrt(13))-1
items =
    13
cost =
   709
average_cost =
       54.538
```

In this case, the number of rolls of tape was set equal to the result of evaluating the expression `round(sqrt(13))-1`.

To see the effect of the `echo` command, add it to the script file and execute:

```
% script M-file example1.m

echo on

erasers = 4;  % Number of each item
pads = 6;
tape = input('Enter the number of rolls of tape purchased > ');

items = erasers + pads + tape
cost = erasers*25 + pads*52 + tape*99
average_cost = cost/items

echo off
```

```
» example1
erasers = 4;  % Number of each item
pads = 6;
tape = input('Enter the number of rolls of tape purchased > ');
Enter the number of rolls of tape purchased > 2

items = erasers + pads + tape
items =
    12
cost = erasers*25 + pads*52 + tape*99
cost =
   610
average_cost = cost/items
average_cost =
       50.833

echo off
```

As you can see in this case, the `echo` command made the result much harder to read. On the other hand, the `echo` command can be very helpful when debugging more complicated script files.

5

# File and Directory Management

## 5.1 THE MATLAB WORKSPACE

When working in the MATLAB workspace, it is often convenient to save or print a copy of your work. The `diary` command opens an ASCII text file named `diary` in the current directory or folder and saves subsequent user input and *Command* window output to the file. `diary` *`fname`* saves the diary to a file named *`fname`*. `diary off` terminates the `diary` command and closes the file. Since this file is a text file, it can be opened by a text editor or word processing program, modified as desired, and then printed.

On the PC and Macintosh platforms, when the *Command* window is active, selecting **Print . . .** from the **File** menu prints a copy of the entire *Command* window. Alternately, if you highlight a portion of the *Command* window using the mouse, selecting **Print Selection . . .** from the **File** menu prints the selected range only.

In addition to remembering variables during a MATLAB session, MATLAB can save and load all variables from files on your computer. On the PC and Macintosh platforms, the

**Save Workspace as . . .** menu item in the **File** menu opens a standard file dialog box for saving all current variables. Similarly, the **Load Workspace . . .** menu item in the **File** menu opens a dialog box for loading variables from a previously saved workspace. Saving variables does not delete them from the MATLAB workspace. Loading variables of the same name as those found in the MATLAB workspace changes the variable values to those loaded from the file.

## 5.2 SAVING, LOADING, AND DELETING VARIABLES

If the **File** menu approach is not available because you are running MATLAB on a workstation platform, or if the **File** menu approach does not meet your needs, MATLAB provides two commands, `save` and `load`, which offer more flexibility. In particular, the `save` command allows you to save one or more variables in the file format of your choice. For example,

```
» save
```

stores all variables in ***MATLAB binary format*** in the file `matlab.mat` in the current directory.

```
» save data
```

saves all variables in MATLAB binary format in the file `data.mat` in the current directory.

```
» save data erasers pads tape
```

saves the variables `erasers`, `pads`, and `tape` in binary format in the file `data.mat` in the current directory.

```
» save data erasers pads tape -ascii
```

saves the variables `erasers`, `pads`, and `tape` in 8-digit ASCII format in the file `data` in the current directory. ASCII-formatted files can be edited using any common text editor. Note that ASCII files do not get the extension `.mat`.

```
» save data erasers pads tape -ascii -double
```

saves the variables `erasers`, `pads`, and `tape` in 16-digit ASCII format in the file `data` in the current directory.

The `load` command uses the same syntax, with the obvious difference of loading variables into the MATLAB workspace. Typing `load` with no additional arguments loads all variables in the file `matlab.mat`. Similarly, entering `load data erasers pads tape` loads the variables `erasers`, `pads`, and `tape` into the MATLAB workspace from the file `data.mat`. All other variables in `data.mat` are ignored.

Sometimes the data file to be created is stored as a character string in a MATLAB variable, such as `fname = 'mydata.dat'`. (Character strings are discussed in Chapter 10.) If this is the case, the preceding approach does not work because

```
» save fname erasers pads tape -ascii
```

saves the four variables `fname`, `erasers`, `pads`, and `tape` in the file `matlab.mat`. To resolve this problem, one must use the ***functional form*** of the `save` command; for example,

```
» save(fname, 'erasers', 'pads')
```

saves the variables `erasers` and `pads` in the file named in the MATLAB character string `fname`. The functional form of MATLAB commands is described in the chapter on function M-files.

After creating a data file using the `save` command, it is possible to query the file for existence as well as the variables it holds. The `exist` function allows one to query the existence of a number of MATLAB items, including data files. For example,

```
» exist('data.mat', 'file')
ans =
     2
```

returns a value of 2 that denotes the existence of the file `data.mat`. For further help using the `exist` command, type `helpwin exist` at the MATLAB prompt.

To see what variables are stored in a particular data file, the MATLAB command `whos` provides an option for querying a file for variable names rather than displaying variable names in the MATLAB workspace. For example,

```
» whos -file data.mat
Name              Size           Bytes  Class

erasers           1x1                8  double array
pads              1x1                8  double array
tape              1x1                8  double array

Grand total is 3 elements using 24 bytes
```

While it is possible to delete data files using standard operating system commands, files can also be deleted from within MATLAB using the `delete` command. For example,

```
» delete('data.mat')
```

deletes the data file `data.mat`.

## 5.3 SPECIAL-PURPOSE FILE I/O

In addition to the built-in commands `load` and `save`, MATLAB offers several special-purpose file I/O commands or functions. `dlmread` and `dlmwrite` are used to read and write ASCII text files, respectively, where numerical values are separated by a user-specified delimiter. While it has not been discussed yet, MATLAB stores arrays as well as scalars in variables. `dlmread` and `dlmwrite` read and write a single array containing any number of rows and columns. For example,

```
» dlmwrite('mydata.dat',ones(3,5),',')
» mydata = dlmread('mydata.dat',',')
mydata =
     1     1     1     1     1
     1     1     1     1     1
     1     1     1     1     1
```

stores an array of all ones, having three rows and five columns, in a file named `mydata.dat`, separating the array elements with commas as indicated by the following file contents:

```
1,1,1,1,1
1,1,1,1,1
1,1,1,1,1
```

The `dlmread` function opens the file and places the array into the variable `mydata`. For more information about `dlmread` and `dlmwrite`, use on-line help.

Business spreadsheet programs store results in a grid or array format of rows and columns. If a spreadsheet is stored in ***Lotus WK1 spreadsheet format,*** `wk1read` can be used to open the file and read its content into a MATLAB variable. Similarly, `wk1write` saves a MATLAB variable in Lotus WK1 spreadsheet format so that it can be opened by a spreadsheet program.

## 5.4 LOW-LEVEL FILE I/O

Because an infinite variety of file types exists, MATLAB provides low-level file I/O functions for reading or writing any binary or formatted ASCII file imaginable. These functions closely resemble their ANSI C programming language counterparts, but do not necessarily match their characteristics exactly. In fact, the special-purpose file I/O commands just described use these commands. The low-level file I/O functions in MATLAB are shown in the following table.

| Category | Function | Description/Syntax Example |
|---|---|---|
| File opening and closing | `fopen` | Open file.<br>`fid = fopen('filename','permission')` |
| | `fclose` | Close file.<br>`status = fclose(fid)` |
| Binary I/O | `fread` | Read part or all of a binary data file.<br>`A = fread(fid,num,precision)` |
| | `fwrite` | Write array to a binary data file.<br>`count = fwrite(fid,array,precision)` |
| Formatted I/O | `fscanf` | Read formatted data from file.<br>`A = fscanf(fid,format,num)` |
| | `fprintf` | Write formatted data to file.<br>`count = fprintf(fid,format,A)` |
| | `fgetl` | Read line from file, discard newline character.<br>`line = fgetl(fid)` |
| | `fgets` | Read line from file, keep newline character.<br>`line = fgets(fid)` |
| String conversion | `sprintf` | Write formatted data to string.<br>`S = sprintf(fid,format,A)` |
| | `sscanf` | Read string under format control.<br>`A = sscanf(string,format,num)` |
| File positioning | `ferror` | Inquire about file I/O status.<br>`message = ferror(fid)` |
| | `feof` | Test for end of file.<br>`TF = feof(fid)` |
| | `fseek` | Set file position indicator.<br>`status = fseek(fid,offset,origin)` |
| | `ftell` | Get file position indicator.<br>`position = ftell(fid)` |
| | `frewind` | Rewind file.<br>`frewind(fid)` |
| Temporary files | `tempdir` | Get temporary directory name. |
| | `tempname` | Get temporary file name. |

In the preceding table, *fid* is a file identifier number and *permission* is a character string identifying the permissions requested. Possible strings include `'r'` for reading only, `'w'` for writing only, `'a'` for appending only, and `'r+'` for both reading and writing. Since the PC distinguishes between text and binary files, a `'b'` must often be appended when working with binary files (e.g., `'rb'`). In the preceding table, *`format`* is a character string defining the desired formatting. *`format`* very closely follows ANSI standard C.

## 5.5 DISK FILE MANIPULATION

MATLAB provides several file and directory manipulation commands that allow you to list file names, view and delete them, and show and change the current working directory or folder. A summary of these commands is shown in the following table.

| File Manipulation Commands | Description |
|---|---|
| `cd` (or) `pwd` | Show present working directory or folder. |
| `p = cd` | Return present working directory in the string `p`. |
| `delete` *`filename.m`* | Delete the M-file *`filename.m`*. |
| `dir` (or) `ls` | Display files in current directory. |
| `d = dir` | Return files in current directory in structure *`d`*. |
| `exist('`*`cow`*`','file')` | Check existence of M-file *`cow.m`*. |
| `exist('`*`dname`*`','dir')` | Check existence of directory *`dname`*. |
| `p = matlabroot` | Return directory path to MATLAB program in the string `p`. |
| `type` *`cow`* | Type the M-file *`cow.m`* in the *Command* window. |
| `what` | Display an organized listing of all MATLAB files in the current directory. |
| `which` *`cow`* | Display the directory path to *`cow.m`*. |

In addition to the preceding commands, MATLAB provides several functions that assist in the construction of directory path character strings. These functions include the following:

| Directory String Creation Functions | Description |
|---|---|
| `filesep` | File separator for this computer platform (e.g., `'\'` on a PC). |
| `fullfile` | Build complete directory path string from parts. |
| `pathsep` | Path separator for this computer platform. |

## 5.6 THE MATLAB SEARCH PATH

MATLAB uses a ***search path*** to find script and function M-files. MATLAB's M-files are organized into numerous directories (or folders) and subdirectories (or subfolders). ***The list of all directories where M-files are found is called the MATLAB search path or MATLABPATH.***

> Use of the MATLAB search path is described as follows. When you enter `cow`, MATLAB does the following:
>
> 1. It checks to see if `cow` is a *variable* in the MATLAB workspace; if not,
> 2. It checks to see if `cow` is a *built-in function;* if not,
> 3. It checks to see if an M-file named `cow.m` exists in the *current directory;* if not,
> 4. It checks to see if `cow.m` exists anywhere on the *MATLAB search path,* by searching in the order in which it is specified.

In reality, the search procedure is more complicated because of several advanced features in MATLAB. However, for the most part, the preceding search procedure is sufficient for most MATLAB work.

When MATLAB starts up, it defines a default MATLAB search path that points to all directories where MATLAB stores its M-files. This search path can be displayed and modified in several ways. The PC and Macintosh platforms have a ***Path Browser*** for graphically viewing and modifying the MATLAB search path. The Path Browser is available by pressing the **Path Browser** button on the *Command* window toolbar, or by selecting **MATLAB Path** from the **Window** menu. On all platforms the command `editpath` brings up the appropriate platform-dependent path browser.

The MATLAB search path can be viewed and modified from the *Command* window as well by using the commands `addpath`, `path`, and `rmpath.path` returns the current search path string. `path(`*`pathstring`*`)` changes the search path to *`pathstring`*. `addpath` *`dirstring`* and `path(path,`*`dirstring`*`)` append the directory path string *`dirstring`* to the current path. `rmpath` *`dirstring`* removes the directory path string *`dirstring`* from the current path.

## 5.7 MATLAB AT STARTUP

When MATLAB starts up, it executes two script M-files, `matlabrc.m` and `startup.m`. Of these, `matlabrc.m` comes with MATLAB and generally should not be modified. The commands in this M-file set the default *Figure* window size and placement as well as a number of other default features. On the PC and workstation platforms, the default MATLAB search path is set by calling the script file `pathdef.m` from `matlabrc.m`. On the

Macintosh, the default MATLAB search path is stored in a file in the Preferences folder within the System Folder.

On all platforms, commands in `matlabrc.m` check for the existence of the script M-file `startup.m` on the MATLAB Search Path. If it exists, the commands in it are executed. This optional M-file `startup.m` typically contains commands that add personal default features to MATLAB. For example, it is very common to put one or more `addpath` or `path` commands in `startup.m` to append additional directories to the MATLAB Search Path. Similarly, the default number display format can be changed (e.g., `format compact`). If you have a grayscale monitor, the command `graymon` is useful for setting default grayscale graphics features. Further still, if you want plots to have different default characteristics, a call to `colordef` could appear in `startup.m`. Since `startup.m` is a standard script M-file, there are no restrictions as to what commands can be placed in it. However, it is probably not wise to include the command `quit` in `startup.m`.

6

# Arrays and Array Operations

All of the computations considered to this point have involved single numbers called scalars. Operations involving scalars are the basis of mathematics. At the same time, when one wishes to perform the same operation on more than one number at a time, repeated scalar operations are time-consuming and cumbersome. To solve this problem, MATLAB defines operations on data arrays.

## 6.1 SIMPLE ARRAYS

Consider the problem of computing values of the sine function over one half of its period—namely, $y = \sin(x)$ over $0 \leq x \leq \pi$. Since it is impossible to compute $\sin(x)$ at all points over this range (there are an infinite number of them), we must choose a finite number of points. In doing so we are sampling the function. To pick some number, let's evaluate $\sin(x)$ every $0.1\pi$ in this range (i.e., let $x = 0, 0.1\pi, 0.2\pi, \ldots, 1.0\pi$). If you were using a scientific calculator to compute these values, you would start by making a list or array of the values of $x$. Then you would enter each value of $x$ into your calculator, find its sine, and write down the result as the second array *y*. Perhaps you would write them in an organized fashion as follows:

| $x$ | 0 | $.1\pi$ | $.2\pi$ | $.3\pi$ | $.4\pi$ | $.5\pi$ | $.6\pi$ | $.7\pi$ | $.8\pi$ | $.9\pi$ | $\pi$ |
|---|---|---|---|---|---|---|---|---|---|---|---|
| $y$ | 0 | .31 | .59 | .81 | .95 | 1.0 | .95 | .81 | .59 | .31 | 0 |

As shown, *x* and *y* are ordered lists of numbers (i.e., the first value or element in *y* is associated with the first value or element in *x*, the second element in *y* is associated with the second element in *x*, and so on). Because of this ordering, it is common to refer to individual values or elements in *x* and *y* with subscripts (e.g., $x_1$ is the first element in *x*, $y_5$ is the fifth element in *y*, $x_n$ is the *n*th element in *x*).

MATLAB handles arrays in a straightforward and intuitive way. Creating arrays is easy—just follow the visual organization given previously:

```
» x = [0 .1*pi .2*pi .3*pi .4*pi .5*pi .6*pi .7*pi .8*pi .9*pi pi]
x =
  Columns 1 through 7
        0    0.3142    0.6283    0.9425    1.2566    1.5708    1.8850
  Columns 8 through 11
    2.1991    2.5133    2.8274    3.1416

» y = sin(x)
y =
  Columns 1 through 7
        0    0.3090    0.5878    0.8090    0.9511    1.0000    0.9511
  Columns 8 through 11
    0.8090    0.5878    0.3090    0.0000
```

To create an array in MATLAB, all you have to do is start with a left bracket, enter the desired values separated by spaces (or commas), and then close the array with a right bracket. Notice that finding the sine of the values in x follows naturally. MATLAB understands that you want to find the sine of each element in x and place the results in an associated array called y. This fundamental capability makes MATLAB different from other computer languages.

Since spaces separate array values, complex numbers entered as array values cannot have embedded spaces unless expressions are enclosed in parentheses. For example, [1 -2i 3 4 5+6i] contains five elements whereas the identical arrays [(1 - 2i) 3 4 5+6i] and [1-2i 3 4 5+6i] contain four.

## 6.2 ARRAY ADDRESSING OR INDEXING

Now since x in the preceding example has more than one element (namely, it has 11 values separated into columns), MATLAB gives you the result back with the columns identified. As shown previously, x is an array having one row and eleven columns, or in mathematical jargon it is a row vector, a one-by-eleven array or simply an array of length 11.

In MATLAB, individual array elements are accessed using subscripts—for example, x(1) is the first element in x, x(2) is the second element in x, and so on. As another example,

```
» x(3)  % The third element of x
ans =
    0.6283
» y(5)  % The fifth element of y
ans =
    0.9511
```

To access a block of elements at one time, MATLAB provides colon notation:

```
» x(1:5)
ans =
        0    0.3142    0.6283    0.9425    1.2566
```

This is the first through fifth elements in x. 1:5 says start with 1 and count up to 5.

```
» x(7:end)
ans =
        1.885       2.1991       2.5133       2.8274       3.1416
```

starts with the seventh element and continues to the last element. Here the word end signifies the last element in the array x.

```
» y(3:-1:1)
ans =
    0.5878    0.3090         0
```

This is the third, second, and first elements in reverse order. 3:-1:1 says start with 3, count down by 1, and stop at 1.

```
» x(2:2:7)
ans =
    0.3142    0.9425    1.5708
```

This is the second, fourth, and sixth elements in x. 2:2:7 says start with 2, count up by 2, and stop when you get to 7. In this case adding 2 to 6 gives 8, which is greater than 7 so the eighth element is not included.

```
» y([8 2 9 1])
ans =
    0.8090    0.3090    0.5878         0
```

Here we used another array [8 2 9 1] to extract the elements of the array y in the order we wanted them. The first element taken is the eighth, the second is the second, the third is the ninth, and the fourth is the first. In reality [8 2 9 1] itself is an array that addresses the desired elements of y. Addressing one array with another works as long as the addressing array contains integers between 1 and the length of the array.

## 6.3 ARRAY CONSTRUCTION

Earlier we entered the values of x by typing each individual element in x. While this is fine when there are only 11 values in x, what if there are 111 values? Using the colon notation, two other ways of entering x are as follows:

```
» x = (0:0.1:1)*pi
x =
  Columns 1 through 7
         0    0.3142    0.6283    0.9425    1.2566    1.5708    1.8850
  Columns 8 through 11
    2.1991    2.5133    2.8274    3.1416

» x = linspace(0,pi,11)
x =
  Columns 1 through 7
         0    0.3142    0.6283    0.9425    1.2566    1.5708    1.8850
  Columns 8 through 11
    2.1991    2.5133    2.8274    3.1416
```

In the first case, the colon notation (0:0.1:1) creates an array that starts at 0, increments or counts by 0.1, and ends at 1. Each element in this array is then multiplied by $\pi$ to create the desired values in x. In the second case, the MATLAB function linspace is used to create x. This function's arguments are described by

```
linspace(first_value,last_value,number_of_values)
```

Both of these array creation forms are common in MATLAB. The colon notation form allows you to specify directly the increment between data points, but not the number of data points. linspace, on the other hand, allows you to specify directly the number of data points, but not the increment between the data points. Both of these array creation forms create arrays where the individual elements are linearly spaced with respect to each other. For the special case where a logarithmically spaced array is desired, MATLAB provides the logspace function:

```
» logspace(0,2,11)
ans =
  Columns 1 through 7
    1.0000    1.5849    2.5119    3.9811    6.3096   10.0000   15.8489
  Columns 8 through 11
   25.1189   39.8107   63.0957  100.0000
```

Here, we created an array starting at $10^0$, ending at $10^2$, containing 11 values. The function arguments are described by

```
logspace(first_exponent,last_exponent,number_of_values)
```

Though it is common to begin and end at integer powers of ten, `logspace` works equally well with nonintegers.

Sometimes an array is required that is not conveniently described by a linearly or logrithmically spaced element relationship. There is no uniform way to create these arrays. However, array addressing and the ability to combine expressions can help eliminate the need to enter individual elements one at a time:

```
» a = 1:5, b = 1:2:9
a =
     1     2     3     4     5
b =
     1     3     5     7     9
```

creates two arrays. Remember that multiple statements can appear on a single line if they are separated by commas or semicolons.

```
» c = [b a]
c =
     1     3     5     7     9     1     2     3     4     5
```

creates an array `c` composed of the elements of `b` followed by those of `a`.

```
» d = [a(1:2:5) 1 0 1]
d =
     1     3     5     1     0     1
```

creates an array `d` composed of the first, third, and fifth elements of `a` followed by three additional elements.

The simple array construction features of MATLAB are summarized in the following table.

| Array Construction Technique | Description |
|---|---|
| `x=[2 2*pi sqrt(2) 2-3j]` | Create row vector `x` containing elements specified. |
| `x=first:last` | Create row vector `x` starting with `first`, counting by one, ending at or before `last`. |
| `x=first:increment:last` | Create row vector `x` starting with `first`, counting by `increment`, ending at or before `last`. |
| `x=linspace(first,last,n)` | Create row vector `x` starting with `first`, ending at `last`, having n elements. |
| `x=logspace(first,last,n)` | Create logarithmically-spaced row vector `x` starting with $10^{first}$, ending at $10^{last}$, having `n` elements. |

## 6.4 ARRAY ORIENTATION

In the preceding examples, arrays contained one row and multiple columns. As a result of this row orientation they are commonly called row vectors. It is also possible for an array to be a column vector having one column and multiple rows. In this case, all of the preceding array manipulation and mathematics apply without change. The only difference is that results are displayed as columns rather than rows.

Since the array creation functions illustrated in the previous section all create row vectors, there must be some way to create column vectors. The most straightforward way to create a column vector is to specify it element by element and separate values with semicolons:

```
» c = [1;2;3;4;5]
c =
     1
     2
     3
     4
     5
```

> Based on this example, separating elements by spaces or commas specifies elements in different columns, whereas separating elements by semicolons specifies elements in different rows.

To create a column vector using the colon notation `start:increment:end`, or the functions `linspace` and `logspace`, one must ***transpose*** the resulting row into a column using the MATLAB transpose operator ( ' ):

```
» a = 1:5
a =
     1     2     3     4     5
```

creates a row vector using the colon notation format.

```
» b = a'
b =
     1
     2
     3
     4
     5
```

uses the transpose operator to change the row vector `a` into the column vector `b`.

```
» c = b'
c =
     1     2     3     4     5
```

applies the transpose again and changes the column back to a row.

In addition to the simple transpose, MATLAB also offers a transpose operator with a preceding dot. In this case the dot-transpose operator is interpreted as the noncomplex conjugate transpose. When an array is complex, the transpose ( ' ) gives the complex conjugate transpose (i.e., the sign on the imaginary part is changed as part of the transpose operation). On the other hand, the dot-transpose ( . ' ) transposes the array but does not conjugate it.

```
» c = a.'
c =
     1
     2
     3
     4
     5
```

shows that .' and ' are identical for real data.

```
» d = a + i*a
d =
  Columns 1 through 4
```

```
 1.0000 + 1.0000i  2.0000 + 2.0000i  3.0000 + 3.0000i  4.0000 + 4.0000i
Column 5
 5.0000 + 5.0000i
```

creates a simple complex row vector from the array a using the default value i $= \sqrt{-1}$.

```
» e = d'
e =
   1.0000 - 1.0000i
   2.0000 - 2.0000i
   3.0000 - 3.0000i
   4.0000 - 4.0000i
   5.0000 - 5.0000i
```

creates a column vector e that is the complex conjugate transpose of d.

```
» f = d.'
f =
   1.0000 + 1.0000i
   2.0000 + 2.0000i
   3.0000 + 3.0000i
   4.0000 + 4.0000i
   5.0000 + 5.0000i
```

creates a column vector f that is the transpose of d.

If an array can be a row vector or a column vector, it makes intuitive sense that arrays could just as well have both multiple rows and multiple columns. That is, arrays can also be in the form of matrices. Creation of matrices follows that of row and column vectors. ***Commas or spaces are used to separate elements in a specific row and semicolons are used to separate individual rows:***

```
» g = [1 2 3 4;5 6 7 8]
g =
     1     2     3     4
     5     6     7     8
```

Here g is an array or matrix having 2 rows and 4 columns (i.e., it is a 2 by 4 matrix or it is a matrix of dimension 2 by 4). The semicolon tells MATLAB to start a new row between the 4 and 5.

```
» g = [1 2 3 4
5 6 7 8
9 10 11 12]
g =
     1     2     3     4
     5     6     7     8
     9    10    11    12
```

In addition to semicolons, pressing the **Return** or **Enter** key while entering a matrix also tells MATLAB to start a new row.

```
» h = [1 2 3;4 5 6 7]
??? All rows in the bracketed expression must have the same
number of columns.
```

MATLAB strictly enforces the fact that all rows must contain the same number of columns.

## 6.5 SCALAR-ARRAY MATHEMATICS

In the first array example in the preceding section, the array x is multiplied by the scalar $\pi$. Other simple mathematical operations between scalars and arrays follow the same natural interpretation. Addition, subtraction, multiplication, and division by a scalar simply apply the operation to all elements of the array:

```
» g - 2
ans =
    -1     0     1     2
     3     4     5     6
     7     8     9    10
```

subtracts 2 from each element in g.

```
» 2*g - 1
ans =
     1     3     5     7
     9    11    13    15
    17    19    21    23
```

multiplies each element in g by 2 and subtracts 1 from each element of the result. Note that scalar-array mathematics uses the same order of precedence used in scalar expressions to determine the order of evaluation.

## 6.6 ARRAY-ARRAY MATHEMATICS

Mathematical operations between arrays are not quite as simple as those between scalars and arrays. Clearly, array operations between arrays of different sizes or dimensions are difficult to define and of even more dubious value. However, when two arrays have the same dimensions, addition, subtraction, multiplication, and division apply on an element-by-element basis in MATLAB. For example,

```
» g  % recall previous array
g =
     1     2     3     4
     5     6     7     8
     9    10    11    12

» h = [1 1 1 1;2 2 2 2;3 3 3 3]  % create new array
h =
     1     1     1     1
     2     2     2     2
     3     3     3     3

» g + h  % add h to g on an element-by-element basis
ans =
     2     3     4     5
     7     8     9    10
    12    13    14    15

» ans - h  % subtract h from the previous answer to get g back
ans =
     1     2     3     4
     5     6     7     8
     9    10    11    12

» 2*g - h  % multiplies g by 2 and subtracts h from the result
ans =
     1     3     5     7
     8    10    12    14
    15    17    19    21
```

Note that array-array mathematics also uses the same order of precedence used in scalar expressions to determine the order of evaluation.

Element-by-element multiplication and division work similarly but use slightly unconventional notation:

```
» g.*h
ans =
     1     2     3     4
    10    12    14    16
    27    30    33    36
```

Here we multiplied the arrays `g` and `h` element by element using the dot multiplication symbol `.*`.

> The dot preceding the standard asterisk multiplication symbol tells MATLAB to perform element-by-element array multiplication. Multiplication without the dot signifies matrix multiplication, which will be discussed later.

For this particular example, matrix multiplication is not defined:

```
» g*h
??? Error using ==> *
Inner matrix dimensions must agree.
```

Array division, or dot division, also requires use of the dot symbol:

```
» g./h
ans =
    1.0000    2.0000    3.0000    4.0000
    2.5000    3.0000    3.5000    4.0000
    3.0000    3.3333    3.6667    4.0000

» h.\g
ans =
    1.0000    2.0000    3.0000    4.0000
    2.5000    3.0000    3.5000    4.0000
    3.0000    3.3333    3.6667    4.0000
```

As with scalars, division is defined using both the forward and backward slashes. In both cases, the array below the slash is divided into the array above the slash.

Division without the dot is the matrix division operation, which is an entirely different operation:

```
» g/h

Warning: Rank deficient, rank = 1 tol = 5.3291e-15.
ans =
         0         0    0.8333
         0         0    2.1667
         0         0    3.5000
» h/g

Warning: Rank deficient, rank = 2 tol = 1.8757e-14.
ans =
   -0.1250         0    0.1250
   -0.2500         0    0.2500
   -0.3750         0    0.3750
```

Matrix division gives results that are not necessarily the same sizes as g and h. Matrix operations are discussed in the numerical linear algebra chapter (Chapter 16).

Array exponentiation is defined in several ways. As with multiplication and division, ^ is reserved for matrix exponentiation and .^ is used to denote element-by-element exponentiation:

```
» g, h  % recalls the arrays used earlier
g =
     1     2     3     4
     5     6     7     8
     9    10    11    12
h =
     1     1     1     1
     2     2     2     2
     3     3     3     3

» g.^2
ans =
     1     4     9    16
    25    36    49    64
    81   100   121   144
```

squares the individual elements of g.

```
» g.^-1
ans =
      1.0000      0.5000      0.3333      0.2500
      0.2000      0.1667      0.1429      0.1250
      0.1111      0.1000      0.0909      0.0833
```

finds the reciprocal of each element in g,

```
» 1./g
ans =
      1.0000      0.5000      0.3333      0.2500
      0.2000      0.1667      0.1429      0.1250
      0.1111      0.1000      0.0909      0.0833
```

just as array division does.

```
» 2.^g
ans =
           2           4           8          16
          32          64         128         256
         512        1024        2048        4096
```

raises 2 to the power of each element in the array g.

```
» g.^h
ans =
           1           2           3           4
          25          36          49          64
         729        1000        1331        1728
```

raises the elements of g to the corresponding elements in h. In this case, the first row is unchanged; the second row is squared, and the third row is cubed.

```
» g.^(h-1)
ans =
     1     1     1     1
     5     6     7     8
    81   100   121   144
```

shows that scalar and array operations can be combined.

The following table summarizes basic array operations.

| Element-by-element Operation | Representative Data $a = [a_1\ a_2\ ...\ a_n]$, $b = [b_1\ b_2\ ...\ b_n]$, c= *<a scalar>* |
|---|---|
| Scalar addition | $a+c = [a_1+c\ a_2+c\ ...\ a_n+c]$ |
| Scalar multiplication | $a*c = [a_1*c\ a_2*c\ ...\ a_n*c]$ |
| Array addition | $a+b = [a_1+b_1\ a_2+b_2\ ...\ a_n+b_n]$ |
| Array multiplication | $a.*b = [a_1*b_1\ a_2*b_2\ ...\ a_n*b_n]$ |
| Array right division | $a./b = [a_1/b_1\ a_2/b_2\ ...\ a_n/b_n]$ |
| Array left division | $a.\backslash b = [a_1\backslash b_1\ a_2\backslash b_2\ ...\ a_n\backslash b_n]$ |
| Array exponentiation | $a.\hat{}c = [a_1\hat{}c\ a_2\hat{}c\ ...\ a_n\hat{}c]$ |
| | $c.\hat{}a = [c\hat{}a_1\ c\hat{}a_2\ ...\ c\hat{}a_n]$ |
| | $a.\hat{}b = [a_1\hat{}b_1\ a_2\hat{}b_2\ ...\ a_n\hat{}b_n]$ |

## 6.7 STANDARD ARRAYS

Because of their general utility, MATLAB provides functions for creating a number of standard arrays. These include arrays containing all ones or all zeros, identity matrices, arrays of random numbers, diagonal arrays, and arrays whose elements are a given constant.

```
» ones(3)
ans =
     1     1     1
     1     1     1
     1     1     1

» zeros(2,5)
ans =
     0     0     0     0     0
     0     0     0     0     0
```

```
» size(g)
ans =
     3     4

» ones(size(g))
ans =
     1     1     1     1
     1     1     1     1
     1     1     1     1
```

When called with a single input argument, `ones(n)` or `zeros(n)`, MATLAB creates an `n`-by-`n` array containing ones or zeros, respectively. When called with two input arguments, `ones(r,c)` or `zeros(r,c)`, MATLAB creates an array having `r` rows and `c` columns. To create an array of ones or zeros the same size as another array, use the `size` function (discussed later in this chapter) in the argument to `ones` and `zeros`.

```
» eye(4)
ans =
     1     0     0     0
     0     1     0     0
     0     0     1     0
     0     0     0     1

» eye(2,4)
ans =
     1     0     0     0
     0     1     0     0

» eye(4,2)
ans =
     1     0
     0     1
     0     0
     0     0
```

As shown, the function `eye` produces identity matrices using the same syntax style as that used to produce arrays of zeros and ones. An identity matrix or array is all zeros except for the elements `A(i,i)`, where `i = 1:min(r,c)` in which `min(r,c)` is the minimum of the number of rows and columns in `A`.

```
» rand(3)
ans =
    0.9501    0.4860    0.4565
    0.2311    0.8913    0.0185
    0.6068    0.7621    0.8214
```

```
» rand(1,5)
ans =
       0.4447    0.6154    0.7919    0.9218    0.7382
» b = eye(3)
b =
     1     0     0
     0     1     0
     0     0     1

» rand(size(b))
ans =
      0.1763    0.9169    0.0579
      0.4057    0.4103    0.3529
      0.9355    0.8937    0.8132
```

The function `rand` produces uniformly distributed random arrays whose elements lie between 0 and 1.

```
» randn(2)
ans =
     -0.4326    0.1253
     -1.6656    0.2877

» randn(2,5)
ans =
      -1.1465    1.1892    0.3273   -0.1867   -0.5883
       1.1909   -0.0376    0.1746    0.7258    2.183
```

On the other hand, the function `randn` produces arrays whose elements are samples from a zero-mean, unit-variance normal distribution.

```
» a = 1:4  % start with a simple vector
a =
     1     2     3     4

» diag(a)  % place elements on the main diagonal
ans =
     1     0     0     0
     0     2     0     0
     0     0     3     0
     0     0     0     4

» diag(a,1)  % place elements 1 place up from diagonal
ans =
     0     1     0     0     0
     0     0     2     0     0
     0     0     0     3     0
     0     0     0     0     4
     0     0     0     0     0
```

```
» diag(a,-2)  % place elements 2 places down from diagonal
ans =
     0     0     0     0     0     0
     0     0     0     0     0     0
     1     0     0     0     0     0
     0     2     0     0     0     0
     0     0     3     0     0     0
     0     0     0     4     0     0
```

As shown, the function `diag` creates diagonal arrays.

Using the preceding standard arrays, there are several ways to create an array whose elements are all the same value. Some of them include the following:

```
» d = pi;  % choose pi for this example

» d*ones(3,4)  % slowest method (scalar multiplication)
ans =
       3.1416        3.1416        3.1416        3.1416
       3.1416        3.1416        3.1416        3.1416
       3.1416        3.1416        3.1416        3.1416

» d+zeros(3,4) % slower method (scalar addition)
ans =
       3.1416        3.1416        3.1416        3.1416
       3.1416        3.1416        3.1416        3.1416
       3.1416        3.1416        3.1416        3.1416

» d(ones(3,4))  % fast method (array addressing)
ans =
       3.1416        3.1416        3.1416        3.1416
       3.1416        3.1416        3.1416        3.1416
       3.1416        3.1416        3.1416        3.1416

» repmat(d,3,4)  % fastest method (optimum array addressing)
ans =
      3.1416         3.1416        3.1416        3.1416
      3.1416         3.1416        3.1416        3.1416
      3.1416         3.1416        3.1416        3.1416
```

For small arrays all of the preceding methods are fine. However, as the array grows in size, the multiplications required in the scalar multiplication approach slow the method down. Since addition is often faster than multiplication, the next best approach is to add the desired scalar to an array of zeros. Although they are not intuitive, the last two methods are fastest for large arrays. They both involve array indexing as described earlier. The function `repmat` stands for replicate matrix and will be discussed in the next section.

## 6.8 ARRAY MANIPULATION

Since arrays and matrices are fundamental to MATLAB, there are many ways to manipulate them in MATLAB. Once arrays are formed, MATLAB provides powerful ways to insert, extract, and rearrange subsets of them by identifying subscripts of interest. Knowledge of these features are a key to using MATLAB efficiently. To illustrate the matrix and array manipulation features of MATLAB, consider the following examples:

```
» A = [1 2 3;4 5 6;7 8 9]
A =
     1     2     3
     4     5     6
     7     8     9

» A(3,3) = 0  % set element in 3rd row, 3rd column to zero
A =
     1     2     3
     4     5     6
     7     8     0
```

changes the element in the third row and third column to zero.

```
» A(2,6) = 1  % set element in 2nd row, 6th column to one
A =
     1     2     3     0     0     0
     4     5     6     0     0     1
     7     8     0     0     0     0
```

places one in the second row, sixth column. Since A does not have six columns, the size of A is increased as necessary and filled with zeros so that the matrix remains rectangular.

```
» A(:,4) = 4
A =
     1     2     3     4     0     0
     4     5     6     4     0     1
     7     8     0     4     0     0
```

sets the fourth column of A equal to 4. Since 4 is a scalar, it is expanded to fill all the elements specified.

```
» A = [1 2 3;4 5 6;7 8 9];  % restore original data
» B = A(3:-1:1,1:3)
B =
     7     8     9
     4     5     6
     1     2     3
```

creates a matrix B by taking the rows of A in reverse order.

```
» B = A(3:-1:1,:)
B =
     7     8     9
     4     5     6
     1     2     3
```

does the same as the preceding example. Here the final single colon means take all columns. That is, : is short for 1:3 in this example because A has three columns.

```
» C = [A B(:,[1 3])]
C =
     1     2     3     7     9
     4     5     6     4     6
     7     8     9     1     3
```

creates C by appending all rows in the first and third columns of B to the right of A.

```
» B = A(1:2,2:3)
B =
     2     3
     5     6
```

creates B by extracting the first two rows and last two columns of A.

```
» C = [1 3]
C =
     1     3

» B = A(C,C)
B =
     1     3
     7     9
```

uses the array C to index the matrix A rather than specifying them directly using the colon notation `start:increment:end` or `start:end`. In this example, B is formed from the first and third rows and first and third columns of A.

```
» B = A(:)
B =
     1
     4
     7
```

```
     2
     5
     8
     3
     6
     9
```

builds B by stretching A into a column vector taking its columns one at a time in order.

```
» B = B.'
B =
     1     4     7     2     5     8     3     6     9
```

illustrates the dot-transpose operation introduced earlier.

```
» B = A
B =
     1     2     3
     4     5     6
     7     8     9

» B(:,2) = []
B =
     1     3
     4     6
     7     9
```

redefines B by throwing away all rows in the second column of original B. When you set something equal to the empty matrix [], it gets deleted, causing the matrix to collapse to what remains. Note that you must delete whole rows or columns so that the result remains rectangular.

```
» B = B.'
B =
     1     4     7
     3     6     9
```

illustrates the transpose of a matrix. In general, the *i*th row becomes the *i*th column of the result, so the original 3-by-2 matrix becomes a 2-by-3 matrix.

```
» B(2,:) = []
B =
     1     4     7
```

throws out the second row of B.

```
» A(2,:) = B
A =
     1     2     3
     1     4     7
     7     8     9
```

replaces the second row of A with B.

```
» B = A(:,[2 2 2 2])
B =
     2     2     2     2
     4     4     4     4
     8     8     8     8
```

creates B by duplicating all rows in the second column of A four times.

```
» A   % show A again
A =
     1     2     3
     1     4     7
     7     8     9

» A(2,2) = []
???  Indexed empty matrix assignment is not allowed.
```

shows that you can only throw out entire rows or columns. MATLAB does not know how to collapse a matrix when partial rows or columns are thrown out.

```
» B = A(4,:)
???  Index exceeds matrix dimensions.
```

Since A does not have a fourth row, MATLAB doesn't know what to do and says so.

```
» B(1:2,:) = A
???  In an assignment A(matrix,:) = B, the number of columns in A and
B must be the same.
```

shows that you cannot squeeze one matrix into another having a different size.

```
» B(3:4,:) = A(2:3,:)
B =
     1     4     7
     0     0     0
     1     4     7
     7     8     9
```

But you can place the second and third columns of A into the same size area of B. Since the second through fourth rows of B did not exist, they are created as necessary. Moreover, the second row of B is unspecified so it is filled with zeros.

```
» A = [1 2 3;4 5 6;7 8 9]  % fresh data
A =
     1     2     3
     4     5     6
     7     8     9

» G(1:6) = A(:,2:3)
G =
     2     5     8     3     6     9
```

creates a row vector G by extracting all rows in the second and third columns of A. Note that the shape of the matrices are different on both sides of the equal sign.

```
» H = ones(6,1);  % create a column array

» H(:) = A(:,2:3)  % fill H in without changing its shape
H =
     2
     5
     8
     3
     6
     9
```

When (:) appears on the left-hand side of the equal sign, it means take elements from the right-hand side and stick them into the array on the left-hand side without changing its shape. In the preceding example, this process extracts the second and third columns of A and packs them into the column vector H.

When the right-hand side of an assignment is a scalar and the left-hand side is an array, ***scalar expansion*** is used. For example,

```
» A(2,:) = 0
A =
     1     2     3
     0     0     0
     7     8     9
```

replaces the second row of A with zeros. The single zero on the right-hand side is expanded to fill all indices specified on the left. This example is equivalent to

```
» A(2,:) = [0 0 0]
A =
     1     2     3
     0     0     0
     7     8     9
```

Sometimes it is desirable to perform some mathematical operation between a vector and a two-dimensional array. For example, consider the following arrays:

```
» A = [1 2 3;4 5 6;7 8 9;10 11 12]
A =
     1     2     3
     4     5     6
     7     8     9
    10    11    12

» r = [3 2 1]
r =
     3     2     1
```

Suppose that we wish to subtract `r(i)` from the *i*th column of `A`. One way of accomplishing this is as follows:

```
» Ar = [A(:,1)-r(1) A(:,2)-r(2) A(:,3)-r(3)]
Ar =
    -2     0     2
     1     3     5
     4     6     8
     7     9    11
```

Alternatively, one can use indexing:

```
» R = r([1 1 1 1],:) % duplicate r to have 4 rows
R =
     3     2     1
     3     2     1
     3     2     1
     3     2     1

» Ar = A - R % now use element by element subtraction
Ar =
    -2     0     2
     1     3     5
     4     6     8
     7     9    11
```

The array `R` can also be computed more generically using the functions `ones` and `size` or using the function `repmat`, all of which are discussed later:

```
» R = r(ones(size(A,1),1),:) % this has been called Tony's trick
R =
     3     2     1
     3     2     1
     3     2     1
     3     2     1
```

```
» R = repmat(r,size(A,1),1)
R =
     3     2     1
     3     2     1
     3     2     1
     3     2     1

» S = R; % delayed copy example
```

A simple assignment such as the preceding does ***not*** immediately create a copy of the right-hand side array into the left-hand side variable. When an array is large, it is advantageous to delay the copy. That is, future references to `S` simply access the associated contents of `R`, so to the user it appears that `S` is equal to `R`. Time is taken to copy the array `R` into the variable `S` only if the contents of `R` are about to change or if the contents of `S` are about to be assigned new values by some MATLAB statement. While the time saved by this ***delayed copy*** feature is insignificant for smaller arrays, it can lead to significant performance improvements for very large arrays.

Sometimes it is more convenient to address matrix elements with a single index. ***When a single index is used in MATLAB, the index counts elements down the columns starting with the first.*** For example,

```
» D = [1 2 3 4;5 6 7 8; 9 10 11 12]  % new data
D =
     1     2     3     4
     5     6     7     8
     9    10    11    12

» D(2)  % second element
ans =
     5
» D(5)  % fifth element (3 in first column plus 2 in second column)
ans =
     6
» D(end) % last element in matrix
ans =
    12
» D(4:7)  % fourth through seventh elements
ans =
     2     6    10     3
```

The MATLAB functions `sub2ind` and `ind2sub` perform the arithmetic to convert to and from single index to row and column subscripts. For example,

```
» sub2ind(size(D),2,4)  % find single index from row and column
ans =
    11
```

```
» [r,c] = ind2sub(size(D),11)  % find row and column from single index
r =
     2
c =
     4
```

The element in the second row, fourth column is the eleventh element. Note that these two functions want to know the size of the array to search, rather than the array itself.

In addition to addressing matrices based on their subscripts, ***logical arrays*** that result from logical operations (to be discussed more thoroughly later) can also be used if the size of the array is equal to that of the array it is addressing. In this case, True (1) elements are retained and False (0) elements are discarded.

```
» x = -3:3  % Create data
x =
    -3    -2    -1     0     1     2     3

» abs(x)>1
ans =
     1     1     0     0     0     1     1
```

returns a logical array with ones (True) where the absolute value of `x` is greater than 1, and zeros (False) elsewhere.

```
» y = x(abs(x)>1)
y =
    -3    -2     2     3
```

creates `y` by taking those values of `x` where its absolute value is greater than 1. Note, however, that

```
» y = x([1 1 0 0 0 1 1])
??? Index into matrix is negative or zero. See release notes on changes to
logical indices.
```

gives an error even though the `abs(x)>1` and `[1 1 0 0 0 1 1]` appear to be the same vector. In this second case, `[1 1 0 0 0 1 1]` is a ***numeric array*** as opposed to a ***logical array.*** As a result, MATLAB tries to address the element numbers specified in `[1 1 0 0 0 1 1]`, and generates an error because there is no element `0`. Naturally, MATLAB provides the function `logical` for converting numerical arrays to logical arrays:

```
» y = x(logical([1 1 0 0 0 1 1]))
y =
    -3    -2     2     3
```

Now we once again have the desired result. Logical arrays are another data type in MATLAB. Up to now we have only considered numerical arrays. To summarize,

**Specifying array subscripts with numerical arrays extracts the elements having the given numerical indices. On the other hand, specifying array subscripts with logical arrays, which are returned by logical expressions and the function `logical`, extracts elements that are logical True (≠0).**

Logical arrays work on matrices as well as vectors:

```
» B = [5 -3;2 -4]
B =
      5    -3
      2    -4

» x = abs(B)>2
x =
      1     1
      0     1
```

Likewise, 0-1 logical array extraction works for matrices as well.

```
» y = b(x)
y =
      5
     -3
      4
```

However, the results are converted to a column vector, since there is no way to define a matrix having only three elements.

The preceding array addressing techniques are summarized in the following table.

| Array Addressing | Description |
|---|---|
| `A(r,c)` | Addresses a subarray within A defined by the index vector of desired rows in r and index vector of desired columns in c. |
| `A(r,:)` | Addresses a subarray within A defined by the index vector of desired rows in r and all columns. |
| `A(:,c)` | Addresses a subarray within A defined by all rows and the index vector of desired columns in c. |
| `A(:)` | Addresses all elements of A as a column vector taken column by column. If A(:) appears on the left-hand side of the equal sign, it means fill A with elements from the right side of the equal sign without changing its shape. |
| `A(i)` | Addresses a subarray within A defined by the single index vector of desired elements in i, as if A was the column vector, A(:). |
| `A(x)` | Addresses a subarray within A defined by the logical array x. x must be the same size as A. |

## 6.9 SUBARRAY SEARCHING

Many times it is desirable to know the indices or subscripts of those elements of an array that satisfy some relational expression. In MATLAB this task is performed by the function `find`, which returns the subscripts where a relational expression is True:

```
» x = -3:3
x =
    -3    -2    -1     0     1     2     3

» k = find(abs(x)>1)  % finds those subscripts where abs(x)>1
k =
     1     2     6     7

» y = x(k)  % creates y using the indexes in k.
y =
    -3    -2     2     3

» y = x(abs(x)>1) % creates the same y vector by logical addressing
y =
    -3    -2     2     3
```

The `find` function also works for general arrays:

```
» A = [1 2 3;4 5 6;7 8 9]
A =
     1     2     3
     4     5     6
     7     8     9

» [i,j] = find(A>5)
i =
     3
     3
     2
     3
j =
     1
     2
     3
     3
```

Here the indices stored in `i` and `j` are the associated row and column indices, respectively, where the relational expression is True. That is, `A(i(1),j(1))` is the first element of `A` where `A>5`, and so on.

> Note that when a MATLAB function returns two or more variables, they are enclosed by square brackets on the left-hand side of the equal sign. This syntax is different from the array manipulation syntax discussed previously, where [i,j] on the right-hand side of the equal sign builds a new array with j appended to the right of i.

The preceding concepts are summarized in the following table:

| Array Searching | Description |
|---|---|
| `i=find(x)` | Return indices of the array x where its elements are nonzero. |
| `[r,c]=find(X)` | Return row and column indices of the array X where its elements are nonzero. |

## 6.10 ARRAY MANIPULATION FUNCTIONS

In addition to the arbitrary array addressing and manipulation capabilities described in the preceding sections, MATLAB provides several functions that implement common array manipulations. Many of these manipulations are easy to follow, such as the following:

```
» A = [1 2 3;4 5 6;7 8 9]  % fresh data
A =
     1     2     3
     4     5     6
     7     8     9

» flipud(A)  % flip array in up-down direction
ans =
     7     8     9
     4     5     6
     1     2     3

» fliplr(A)  % flip array in the left-right direction
ans =
     3     2     1
     6     5     4
     9     8     7

» rot90(A)  % rotate array 90 degrees counterclockwise
ans =
     3     6     9
     2     5     8
     1     4     7
```

```
» rot90(A,2)  % rotate array 2*90 degrees counterclockwise
ans =
     9     8     7
     6     5     4
     3     2     1

» B = 1:12  % more data
B =
     1     2     3     4     5     6     7     8     9    10    11    12

» reshape(B,2,6)  % reshape to 2 rows, 6 columns, fill by columns
ans =
     1     3     5     7     9    11
     2     4     6     8    10    12

» reshape(B,3,4)  % reshape to 3 rows, 4 columns, fill by columns
ans =
      1    4     7    10
      2    5     8    11
      3    6     9    12

» reshape(A,3,2)  % A has more than 3*2 elements, OOPS!
??? To RESHAPE the number of elements must not change.

» reshape(A,1,9)  % stretch A into a row vector
ans =
     1     4     7     2     5     8     3     6     9
```

The following functions extract parts of an array to create another array:

```
» A  % remember what A is
A =
     1     2     3
     4     5     6
     7     8     9

» diag(A)  % extract diagonal using diag
ans =
     1
     5
     9

» diag(ans)  % remember this? same function, different action
ans =
     1     0     0
     0     5     0
     0     0     9
```

```
» triu(A)  % extract upper triangular part
ans =
     1     2     3
     0     5     6
     0     0     9

» tril(A)  % extract lower triangular part
ans =
     1     0     0
     4     5     0
     7     8     9

» tril(A) - diag(diag(A)) % lower triangular, no diagonal
ans =
     0     0     0
     4     0     0
     7     8     0
```

The following functions create arrays from other arrays:

```
» a = [1 2;3 4]  % a smaller data array
a =
     1     2
     3     4

» b = [0 1;-1 0]  % another smaller data array
b =
     0     1
    -1     0

» kron(a,b)  % the Kronecker tensor product of a and b
ans =
     0     1     0     2
    -1     0    -2     0
     0     3     0     4
    -3     0    -4     0

» kron(b,a)  % the Kronecker tensor product of b and a
ans =
     0     0     1     2
     0     0     3     4
    -1    -2     0     0
    -3    -4     0     0

» repmat(a,1,3)  % replicate a one-by-three times
ans =
     1     2     1     2     1     2
     3     4     3     4     3     4
```

```
» repmat(b,2,2)  % replicate b two-by-two
ans =
     0     1     0     1
    -1     0    -1     0
     0     1     0     1
    -1     0    -1     0
```

## 6.11 ARRAY SIZE

In those cases where the size of an array or vector is unknown and is needed for some mathematical manipulation, MATLAB provides two utility functions `size` and `length`:

```
» A = [1 2 3 4;5 6 7 8]
A =
     1     2     3     4
     5     6     7     8

» s = size(A)
s =
     2     4
```

With one output argument, the `size` function returns a row vector whose first element is the number of rows and whose second element is the number of columns.

```
» [r,c] = size(A)
r =
     2
c =
     4
```

With two output arguments, `size` returns the number of rows in the first variable and the number of columns in the second variable.

```
» r = size(A,1)  % number of rows
r =
     2
» c = size(A,2)  % number of columns
c =
     4
```

Called with two arguments, `size` returns either the number of rows or columns.

```
» length(A)
ans =
     4
```

returns the number of rows or the number of columns, whichever is larger.

```
» B = pi:0.01:2*pi;

» size(B)
ans =
      1    315
```

shows that B is a row vector, and

```
» length(B)
ans =
    315
```

returns the length of the vector.

```
» size([])
ans =
      0      0
```

shows that the empty matrix does indeed have zero size.

The preceding concepts are summarized in the following table:

| Array Size | Description |
|---|---|
| s=size(A) | Returns a row vector s, whose first element is the number of rows in A and whose second element is the number of columns in A. |
| [r,c]=size(A) | Returns two scalars r and c containing the number of rows and columns in A, respectively. |
| r=size(A,1) | Returns the number of rows A in the variable r. |
| c=size(A,2) | Returns the number of columns in A in the variable c. |
| n=length(A) | Returns max(size(A)) in the variable n. |

7

# Multidimensional Arrays

Prior versions of MATLAB were limited to two-dimensional arrays. For many applications this was sufficient. However, for some applications it is very convenient to utilize arrays having higher dimensions. As a result, MATLAB 5 supports arrays having an arbitrary number of dimensions.

For the most part, MATLAB 5 supports multidimensional arrays using the same functions and addressing techniques that apply to one- and two-dimensional arrays. In general, the third dimension is numbered by ***pages,*** and higher dimensions have no generic name. Thus, a three-dimensional array has rows, columns, and pages. Each page contains a two-dimensional array of rows and columns. In addition, just as all columns of a two-dimensional array must have the same number of rows and vice versa, all pages of a three-dimensional array must have the same number of rows and columns. One way to visualize three-dimensional arrays is to think of the residential listings (white pages) in a phone book. Each page has the same number of columns and the same number of names (rows) in each column. Even though there is no limit to the number of dimensions, three-dimensional arrays are used predominately in this chapter because they are more easily visualized and displayed.

## 7.1 ARRAY CONSTRUCTION

Multidimensional arrays can be created in several ways. For example,

```
» A = zeros(2,3)  % start with a 2-D array
A =
     0     0     0
     0     0     0

» A(:,:,2) = ones(2,3) % add a second page to go 3-D!
A(:,:,1) =
     0     0     0
     0     0     0

A(:,:,2) =
     1     1     1
     1     1     1

» A(:,:,3) = 4  % add a third page
A(:,:,1) =
     0     0     0
     0     0     0

A(:,:,2) =
     1     1     1
     1     1     1

A(:,:,3) =
     4     4     4
     4     4     4
```

The preceding approach starts with a two-dimensional array that is the first page of a three-dimensional array. Then other pages are added by straightforward array addressing. The array creation functions `zeros`, `ones`, `rand`, `randn`, and `repmat` can all be used to create multidimensional arrays. For example,

```
» B = ones(1,3,4)  % one row, three columns, four pages
B(:,:,1) =
     1     1     1
B(:,:,2) =
     1     1     1
B(:,:,3) =
     1     1     1
B(:,:,4) =
     1     1     1

» repmat(eye(3),[1 1 2])  % replicate eye(3) to two pages
ans(:,:,1) =
     1     0     0
     0     1     0
     0     0     1
```

```
ans(:,:,2) =
     1     0     0
     0     1     0
     0     0     1
```

The `cat` function creates multidimensional arrays from lower-dimensional arrays:

```
» a = zeros(2);
» b = ones(2);
» c = repmat(2,2,2);
» D = cat(3,a,b,c)  % conCATenate a,b,c along the 3rd dimension
D(:,:,1) =
     0    0
     0    0
D(:,:,2) =
     1    1
     1    1
D(:,:,3) =
     2    2
     2    2

» D = cat(4,a,b,c)  % try the 4th dimension!
D(:,:,1,1) =
     0     0
     0     0
D(:,:,1,2) =
     1     1
     1     1
D(:,:,1,3) =
     2     2
     2     2

» size(D)
ans =
     2     2     1     3
```

D has two rows, two columns, one page, and three fourth dimension parts.

## 7.2 ARRAY MATHEMATICS AND MANIPULATION

As additional dimensions are created, array mathematics and manipulation becomes more complicated. Scalar-array arithmetic remains straightforward, but array-array arithmetic requires that the two arrays have the same size in all dimensions.

MATLAB provides several functions for the manipulation of multidimensional arrays. The function `squeeze` eliminates ***singleton dimensions*** (i.e., eliminates dimensions of size 1). For example,

```
» E = squeeze(D)  % squeeze dimension 4 down to dimension 3
E(:,:,1) =
     0     0
     0     0
E(:,:,2) =
     1     1
     1     1
E(:,:,3) =
     2     2
     2     2
» size(E)
ans =
     2     2     3
```

E has two rows, two columns, and three pages.

The function reshape allows you to change the row, column, page, and higher-order dimensions while not changing the total number of elements. For example,

```
» F = cat(3,2+zeros(2,4),ones(2,4),zeros(2,4)) % new 3-D array
F(:,:,1) =
     2     2     2     2
     2     2     2     2
F(:,:,2) =
     1     1     1     1
     1     1     1     1
F(:,:,3) =
     0     0     0     0
     0     0     0     0

» G = reshape(F,[3 2 4]) % change it to 3 rows, 2 columns, 4 pages
G(:,:,1) =
     2     2
     2     2
     2     2
G(:,:,2) =
     2     1
     2     1
     1     1
G(:,:,3) =
     1     1
     1     0
     1     0
G(:,:,4) =
     0     0
     0     0
     0     0
```

```
» H = reshape(F,[4 3 2]) % or 4 rows, 3 columns, 2 pages
H(:,:,1) =
     2     2     1
     2     2     1
     2     2     1
     2     2     1
H(:,:,2) =
     1     0     0
     1     0     0
     1     0     0
     1     0     0
```

The preceding reshaping is confusing until you get comfortable visualizing arrays in multidimensional space. The reshaping process follows the same pattern as that for two-dimensional arrays. Data are gathered first by rows, then by columns, then by pages, and so on into higher dimensions. That is, all rows in the first column are gathered, then all rows in the second column, etc. Then when the first page has been gathered, one moves on to the second page and starts over with all rows in the first column.

The order in which array elements are gathered is the order in which the functions `sub2ind` and `ind2sub` consider single index addressing:

```
» sub2ind(size(F),1,1,1) % 1st row, 1st column, 1st page is element 1
ans =
     1

» sub2ind(size(F),1,2,1) % 1st row, 2nd column, 1st page is element 3
ans =
     3

» sub2ind(size(F),1,2,3) % 1st element, 2nd column, 3rd page is element 19
ans =
    19
```

The multidimensional equivalent to `flipud` and `fliplr` is `flipdim`. For example,

```
» a = [1 2 3;4 5 6];
» b = [7 8 9;10 11 12];
» c = [13 14 15;16 17 18];
» M = cat(3,a,b,c) % new data
M(:,:,1) =
     1     2     3
     4     5     6
M(:,:,2) =
     7     8     9
    10    11    12
```

```
M(:,:,3) =
    13    14    15
    16    17    18

» flipdim(M,1) % flip rows on all pages
ans(:,:,1) =
     4     5     6
     1     2     3
ans(:,:,2) =
    10    11    12
     7     8     9
ans(:,:,3) =
    16    17    18
    13    14    15

» flipdim(M,2) % flip columns on all pages
ans(:,:,1) =
     3     2     1
     6     5     4
ans(:,:,2) =
     9     8     7
    12    11    10
ans(:,:,3) =
    15    14    13
    18    17    16

» flipdim(M,3) % flip pages, page 3 becomes page 1
ans(:,:,1) =
    13    14    15
    16    17    18
ans(:,:,2) =
     7     8     9
    10    11    12
ans(:,:,3) =
     1     2     3
     4     5     6
```

The function `shiftdim` shifts the dimensions of an array. That is, if an array has `r` rows, `c` columns, and `p` pages, a shift by one dimension creates an array with `c` rows, `p` columns, and `r` pages. For example,

```
» M % recall data: 2 rows, 3 columns, 3 pages
M(:,:,1) =
     1     2     3
     4     5     6

M(:,:,2) =
     7     8     9
    10    11    12
```

```
M(:,:,3) =
    13    14    15
    16    17    18

» shiftdim(M,1) % shift by 1: 3 rows, 3 columns, 2 pages
ans(:,:,1) =
     1     7    13
     2     8    14
     3     9    15
ans(:,:,2) =
     4    10    16
     5    11    17
     6    12    18

» shiftdim(M,2) % shift by 2: 3 rows, 2 columns, 3 pages
ans(:,:,1) =
     1     4
     7    10
    13    16
ans(:,:,2) =
     2     5
     8    11
    14    17
ans(:,:,3) =
     3     6
     9    12
    15    18

» shiftdim(M,3) % shift by 3: back to the original M
ans(:,:,1) =
     1     2     3
     4     5     6
ans(:,:,2) =
     7     8     9
    10    11    12
ans(:,:,3) =
    13    14    15
    16    17    18
```

A generalization of `shiftdim` allowing arbitrary reorientation of an array is accomplished with the functions `permute` and `ipermute`. For example,

```
» permute(M,[2 3 1]) % same as shiftdim(M,1)
ans(:,:,1) =
     1     7    13
     2     8    14
     3     9    15
```

```
ans(:,:,2) =
     4    10    16
     5    11    17
     6    12    18

» shiftdim(M,1)
ans(:,:,1) =
     1     7    13
     2     8    14
     3     9    15
ans(:,:,2) =
     4    10    16
     5    11    17
     6    12    18
```

In the preceding, `[2 3 1]` says let the second dimension become first, the third become the second, and the first become the third. The function `ipermute` performs the inverse permutation using the same (rather than inverse) permutation vector:

```
» N = permute(M,[3 2 1]) % permute M
N(:,:,1) =
     1     2     3
     7     8     9
    13    14    15
N(:,:,2) =
     4     5     6
    10    11    12
    16    17    18

» ipermute(N,[3 2 1]) % undo it
ans(:,:,1) =
     1     2     3
     4     5     6
ans(:,:,2) =
     7     8     9
    10    11    12
ans(:,:,3) =
    13    14    15
    16    17    18
```

## 7.3 ARRAY SIZE

As shown earlier in this chapter, the function `size` returns the size of an array along each of its dimensions. For example,

```
» M  % recall 3-D array M
M(:,:,1) =
     1     2     3
     4     5     6
```

```
M(:,:,2) =
     7     8     9
    10    11    12
M(:,:,3) =
    13    14    15
    16    17    18

» size(M) % return size in all dimensions as a row vector
ans =
     2     3     3

» [r,c,p] = size(M) % return size in individual variables
r =
     2
c =
     3
p =
     3

» r = size(M,1) % return just dimension 1 size (rows)
r =
     2
» c = size(M,2) % return just dimension 2 size (columns)
c =
     3
» p = size(M,3) % return just dimension 3 size (pages)
p =
     3
```

When the number of dimensions is unknown or variable, the function `ndims` is useful:

```
» ndims(M)
ans =
     3
» ndims(M(:,:,1)) % just the 2-D first page of M
ans =
     2
```

In this last example, `M(:,:,1)` is a two-dimensional array because it has only one page. That is, it has a singleton third dimension.

To summarize, this chapter illustrates the multidimensional array capabilities of MATLAB 5. It should be clear that added dimensions increase computational and addressing complexity. As a result, multidimensional arrays should only be used where they make sense, such as storing data that are a function of more than two variables.

# 8

# Relational and Logical Operations

In addition to traditional mathematical operations, MATLAB supports relational and logical operations. You may be familiar with these if you have had some experience with other programming languages. The purpose of these operators and functions is to provide answers to True/False questions. One important use of this capability is to control the flow or order of execution of a series of MATLAB commands (usually in an M-file) based on the results of True/False questions.

As inputs to all relational and logical expressions, MATLAB considers any nonzero number to be True and zero to be False. The output of all relational and logical expressions produces logical arrays with One for True and Zero for False. Logical arrays are a special type of numerical array that can be used for logical array addressing, as shown in the chapter on arrays (Chapter 6), as well as in any numerical expression.

## 8.1 RELATIONAL OPERATORS

MATLAB relational operators include all common comparisons:

| Relational Operator | Description |
|---|---|
| < | less than |
| <= | less than or equal to |
| > | greater than |
| >= | greater than or equal to |
| == | equal to |
| ~= | not equal to |

MATLAB relational operators can be used to compare two arrays of the same size, or to compare an array to a scalar. In the latter case, scalar expansion is used to compare the scalar to each array element and the result has the same size as the array. Examples include the following:

```
» A = 1:9, B = 9-A
A =
     1     2     3     4     5     6     7     8     9
B =
     8     7     6     5     4     3     2     1     0

» tf = A>4
tf =
     0     0     0     0     1     1     1     1     1
```

finds elements of A that are greater than 4. Zeros appear in the result where $A \leq 4$ and ones appear where $A > 4$.

```
» tf = (A==B)
tf =
     0     0     0     0     0     0     0     0     0
```

finds elements of A that are equal to those in B.

> Note that = and == mean two different things: == compares two variables and returns ones where they are equal and zeros where they are not; on the other hand, = is used to assign the output of an operation to a variable.

```
» tf = B - (A>2)
tf =
     8     7     5     4     3     2     1     0    -1
```

finds where A > 2 and subtracts the resulting vector from B. This example shows that since the output of logical operations are numerical arrays of ones and zeros, they can be used in mathematical operations, too.

```
» B = B + (B==0)*eps
B =
  Columns 1 through 7
    8.0000    7.0000    6.0000    5.0000    4.0000    3.0000    2.0000
  Columns 8 through 9
    1.0000    0.0000
```

is a demonstration of how to replace zero elements in an array with the special MATLAB number eps, which is approximately 2.2e-16. This particular expression is sometimes useful to avoid dividing by zero, as in

```
» x = (-3:3)/3
x =
   -1.0000  -0.6667 -0.3333 0 0.3333 0.6667 1.0000

» sin(x)./x

Warning: Divide by zero
ans =
    0.8415    0.9276    0.9816       NaN    0.9816    0.9276    0.8415
```

Computing the function sin(x)/x gives a warning because the fifth data point is zero. Since sin(0)/0 is undefined, MATLAB returns NaN (meaning Not-a-Number) at that location in the result. Try again, after replacing the zero with eps:

```
» x = x + (x==0)*eps;
» sin(x)./x
ans =
    0.8415    0.9276    0.9816    1.0000    0.9816    0.9276    0.8415
```

Now sin(x)/x for x=0 gives the correct limiting answer.

## 8.2 LOGICAL OPERATORS

Logical operators provide a way to combine or negate relational expressions. MATLAB logical operators include the following:

| Logical Operator | Description |
|---|---|
| & | AND |
| \| | OR |
| ~ | NOT |

Some examples of the use of logical operators are as follows:

```
» A = 1:9; B = 9-A;
» tf = A>4
tf =
     0    0    0    0    1    1    1    1    1
```

finds where A is greater than 4.

```
» tf = ~(A<4)
tf =
     1    1    1    1    0    0    0    0    0
```

negates the preceding result (i.e., swaps where the ones and zeros appear).

```
» tf = (A>2) & (A<6)
tf =
     0    0    1    1    1    0    0    0    0
```

returns ones where A is greater than 2 AND less than 6.

Finally, the preceding capabilities make it easy to generate arrays representing signals with discontinuities or signals that are composed of segments of other signals. The basic idea is to multiply those values in an array that you wish to keep with ones, and multiply all other values with zeros. For example,

```
» x = linspace(0,10,100);    % create data
» y = sin(x);                % compute sine
» z = (y>=0).*y;             % set negative values of sin(x) to zero
» z = z + 0.5*(y<0);         % where sin(x) is negative add 1/2
» z = (x<=8).*z;             % set values past x=8 to zero

» plot(x,z)
» xlabel('x'), ylabel('z=f(x)'),
» title('Figure 8.1: A Discontinuous Signal')
```

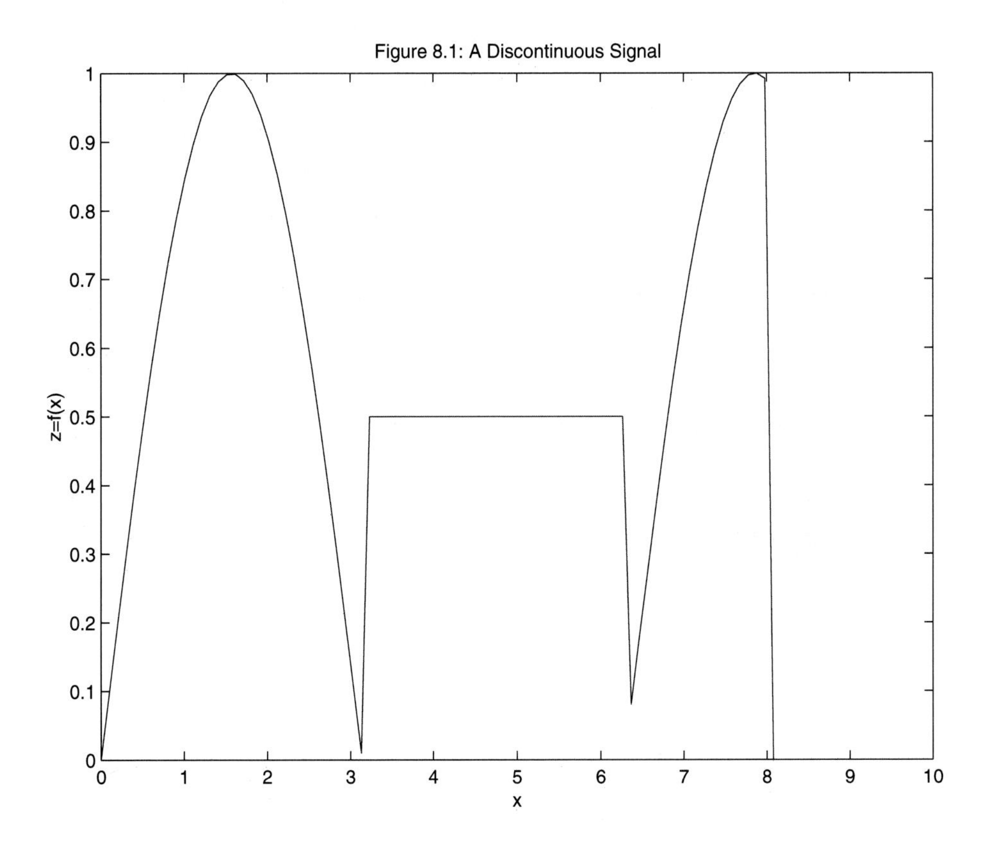


## 8.3 RELATIONAL AND LOGICAL FUNCTIONS

In addition to the preceding basic relational and logical operators, MATLAB provides a number of additional relational and logical functions, including the following:

| Function | Description |
|---|---|
| `xor(x,y)` | Exclusive OR operation. Return True (1) for each element where either `x` or `y` is nonzero (True). Return False (0) where both `x` and `y` are zero (False) or both are nonzero (True). |
| `any(x)` | Return True (1) if any element in a vector `x` is nonzero. Return True (1) for each column in a matrix `x` that has nonzero elements. |
| `all(x)` | Return True (1) if all elements in a vector `x` are nonzero. Return True (1) for each column in a matrix `x` that has all nonzero elements. |

In addition to these functions, MATLAB provides numerous functions that test for the existence of specific values or conditions and return logical results:

| Function | Description |
| --- | --- |
| `isa(X,'name')` | True if `X` has object class `'name'`. |
| `iscell(X)` | True if argument is a cell array. |
| `iscellstr(S)` | True if argument is a cell array of strings. |
| `ischar(S)` | True if argument is a character string. |
| `isempty(X)` | True if argument is empty. |
| `isequal(A,B)` | True if `A` and `B` have identical contents and dimensions. |
| `isfield(S,'name')` | True if `'name'` is a field of structure `S`. |
| `isfinite(X)` | True where elements are finite. |
| `isglobal(X)` | True if argument is a global variable. |
| `ishandle(h)` | True if argument is a valid object handle. |
| `ishold` | True if current plot hold state in ON. |
| `isieee` | True if computer performs IEEE arithmetic. |
| `isinf(X)` | True where elements are infinite. |
| `isletter(S)` | True where elements are letters of the alphabet. |
| `islogical(X)` | True if argument is a logical array. |
| `ismember(A,B)` | True where elements of `A` are also in `B`. |
| `isnan(X)` | True where elements are `NAN`s. |
| `isnumeric(X)` | True if argument is a numeric array. |
| `isppc` | True for Macintosh with PowerPC processor. |
| `isprime(X)` | True where elements are prime. |
| `isreal(X)` | True if argument has no imaginary part. |
| `isspace(S)` | True where elements are whitespace characters. |
| `issparse(A)` | True if argument is a sparse matrix. |
| `isstruct(S)` | True if argument is a structure. |
| `isstudent` | True if Student Edition of MATLAB. |
| `isunix` | True if computer is UNIX. |
| `isvms` | True if computer is VMS. |

## 8.4 NaNs AND EMPTY ARRAYS

NaNs (Not-a-Numbers) and empty arrays ([ ]) require special treatment in MATLAB, especially when used in logical or relational expressions. According to IEEE mathematical standards, almost all operations on NaNs result in NaNs. For example,

```
» a = [1 2 nan inf nan]
a =
     1     2   NaN   Inf   NaN

» b = 2*a
b =
     2     4   NaN   Inf   NaN

» c = sqrt(a)
c =
    1.0000    1.4142       NaN       Inf       NaN

» d = (a==nan)
d =
     0     0     0     0     0

» f = (a~=nan)
f =
     1     1     1     1     1
```

The first two computations give NaN results for NaN inputs. However, the final two relational computations produce somewhat surprising results. (a==nan) produces all zeros or False results even when NaN is compared to NaN. At the same time, (a~=nan) produces all ones or True results. Thus, individual NaNs are not equal to each other. As a result of this property of NaNs, MATLAB has a built-in logical function for finding NaNs:

```
» g = isnan(a)
g =
     0     0     1     0     1
```

This function makes it possible to find the indices of NaNs using the find command. For example,

```
» i = find(isnan(a))  % find indices of NaNs
i =
     3     5
» a(i) = zeros(size(i))  % changes NaNs in a to zeros
a =
     1     2     0   Inf     0
```

While NaNs are well defined mathematically by IEEE standards, empty arrays are defined by the creators of MATLAB and have their own interesting properties. Empty arrays are simply that: They are MATLAB variables having zero length in one or more dimensions:

```
» size([])  % simplest empty array
ans =
     0     0
» c = zeros(0,5)  % how about an empty array with multiple columns!
c =
   Empty matrix: 0-by-5
» size(c)
ans =
     0     5

» d = ones(4,0)  % an empty array with multiple rows!
d =
   Empty matrix: 4-by-0
» size(d)
ans =
     4     0
» length(d)  % it's length is zero even though it has 4 rows
ans =
     0
```

The preceding may seem strange, but allowing an empty array to have zero length in any dimension is sometimes useful. [] is just the simplest empty array.

In MATLAB, many functions return empty arrays when no other result is appropriate. Perhaps the most common example is the find function:

```
» x = -2:2  % new data
x =
    -2    -1     0     1     2

» y = find(x>2)
y =
     []
```

In this example, x contains no values greater than 2, so there are no indices to return. To test for empty results, MATLAB provides the logical function isempty:

```
» isempty(y)
ans =
     1
```

When performing relational tests for the empty array, it is important to use isempty because of ambiguities such as

```
» c==[] % compare 0-by-5 to 0-by-0 empty arrays
Warning: X == [] is technically incorrect. Use isempty(X) instead.
ans =
     0

» a = [] % create an empty variable
a =
     []
» a==[] % equal, but still not good
Warning: X == [] technically incorrect. Use isempty(X) instead.
ans =
     1
```

In MATLAB, the empty array is not equal to any nonzero array or scalar. This fact leads to the following example:

```
» y = [];
» a = (y==0)
Warning: Future versions will return empty for empty == scalar comparisons.
a =
     0
» a = (y==1)
Warning: Future versions will return empty for empty == scalar comparisons.
a =
     0
» b = (y~=0)
Warning: Future versions will return empty for empty ~= scalar comparisons.
b =
     1
```

which shows that an empty array is not equal to a scalar and that at some point MATLAB 5 will define empty == scalar as being empty. Until then one must be careful in situations like the following:

```
» find(y==0)
Warning: Future versions will return empty for empty == scalar comparisons.
ans =
     []
```

says there are no indices to return. Likewise,

```
» b = (y~=0)
Warning: Future versions will return empty for empty ~= scalar comparisons.
b =
     1
```

An empty matrix is not equal to a scalar yet again. But

```
» j = find(y~=0)
Warning: Future versions will return empty for empty ~= scalar comparisons.
j =
     1
```

Now there is an index even though `y` has zero size. Therefore, until the promised changes in empty array treatment are made, one must check variables with `isempty` before conducting relational tests. The current interpretation usually leads to problems since `y(find(y~=0))` does not exist. But the new treatment will lead to

```
» find([])
ans =
     []
```

which is generally what is desired.

# 9

# Set, Bit, and Base Functions

## 9.1 SET FUNCTIONS

Since arrays are ordered collections of values, they can be thought of as sets. With that understanding, MATLAB provides several basic functions for testing and comparing sets. The simplest test is for equality:

```
» a = rand(2,5);  % random array
» b = randn(2,5); % a different random array

» isequal(a,b) % a and b are not equal
ans =
     0
» isequal(a,a) % but a is certainly equal to a
ans =
     1
```

For two arrays to be equal they must have the same dimensions and the same contents. The function `unique` removes duplicate items from a given set:

```
» a = [2:2:8;4:2:10] % new data
a =
     2     4     6     8
     4     6     8    10

» unique(a) % unique elements sorted into a column
ans =
     2
     4
     6
     8
    10
```

`unique` returns a sorted column vector because removal of duplicate values makes it impossible to maintain the array dimensions. Set membership is determined with the function `ismember`:

```
» a = 1:9
a =
     1     2     3     4     5     6     7     8     9
» b = 2:2:9
b =
     2     4     6     8

» ismember(a,b)
ans =
     0     1     0     1     0     1     0     1     0
» ismember(b,a)
ans =
     1     1     1     1
```

For vector arguments, `ismember` returns a logical array the same size as its first argument, with ones appearing at those indices where the two vectors share common values.

```
» A = eye(3);  % new data
» B = ones(3);

» ismember(A,B) % those elements in A that are also in B
ans =
     1
     0
     0
     0
     1
     0
```

```
     0
     0
     1

» ismember(B,A)
ans =
     1
     1
     1
     1
     1
     1
     1
     1
     1
```

For two-dimensional arrays `ismember` returns a column vector, with ones appearing at those *single* indices of the first argument where the two arguments share common values.

Set arithmetic is accomplished with the functions `union`, `intersect`, `setdiff`, and `setxor`. Examples of the use of these functions include the following:

```
» a % recall prior data
a =
     1     2     3     4     5     6     7     8     9
» b
b =
     2     4     6     8

» union(a,b) % union of a and b
ans =
     1     2     3     4     5     6     7     8     9

» intersect(a,b) % intersection of a and b
ans =
     2     4     6     8

» setxor(a,b) % set exclusive or
ans =
     1     3     5     7     9

» setdiff(a,b) % values in a that are not in b
ans =
     1     3     5     7     9

» setdiff(b,a) % values in b that are not in a
ans =
     []
```

```
» union(A,B) % matrix inputs produce sorted column arrays
ans =
     0
     1
     2
     3
     4
     5
     6
     7
     8
```

## 9.2 BIT FUNCTIONS

In addition to the logical operators discussed in the last chapter, MATLAB provides functions that allow logical operations on individual bits of floating-point integers. Because MATLAB represents all numeric values as double-precision floating-point numbers (IEEE arithmetic on most platforms), integers are represented as *flints*—floating-point integers. As a result, IEEE arithmetic compliant computers can uniquely represent integers in the range of 0 to $2^{53} - 1$. The MATLAB bitwise functions `bitand`, `bitcmp`, `bitor`, `bitxor`, `bitset`, `bitget`, and `bitshift` work on integers in this range.

The maximum integer representable using bit operations is `bitmax`, which is $2^{53} - 1$ on IEEE arithmetic computers.

```
» format hex % change format to hexadecimal
» bitmax
ans =
   433fffffffffffff
```

Examples of bit operations include the following:

```
» a = (2^24)-1 % data
a =
    416fffffe0000000
» b = 123456789 % data
b =
    419d6f3454000000

» bitand(a,b) %  (a & b)
ans =
   4156f34540000000

» bitor(a,b) %  (a | b)
ans =
   419ffffffc000000
```

```
» bitxor(a,b) % xor(a,b)
ans =
   419e90cba8000000

» bitcmp(0,10) % complement 0 as a 10 bit number
ans =
   408ff80000000000

» bitget(a,1) % get first bit of a
ans =
   3ff0000000000000

» bitset(a,30) % set 30th bit of a to 1
ans =
   41c07fffff800000

» format short g % reset display format
```

## 9.3 BASE CONVERSIONS

MATLAB provides a number of utility functions for converting decimal numbers to other bases in the form of character strings. Conversions between decimals and binary numbers are performed by the functions `dec2bin` and `bin2dec`:

```
» a = dec2bin(17) % find binary representation of 17
a =
10001
» class(a) % result is a character string
ans =
char
» bin2dec(a) % convert a back to decimal
ans =
    17
» class(ans) % result is a double precision decimal
ans =
double
```

Conversions between decimals and hexadecimals are performed by `dec2hex` and `hex2dec`:

```
» a = dec2hex(2047) % hex representation of 2047
a =
7FF
» class(a) % result is a character string
ans =
char
```

```
» hex2dec(a) % convert a back to decimal
ans =
        2047
» class(ans) % result is a double precision decimal
ans =
double
```

Conversions between decimals and any base between 2 and 36 are performed by `dec2base` and `base2dec`:

```
» a = dec2base(26,3)
a =
222
» class(a)
ans =
char
» base2dec(a,3)
ans =
    26
```

Base 36 is the maximum usable base because it uses the numbers `0` through `9` and letters `A` through `Z` to represent the 36 distinct digits of a base 36 number.

# 10

# Character Strings

MATLAB's true power is in its ability to crunch numbers. However, there are times when it is desirable to manipulate text, such as when putting labels and titles on plots. In MATLAB, text is referred to as character strings or simply strings.

## 10.1 STRING CONSTRUCTION

Character strings in MATLAB are special numerical arrays of ASCII values that are displayed as their character string representation. For example,

```
» t = 'How about this character string?'
t =
How about this character string
» size(t)
ans =
     1    32
```

```
» whos
  Name      Size         Bytes  Class
  ans       1x2             16  double array
  t         1x32            64  char array

Grand total is 34 elements using 80 bytes
```

A character string is simple text surrounded by single quotes. Each character in a string is one element in an array that requires 2 bytes per character for storage. This is different from the 8 bytes per element required for numerical or double arrays, as shown previously.

To see the underlying ASCII representation of a character string, one need only perform some arithmetic operation on the string or use the dedicated function `double`. For example,

```
» u = double(t)
u =
  Columns 1 through 12
    72   111   119    32    97    98   111   117   116    32   116   104
  Columns 13 through 24
   105   115    32    99   104    97   114    97    99   116   101   114
  Columns 25 through 32
    32   115   116   114   105   110   103    63
» abs(t)
ans =
  Columns 1 through 12
    72   111   119    32    97    98   111   117   116    32   116   104
  Columns 13 through 24
   105   115    32    99   104    97   114    97    99   116   101   114
  Columns 25 through 32
    32   115   116   114   105   110   103    63
```

The function `char` performs the inverse transformation:

```
» char(u)
ans =
How about this character string?
```

Numerical values less than 0 produce a warning message when converted to character; values greater than 255 are converted after finding their remainder—that is, `rem(n,256)` is computed. For example,

```
» a = double('a')
a =
    97
» char(a)
ans =
a
```

```
» char(a+256) % adding 256 does not change result
ans =
a
» char(a-256) % negative arguments are flagged
Warning: Out of range or non-integer values truncated during
conversion from double to character.
ans =
a
```

Since strings are arrays, they can be manipulated with all the array manipulation tools available in MATLAB. For example,

```
» u = t(16:24)
u =
character
```

Strings are addressed just like arrays. Here elements 16 through 24 contain the word `character`.

```
» u = t(24:-1:16)
u =
retcarahc
```

This is the word `character` spelled backward.

```
» u = t(16:24)'
u =
c
h
a
r
a
c
t
e
r
```

Using the transpose operator changes the word `character` to a column.

```
» v = 'I can''t find the manual!'
v =
I can't find the manual!
```

Single quotes within a character string are symbolized by two consecutive quotes.

String concatenation follows directly from array concatenation:

```
» u = 'If a woodchuck could chuck wood,';
```

```
» v = ' how much wood could a woodchuck chuck?';

» w = [u v]
w =
If a woodchuck could chuck wood, how much wood could a woodchuck chuck?
```

The function `disp` allows you to display a string without printing its variable name. For example,

```
» disp(u)
If a woodchuck could chuck wood,
```

Note that the `u =` statement is suppressed. This is useful for displaying help text within a script file.

> As with other arrays, character strings can have multiple rows, but each row must have an equal number of columns. Therefore, blanks are explicitly required to make all rows the same length. For example,

```
» v = ['Character strings having more than'
       'one row must have the same number '
       'of columns just like arrays!      ']
v =
Character strings having more than
one row must have the same number
of columns just like arrays!
```

The functions `char`, `str2mat`, and `strvcat` create multiple-row string arrays from individual strings of varying lengths:

```
» lawyers = char('Cochran','Shapiro','Clark','Darden')
lawyers =
Cochran
Shapiro
Clark
Darden
» lawyers = str2mat('Cochran','Shapiro','Clark','Darden')
lawyers =
Cochran
Shapiro
Clark
Darden
» lawyers = strvcat('Cochran','Shapiro','Clark','Darden')
lawyers =
Cochran
Shapiro
```

```
Clark
Darden
» size(lawyers)
ans =
     4     7
```

The functions `char` and `str2mat` implement the same algorithm, whereas `strvcat` is slightly different. `strvcat` ignores empty string inputs, whereas `char` and `str2mat` insert blank rows for empty strings. For example,

```
» str2mat('one','','two','three')
ans =
one

two
three
» strvcat('one','','two','three')
ans =
one
two
three
```

Horizontal concatenation of string arrays having the same number of rows is accomplished by the function `strcat`:

```
» a = char('apples','bananas')
a =
apples
bananas
» b = char('oranges','grapefruit')
b =
oranges
grapefruit

» strcat(a,b)
ans =
applesoranges
bananasgrapefruit
```

Once a string array is created with padded blanks, the function `deblank` is useful for eliminating the extra blanks from individual rows extracted from the array:

```
» c = lawyers(3,:)
c =
Clark
```

```
» size(c)
ans =
     1     7
» d = deblank(lawyers(3,:))
d =
Clark
» size(d)
ans =
     1     5
```

## 10.2 NUMBERS TO STRINGS TO NUMBERS

There are numerous contexts where it is desirable to convert numerical results into character strings and to extract numerical data from character strings. MATLAB provides the functions `int2str`, `num2str`, `mat2str`, `sprintf`, and `fprintf` for converting numerical results into character strings. Examples of the first three functions include the following:

```
» int2str(eye(3)) % convert integer arrays
ans =
1  0  0
0  1  0
0  0  1
» size(ans) % it's a character array, not a numerical matrix
ans =
     3     7
» num2str(rand(2,4)) % convert noninteger arrays
ans =
0.95013     0.60684      0.8913     0.45647
0.23114     0.48598      0.7621     0.01850
» size(ans) % again its a character array
ans =
     2    40
» mat2str(pi*eye(2)) % convert to MATLAB input syntax form!
ans =
[3.14159265358979 0; 0 3.14159265358979]
» size(ans)
ans =
     1    40
» fprintf('%.4g\n',sqrt(2)) % display in Command window
1.414
» sprintf('%.4g',sqrt(2)) % create character string
ans =
1.414
» size(ans)
ans =
     1     5
```

The last two functions are general-purpose conversion functions that closely resemble their ANSI C language counterparts. As a result, they offer the most flexibility. Normally, `fprintf` is used to convert numerical results into ASCII format and append it to a data file. However, if no file identifier is provided as the first argument to `fprintf` or if a file identifier of 1 is used, the resulting output is displayed in the *Command* window. `sprintf` is identical to `fprintf` except that it simply creates a character array that can be displayed, passed to a function, or modified like any other character array. Because `sprintf` and `fprintf` are nearly identical, consider the usage of `sprintf` in the following example:

```
» radius = sqrt(2);
» area = pi * radius^2;

» s = sprintf('A circle of radius %.5g has an area of %.5g.',radius,area)
s =
A circle of radius 1.4142 has an area of 6.2832.
```

Here `%.5g`, the format specification for the variable `radius`, indicates that five significant digits in general conversion format are desired. The most common usage of `sprintf` is to create a character string for annotating a graph, for displaying numerical values in a graphical user interface, or for creating a sequence of data file names. A rudimentary example of this last usage is as follows:

```
» i = 3;
» fname = sprintf('mydata%.0f.dat',i)
fname =
mydata3.dat
```

In the past, the functions `int2str` and `num2str` were nothing more than a simple call to `sprintf` with `%.0f` and `%.4g` format specifiers, respectively. In MATLAB 5, `int2str` and `num2str` were enhanced to work with numerical arrays, as illustrated previously. As a result of their former simplicity, it was common in prior versions of MATLAB to create the preceding example as

```
» s = ['A circle of radius ' num2str(radius) ' has an area of ' ...
num2str(area) '.']
s =
A circle of radius 1.4142 has an area of 6.2832.
```

While the result is the same as that shown previously, this latter form requires more computational effort, is more prone to typographical errors such as missing spaces or single quotes, and requires more effort to read. As a result, it is suggested that usage of `int2str` and `num2str` be limited to conversion of arrays, as illustrated earlier in this section. In almost all other cases, it is more productive to use `sprintf` directly.

The help text for `sprintf` concisely describes its use:

```
» help sprintf
SPRINTF Write formatted data to string.
[S,ERRMSG] = SPRINTF(FORMAT,A,...) formats the data in the real
part of matrix A (and in any additional matrix arguments), under
control of the specified FORMAT string, and returns it in the
MATLAB string variable S. ERRMSG is an optional output argument
that returns an error message string if an error occurred or an
empty matrix if an error did not occur. SPRINTF is the same as
FPRINTF except that it returns the data in a MATLAB string
variable rather than writing it to a file.

FORMAT is a string containing C language conversion specifications.
Conversion specifications involve the character %, optional flags,
optional width and precision fields, optional subtype specifier, and
conversion characters d, i, o, u, x, X, f, e, E, g, G, c, and s.
See the Language Reference Guide or a C manual for complete details.

The special formats \n, \r, \t, \b, \f can be used to produce linefeed,
carriage return, tab, backspace, and formfeed characters respectively.
Use \\ to produce a backslash character and %% to produce the percent
character.

SPRINTF behaves like ANSI C with certain exceptions and extensions.
These include:
1. The following non-standard subtype specifiers are supported for
conversion characters o, u, x, and X.
t - The underlying C datatype is a float rather than an
    unsigned integer.
b - The underlying C datatype is a double rather than an
    unsigned integer.
For example, to print out in hex a double value use a format like
'%bx'.

2. SPRINTF is "vectorized" for the case when A is nonscalar. The
format string is recycled through the elements of A (columnwise)
until all the elements are used up. It is then recycled in a similar
manner through any additional matrix arguments.

Examples
   sprintf('%0.5g',(1+sqrt(5))/2)       1.618
   sprintf('%0.5g',1/eps)               4.5036e+15
   sprintf('%15.5f',1/eps)              4503599627370496.00000
   sprintf('%d',round(pi))              3
   sprintf('%s','hello')                hello
   sprintf('The array is %dx%d.',2,3)   The array is 2x3.
   sprintf('\n') is the line termination character on all platforms.

See also FPRINTF, SSCANF, NUM2STR, INT2STR.
```

The following table shows how `pi` is displayed under a variety of conversion specifications:

| Command | Result |
|---|---|
| `sprintf('%.0e\n',pi)` | `3e+00` |
| `sprintf('%.1e\n',pi)` | `3.1e+00` |
| `sprintf('%.3e\n',pi)` | `3.142e+00` |
| `sprintf('%.5e\n',pi)` | `3.14159e+00` |
| `sprintf('%.10e\n',pi)` | `3.1415926536e+00` |
| `sprintf('%.0f\n',pi)` | `3` |
| `sprintf('%.1f\n',pi)` | `3.1` |
| `sprintf('%.3f\n',pi)` | `3.142` |
| `sprintf('%.5f\n',pi)` | `3.14159` |
| `sprintf('%.10f\n',pi)` | `3.1415926536` |
| `sprintf('%.0g\n',pi)` | `3` |
| `sprintf('%.1g\n',pi)` | `3` |
| `sprintf('%.3g\n',pi)` | `3.14` |
| `sprintf('%.5g\n',pi)` | `3.1416` |
| `sprintf('%.10g\n',pi)` | `3.141592654` |
| `sprintf('%8.0g\n',pi)` | `       3` |
| `sprintf('%8.1g\n',pi)` | `       3` |
| `sprintf('%8.3g\n',pi)` | `    3.14` |
| `sprintf('%8.5g\n',pi)` | `  3.1416` |
| `sprintf('%8.10g\n',pi)` | `3.141592654` |

In this table, the format specifier `e` signifies exponential notation, `f` signifies fixed-point notation, and `g` signifies the use of `e` or `f`, whichever is shorter. Note that for the `e` and `f` formats, the number to the right of the decimal point says how many digits to the right of the decimal point to display. On the other hand, in the `g` format, the number to the right of the decimal specifies the total number of digits to display. In addition, note that in the last five entries, a width of eight characters is specified for the result, and the result is right justified. In the very last case, the 8 is ignored because more than eight digits were specified.

Though it is not as common, sometimes it is necessary to convert or extract a numerical value from a character string. The MATLAB functions `eval`, `sscanf`, and `str2num` provide this capability. For example,

```
» s = num2str(pi*eye(2)) % convert array to string
s =
3.1416          0
     0     3.1416
» size(s) % it's a 2-by-17 string array
ans =
     2    17
» m = str2num(s) % str2num is the inverse of num2str
m =
         3.1416             0
              0        3.1416
» size(m) % it's a 2-by-2 matrix
ans =
     2     2
» pi*eye(2) - m % resolution is lost in this process!
ans =
  -7.3464e-06             0
            0   -7.3464e-06
```

The function `str2num` only handles numerically valued string arrays. That is, the input to `str2num` cannot contain expressions or variables. On the other hand, the function `eval` calls in the entire MATLAB interpreter to interpret and evaluate a string, such as

```
» s = '[sqrt(2) eps; 3*randn(1,2)]' % a string that requires evaluation
s =
[sqrt(2) eps;3*randn(1,2)]
» eval(s) % evaluate what's in the string s
ans =
      1.4142   2.2204e-16
     -1.2977      -4.9968
```

The function `eval` gives MATLAB macro capability. The argument to `eval` can be any MATLAB statement and it can include a variable assignment:

```
» ss = ['t=' s] % prepend a variable assignment
ss =
t = [sqrt(2) eps;3*randn(1,2)]
» eval(ss)
t =
      1.4142   2.2204e-16
       0.376      0.86303
```

In addition, `eval` can be called as `eval('try', 'catch')`, where the string `'catch'` is evaluated if the string `'try'` cannot be evaluated because of an error. For example,

```
» eval('a=sqrtt(2)','a=[]')
a =
     []
```

Here the second argument is executed since the first has an error—namely, `sqrtt` is not a valid MATLAB function.

In those cases where it is not necessary to bring in the entire MATLAB interpreter to convert a string into one or more numerical values, the function `sscanf` is preferred:

```
» version % get MATLAB version number string
ans =
5.0.0.4064
» sscanf(version,'%g') % scan the string for numbers
ans =
            5
            0
       0.4064
» sscanf(version,'%g',1) % get just the first number: version 5
ans =
     5
```

In a sense, `sscanf` is the inverse of `sprintf`. For simple numerical conversions, `sscanf` is much faster than using `eval`.

## 10.3 STRING FUNCTIONS

MATLAB provides a number of string testing and manipulation functions, including the following:

| Function | Description |
|---|---|
| `ischar(S)` | True if argument is a character string. |
| `isletter(S)` | True for indices that contain letters. |
| `isspace(S)` | True for indices that contain white space. |
| `findstr(S1,S2)` | Find indices where one string appears in the other. |
| `strcmp(S1,S2)` | True if two strings are identical. |
| `strncmp(S1,S2,n)` | True if first `n` characters of two strings are identical. |
| `strjust(S)` | Right justify string array. |
| `strmatch(S1,S2)` | Return row indices of `S2` that contain `S1` as first elements. |

| Function | Description |
|---|---|
| `strrep(S1,S2,S3)` | Replace all occurrences of S2 in S1 with S3. |
| `strtok(S1,D)` | Return first token in S1 delimited by characters in D. |
| `lower(S)` | Convert to lowercase. |
| `upper(S)` | Convert to uppercase. |

A number of the string functions listed in the preceding table provide basic string parsing capabilities. For example, `findstr` returns the starting indices of one string within another:

```
» b='Peter Piper picked a peck of pickled peppers';
» findstr(b,' ')  % find spaces
ans =
     6    12    19    21    26    29    37
» findstr(b,'p')  % find the letter p
ans =
     9    13    22    30    38    40    41
» find(b=='p')  % for single character searches the find command works too
ans =
     9    13    22    30    38    40    41
» findstr(b,'cow')  % find the word cow
ans =
     []
» findstr(b,'pick')  % find the string pick
ans =
    13    30
```

Note that this function is case sensitive and returns the empty matrix when no match is found. `findstr` does not work on string arrays with multiple rows.

Tests on character strings include the following:

```
» c = 'a2 : b_c'
c =
a2 : b_c
» ischar(c) % it is a character string
ans =
     1
» isletter(c) % where are the letters?
ans =
     1     0     0     0     0     1     0     1
» isspace(c) % where are the spaces?
ans =
     0     0     1     0     1     0     0     0
```

To illustrate string comparison, consider the situation where a user types a string (perhaps into an editable text `uicontrol`) that must match at least in part one of a list of strings. The function `strmatch` provides this capability:

```
» S = char('apple','banana','peach','mango','pineapple')
S =
apple
banana
peach
mango
pineapple
» strmatch('pe',S) % pe is in 3rd row
ans =
     3
» strmatch('p',S) % p is in 3rd and 5th rows
ans =
     3
     5
» strmatch('banana',S) % banana is in 2nd row
ans =
     2
» strmatch('Banana',S) % but Banana is nowhere
ans =
     []
» strmatch(lower('Banana'),S) % changing B to b finds banana
ans =
     2
```

## 10.4 CELL ARRAYS OF STRINGS

The fact that all rows in string arrays must have the same number of columns is sometimes cumbersome, especially when the nonblank portions vary significantly from row to row. This cumbersome issue is eliminated by using cell arrays, which are a data type discussed in a later chapter (Chapter 12). All data forms can be placed in cell arrays, but their most frequent use is with character strings. A cell array is a data type that simply allows one to name a group of data of various sizes and types. For example,

```
» C = {'How';'about';'this for a';'cell array of strings?'}
C =
    'How'
    'about'
    'this for a'
    'cell array of strings?'
» size(C)
ans =
     4     1
```

Note that curly brackets { } are used to create cell arrays and that the quotes around each string are displayed. In this example, the cell array C has four rows and one column. However, each element of the cell array contains a character string of different lengths.

Cell arrays are addressed just like other arrays:

```
» C(2:3)
ans =
    'about'
    'this for a'
» C([4 3 2 1])
ans =
    'cell array of strings?'
    'this for a'
    'about'
    'How'
» C(1)
ans =
    'How'
```

Here the results are still cell arrays. That is, C(indices) addresses given cells, but not the contents of those cells. To retrieve the contents of a particular cell, use curly brackets:

```
» s = C{4}
s =
cell array of strings?
» size(s)
ans =
     1    22
```

To extract more than one cell, use the function deal:

```
» [a,b,c,d] = deal(C{:})
a =
How
b =
about
c =
this for a
d =
cell array of strings?
```

Here C{:} denotes all the cells as a list. That is, it is the same as

```
» [a,b,c,d] = deal(C{1},C{2},C{3},C{4})
a =
How
```

```
b =
about
c =
this for a
d =
cell array of strings?
```

Partial cell array contents can also be dealt:

```
» [a,b] = deal(C{2:2:4}) % get 2nd and 4th cell contents
a =
about
b =
cell array of strings?
```

Some of the contents of a particular cell array can also be addressed:

```
» C{4}(1:10) % 4th cell, elements 1 through 10
ans =
cell array
```

The multipurpose function `char` converts the contents of a cell array to a conventional string array:

```
» s = char(C)
s =
How
about
this for a
cell array of strings?
» size(s)  % result is a standard array with blanks
ans =
     4    22
» ss = char(C(1:2))  % naturally you can convert subsets
ss =
How
about
» size(ss)  % result is a standard string array with blanks
ans =
     2    5
```

The inverse conversion is performed by the function `cellstr`:

```
» cellstr(s)
ans =
    'How'
    'about'
    'this for a'
    'cell array of strings?'
```

One can test if a particular variable is a cell array of strings by using the function `iscellstr`:

```
» iscellstr(C) % True for cell arrays of strings
ans =
     1
» ischar(C) % True for string arrays not cell arrays of strings
ans =
     0
» ischar(C{3}) % Contents of 3rd cell is a string array
ans =
     1
» iscellstr(C(3)) % but 3rd cell itself is a cell
ans =
     1
» ischar(C(3)) % and not a string array
ans =
     0
```

Most of the string functions in MATLAB work with either string arrays or cell arrays of strings. In particular, the functions `deblank`, `strcat`, `strcmp`, `strncmp`, `strmatch`, and `strrep` all work with either string arrays or cell arrays of strings. Further information regarding cell arrays in general can be found in the chapter on cell arrays and structures (Chapter 12).

# 11

# Time Computations

MATLAB offers a number of functions to manipulate time. You can do arithmetic with dates and times, print calendars, and find specific days. MATLAB does this by storing the date and time as a double-precision number representing the number of days since the beginning of year zero. For example, January 1, 1997 at midnight is represented as 729391 and the same day at noon is 729391.5. This format may make calculations easier for the computer, but it can be difficult to interpret visually. That is why MATLAB supplies a number of functions to convert between date numbers and character strings and to manipulate dates and times.

## 11.1 CURRENT DATE AND TIME

The function `clock` returns the current date and time in an array. For example,

```
» T = clock
T =
        1997          5          18          15          15          1.5205
```

The preceding data are organized as `T = [year month day hour minute seconds]`. So the time is the year 1997, the fifth month, eighteenth day, fifteenth hour, fifteenth minute, 1.5205 seconds.

The function `now` returns the current date and time as a double-precision date number or simply a date number.

```
» format long
» t = now
t =
        7.295286347238825e+05
```

Both `T` and `t` represent the same information.

The function `date` returns the current date as a string in the dd-mmm-yyyy format.

```
» date
ans =
18-May-1997
```

## 11.2 DATE FORMAT CONVERSIONS

In general, mathematics with time involves converting times to date number format, performing standard mathematical operations on the times, and then converting the result back to a format that makes human sense. As a result, converting time among different formats is very important.

MATLAB supports three formats for dates: (1) double-precision date number, (2) date (character) strings in a variety of styles, and (3) numerical date vector, where each element contains a different date component; that is,

```
[year,month,day,hour,minute,seconds].
```

You can convert the date number to a string using the `datestr` function. The syntax for using `datestr` is `datestr(date,dateform)`, where `dateform` is described by the help text for `datestr`:

```
» help datestr
DATESTR String representation of date.
   DATESTR(D,DATEFORM) converts a serial data number D (as returned
   by DATENUM) into a date string. The string is formatted according to
   the format number or string DATEFORM (see table below). By default,
   DATEFORM is 1, 16, or 0 depending on whether D contains
   dates, times or both.
```

```
DATEFORM number   DATEFORM string          Example
   0             'dd-mmm-yyyy HH:MM:SS'   01-Mar-1995 15:45:17
   1             'dd-mmm-yyyy'            01-Mar-1995
   2             'mm/dd/yy'               03/01/95
   3             'mmm'                    Mar
   4             'm'                      M
   5             'mm'                     3
   6             'mm/dd'                  03/01
   7             'dd'                     1
   8             'ddd'                    Wed
   9             'd'                      W
  10             'yyyy'                   1995
  11             'yy'                     95
  12             'mmmyy'                  Mar95
  13             'HH:MM:SS'               15:45:17
  14             'HH:MM:SS PM'             3:45:17 PM
  15             'HH:MM'                  15:45
  16             'HH:MM PM'                3:45 PM
  17             'QQ-YY'                  Q1-96
  18             'QQ'                     Q1

See also DATE, DATENUM, DATEVEC.
```

Examples of datestr usage include the following:

```
» t = now;
» datestr(t)
ans =
18-May-1997 15:23:59
» datestr(t,14)
ans =
 3:23:59 PM
```

The function datenum is the inverse of datestr. That is, datenum converts a date string to a date number using the form datenum(str). Alternatively, it converts individual date specifications using the forms datenum(year,month,day) or datenum(year,month,day,hour,minute,second). For example,

```
» t = now
t =
     7.295286656729241e+05
» datestr(t)
ans =
18-May-1997 15:58:34
```

```
» datenum(ans)
ans =
     7.295286656712963e+05
» datenum(1997,5,14)
ans =
     729524
» datenum(1997,5,17,15,58,34)
ans =
     7.295276656712963e+05
```

The `datevec` function converts a date string using `datestr` (formats 0, 1, 2, 6, 13, 14, 15, or 16) to a numerical vector containing the date components. Alternatively, it converts a date number into a numerical vector of date components. For example,

```
» c = datevec('12/24/1984')
c =
    1984     12     24     0     0     0
» [yr,mo,day,hr,min,sec] = datevec('24-Dec-1984 08:22')
yr =
    1984
mo =
    12
day =
    24
hr =
    8
min =
    22
sec =
    0
```

## 11.3 DATE FUNCTIONS

The numerical day of the week can be found from a date string or a date number using the function `weekday`. ***MATLAB uses the convention where Sunday is day one and Saturday is day seven.*** For example,

```
» [d,w] = weekday(728647)
d =
    2
w =
    Mon
» [d,w] = weekday('21-Dec-1994')
d =
    4
```

```
w =
    Wed
```

The last day of any month can be found using the function `eomday`. Because of leap year, both the year and month are required.

```
» eomday(1996,2)  % 1996 was a leap year
ans =
    29
```

MATLAB can generate a calendar for any month you request and display it in the *Command* window or place it in a 6-by-7 matrix using the function `calendar`:

```
» calendar('7/17/95') % display calendar
                    Jul 1995
       S      M     Tu      W     Th      F      S
       0      0      0      0      0      0      1
       2      3      4      5      6      7      8
       9     10     11     12     13     14     15
      16     17     18     19     20     21     22
      23     24     25     26     27     28     29
      30     31      0      0      0      0      0
» S = calendar(1994,12) % return calendar as a numerical array
S =
       0      0      0      0      1      2      3
       4      5      6      7      8      9     10
      11     12     13     14     15     16     17
      18     19     20     21     22     23     24
      25     26     27     28     29     30     31
       0      0      0      0      0      0      0
» size(S)
ans =
     6     7
```

## 11.4 TIMING FUNCTIONS

The functions `tic` and `toc` are used to time a sequence of MATLAB operations. `tic` starts a stopwatch; `toc` stops the stopwatch and displays the elapsed time:

```
» tic; plot(rand(5)); toc
elapsed_time =
   0.17444700000000
```

```
» tic; plot(rand(5)); toc
elapsed_time =
   0.03601500000000
```

Note the difference in elapsed times for identical `plot` commands. The second `plot` was significantly faster because MATLAB had already created the *Figure* window and compiled the functions it needed into memory.

The function `cputime` returns the amount of central processing unit (CPU) time in seconds that MATLAB has used since the current session was started. The function `etime` calculates the elapsed time between two time vectors in 6-element row vector form such as that returned by the functions `clock` and `datevec`. At the present time, `etime` does not work across month and year boundaries. Both `cputime` and `etime` can be used to compute the time it takes for an operation to complete. In fact, the functions `tic` and `toc` just automate the use of `clock` and `etime` to compute elapsed time. Usage of `cputime` and `etime` are demonstrated by the following examples, where `myoperation` is a script file containing a number of MATLAB commands:

```
» t0 = cputime; myoperation; cputime-t0
ans =
    0.149999999999991
» t1 = clock; myoperation; etime(clock,t1)
ans =
   11.284853
```

## 11.5 PLOT LABELS

Sometimes it is useful to plot data and use dates or time strings for one or more of the axis labels. The `datetick` function automates this task. ***Use of this function requires that the axis to be marked and plotted with a vector of date numbers (e.g., the output of the*** `datenum` ***function).*** Consider the following examples:

```
» t = (1900:10:1990)';
» p = [ 75.995;  91.972; 105.711; 123.203; 131.669;
        150.697; 179.323; 203.212; 226.505; 249.633];
» plot(datenum(t,1,1),p)

» datetick('x','yyyy')      % use 4-digit year on the x-axis
» title('Figure 11.1: Population by Year')
```

Here we create a bar chart of company sales from November 1994 to December 1995:

```
» y = [1994 1994 1995*ones(1,12)]';
» m = [11 12 (1:12)]';
» s = [1.1 1.3 1.2 1.4 1.6 1.5 1.7 1.6 1.8 1.3 1.9 1.7 1.6 1.95]';
```

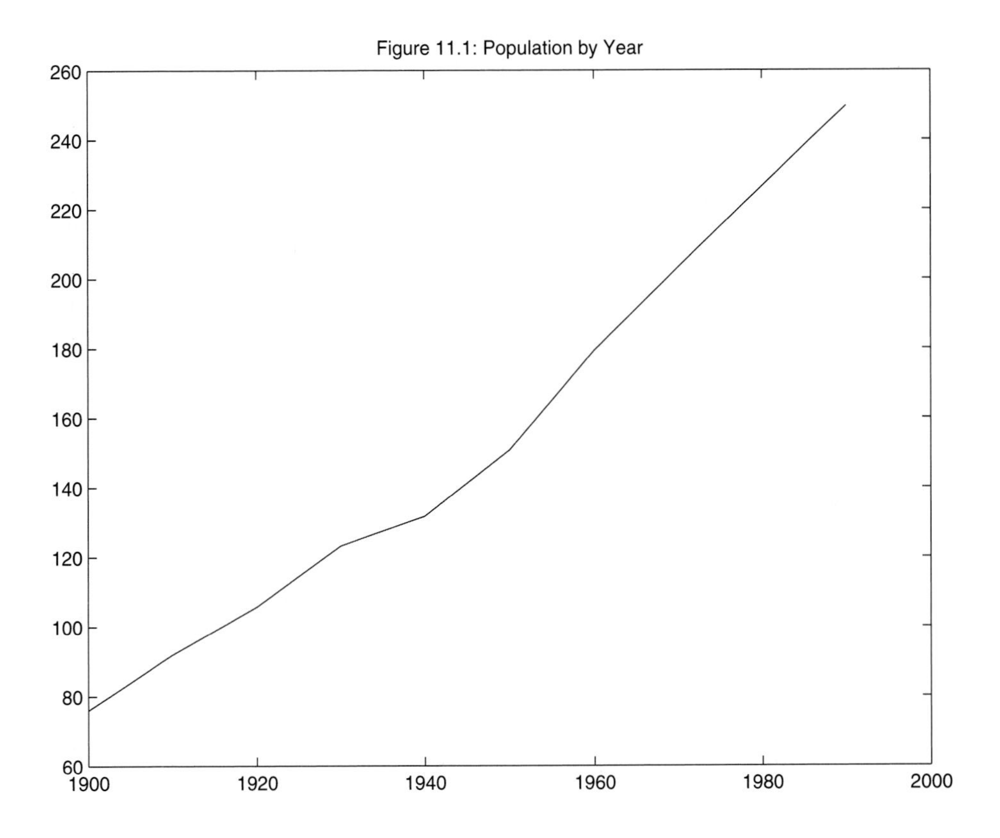


```
» bar(datenum(y,m,1),s)

» datetick('x','mmmyy')
» ylabel('$ Million')
» title('Figure 11.2: Monthly Sales')
```

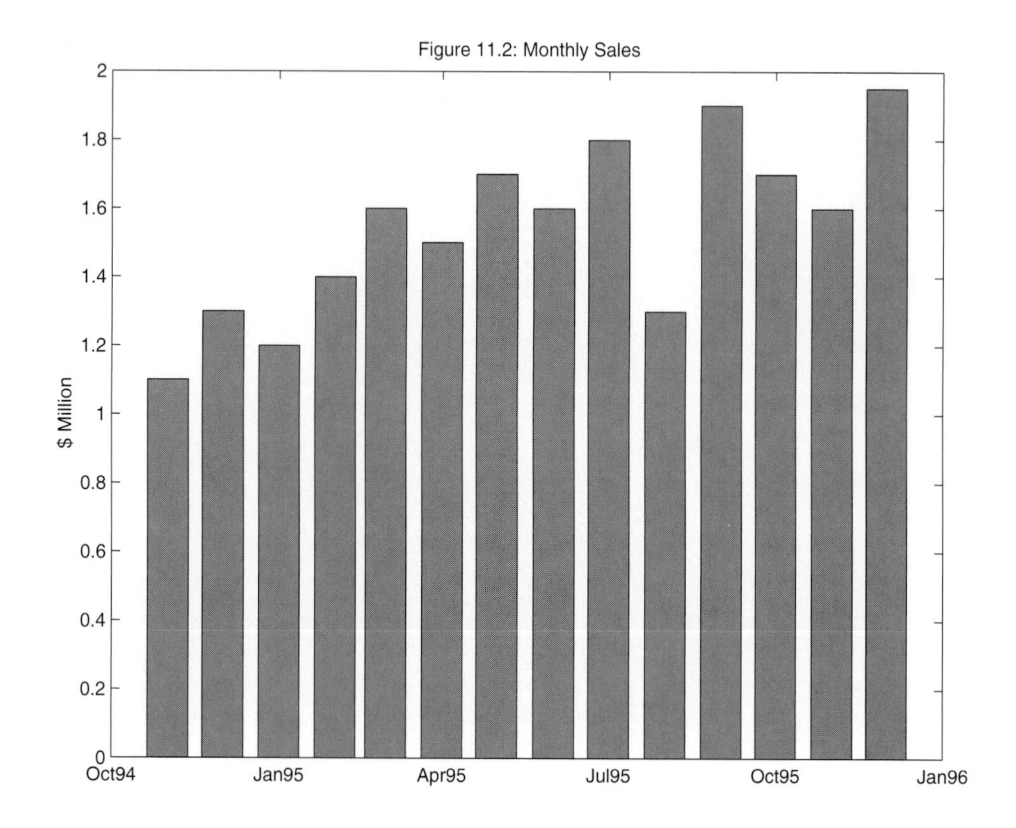
Figure 11.2: Monthly Sales
$ Million
2
1.8
1.6
1.4
1.2
1
0.8
0.6
0.4
0.2
0
Oct94
Jan95
Apr95
Jul95
Oct95
Jan96

12

# Cell Arrays and Structures

MATLAB 5 introduces two new data types called ***cell arrays*** and ***structures***. These data types allow one to group dissimilar but related arrays into a single variable. In doing so, data management becomes easier since groups of related data can be organized and accessed through a cell array or structure. Since cell arrays and structures are containers for other data types, mathematical operations on them are not defined. One must address the contents of a cell or structure to perform mathematical operations.

One way to visualize cell arrays is to consider a collection of post office boxes covering a wall at the post office. The collection of post office boxes is the cell array, with each post office box being one cell in the cell array. The contents of each post office box are different, just as the contents of each cell in a cell array are different types or sizes of MATLAB data, such as character strings or numerical arrays of varying dimensions. Just as each post office box is identified by a number, each cell in a cell array is indexed by a number. When you send mail to a post office box, you identify the box number you want it put in. When you want to put data in a particular cell, you identify the cell number you want it put in. The number is also used to identify which box or cell to take data out of.

Structures are almost identical to cell arrays, except that the individual post office boxes or data storage locations are not identified by number. Instead they are identified by name. Using the post office box analogy, the collection of post office boxes is the structure and each post office box is identified by its owner's name rather than by a number. To send mail to a post office box, you identify the name of the box you want it put in. To place data in a particular structure element, you identify the name (i.e., field) of the structure element to put the data in.

## 12.1 CREATING AND DISPLAYING CELL ARRAYS

Cell arrays are MATLAB arrays whose elements are cells. Each cell in a cell array can hold any MATLAB data type, including numeric arrays, text, symbolic objects, other cell arrays, and structures. For example, one cell of a cell array might contain a numeric array, another an array of text strings, and another a vector of complex values. Cell arrays can be created with any number of dimensions just as numerical arrays can. However, in most cases cell arrays are created as a simple vector of cells.

Cell arrays can be created using assignment statements or by preallocating the array using the `cell` function and then assigning data to the cells. If you have trouble with these examples, it is very likely that you have another variable in the workspace of the same name. If you assign a cell to an existing numeric array, MATLAB will report an error. If any of the following examples give unexpected results, clear the array from the workspace and try again.

Like other kinds of arrays, cell arrays can be built by assigning data to individual cells, one at a time. There are two different ways to access cells. If you use standard array syntax to index the array, you must enclose the cell contents in curly braces, { }. For example,

```
» A(1,1) = { [1 2 3; 4 5 6; 7 8 9] };
» A(1,2) = { 2+3i };
» A(2,1) = { 'A text string' };
» A(2,2) = { 12:-2:0 };
```

The curly braces on the right side of the equal sign indicate that the expression represents a cell rather than a numerical value. This is called ***cell indexing.*** Alternately, the following statements create the same cell array:

```
» A{1,1} = [1 2 3; 4 5 6; 7 8 9];
» A{1,2} = 2+3i;
» A{2,1} = 'A text string';
» A{2,2} = 12:-2:0;
```

Here the curly braces on the left side of the assignment indicate that `A` is a cell array rather than a numerical array and the expression on the right is put inside the specified cell. This is called ***content addressing.*** Both methods can be used interchangeably.

> Based on the preceding, curly braces { } are used to access the contents of cells, whereas parentheses ( ) are used to identify cells, but not their contents. To use the post office box analogy, curly braces are used to look at the contents of post office boxes whereas parentheses are used to identify one or more post office boxes without looking at their contents.

MATLAB displays the cell array A as follows:

```
» A
A =
       [3x3 double]    [2.0000+ 3.0000i]
    'A text string'         [1x7 double]
```

Here [3x3 double] indicates that A(1,1) is a cell that holds a 3-by-3 double array. When a cell holds a single numerical value or character string, it is displayed. However, when a cell holds a numerical array, the array size and type is shown. To display the contents of each cell in a cell array, use the celldisp function:

```
» celldisp(A)
A{1,1} =
    1     2     3
    4     5     6
    7     8     9
A{2,1} =
    A text string
A{1,2} =
    2.0000+ 3.0000i
A{2,2} =
    12    10     8     6     4     2     0
```

To display the contents of a single cell, access the cell using curly braces.

```
» A{2,2}
ans =
    12    10     8     6     4     2     0
```

MATLAB displays a graphical structure map of a cell array in a *Figure* window using the cellplot function. Filled rectangles are used to represent the number of elements in each cell. Scalars and string data are displayed as values as well. For example,

```
» cellplot(A)
```

Figure 12.1: A Cell Plot

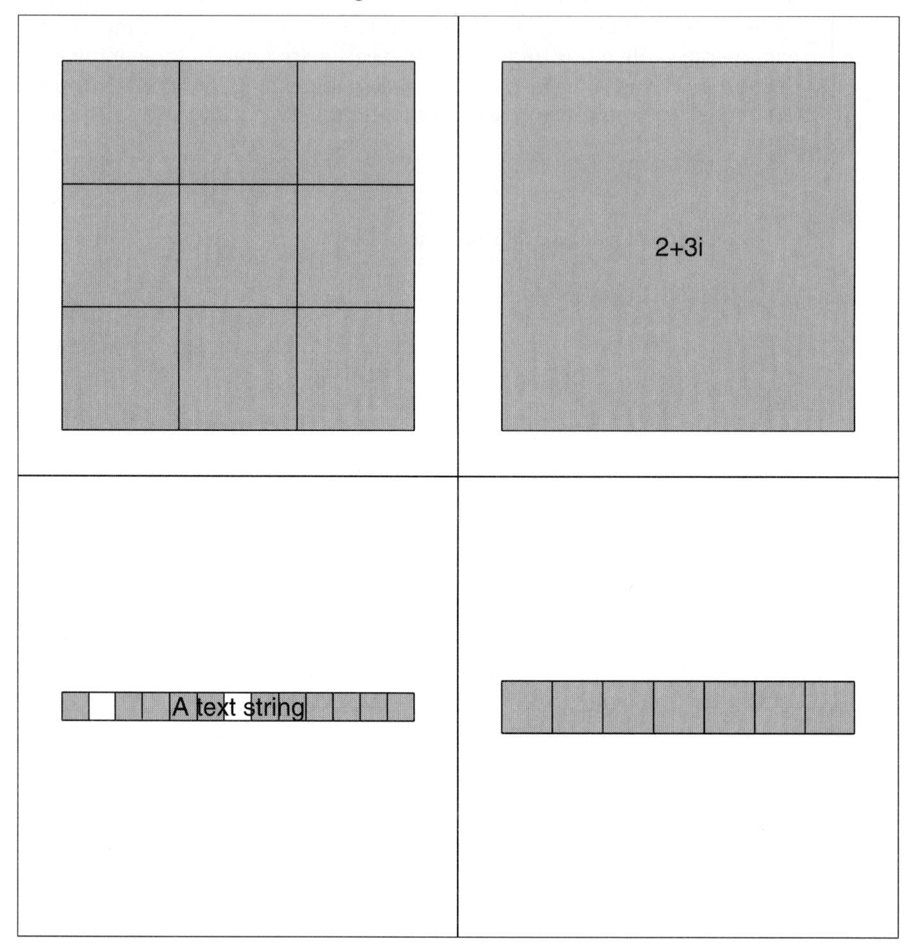


Curly braces work the same way as square brackets for numeric arrays, except that cells can be nested. Commas and semicolons provide column and row breaks, just as they do for numerical and character arrays:

```
» B = { [1 2], 'John Smith'; 2+3i, 5}
B =
          [1x2 double]    'John Smith'
    [2.0000+  3.0000i]    [         5]
```

The `cell` function makes working with cell arrays more efficient by preallocating an empty array of the specified size. `cell` actually creates the cell array and fills it with empty numeric matrices.

```
» C = cell(2,3)
C =
    []     []     []
    []     []     []
```

Once the cell array has been defined, use the assignment syntax illustrated earlier to populate the cells.

```
» C(1,1) = 'This doesn''t work' % because there are no curly brackets
??? Conversion to cell from char is not possible.
» C(1,1) = { 'This does work' }
C =
    'This does work'     []     []
                  []     []     []
» C{2,3} = 'This works too'
C =
    'This does work'     []                   []
                  []     []     'This works too'
```

## 12.2 COMBINING AND RESHAPING CELL ARRAYS

If you assign data to a cell that is outside the dimensions of the current array, MATLAB automatically expands the array and fills the intervening cells with the empty numeric matrix [ ]. Note that { } represents an empty cell array, just as [ ] represents an empty array.

Square brackets are used to combine cell arrays into larger cell arrays, just as they are used to construct larger numerical or character arrays. For example,

```
» C = [A B]
C =
      [3x3 double]    [2.0000+ 3.0000i]          [1x2 double]   'John Smith'
   'A text string'         [1x7 double]    [2.0000+ 3.0000i]   [          5]
» C = [A;B]
C =
         [3x3 double]   [2.0000+ 3.0000i]
    'A text string'          [1x7 double]
         [1x2 double]   'John Smith'
    [2.0000+  3.0000i]     [             5]
```

A subset of cells can be extracted to create a new cell array by conventional array addressing techniques. For example,

```
» D = C([1 3],:) % all columns of C in rows 1 and 3
D =
    [3x3 double]    [2.0000+ 3.0000i]
    [1x2 double]    'John Smith'
```

An entire row or column of a cell array can be deleted using the empty matrix:

```
» C(3,:) = [];
```

Note that curly braces do not appear in either of the preceding expressions because we are working with the cell array itself. Curly braces are used to address the contents of cells, whereas parentheses are used to identify the cells, not their contents.

The `reshape` function can be used to change the configuration of a cell array but cannot be used to add or remove cells:

```
» X = cells(3,4);
» size(X)
ans =
    3    4
» Y = reshape(X,6,2);
» size(Y)
ans =
    6    2
```

## 12.3 RETRIEVING CELL ARRAY CONTENTS

Curly braces are used to retrieve data contained in cell array, whereas parentheses index the cells themselves. For example,

```
» x = B{2,2} % access the contents of the cell
x =
      5
» class(x) % class returns the data type of its argument
ans =
      double
» y = B(2,2) % access the cell but not the contents
y =
      [5]
» class(y)
ans =
      cell
» iscell(y) % True = 1 when argument is a cell array
ans =
     1
» isa(y,'cell') % more general test for cell arrays
ans =
     1
```

While you can display the contents of more than one cell at a time, it is not possible to assign more than one at a time to a variable using a standard assignment statement. For example,

```
» B{:,2} % display the second column
ans =
John Smith
ans =
     5
» x = B{:,1} % assign first column to the variable x
??? Illegal right hand side in assignment. Too many elements.
```

When addressing the contents of a cell, you can also address a subarray within the contents of a cell by appending the desired subscript range:

```
» B{1,1} % contents of cell (1,1)
ans =
     1     2
» B{1,1}(1) % contents first element of cell (1,1)
ans =
     1
» B{1,1}(2) % contents second element of cell (1,1)
ans =
     2
» B{1,2} % contents of cell (1,2)
ans =
John Smith
» B{1,2}(1:4) % first four elements of cell (1,2)
ans =
John
```

## 12.4 COMMA-SEPARATED LISTS

Assigning the contents of multiple cells to variables with one command requires the `deal` function. For example,

```
» [a,b,c,d] = deal(B{:}) % extract all four cells
a =
     1     2
b =
     2.0000+    3.0000i
c =
John Smith
d =
     5
```

```
» [g,h] = deal(B{1:2}) % extract first two elements
g =
     1     2
h =
    2.0000+    3.0000i
```

To access elements of a range of cells in a cell array, use the `deal` function:

```
» [a,b] = deal(B{2,:})
a =
     2.0000+    3.0000i
b =
     5
» [a,b,c,d] = deal(B{1},B{2},B{3},B{4})
a =
    1     2
b =
    2.0000+    3.0000i
c =
John Smith
d =
    5
```

Based on the preceding example, the syntax `deal(B{:})` and `deal(B{1},B{2},B{3},B{4})` are identical. That is, `B{:}` is equivalent to the comma-separated list `B{1},B{2},B{3},B{4}`. This feature enhances the use of cell arrays (and structures) in the calling syntax of functions and in horizontal concatenation. Simply put, any indexing expression that attempts to extract more than one element produces a comma-separated list. In MATLAB, comma-separated lists are described in the `lists` help text:

```
» help lists
 LISTS Comma separated lists.
    Extracting multiple values from a cell array or structure results
    in a list of values (as if separated by commas).  The comma-separated
    lists are valid:
       1) by themselves on the command line to display the values,
            C{:} or S.name
       2) within parentheses as part of a function call,
            myfun(x,y,C{:}) or myfun(x,y,S.name)
       3) within square brackets as part of a horizontal concatenation
            [C{:}] or [S.name]
       4) within square brackets as part of a function output list,
            [C{:}] = myfun or [S.name] = myfun
       5) within braces as part of a cell array construction.
            {C{:}} or {S.name}
    In all these uses, C{:} is the same as C{1},C{2},...,C{end} and
    S.name is the same as S(1).name,S(2).name,...,S(end).name.  If
    C or S is 1-by-1 then these expressions produce the familiar
```

```
    single element extraction.  Any indexing expression that attempts
    to extract more than one element produces a comma separated list.
    Hence C{1:5} and S(2:3).name are also comma separated lists.

    Comma separated lists are very useful when dealing with variable
    input or output argument lists or for converting the contents
    of cell arrays and structures into matrices.

    Examples
       C = {1 2 3 4};
       A = [C{:}];
       B = cat(2,C{:});
       [S(1:3).FIELD] = deal(5);

    See also VARARGIN, VARARGOUT, DEAL, CELL2STRUCT, STRUCT2CELL,
             NUM2CELL, CAT.
```

## 12.5 CELL ARRAYS OF CHARACTER STRINGS

One of the most common applications of cell arrays is the creation of string arrays. Standard arrays of character strings require that all strings have the same length. Because cell arrays can contain different types of data in each element, text strings in cell arrays do not have this limitation. For example,

```
» T = {'Tom'; 'Dick'; 'Harry Smith'; 'Mohamad'; 'Suzanne'}
T =
     'Tom'
     'Dick'
     'Harry Smith'
     'Mohamad'
     'Suzanne'
» iscellstr(T) % logical test for cell array of strings
ans =
     1
» ischar(T) % logical test for a string array
ans =
     0
```

T and its elements can now be used anywhere a cell array of strings or string array is desired. For more information on cell arrays of strings, see the chapter on character strings (Chapter 10).

## 12.6 CREATING AND DISPLAYING STRUCTURES

Structures are like cell arrays in that they allow one to group collections of dissimilar data into a single variable. However, instead of addressing elements by number, structure elements are addressed by names called ***fields***. As with cell arrays, structures can have any number of dimensions, but a simple vector array is most common.

Structures use dot notation to access fields. Creating a structure can be as simple as assigning data to individual fields. This example builds a client record for a testing laboratory:

```
» client.name = 'John Doe';
» client.cost = 86.50;
» client.test.A1C = [6.3 6.8 7.1 7.0 6.7 6.5 6.3 6.1 6.4];
» client.test.CHC = [2.8 3.4 3.6 4.1 3.5];
» client
client =
    name: 'John Doe'
    cost: 86.50
    test: [1x1 struct]
» client.test
ans =
    A1C: 6.3000 6.8000 7.1000 7.0000 6.7000 6.5000 6.3000 6.1000 6.4000
    CHC: 2.8000 3.4000 3.6000 4.1000 3.5000
```

Now create a second client record to create a structure array.

```
» client(2).name = 'Alice Smith';
» client(2).cost = 112.35;
» client(2).test.A1C = [5.3 5.8 7.0 6.5 6.7 5.5 6.0 5.9 6.1];
» client(2).test.CHC = [3.8 3.6 3.2 3.1 2.5];
» client
client =
1x2 struct array with fields
    name
    cost
    text
```

In this example, the field named `test` is itself a structure. That is why `client.test.A1C` has two periods in it. `client` is the overall structure having three fields named `name`, `cost`, and `test`. The field named `name` is a simple character array, the field named `cost` is a scalar, and the field named `test` is a structure having field names `A1C` and `CHC`.

As you expand the structure array, MATLAB fills in unspecified fields with the empty numeric matrix so that all structures in the array have the same number of fields and the same field names. Note that data stored in fields do not have to match in all indexed elements of a structure array. For example, `'John Doe'` has a different length than `'Alice Smith'`.

Structures can also be created using the `struct` function to preallocate an array of structures. The syntax is `struct('`*`field1`*`',`*`V1`*`,'`*`field2`*`',`*`V2`*`,...)`, where *`field1`*, *`field2`*, etc. are field names and arrays *`V1`*, *`V2`*, etc. must be cell arrays of the same size, scalar cells, or single values. For example, a structure array can be created as follows:

```
» N = {'John Doe', 'Alice Smith'};
» C = {86.50, 112.35};
» P = {[10.00 20.00 45.00],[100.00 12.35]};
» bills = struct('name',N,'cost',C,'payment',P)
bills =
1x2 struct array with fields
      name
      cost
      payment
```

## 12.7 RETRIEVING STRUCTURE FIELD CONTENTS

Because structure elements are named rather than indexed by number, the names of the fields in a structure must be known to retrieve the data contained within them. Field names can be found in the *Command* window simply by typing the name of the structure. In an M-file, however, a function is needed to obtain the field names.

The `fieldnames` function returns a cell array containing the names of the fields in a structure. For example,

```
» T = fieldnames(bills)
T =
    'name'
    'cost'
    'payment'
```

There are two methods for retrieving the contents of structure fields. Direct indexing uses a period to access a structure field and an optional appended index range to specify a subarray range. Consider the following example based on the preceding `bills` and `client` structures:

```
» bills.name % get both names
ans =
    John Doe
ans =
    Alice Smith
» bills(2).cost % get just the 2nd cost
ans =
    112.3500
» bills(1) % get all fields of the first bills
ans =
     name:  'John Doe'
     cost:  86.5000
  payment:  10.0000   20.0000   45.0000
```

```
» baldue = bills(1).cost - sum(bills(1).payment) % perform math
baldue =
     6.5000
» bills(2).payment(2) % get 2nd element of bills(2).payment
ans =
     12.3500
» client(2).test.A1C(3) % this is getting complicated!
ans =
     7.0000
```

Direct indexing is usually the most efficient way to retrieve field values. However, in an M-file where the field names are obtained from the `fieldnames` function, the `getfield` and `setfield` functions can be used to retrieve structure data. Each of these functions let you build up a string from individual fields. For example,

```
» getfield(bills,{1},'name')    % same as bills(1).name
ans =
    John Doe
» T = fieldnames(bills);
» getfield(bills,{2},T{3},{2})  % same as bills(2).payment(2)
ans =
     12.3500
```

The following example returns a structure containing the same data as the original structure with one value changed. The direct addressing equivalent is `client(2).test.A1C(3) = 7.1`.

```
» client = setfield(client,{2},'test','A1C',{3},7.1)
client =
1x2 struct array with fields
    name
    cost
    test
» client(2).test.A1C(3)
ans =
    7.1000
```

Fields can be added to a structure or structure array by simply assigning a value to a new structure field:

```
» client(1).addr = {'MyStreet'; 'MyCity'}
client =
1x2 struct array with fields
    name
    cost
    test
    addr
```

Fields are removed from a structure or structure array by the `rmfield` command. `S=rmfield(S,'field')` removes the field `'field'` from the structure `S`. Alternatively, `S=rmfield(S,F)`, where `F` is a cell array of field names, removes more than one field at a time from the structure `S`. For example,

```
» client = rmfield(client,'addr')
client =
1x2 struct array with fields
    name
    cost
    test
```

## 12.8 CONVERSION AND TEST FUNCTIONS

Because cell arrays and structures are so similar, it is possible to convert from one to the other using the functions `struct2cell` and `cell2struct`. Field names must be supplied to `cell2struct` and are lost when converting from a structure to a cell array. Conversions from numeric arrays and character arrays to cell arrays are provided by the functions `num2cell` and `cellstr`, respectively. As described in the chapter on character strings (Chapter 10), conversion from a cell array to a character array uses the `char` function.

Given a MATLAB variable, it is often necessary to determine the type or ***class*** of the variable. As shown earlier in this chapter, the function `class` returns a character string identifying the type of data stored in a variable. In addition, the function `isa` returns the logical result of testing a variable for membership in a particular class of data. For example,

```
» A % recall a cell array
A =
       [3x3 double]    [2.0000+    3.0000i]
    'A text string'            [1x7 double]
» class(A) % string identifying what A is
ans =
cell
» isa(A,'double') % A is not a double, i.e., a numerical array
ans =
     0
» isa(A,'cell') % A is a cell
ans =
     1
» isa(A,'struct') % but not a structure
ans =
     0
» class(client) % test client structure
ans =
struct
```

```
» isa(client,'cell') % it's not a cell
ans =
     0
» isa(client,'struct') % but it is a structure
ans =
     1
```

In addition to the `class` and `isa` functions, there are specific logical testing functions for all MATLAB data types or classes. For example,

```
» isstruct(client) % logical test for structures
ans =
     1
» isfield(client,'name') % logical test for field names in a structure
ans =
     1
» isfield(client,'home') % 'home' is not a field name!
ans =
     0
» isnumeric(pi) % test for numeric or double
ans =
     1
» ischar('John Doe') % test for character strings
ans =
     1
```

# 13

# Control Flow

Computer programming languages and programmable calculators offer features that allow you to control the flow of command execution based on decision-making structures. If you have used these features before, this section will be very familiar to you. On the other hand, if control flow is new to you, this material may seem complicated the first time through.

Control flow is extremely powerful since it lets past computations influence future operations. MATLAB offers four decision-making or control flow structures: For Loops, While Loops, If-Else-End constructions, and Switch-Case constructions. Because these constructions often encompass numerous MATLAB commands, they often appear in M-files rather than being typed directly at the MATLAB prompt.

## 13.1 FOR LOOPS

For Loops allow a group of commands to be repeated a fixed, predetermined number of times. The general form of a For Loop is

```
for x = array
    (commands)
end
```

The `(commands)` between the `for` and `end` statements are executed once for every column in `array`. At each iteration, `x` is assigned to the next column of `array`—that is, during the `n`th time through the loop, `x=array(:,n)`. For example,

```
» for n = 1:10
        x(n) = sin(n*pi/10);
end

» x
x =
  Columns 1 through 7
    0.3090    0.5878    0.8090    0.9511    1.0000    0.9511    0.8090
Columns 8 through 10
    0.5878    0.3090    0.0000
```

In words, the first statement says, for `n` equals one to ten evaluate all statements until the next `end` statement. The first time through the For Loop `n=1`, the second time `n=2`, and so on through the `n=10` case. After the `n=10` case, the For Loop ends and any commands after the `end` statement are evaluated, which in this case is to display the computed elements of `x`.

A For Loop cannot be terminated by reassigning the loop variable `n` within the For Loop:

```
» for n = 1:10
        x(n) = sin(n*pi/10);
        n = 10;
  end
» x
x =
  Columns 1 through 7
    0.3090    0.5878    0.8090    0.9511    1.0000    0.9511    0.8090
Columns 8 through 10
    0.5878    0.3090    0.0000
```

The statement `1:10` is a standard MATLAB array creation statement. Any valid MATLAB array is acceptable in the For Loop:

```
» data = [3 9 45 6; 7 16 -1 5]
data =
     3     9    45     6
     7    16    -1     5
```

```
» for n = data
      x = n(1) - n(2)
  end
x =
    -4
x =
    -7
x =
    46
x =
     1
```

Naturally, For Loops can be nested as desired:

```
» for n = 1:5
      for m = 5:-1:1
           A(n,m) = n^2 + m^2;
      end
      disp(n)
end
   1
   2
   3
   4
   5
» A
A =
   2    5   10   17   26
   5    8   13   20   29
  10   13   18   25   34
  17   20   25   32   41
  26   29   34   41   50
```

For Loops should be avoided whenever there is an equivalent array approach to solving a given problem. For example, the first example can be rewritten as

```
» n = 1:10;
» x = sin(n*pi/10)
x =
  Columns 1 through 6
      0.30902      0.58779      0.80902      0.95106      1      0.95106
  Columns 7 through 10
      0.80902      0.58779      0.30902            0
```

While both approaches lead to identical results, the latter approach executes faster, is more intuitive, and requires less typing.

To maximize speed, arrays should be preallocated before a For Loop (or While Loop) is executed. For example, in the first case considered, every time the commands within the For Loop are executed, the size of the variable x is increased by 1. This forces MATLAB to take the time to allocate more memory for x every time through the loop. To eliminate this step, the For Loop example should be rewritten as

```
» x = zeros(1,10); % preallocated memory for x
» for n = 1:10
      x(n) = sin(n*pi/10);
  end
```

Now only the values of x(n) need to be changed.

## 13.2 WHILE LOOPS

As opposed to a For Loop that evaluates a group of commands a fixed number of times, a While Loop evaluates a group of statements an indefinite number of times. The general form of a While Loop is

```
while expression
    (commands)
end
```

The *(commands)* between the while and end statements are executed as long as **all** elements in expression are True. Usually, evaluation of expression gives a scalar result, but array results are also valid. In the array case, all elements of the resulting array must be True. Consider the following example:

```
» num = 0; EPS = 1;

» while (1+EPS)>1
      EPS = EPS/2;
      num = num+1;
end

» num
num =
    53
» EPS = 2*EPS
EPS =
   2.2204e-16
```

This example shows one way of computing the special MATLAB value eps, which is the smallest number that can be added to 1 such that the result is greater than 1 using finite precision. Here we used uppercase EPS so that the MATLAB value eps is not overwritten. In this example EPS starts at 1. As long as (1+EPS)>1 is True (nonzero), the commands inside the While Loop are evaluated. Since EPS is continually divided in two, EPS eventually gets so small that adding EPS to 1 is no longer greater than 1. (Recall that this happens because a computer uses a fixed number of digits to represent numbers. MATLAB uses 16 digits so one would expect eps to be near $10^{-16}$.) At this point, (1+EPS)>1 is False (zero) and the While Loop terminates. Finally, EPS is multiplied by 2 because the last division by 2 made it too small by a factor of 2.

## 13.3 IF-ELSE-END CONSTRUCTIONS

Many times, sequences of commands must be conditionally evaluated based on a relational test. In programming languages this logic is provided by some variation of an If-Else-End construction. The simplest If-Else-End construction is

```
if expression
    (commands)
end
```

The *(commands)* between the if and end statements are evaluated if **all** elements in *expression* are True (nonzero). In those cases where *expression* involves several logical subexpressions, only the minimum number required to determine the final logical state are evaluated. For example, if *expression* is *(expression1 | expression2)*, then *expression2* is only evaluated if *expression1* is False. Similarly, if *expression* is *(expression1 & expression2)*, then *expression2* is not evaluated if *expression1* is False. An example of the If-Else-End construction is as follows:

```
» apples = 10;                 % number of apples
» cost = apples*25             % cost of apples
cost =
   250
» if apples>5                  % give 20% discount for larger purchases
      cost = (1-20/100)*cost;
  end
» cost
cost =
   200
```

In cases where there are two alternatives, the If-Else-End construction is

```
if expression
    (commands evaluated if True)
else
    (commands evaluated if False)
end
```

Here the first set of commands is evaluated if `expression` is True; the second set is evaluated if `expression` is False.

When there are three or more alternatives, the If-Else-End construction takes the form

```
if expression1
    (commands evaluated if expression1 is True)
elseif expression2
    (commands evaluated if expression2 is True)
elseif expression3
    (commands evaluated if expression3 is True)
elseif expression 4
    (commands evaluated if expression 4 is True)
elseif ...
    .
    .
    .
else
    (commands evaluated if no other expression is True)
end
```

In this last form only the commands associated with the first True expression encountered are evaluated; ensuing relational expressions are not tested and the rest of the If-Else-End construction is skipped. Furthermore, the final `else` command may or may not appear.

Now that we know how to make decisions with If-Else-End structures, it is possible to show a legal way for jumping or breaking out of For Loops and While Loops:

```
» EPS = 1;
» for num = 1:1000
     EPS = EPS/2;
     if (1+EPS)<=1
         EPS = EPS*2
         break
      end
  end
EPS =
   2.2204e-16
» num
num =
    53
```

This example demonstrates another way of estimating `eps`. In this case, the For Loop is instructed to run some sufficiently large number of times. The If-Else-End structure tests to see if `EPS` has gotten small enough. If it has, `EPS` is multiplied by 2 and the `break` command forces the For Loop to end prematurely (i.e., at `num=53` in this case).

In this example, when the `break` statement is executed, MATLAB jumps to the next statement outside the Loop in which it appears. In this case, it returns to the MATLAB prompt and displays `EPS`. If a `break` statement appears in a nested For Loop or While Loop structure, MATLAB only jumps out of the Loop in which it appears. It does not jump all the way out of the entire nested structure.

## 13.4 SWITCH-CASE CONSTRUCTIONS

When sequences of commands must be conditionally evaluated based on repeated use of an equality test with one common argument, a Switch-Case construction is often easier. Switch-Case constructions have the form

```
switch expression
   case test_expression1
      (commands1)
   case {test_expression2,test_expression3,test_expression4}
      (commands2)
   otherwise
      (commands3)
end
```

Here `expression` must be either a scalar or a character string. If `expression` is a scalar, `expression==`*`test_expressionN`* is tested by each case statement. If `expression` is a character string, `strcmp(expression,`*`test_expression`*`)` is tested. In the preceding example, *`expression`* is compared with *`test_expression1`* at the first case statement. If they are equal, `(`*`commands1`*`)` are evaluated and the rest of the statements before the `end` statement are skipped. If the first comparison is not true, the second is considered. In the preceding example, `expression` is compared with *`test_expression2`*, *`test_expression3`*, and *`test_expression4`*, which are contained in a cell array. If any of these are equal to `expression`, *`(commands2)`* are evaluated and the rest of the statements before the `end` are skipped. If all `case` comparisons are false, *`(commands3)`* following the optional `otherwise` statement are executed. ***Note that this implementation of the Switch-Case construction allows at most one of the command groups to be executed.***

A simple example demonstrating the Switch-Case construction is as follows:

```
x = 2.7;
units = 'm';
switch units  % convert x to centimeters
   case {'inch','in'}
      y = x*2.54;
   case {'feet','ft'}
      y = x*2.54*12;
```

```
case {'meter','m'}
      y = x/100;
   case {'millimeter','mm'}
      y = x*10;
   case {'centimeter','cm'}
      y = x;
   otherwise
      disp(['Unknown Units: ' units])
      y = nan;
end
```

Executing the preceding example gives a final value of `y=0.027`.

14

# Function M-files

When you use MATLAB functions such as `inv`, `abs`, `angle`, and `sqrt`, MATLAB takes the variables you pass it, computes the required results using your input, and then passes those results back to you. The commands evaluated by the function as well as any intermediate variables created by those commands are hidden. All you see is what goes in and what comes out (i.e., a function is a black box).

These properties make functions very powerful tools for evaluating commands that encapsulate useful mathematical functions or sequences of commands that often appear when solving some larger problem. Because of this power, MATLAB provides a structure for creating functions of your own in the form of text M-files stored on your computer. The MATLAB function `flipud` is a good example of an M-file function:

```
function y = flipud(x)
%FLIPUD Flip matrix in up/down direction.
%  FLIPUD(X) returns X with columns preserved and rows flipped
%  in the up/down direction. For example,
%
%  X = 1 4       becomes 3 6
%      2 5               2 5
%      3 6               1 4
%
%  See also FLIPLR, ROT90, FLIPDIM.
%  Copyright (c) 1984-96 by The MathWorks, Inc.
%  $Revision: 5.3 $ $Date: 1996/10/24 18:41:14 $

if ndims(x)~=2, error('X must be a 2-D matrix.'); end
[m,n] = size(x);
y = x(m:-1:1,:);
```



A function M-file is similar to a script file in that it is a text file having a `.m` extension. As with script M-files, function M-files are not entered in the *Command* window, but rather are external text files created with a text editor. A function M-file is different from a script file in that a function communicates with the MATLAB workspace only through the variables passed to it and through the output variables it creates. Intermediate variables within the function do not appear in or interact with the MATLAB workspace. As can be seen in the preceding example, the first line of a function M-file defines the M-file as a function and specifies its name, which is the same as its file name without the `.m` extension. It also defines its input and output variables. The next continuous sequence of comment lines is the text displayed in response to the help commands: `help flipud` or `helpwin flipud`. The first help line, called the H1 line, is the line searched by the `lookfor` command. Finally, the remainder of the M-file contains MATLAB commands that create the output variables. Note that there is no `return` command in `flipud`; the function simply terminates after it executes the last command.

## 14.1 M-FILE CONSTRUCTION RULES

Function M-files must satisfy a number of criteria. In addition, there are a number of desirable features that they should have.

1. The function M-file name and the function name (e.g., `flipud`) that appears in the first line of the file should be identical. In reality, MATLAB ignores the function name in the first line and executes functions based on the file name stored on disk.

2. Function M-file names can have up to 31 characters. This maximum may be limited by the operating system, in which case the lower limit applies. MATLAB ignores characters beyond the thirty-first or the operating system limit, so longer names can be used, provided the legal characters point to a unique file name.

3. Function names must begin with a letter. Any combination of letters, numbers, and underscores can appear after the first character. This naming rule is identical to that for variables.

4. The first line of a function M-file is called the ***function declaration line*** and must contain the word `function` followed by the calling syntax for the function in its most general form. The input and output variables identified in the first line are variables local to the function. The input variables contain data passed to the function and the output variables contain data passed back from the function. It is not possible to pass data back through the input variables.

5. The first set of contiguous comment lines after the function declaration line are the help text for the function. The first comment line is called the H1 line and is the line searched by the `lookfor` command. The H1 line typically contains the function name in uppercase characters and a concise description of the function's purpose. Comment lines after the first describe possible calling syntaxes, algorithms used, and simple examples, if appropriate.

6. All statements after the first set of contiguous comment lines compose the body of the function. The body of a function contains MATLAB statements that operate on the input arguments and produce results in the output arguments.

7. A function M-file terminates after the last line in the file is executed or whenever a `return` statement is encountered.

8. A function can abort operation and return control to the *Command* window by calling the function `error`. This function is useful for flagging improper function usage, as shown in the following file fragment:

```
if length(val) > 1
     error('VAL must be a scalar.')
end
```

   If the function `error` is executed as shown, the string `'VAL must be a scalar.'` is displayed in the *Command* window after a line identifying the file the error message originated from. If the function `error` is given a zero-length character string, it takes no action.

9. A function can report a warning and then continue operation by calling the function `warning`. This function is useful for reporting exceptions and other anomalous behavior. `warning('`*`some message`*`')` simply displays a character string in the *Command* window. The difference between the function `warning` and the function `disp` is that warning messages can be turned on or off globally by issuing the commands `warning on` and `warning off`, respectively.

10. Function M-files can contain calls to script files. When a script file is encountered, it is evaluated in the function's workspace, not the MATLAB workspace.
11. Multiple functions can appear in a single function M-file. Additional functions, called subfunctions or local functions, are simply appended to the end of the primary function. Subfunctions begin with a standard function statement line and follow function construction rules.
12. Subfunctions can be called by the primary function in the M-file as well as by other subfunctions in the same M-file. As with all functions, subfunctions have their own individual workspaces.
13. Subfunctions can appear in any order after the primary function in an M-file. Help text for subfunctions is not returned by the `help` command.
14. It is suggested that subfunction names begin with the word `local` (e.g., `local_myfun`). Doing so improves the readability of the primary function because calls to local functions are clearly identifiable. All local function names can have up to 31 characters.
15. In addition to subfunctions, M-files can call private M-files, which are standard function M-files that reside in a subdirectory of the calling function that is entitled `private`. Only functions in the immediate parent directory of private M-files have access to private M-files.
16. It is suggested that private M-file names begin with the word `private` (e.g., `private_myfun`). Doing so improves the readability of the primary function because calls to private functions are clearly identifiable. As with other function names, all private M-file names can have up to 31 characters.

## 14.2 INPUT AND OUTPUT ARGUMENTS

MATLAB functions can have any number of input and output arguments. The features and criteria of these arguments are as follows:

1. Function M-files can have zero input and zero output arguments.
2. Functions can be called with fewer input and output arguments than the function definition line in the M-file specifies. Functions cannot be called with more input or output arguments than the M-file specifies.
3. The number of input and output arguments used in a function call can be determined by calls to the functions `nargin` and `nargout`, respectively. Since `nargin` and `nargout` are functions, not variables, one cannot reassign them with statements such as `nargin = nargin - 1`. The function `mmdigit` in the *Mastering MATLAB Toolbox* illustrates the use of `nargin`:

can be called recursively and each call has a separate workspace. In addition to input and output arguments, MATLAB provides several techniques for communicating among function workspaces and the MATLAB or base workspace.

1. Functions can share variables with other functions, the MATLAB workspace, and recursive calls to themselves if the variables are declared `global`. To gain access to a global variable within a function or the MATLAB workspace, the variable must be declared `global` within each desired workspace. The MATLAB functions `tic` and `toc` illustrate the use of global variables:

```
function tic
%TIC Start a stopwatch timer.
%   The sequence of commands
%       TIC, operation, TOC
%   prints the time required for the operation.
%

%   See also TOC, CLOCK, ETIME, CPUTIME.

%   Copyright (c) 1984-96 by The MathWorks, Inc.
%   $Revision: 5.2 $ $Date: 1996/03/29 20:29:26 $

% TIC simply stores CLOCK in a global variable.
global TICTOC
TICTOC = clock
```



```
function t = toc
%TOC Read the stopwatch timer.
%  TOC, by itself, prints the elapsed time since TIC was used.
%  t = TOC; saves the elapsed time in t, instead of printing it out.
%
%  See also TIC, ETIME, CLOCK, CPUTIME.

%  Copyright (c) 1984-96 by The MathWorks, Inc.
%  $Revision: 5.2 $ $Date: 1996/03/29 20:30:02 $

% TOC uses ETIME and the value of CLOCK saved by TIC.
global TICTOC
if nargout < 1
   elapsed_time = etime(clock,TICTOC)
else
   t = etime(clock,TICTOC);
end
```

The functions `tic` and `toc` form a simple stopwatch for timing MATLAB operations. When `tic` is called, it declares the variable `TICTOC` global, assigns the current time to it, and then terminates. Later when `toc` is called, `TICTOC` is declared global in the `toc` workspace, thereby providing access to its contents, and the elapsed time is computed. It is important to note that the variable `TICTOC` exists only in the workspaces of the functions `tic` and `toc`; it does not exist in the MATLAB workspace unless `global TICTOC` is issued there.

> As a matter of programming practice, the use of global variables is discouraged whenever possible. However, if they are used, it is suggested that global variable names be long, contain all capital letters, and optionally start with the name of the M-file where they appear (e.g., `MYFUN_ALPHA`). If followed, these suggestions will minimize unintended conflicts among global variables.

2. In addition to sharing data through global variables, MATLAB provides the function `evalin`, which allows one to reach into another workspace, evaluate an expression, and return the result to the current workspace. The function `evalin` is similar to `eval`, except that the string is evaluated in either the ***caller*** or ***base*** workspace. The caller workspace is the workspace from which the current function was called. The base workspace is the MATLAB workspace in the *Command* window. For example, `A=evalin('caller','expression')` evaluates *'expression'* in the caller workspace and returns the results to the variable `A` in the current workspace. Alternatively, `A=evalin('base','expression')` evaluates *'expression'* in the MATLAB workspace and returns the results to the variable `A` in the current workspace. `evalin` also provides error trapping with the syntax `evalin('workspace','try','catch')`, where *'workspace'* is either `'caller'` or `'base'`, *'try'* is the first expression evaluated, and *'catch'* is an expression that is evaluated in the ***current*** workspace if the evaluation of *'try'* produces an error.
3. Since one can evaluate an expression in another workspace, it makes sense that one can also assign the results of some expression in the current workspace to a variable in another workspace. The function `assignin` provides this capability. `assignin('workspace','vname',X)`, where *'workspace'* is either `'caller'` or `'base'`, assigns the contents of the variable *X* in the current workspace to a variable in the `'caller'` or `'base'` workspace named *'vname'*.
4. The function `inputname` provides a way to determine the variable names used when a function is called. For example, suppose a function is called as

```
» y = myfunction(xdot,time,sqrt(2))
```

Issuing `inputname(1)` inside of `myfunction` returns the character string `'xdot'`, `inputname(2)` returns `'time'`, and `inputname(3)` returns an empty array because `sqrt(2)` is not a variable, but rather an expression that produces an unnamed temporary result.

5. The name of the M-file being executed is available within a function in the variable `mfilename`. For example, when the M-file `myfunction.m` is being executed, the workspace of the function contains the variable `mfilename`, which contains the character string `'myfunction'`. This variable also exists within script files, in which case it contains the name of the script file being executed.

## 14.4 FUNCTION M-FILES AND THE MATLAB SEARCH PATH

Function M-files are one of the fundamental strengths of MATLAB. They allow one to encapsulate sequences of useful commands and apply them over and over. Since M-files exist as text files on disk, it is important that MATLAB maximize the speed at which the files are found, opened, and executed. The techniques that MATLAB uses to maximize speed are as follows:

1. The first time MATLAB executes a function M-file, it opens the corresponding text file and ***compiles*** the commands into an internal pseudocode representation in memory that speeds execution for all later calls to the function. If the function contains references to other M-file functions and script M-files, they too are compiled into memory.
2. The function `inmem` returns a cell array of strings containing a list of functions and script files currently compiled into memory.
3. It is possible to store the compiled or P-Code version of a function M-file to disk using the `pcode` command. When this is done, MATLAB loads the compiled function into memory rather than the M-file. For most functions this step does not significantly shorten the amount of time required to execute a function the first time. However, it can speed up large M-files associated with complex graphical user interface functions. P-Code files are created by issuing

```
» pcode myfunction
```

   where *`myfunction.m`* is the M-file name to be compiled. P-Code files are platform-independent binary files that have the same name as the original M-file, but end in `.p` rather than `.m`. P-Code files also provide a level of security, since they are visually undecipherable and can be run without the corresponding M-file.
4. As discussed in the chapter on file and directory management (Chapter 5), when MATLAB encounters a name that it does not recognize, it follows a set of rules to determine what to do. Given the information presented here, the set of rules can be updated as follows:

When you enter `cow` at the MATLAB prompt or if MATLAB encounters a reference to `cow` in a script or function M-file,

1. it checks to see if `cow` is a *`variable`* in the current workspace; if not,
2. it checks to see if `cow` is a *`built-in function`*; if not,
3. it checks to see if `cow` is a subfunction in the M-file in which `cow` appears; if not,
4. it checks to see if `cow.p` and then `cow.m` is a *`private`* function to the M-file in which `cow` appears; if not,
5. it checks to see if `cow.p` and then `cow.m` exists in the *`current directory`*; if not,
6. it checks to see if `cow.p` and then `cow.m` exists in each directory specified on the *MATLAB `search path`*, by searching in the order in which the search path is specified.

MATLAB uses the first match it finds.

5. When MATLAB is started, it ***caches*** the name and location of all M-files stored within the `toolbox` subdirectory and in all subdirectories of the `toolbox` directory. This allows MATLAB to find and execute function M-files much faster. It also makes the command `lookfor` work faster.

M-file functions that are cached are considered read-only. If they are executed and then later altered, MATLAB will simply execute the function that was previously compiled into memory, ignoring the changed M-files. Moreover, if new M-files are added within the `toolbox` directory after MATLAB is running, their presence will not be noted in the cache, and thus they will be unavailable for use.

As a result, in the development of M-file functions, it is best to store them outside the `toolbox` directory, perhaps in the `MATLAB` directory, until they are considered complete. When they are complete, move them to a subdirectory inside the read only `toolbox` directory. Finally, make sure the MATLAB search path is changed to recognize their existence.

6. When new M-files are added to a cached location, MATLAB will find them only if the cache is refreshed by issuing the command `path(path)`. On the other hand, when cached M-files are modified, MATLAB will recognize the changes only if a previously compiled version is dumped from memory by issuing the `clear` command (e.g., » `clear` *`myfun`* clears the M-file function *`myfun`* from memory, and » `clear functions` clears all compiled functions from memory).
7. MATLAB keeps track of the modification date of M-files outside the `toolbox` directory. As a result, when an M-file function is encountered that was previously compiled into memory, MATLAB compares the modification dates of the compiled M-file with that of the M-file on disk. If the dates are the same, MATLAB executes the com-

piled M-file. On the other hand, if the M-file on disk is newer, MATLAB dumps the previously compiled M-file and compiles the newer, revised M-file for execution.

## 14.5 CREATING YOUR OWN TOOLBOXES

It is common to organize a group of M-files into a subdirectory on the MATLAB search path. If the M-files are considered complete, the subdirectory should be placed in the `toolbox` directory so that the M-file names are cached. When a `toolbox` subdirectory is created, it is beneficial to include two additional script M-files that contain only MATLAB comments (i.e., lines that begin with a percent sign %). These M-files, named `Readme.m` and `Contents.m`, are described next.

1. The script file `Readme.m` typically contains comment lines that describe late-breaking changes or descriptions of undocumented features. Issuing the command `» whatsnew MyToolbox`, where *MyToolbox* is the name of the directory containing the group of M-files, displays the comment lines in the file. If the Toolbox is posted to *The MathWorks* ftp site or is redistributed by *The MathWorks*, the `Readme.m` file should include a disclaimer such as the following:

```
% These M-files are User Contributed Routines that are being redistributed
% by The MathWorks, upon request, on an "as is" basis. A User Contributed
% Routine is not a product of The MathWorks, Inc. and The MathWorks assumes
% no responsibility for any errors that may exist in these routines.
```

2. The script file `Contents.m` contains comment lines that list all M-files in the Toolbox. Issuing the command `» help MyToolbox` or `» helpwin MyToolbox`, where *MyToolbox* is the name of the directory containing the group of M-files, displays the file listing in the *Command* window or *Help* window, respectively. The first line in the `Contents.m` file should specify the name of the Toolbox, and the second line should state the Toolbox version and date, as shown next for the *Mastering MATLAB Toolbox.*

```
% Mastering MATLAB Toolbox
% Version 5.0 31-Dec-1997
```

When this is done, the `ver` command in MATLAB decodes the first two lines and includes this information in its list of version information. For example,

```
» ver
-------------------------------------------------
MATLAB version 5.0.0.4064 on MAC2
MATLAB License Identification Number: 51483
-------------------------------------------------
MATLAB Toolbox                          Version 5.0      05-Nov-1996
Control System Toolbox.                 Version 4.0      15-Nov-1996
Mastering MATLAB Toolbox                Version 5.0      31-Dec-1997
Optimization Toolbox.                   Version 1.5.0    31-Oct-1996
Signal Processing Toolbox.              Version 4.0      15-Nov-1996
SIMULINK                                Version 2.0      15-Nov-1996
```

## 14.6 COMMAND–FUNCTION DUALITY

In addition to creating function M-files, it is also possible to create MATLAB ***commands***. Examples of MATLAB commands include `clear`, `who`, `dir`, `ver`, `help`, and `whatsnew`. MATLAB commands are very similar to functions. In fact, there are only two differences between commands and functions: (1) commands do not have output arguments, and (2) input arguments to commands are not enclosed by parentheses. For example, `clear functions` is a command that accepts the input argument `functions` without parentheses, performs the action of clearing all compiled functions from memory, and produces no output. A function, on the other hand, usually places data in one or more output arguments and must have its input arguments separated by commas and enclosed by parentheses—for example, `a=atan2(x,y)`.

In reality, MATLAB commands are function calls that obey the two differences just mentioned. For example, the command `whatsnew` is a function M-file. When called from the MATLAB prompt as

```
» whatsnew MyToolbox
```

MATLAB interprets the command as a call to the function `whatsnew` with the syntax:

```
» whatsnew('MyToolbox')
```

In other words, as long as there are no output arguments requested, MATLAB interprets command arguments as character strings, places them in parentheses, and then calls the requested function. This interpretation applies to all commands such as the following:

| Command | Function |
|---|---|
| `format short g` | `format('short','g')` |
| `save x y z` | `save('x','y','z')` |
| `hold on` | `hold('on')` |
| `axis square equal` | `axis('square','equal')` |
| `which` *`fname`* | `which('`*`fname`*`')` |

Both command and function forms can be entered at the MATLAB prompt, although the command form generally requires less typing.

A command M-file can also be interpreted as a function if it obeys the rules for calling functions. For example, the command

```
» which fname
```

displays the directory path string to the M-file *fname* and the function call

```
» s = which('fname')
```

returns the directory path string in the variable s. At the same time,

```
» s = which fname
??? s = which fname
                |
Missing operator, comma, or semi-colon.
```

causes an error because it mixes function and command syntaxes. ***Whenever MATLAB sees an equal sign, it interprets the rest of the statement as a function, which requires comma-separated arguments enclosed in parentheses.***

To summarize, commands and functions both call functions. Commands are translated into function calls by interpreting command arguments as character strings, placing them in parentheses, and then calling the requested function. Any function call can be made in the form of a command if it produces no output arguments and if it requires only character string input.

## 14.7 IN-LINE FUNCTIONS AND feval

There are a number of occasions where the character string name of a function is passed into a function for evaluation. For example, many of the numerical analysis functions in MATLAB require the name of a function to be evaluated. MATLAB provides the function feval to facilitate this evaluation. For example,

```
» a = feval('myfunction',x)
```

is equivalent to

```
» a = myfunction(x)
```

Now you might be thinking that feval is not necessary because the function eval can also be used to evaluate a character string; that is,

```
» a = eval('myfunction(x)')
```

The difference between these two functions is that eval calls in the entire MATLAB interpreter to evaluate the string, whereas feval does just what it says it does. As a result, feval is much more efficient, especially if the function must be evaluated many times as part of some iterative procedure.

feval also supports multiple input and output arguments, such as

```
» [a,b] = feval('myfunction',x,y,x,t)
```

which is equivalent to [a,b] = *myfunction*(x,y,z,t).

Normally, `myfunction` in the preceding examples is the name of an M-file function. Sometimes, however, it is convenient to express the entire function as a character string. For example,

```
» myfun = '100*(y -x^2)^2 + (1-x)^2';
```

In this case, `feval` cannot be used, but `eval` still works. That is,

```
» a = eval(myfun);
```

evaluates the function provided `x` and `y` have been assigned the desired values.

To make both of these function definitions work with `feval`, MATLAB created an object called an ***in-line*** function that overloads `feval` with the capability to handle cases such as `myfun`. In-line functions are created and manipulated by the functions `inline`, `fcnchk`, `argnames`, and `formula`. For example,

```
» x = 1.2; % pick some values for x and y
» y = 2;
» myfun    % show function
myfun =
100*(y -x^2)^2 + (1-x)^2
» a = eval(myfun) % traditional evaluation
a =
          31.4
» myfuni = inline(myfun,'x','y') % convert to inline function
myfuni =
     Inline function:
     myfuni(x,y) = 100*(y -x^2)^2 + (1-x)^2
» a = feval(myfuni,x,y) % now feval works
a =
          31.4
» b = feval(myfuni,-1.2,0) % and works for any arguments
b =
         212.2
» argnames(myfuni) % what are the function arguments?
ans =
    'x'
    'y'
» formula(myfuni) % what is the formula for this function?
ans =
100*(y -x^2)'2 + (1-x)^2
```

As shown, the `inline` object type overloads the function `feval` with the function `@inline/feval.m`, which takes the `inline` function, evaluates it, and returns results just as the standard or built-in function `feval` does. For further information on this topic, see the chapter on object-oriented programming (Chapter 25).

When the M-file name or string function definition is passed to a function for evaluation, the function must determine whether the string is just the name of a function M-file, like `myfunction`, or is the string function definition, like `myfun`. Because most of the numerical analysis functions in MATLAB support both function definition forms, the utility function fcnchk accepts an input string and does the right thing. That is, if the input string is just an M-file name, it returns it for use with the built-in feval. However, if the string is an entire function definition, it creates and returns an in-line function that calls the overloaded feval for evaluation. For example,

```
» FUN = 'humps' % M-file function name
FUN =
humps
» fcnchk(FUN) % inline function needed? No, so return input
ans =
humps
» FUN = '100*(y -x^2)^2 + (1-x)^2' % function definition
FUN =
100*(y -x^2)^2 + (1-x)^2
» fcnchk(FUN) % inline function needed? Yes
ans =
     Inline function:
     ans(x) = 100*(y -x^2)^2 + (1-x)^2
» argnames(ans)
ans =
    'x'
```

The function argnames makes a number of assumptions regarding what the variable or argument of a character string function is. In general, it looks for only one argument, although that argument can be a vector. So, in the preceding example, rather than have x and y as the variables, let the variables be x(1) and x(2), respectively. For example,

```
» FUN % recall FUN
FUN =
100*(y -x^2)^2 + (1-x)^2
» FUN = strrep(FUN,'x','x(1)') % change x to x(1)
FUN =
100*(y -x(1)^2)^2 + (1-x(1))^2
» FUN = strrep(FUN,'y','x(2)') % change y to x(2)
FUN =
100*(x(2) -x(1)^2)^2 + (1-x(1))^2
» fcnchk(FUN) % create inline function with vector argument x
ans =
     Inline function:
     ans(x) = 100*(x(2) -x(1)^2)^2 + (1-x(1))^2
» argnames(ans)
ans =
    'x'
```

The utility of in-line functions is that feval works for functions defined by M-file names as well as for in-line functions. The utility function fcnchk deciphers whether a string contains just an M-file name or the string function description. For further help with fcnchk, type helpwin fcnchk.

# 15

# Debugging and Profiling Tools

In the process of developing function M-files, it is inevitable that errors (i.e., *bugs*) appear. MATLAB provides a number of approaches and functions to assist in *debugging* M-files. MATLAB also provides a tool to help you improve the performance of your M-files. *Profiling* is a method of determining where your M-file spends most of its time.

## 15.1 DEBUGGING TOOLS

Two types of errors can appear in MATLAB expressions: syntax errors and run-time errors. Syntax errors (such as misspelled variables or function names or missing quotes or parentheses) are found when MATLAB evaluates an expression or when a function is compiled into memory. MATLAB flags these errors immediately and provides feedback about the type of error encountered and the line number in the M-file where it occurs. Given this feedback, these errors are usually easy to spot. Note, however, that syntax errors within GUI callback strings are not detected until the strings themselves are evaluated by activating the callback at run-time.

Run-time errors, on the other hand, are generally more difficult to find, even though MATLAB flags them also. When a run-time error is found, MATLAB returns control to the *Command* window and the MATLAB workspace. Access to the function workspace where the error occurred is lost, so one cannot interrogate the contents of the function workspace in an effort to isolate the problem.

Based on the authors' experience, the most common run-time errors occur when the result of some operation leads to empty arrays or `NaN`s. All operations on `NaN`s return `NaN`s, so if `NaN`s are a possible result, it is good to use the logical function `isnan` to perform some default action when `NaN`s occur. Addressing arrays that are empty always leads to an error since empty matrices have a zero dimension. The `find` function represents a common situation where an empty array may result. If the empty array output of the `find` function is used to index some other array, the result returned will also be empty. That is, empty matrices tend to propagate empty matrices. For example,

```
» x = pi*(1:4)  % example data
x =
    3.1416    6.2832    9.4248    12.5664
» i = find(x>20)  % use find function
i =
     []
» y = 2*x(i)  % propagate the empty matrix
y =
     []
```

Clearly, when `y` is expected to have a finite dimension and value(s), a run-time error is likely to occur. When performing operations or using functions that can return empty results, the logical function `isempty` is useful to define a default result for the empty matrix case, thereby avoiding a run-time error.

There are several approaches to debugging function M-files. For simple problems, it is straightforward to use a combination of the following:

1. Remove semicolons from selected lines within the function so that intermediate results are displayed in the *Command* window.
2. Add statements that display variables of interest within the function.
3. Place the `keyboard` command at selected places in the M-file to give temporary control to the keyboard. By doing so, the function workspace can be interrogated and values changed as necessary. Resume function execution by issuing a `return` command at the keyboard prompt (i.e., `K» return`).
4. Change the function M-file into a script M-file by placing a `%` before the function definition statement at the beginning of M-file. When executed as a script file, the workspace is the MATLAB workspace and thus it can be interrogated after the error occurs.

When the M-file is large, recursive, or highly nested (i.e., it calls other M-file functions that call still other functions, etc.), it may be more convenient to use MATLAB debugging functions. As opposed to the approaches listed previously, these functions do not require one to edit the M-file in question. These functions, shown in the following table, are similar to those found in other high-level programming languages.

| Debugging Command | Description |
|---|---|
| `dbstop in` *mfile*<br>`dbstop in` *mfile* `at` *lineno* | Set a breakpoint in *mfile* (at *lineno*). |
| `dbstop if warning`<br>`dbstop if error`<br>`dbstop if naninf`<br>`dbstop if infnan` | Stop on any warning, run-time error, or when a `NaN` or `Inf` is generated (`naninf` and `infnan` are equivalent forms). |
| `dbclear all`<br>`dbclear all in` *filename*<br>`dbclear in` *filename*<br>`dbclear if warning`<br>`dbclear if error`<br>`dbclear if infnan (or naninf)` | Remove breakpoint(s). |
| `dbstatus`<br>`dbstatus` *filename* | List all breakpoints (in *filename*). |
| `dbtype` *mfile*<br>`dbtype` *mfile m:n* | List *mfile* with line numbers (between line numbers *m* and *n*. |
| `dbstep`<br>`dbstep` *n* | Execute one line (or *n* lines) and stop. |
| `dbcont` | Resume execution. |
| `dbstack` | List who called whom. |
| `dbup` | Move up one workspace level. |
| `dbdown` | Move down one workspace level. |
| `dbquit` | Quit debug mode. |

The usual debugging process is first to set breakpoints using `dbstop` in the M-file(s) of interest. Use `dbtype` to determine the appropriate line numbers if necessary. Run the function until a breakpoint is reached. Examine variables at the keyboard prompt (`K»`) and then use `dbstep` to single step through the M-file or `dbcont` to resume execution until another breakpoint is encountered. Use `dbstack` to determine the calling sequence, and use `dbup` and `dbdown` to enter the various workspaces to examine variables. Set and clear breakpoints, step through the code, and examine variables until the problem has been found. Finally, use `dbclear all` and `dbquit` to end the debugging session.

Both PC and Macintosh versions of MATLAB now feature a graphical debugger for M-files. The PC version has an integrated M-file Editor/Debugger. If you have just created or edited the M-files using the MATLAB Editor, you are ready to start. Otherwise, choose **Debug** from the **File** menu to open the Editor/Debugger window. The Macintosh version has a debugger icon on the toolbar of the M-file editor. You can also choose **M-File Debugger** from the **Window** menu and then select a file to debug by choosing **Open** from the **File** menu.

The debugging icons on the toolbar of the debugger window include buttons to single step, continue, and quit debugging. The PC has buttons to set or clear breakpoints as well. On both platforms you can set or clear breakpoints by clicking on the dash or red stop sign next to a line or by selecting **Set/Clear Breakpoints** from the **Debug** menu on the PC or the **Breakpoints** menu on the Macintosh. To view the value of a variable, highlight the variable name and choose **Evaluate Selection** from the **View** menu on the PC, or press the **Return** or **Enter** key on the Macintosh. The **Debug** menu also includes commands to single step or to continue execution. The current workspace can be selected using the **Stack** menu. The **Base** workspace is the top-level *Command* window workspace.

## 15.2 PROFILING M-files

Even when a function M-file works correctly, there may be ways to fine-tune the code to avoid unnecessary calculations or function calls. Performance improvements may be obtained by simply storing the result of a calculation to avoid a complex recalculation or by using vectorization techniques to avoid a For Loop. Do not try to guess where most execution time is spent, however. With today's high-speed processors with integrated floating-point units, it may turn out to be faster to calculate a result more than once than to store it in a variable and recall it again later. The recommended method is to write the code as simply as possible, analyze the code, and then optimize as necessary.

M-file performance can be analyzed through the use of ***profiling.*** A profiler examines a running program to determine where the program spends most of its time. Use a profiler to identify time-consuming function calls or calculations and try to minimize the use of these functions. Remember, the most expensive function in terms of execution time may be several layers down in your code.

MATLAB uses the `profile` command to determine which lines of code in an M-file take the most time to execute. The syntax is shown in the following table.

| Profiling Command | Description |
|---|---|
| `profile` *`function`* | Begin profiling for the *`function`*`.m` M-file. |
| `profile report` | Display a profile report for the selected M-file. |
| `profile plot` | Plot the profile report using a `pareto` plot. |
| `profile reset` | Reset elapsed time data to zero. |
| `profile off` | Disable profiling. |
| `profile on` | Enable profiling after a `profile off` command. |
| `profile done` | Turn off the profiler and clear its data. |

Only one function may be profiled at any one time. Use the `profile` *`function`* syntax to reset the data and select a new M-file to examine.

Here is an example of a profiling session to analyze the `cellstr` function.

```
» g = char('Hi','There','Matlab','Simulink','Testing 1 2 3');
» profile cellstr
» for i=1:1000
    h = cellstr(g);
  end

» profile report
Total time in "/matlab/toolbox/matlab/strfun/cellstr.m": 9.35 seconds

100% of the total time was spent on lines:
        [29 17 28 27 20 18 30 1 31]

 0.06s,  1%  1: function c = cellstr(s)
             2: %CELLSTR Create cell array of strings from character array.

            16:
 0.46s,  5% 17: if iscellstr(s), c = s; return, end
 0.08s,  1% 18: if ndims(s)~=2, error('S must be 2-D.'); end
            19:
 0.08s,  1% 20: if isempty(s)
            21: if isstr(s)

            26: else
 0.17s,  2% 27: c = cell(size(s,1),1);
 0.27s,  3% 28: for i=1:size(s,1)
 8.15s, 87% 29:   c{i} = deblank(s(i,:));
 0.07s,  1% 30: end
 0.01s,  0% 31: end

» profile done
```

As you can see, the greatest amount of time by far was spent in the `deblank` function given the input data used in this example. Your mileage may vary.

# 16

# Numerical Linear Algebra

MATLAB was originally written to provide an easy-to-use interface to professionally developed numerical linear algebra subroutines. As MATLAB has evolved over the years, other features, such as graphics and graphical user interfaces, have made the numerical linear algebra routines less prominent. Nevertheless, MATLAB offers a wide range of valuable numerical linear algebra functions.

## 16.1 SETS OF LINEAR EQUATIONS

One of the most common linear algebra problems is the solution of a linear set of equations. For example, consider the set of equations

$$\begin{bmatrix} 1 & 2 & 3 \\ 4 & 5 & 6 \\ 7 & 8 & 0 \end{bmatrix} \cdot \begin{bmatrix} x_1 \\ x_2 \\ x_3 \end{bmatrix} = \begin{bmatrix} 366 \\ 804 \\ 351 \end{bmatrix}$$

$$\mathbf{A} \cdot \mathbf{x} = \mathbf{b}$$

where the mathematical multiplication symbol ($\cdot$) is now defined in the matrix sense as opposed to the array sense discussed earlier. In MATLAB, this matrix multiplication is denoted with the asterisk notation *. The preceding equations define the product of the matrix **A** and the vector **x** as being equal to the vector **b.** The existence of solutions to the preceding equation is a fundamental problem in linear algebra. Moreover, when a solution does exist, there are numerous approaches to finding the solution, such as Gaussian elimination, LU factorization, or direct use of $\mathbf{A}^{-1}$. It is clearly beyond the scope of this text to discuss the many analytical and numerical issues of linear algebra. We only wish to demonstrate how MATLAB can be used to solve problems like the foregoing.

To solve the preceding problem it is necessary to enter **A** and **b:**

```
» A = [1 2 3;4 5 6
7 8 0]
A =
     1     2     3
     4     5     6
     7     8     0
» b = [366;804;351]
b =
   366
   804
   351
```

As discussed earlier, the entry of the matrix A shows the two ways that MATLAB distinguishes between rows. The semicolon between the 3 and 4 signifies the start of a new row, as does the new line between the 6 and 7. The vector b is a column because each semicolon signifies the start of a new row.

With a background in linear algebra, it is easy to show that this problem has a unique solution if the rank of A and the rank of the augmented matrix [A b] are both equal to 3:

```
» rank(A)
ans =
     3
» rank([A b])
ans =
     3
```

Since this is true, MATLAB can find the solution of $\mathbf{A} \cdot \mathbf{x} = \mathbf{b}$ in two ways, one of which is preferred. The less favorable but more straightforward method is to take $\mathbf{x} = \mathbf{A}^{-1} \cdot \mathbf{b}$ literally:

```
» x = inv(A)*b
x =
   25.0000
   22.0000
   99.0000
```

Here `inv(A)` is a MATLAB function that computes $A^{-1}$ and the multiplication operator `*` is matrix multiplication. The preferable solution is found using the matrix left division operator or backward slash:

```
» x = A\b
x =
    25.0000
    22.0000
    99.0000
```

This equation utilizes an LU factorization approach and expresses the answer as the left division of A into b. The left division operator \ has no preceding dot as this is a matrix operation, not an element-by-element array operation. There are many reasons why this second solution is preferable. Of these, the simplest is that the latter method requires fewer multiplications and divisions and as a result is significantly faster. In addition, this solution is generally more accurate, especially for larger problems. In either case, if MATLAB cannot find a solution or cannot find it accurately, it displays an warning message.

If the transpose of the preceding set of linear equations is taken—that is, $(\mathbf{A} \cdot \mathbf{x})' = \mathbf{b}'$—then the preceding set of linear equations can be written as $\mathbf{x}' \cdot \mathbf{A}' = \mathbf{b}'$, where $\mathbf{x}'$ and $\mathbf{b}'$ are now row vectors. As a result, it is equally valid to express a set of linear equations in terms of the product of a row vector and a matrix being equal to another row vector (e.g., $\mathbf{y} \cdot \mathbf{D} = \mathbf{c}$). In MATLAB, this case is solved by the same internal algorithms using the matrix right division operator or forward slash as `y = c/D`.

It is important to note that when MATLAB encounters a forward (/) or backward (\) slash, it checks the structure of the coefficient matrix to determine what internal algorithm to use to find the solution. In particular, if the matrix is upper or lower triangular or a permutation of an upper or lower triangular matrix, MATLAB does not refactor the matrix, but rather just performs the forward or backward substitution steps required to find the solution. As a result, MATLAB makes use of the properties of the coefficient matrix to compute the solution as quickly as possible.

If you have studied linear algebra rigorously, you know that when the number of equations and number of unknowns differ, a single unique solution usually does not exist. However, with further contraints a practical solution usually can be found. In MATLAB, when $\text{rank}(A) = \min(r,c)$, where $r$ and $c$ are the number of rows and columns in $A$, respectively, and there are more equations than unknowns ($r > c$) (i.e., the overdetermined case), use of a division operator / or \ automatically finds the solution that minimizes the squared error in $\mathbf{A} \cdot \mathbf{x} - \mathbf{b}$. This solution is of great practical value and is called the ***least squares solution.*** Consider the following example:

```
» A = [1 2 3;4 5 6;7 8 0;2 5 8]  % 4 equations in 3 unknowns
A =
     1     2     3
     4     5     6
     7     8     0
     2     5     8
» b = [366 804 351 514]'  % a new r.h.s. vector
```

```
b =
   366
   804
   351
   514
» x = A\b  % compute the least squares solution
x =
  247.9818
 -173.1091
  114.9273
» res = A*x-b % this residual has the smallest norm.
res =
 -119.4545
   11.9455
    0.0000
   35.8364
```

In addition to the least squares solution computed with the left and right division operators, MATLAB also offers the functions `lscov` and `nnls`. The function `lscov` solves the weighted least squares problem when the covariance matrix of the data is known, and `nnls` finds the nonnegative least squares solution where all solution components are constrained to be positive.

When there are fewer equations than unknowns ($r < c$) (i.e., the underdetermined case), an infinite number of solutions exist. Of these solutions, MATLAB computes two in a straightforward way. Use of the division operator gives a solution that has a maximum number of zeros in the elements of $x$. Alternatively, computing `x=pinv(A)*b` gives a solution where the length or norm of `x` is smaller than all other possible solutions. This solution, based on the pseudoinverse, also has great practical value and is called the ***minimum norm solution.*** Consider the following example:

```
» A = A'  % create 3 equations in 4 unknowns
A =
     1     4     7     2
     2     5     8     5
     3     6     0     8
» b = b(1:3) % new r.h.s vector
b =
   366
   804
   351
» x = A\b  % solution with maximum zero elements
x =
         0
 -165.9000
   99.0000
  168.3000
» xn = pinv(A)*b % find minimum norm solution
xn =
```

```
   30.8182
 -168.9818
   99.0000
  159.0545
» norm(x) % norm of solution with zero elements
ans =
  256.2200
» norm(xn) % minimum norm solution has smaller norm!
ans =
  254.1731
```

## 16.2 MATRIX FUNCTIONS

In addition to the solution of linear sets of equations, MATLAB offers numerous matrix functions that are useful for solving numerical linear algebra problems. Thorough discussion of these functions is beyond the scope of this text. A brief description of many of the matrix functions is given in the following table.

| Function | Description |
|---|---|
| `A^n` | Exponentiation, e.g., `A^3 = A · A · A`. |
| `balance(A)` | Scale to improve eigenvalue accuracy. |
| `cdf2rdf(A)` | Complex diagonal form to real block diagonal form. |
| `chol(A)` | Cholesky factorization. |
| `cholinc(A,DropTol)` | Incomplete Cholesky factorization. |
| `cond(A)` | Matrix condition number. |
| `condest(A)` | 1-norm matrix condition number estimate. |
| `condeig(A)` | Condition number with respect to repeated eigenvalues. |
| `det(A)` | Determinant. |
| `eig(A)` | Vector of eigenvalues. |
| `[V,D]=eig(A)` | Matrix of eigenvectors `V`, and diagonal matrix of eigenvalues `D`. |
| `[V,D]=eigs(A)` | A few eigenvectors and eigenvalues. |
| `expm(A)` | Matrix exponential (preferred method). |
| `expm1(A)` | M-file implementation of `expm(A)`. |
| `expm2(A)` | Matrix exponential via Taylor series. |
| `expm3(A)` | Matrix exponential via eigenvalues and eigenvectors. |
| `funm(A,'fun')` | Compute general matrix function. |
| `hess(A)` | Hessenberg form. |

| Function | Description |
|---|---|
| inv(A) | Matrix inverse. |
| logm(A) | Matrix logarithm. |
| lscov(A,b,V) | Least squares with known covariance. |
| lu(A) | Factors from Gaussian elimination. |
| luinc(A,DropTol) | Incomplete LU factorization. |
| nnls(A,b) | Nonnegative least squares. |
| norm(A)<br>norm(A,1)<br>norm(A,2)<br>norm(A,inf)<br>norm(A,p)<br>norm(A,'fro') | Matrix and vector 2-norm,<br>1-norm,<br>2-norm<br>infinity norm,<br>P-norm (vectors only),<br>Frobenius norm, i.e., norm(A(:)). |
| normest(A) | Estimate of matrix 2-norm. |
| null(A) | Null space. |
| orth(A) | Orthogonalization. |
| pinv(A) | Pseudoinverse. |
| planerot(x) | Given's plane rotation. |
| poly(A) | Characteristic polynomial. |
| polyeig(AØ,A1,...AP) | Solve polynomial eigenvalue problem. |
| polyvalm(A) | Evaluate matrix polynomial. |
| qr(A) | Orthogonal-triangular decomposition. |
| qrdelete(Q,R,j) | Delete column from qr factorization. |
| qrinsert(Q,R,j,x) | Insert column in qr factorization. |
| qz(A,B) | Generalized eigenvalues. |
| rank(A) | Number of linearly independent rows or columns. |
| rref(A) | Reduced row echelon form. |
| rsf2csf(U,T) | Real schur form to complex schur form. |
| schur(A) | Schur decomposition. |
| sqrtm(A) | Matrix square root. |
| subspace(A,B) | Angle between two subspaces. |
| svd(A) | Singular value decomposition. |
| svds(A,k) | A few singular values. |
| trace(A) | Sum of diagonal elements. |

## 16.3 SPECIAL MATRICES

MATLAB offers a number of special matrices; some of them are general utilities, while others are matrices of interest to specialized disciplines. These and other special matrices include those given in the following table. Use on-line help to learn more about these matrices.

| Matrix | Description |
|---|---|
| `[]` | The empty matrix. |
| `compan` | Companion matrix. |
| `eye` | Identity matrix. |
| `gallery` | Over 50 test matrices. |
| `hadamard` | Hadamard matrix. |
| `hankel` | Hankel matrix. |
| `hilb` | Hilbert matrix. |
| `invhilb` | Inverse Hilbert matrix. |
| `magic` | Magic square. |
| `ones` | Matrix containing all ones. |
| `pascal` | Pascal triangle matrix. |
| `rand` | Uniformly distributed random matrix with elements between 0 and 1. |
| `randn` | Normally distributed random matrix with elements having zero mean and unit variance. |
| `rosser` | Symmetric eigenvalue test matrix. |
| `toeplitz` | Toeplitz matrix. |
| `vander` | Vandermonde matrix. |
| `wilkinson` | Wilkinson eigenvalue test matrix. |
| `zeros` | Matrix containing all zero elements. |

## 16.4 SPARSE MATRICES

In many practical applications, matrices are generated that contain only a few nonzero elements. As a result, these matrices are said to be sparse. For example, circuit simulation and finite element analysis programs routinely deal with matrices containing fewer than 1% nonzero elements. If a matrix is large—for example, max(size(A))>100—and has a high percentage of zero elements, it is both wasteful of computer storage to store the zero elements and wasteful of computational power to perform arithmetic operations using the zero elements. To eliminate the storage of zero elements, it is common to store only the nonzero elements of a sparse matrix and two sets of indices that identify the row and column positions of those elements. Similarly, to eliminate arithmetic operations on the zero elements, special algorithms have been developed to solve typical matrix problems such as solving a set of linear equations wherein operations involving zeros are eliminated as much as possible.

The techniques used to optimize sparse matrix computations are complex in implementation as well as in theory. Fortunately, MATLAB hides this complexity. In MATLAB, sparse matrices are stored in variables just as regular full matrices are. Moreover, most computations with sparse matrices use the same syntax as that used for full matrices. In particular, all the array manipulation capabilities of MATLAB work equally well on sparse matrices. For example, s(i,j) = *value* adds a nonzero element to the ith row and jth column of the sparse matrix s.

In this text, only the creation of sparse matrices and the conversion to and from sparse matrices will be illustrated. In general, operations on full matrices produce full matrices and operations on sparse matrices produce sparse matrices. In addition, operations on a mixture of full and sparse matrices generally produce sparse matrices, unless the operation makes the result too densely populated with nonzeros to make sparse storage efficient.

Sparse matrices are created using the MATLAB function sparse. For example,

```
» As = sparse(1:10,1:10,ones(1,10))
As =
   (1,1)        1
   (2,2)        1
   (3,3)        1
   (4,4)        1
   (5,5)        1
   (6,6)        1
   (7,7)        1
   (8,8)        1
   (9,9)        1
  (10,10)       1
```

creates a 10 by 10 identity matrix. In this usage sparse(i,j,s) creates a sparse matrix whose kth nonzero element is s(k), and s(k) appears in the row i(k) and column j(k). Note the difference in how sparse matrices are displayed.

```
» As = sparse(eye(10))
As =
   (1,1)        1
   (2,2)        1
   (3,3)        1
```

```
    (4,4)        1
    (5,5)        1
    (6,6)        1
    (7,7)        1
    (8,8)        1
    (9,9)        1
   (10,10)       1
```

creates the 10 by 10 identity matrix again, this time by converting the full matrix `eye(10)` to sparse format. While this method of creating a sparse matrix works, it is seldom in used in practice because the initial full matrix wastes a great deal of memory.

```
» A = full(As)
A =
     1     0     0     0     0     0     0     0     0     0
     0     1     0     0     0     0     0     0     0     0
     0     0     1     0     0     0     0     0     0     0
     0     0     0     1     0     0     0     0     0     0
     0     0     0     0     1     0     0     0     0     0
     0     0     0     0     0     1     0     0     0     0
     0     0     0     0     0     0     1     0     0     0
     0     0     0     0     0     0     0     1     0     0
     0     0     0     0     0     0     0     0     1     0
     0     0     0     0     0     0     0     0     0     1
```

converts the sparse matrix back to its full form.

To compare sparse matrix storage to full matrix storage, consider the following example:

```
» B = eye(200);    % FULL   200 by 200 identity matrix
» Bs = sparse(B); % Sparse 200 by 200 identity matrix
» whos
  Name        Size          Bytes  Class


B           200x200       320000  double array
Bs          200x200         3204  sparse array

Grand total is 40200 elements using 323204 bytes
```

Here the sparse matrix `Bs` contains only 0.5% nonzero elements and requires 3204 bytes of storage. On the other hand, `B`, the same matrix in full matrix form, requires two orders of magnitude more bytes of storage.

## 16.5 SPARSE MATRIX FUNCTIONS

MATLAB provides numerous sparse matrix functions. Many involve different aspects of and techniques for the solution of sparse simultaneous equations. The functions available are listed in the following table.

| Sparse Matrix Function | Description |
|---|---|
| `bic` | Biconjugate gradients method. |
| `bicstab` | Biconjugate gradients stabilized method. |
| `cgs` | Preconditioned conjugate gradients method. |
| `cholinc` | Incomplete Choleski factorization. |
| `colmmd` | Reorder by column minimum degree. |
| `colperm` | Reorder by ordering columns based on nonzero count. |
| `condest` | Estimate 1-norm matrix condition. |
| `dmperm` | Reorder by Dulmage-Mendelsohn decomposition. |
| `eigs` | A few eigenvalues and eigenvectors. |
| `etree` | Elimination tree. |
| `etreeplot` | Plot elimination tree. |
| `find` | Find indices of nonzero entries. |
| `full` | Convert to full matrix. |
| `gmres` | Generalized minimum residual method. |
| `gplot` | Graph theory plot of sparse matrix. |
| `issparse` | Return True for sparse input. |
| `luinc` | Incomplete LU factorization. |
| `nnz` | Number of nonzero entries. |
| `nonzeros` | Nonzero entries. |
| `normest` | Estimate 2-norm. |
| `nzmax` | Storage allocated for nonzeros. |
| `qmr` | Quasi-minimal residual method. |
| `randperm` | Random permutation. |
| `spalloc` | Allocate memory for nonzeros. |

| Sparse Matrix Function | Description |
|---|---|
| `sparse` | Create sparse matrix. |
| `spaugment` | Form least squares augmented system. |
| `spconvert` | Convert sparse to external format. |
| `spdiags` | Sparse matrix formed from diagonals. |
| `speye` | Sparse identity matrix. |
| `spfun` | Apply function to nonzero elements. |
| `spones` | Replace nonzeros with ones. |
| `spparms` | Set sparse matrix routine parameters. |
| `sprand` | Sparse uniformly distributed random matrix. |
| `sprandn` | Sparse normally distributed random matrix. |
| `sprandsym` | Sparse symmetric random matrix. |
| `sprank` | Structural rank. |
| `spy` | Visualize sparse structure. |
| `svds` | A few singular values. |
| `symbfact` | Symbolic factorization analysis. |
| `symmd` | Reorder by symmetric minimum degree. |
| `symrcm` | Reorder by reverse Cuthill-McKee algorithm. |
| `treelayout` | Lay out tree or forest. |
| `treeplot` | Plot picture of tree. |

# 17

# Data Analysis

Because of its array orientation, MATLAB readily performs statistical analyses on data sets. While MATLAB by default considers data sets stored in column-oriented arrays, data analysis can be conducted along any specified dimensions.

> That is, unless specified otherwise, the each column of an array represents a different measured variable and each row represents individual samples or observations.

## 17.1 BASIC STATISTICAL ANALYSIS

For example, let us assume that the daily high temperature (in Celsius) of three cities over a 31-day month was recorded and assigned to the variable `temps` in a script M-file. Running the M-file puts the variable `temps` in the MATLAB workspace. Doing this work, the variable `temps` contains the following:

```
» temps
temps =
    12     8    18
    15     9    22
    12     5    19
    14     8    23
    12     6    22
    11     9    19
    15     9    15
     8    10    20
    19     7    18
    12     7    18
    14    10    19
    11     8    17
     9     7    23
     8     8    19
    15     8    18
     8     9    20
    10     7    17
    12     7    22
     9     8    19
    12     8    21
    12     8    20
    10     9    17
    13    12    18
     9    10    20
    10     6    22
    14     7    21
    12     5    22
    13     7    18
    15    10    23
    13    11    24
    12    12    22
```

Each row contains the high temperatures for a given day. Each column contains the high temperatures for a different city. To visualize the data, plot them:

```
» d = 1:31;   % number the days of the month
» plot(d,temps)
» xlabel('Day of Month'), ylabel('Celsius')
» title('Figure 17.1: Daily High Temperatures in Three Cities')
```

The `plot` command illustrates yet another form of `plot` command usage. The variable `d` is a vector of length 31; whereas `temps` is a 31 by 3 matrix. Given this data, the `plot` command plots each column of `temps` versus `d`.

To illustrate some of the data analysis capabilities of MATLAB, consider the following commands based on the preceding temperature data:

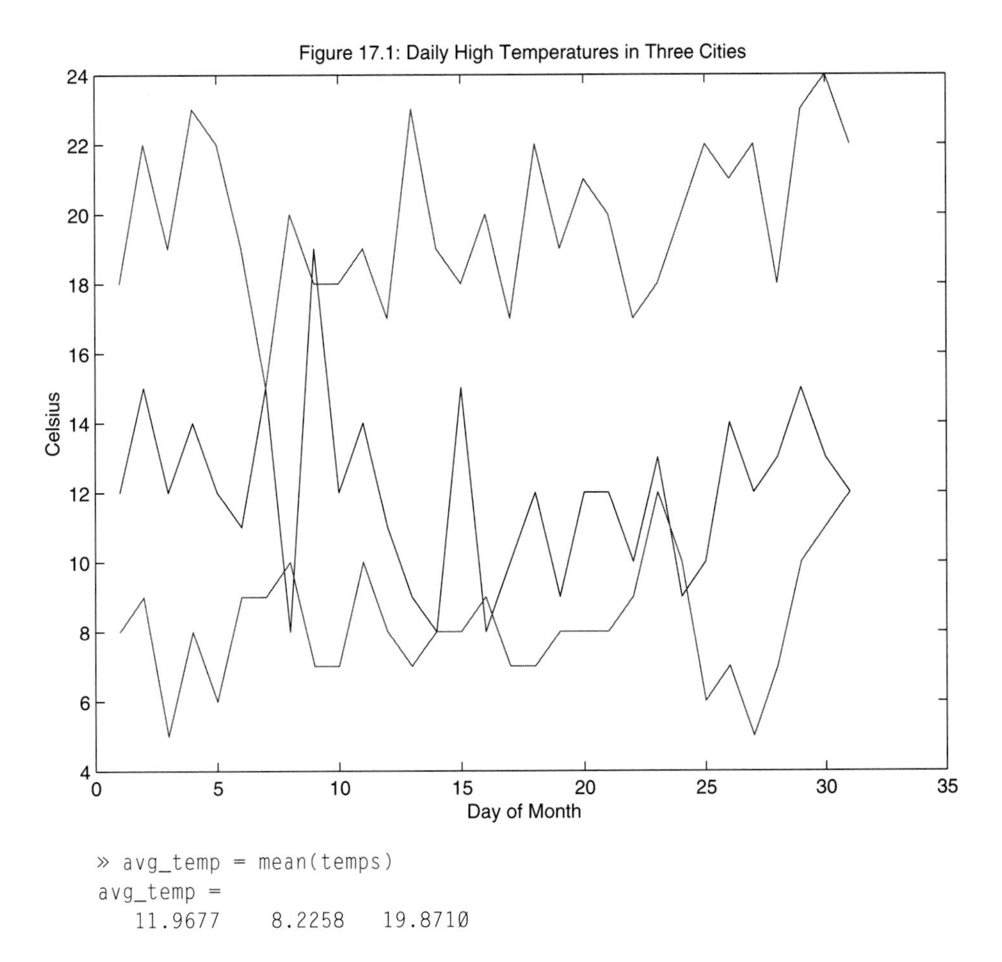


```
» avg_temp = mean(temps)
avg_temp =
   11.9677    8.2258   19.8710
```

shows that the third city has the highest average temperature. Here MATLAB found the average of each column individually. Taking the average again gives

```
» avg_avg = mean(avg_temp)
avg_avg =
   13.3548
```

This finds the overall average temperature of the three cities. When the input to a data analysis function is a row or column vector, MATLAB simply performs the operation on the vector, returning a scalar result.

Alternately, you can specify the dimension to work on:

```
» avg_temp = mean(temps,1)  % same as above, work down the rows
avg_temp =
        11.968        8.2258        19.871
```

```
» avg_tempr = mean(temps,2)  % compute means across columns
avg_tempr =
       12.667
       15.333
           12
           15
       13.333
           13
           13
       12.667
       14.667
       12.333
       14.333
           12
           13
       11.667
       13.667
       12.333
       11.333
       13.667
           12
       13.667
       13.333
           12
       14.333
           13
       12.667
           14
           13
       12.667
           16
           16
       15.333
```

This is the three-city average temperature on each day. The scalar second argument to `mean` dictates the dimension to be analyzed (e.g., 1 is the row dimension number, so perform analysis accumulating information down the rows; 2 is the column dimension number, so perform the analysis accumulating information across the columns).

Consider the problem of finding the daily deviation from the mean of each city. That is, `avg_temp(i)` must be subtracted from column `i` of `temps`. One cannot simply issue the statement

```
» temps - avg_temp
??? Error using ==> -
Matrix dimensions must agree.
```

because the operation is not a defined array operation (`temps` is 31 by 3 and `avg_temp` is 1 by 3). Perhaps the most straightforward approach is to use a For Loop:

```
» for c=1:3
   tdev(:,c) = temps(:,c) - avg_temp(c);
end
```

While the preceding approach works, it is slower than using the array manipulation features of MATLAB. It is much faster to duplicate `avg_temp` to make it the size of `temps`, and then do the subtraction:

```
» tdev = temps - avg_temp(ones(31,1),:)
tdev =
    0.0323   -0.2258   -1.8710
    3.0323    0.7742    2.1290
    0.0323   -3.2258   -0.8710
    2.0323   -0.2258    3.1290
    0.0323   -2.2258    2.1290
   -0.9677    0.7742   -0.8710
    3.0323    0.7742   -4.8710
   -3.9677    1.7742    0.1290
    7.0323   -1.2258   -1.8710
    0.0323   -1.2258   -1.8710
    2.0323    1.7742   -0.8710
   -0.9677   -0.2258   -2.8710
   -2.9677   -1.2258    3.1290
   -3.9677   -0.2258   -0.8710
    3.0323   -0.2258   -1.8710
   -3.9677    0.7742    0.1290
   -1.9677   -1.2258   -2.8710
    0.0323   -1.2258    2.1290
   -2.9677   -0.2258   -0.8710
    0.0323   -0.2258    1.1290
    0.0323   -0.2258    0.1290
   -1.9677    0.7742   -2.8710
    1.0323    3.7742   -1.8710
   -2.9677    1.7742    0.1290
   -1.9677   -2.2258    2.1290
    2.0323   -1.2258    1.1290
    0.0323   -3.2258    2.1290
    1.0323   -1.2258   -1.8710
    3.0323    1.7742    3.1290
    1.0323    2.7742    4.1290
    0.0323    3.7742    2.1290
```

Here `avg_temp(ones(31,1),:)` duplicates the first (and only) row of `avg_temp` 31 times, creating a 31 by 3 matrix whose *i*th column is `avg_temp(i)`. Alternatively, `avg_temp(ones(31,1),:)` can be replaced by `repmat(avg_temp,31,1)`.

MATLAB can also find minima and maxima. That is,

```
» max_temp = max(temps)
max_temp =
    19    12    24
```

finds the maximum high temperature of each city over the month.

```
» [max_temp,maxday] = max(temps)
max_temp =
    19    12    24
maxday =
     9    23    30
```

finds the maximum high temperature of each city and the row index `maxday` where the maximum appears. For this example, `maxday` identifies the day of the month when the highest temperature occurred.

```
» min_temp = min(temps)
min_temp =
     8     5    15
```

finds the minimum high temperature of each city.

```
» [min_temp,minday] = min(temps)
min_temp =
     8     5    15
minday =
     8     3     7
```

finds the minimum high temperature of each city and the row index `minday` where the minimum appears. For this example, `minday` identifies the day of the month when the lowest high temperature occurred.

Other standard statistical measures are also provided in MATLAB. For example,

```
» s_dev = std(temps) % standard deviation in each city
s_dev =
    2.5098    1.7646    2.2322

» median(temps) % median temperature in each city
ans =
    12     8    20
```

```
» cov(temps) % covariance
ans =
       6.2989       0.04086      -0.13763
      0.04086         3.114      0.063441
     -0.13763      0.063441        4.9828

» corrcoef(temps) % correlation coefficients
ans =
            1     0.0092259     -0.024567
    0.0092259             1      0.016106
    -0.024567      0.016106             1
```

You can also compute differences from day to day by using the function `diff`:

```
» daily_change = diff(temps)
daily_change =
     3     1     4
    -3    -4    -3
     2     3     4
    -2    -2    -1
    -1     3    -3
     4     0    -4
    -7     1     5
    11    -3    -2
    -7     0     0
     2     3     1
    -3    -2    -2
    -2    -1     6
    -1     1    -4
     7     0    -1
    -7     1     2
     2    -2    -3
     2     0     5
    -3     1    -3
     3     0     2
     0     0    -1
    -2     1    -3
     3     3     1
    -4    -2     2
     1    -4     2
     4     1    -1
    -2    -2     1
     1     2    -4
     2     3     5
    -2     1     1
    -1     1    -2
```

This computes the difference between daily high temperatures, which describes how much the daily high temperature varied from day to day. For example, between the first and second day of the month, the first row of daily_change is the amount the daily high changed.

In prior versions of MATLAB, the existence of NaNs in a data array to signify missing data caused the functions min and max to return NaNs for results. As a result, it was necessary to eliminate NaNs from data before performing analysis. In version 5, NaNs are ignored by the functions min and max. However, mean, median, and other functions must still be protected from NaNs.

## 17.2 BASIC DATA ANALYSIS

In addition to statistical data analysis, MATLAB offers a variety of general-purpose data analysis functions. For example, the preceding temperature data can be filtered using the function filter(b,a,data):

```
» filter(ones(1,4),4,temps)
ans =
          3          2        4.5
       6.75       4.25         10
       9.75        5.5      14.75
      13.25        7.5       20.5
      13.25          7       21.5
      12.25          7      20.75
         13          8      19.75
       11.5        8.5         19
      13.25       8.75         18
       13.5       8.25      17.75
      13.25        8.5      18.75
         14          8         18
       11.5          8      19.25
       10.5       8.25       19.5
      10.75       7.75      19.25
         10          8         20
      10.25          8       18.5
      11.25       7.75      19.25
       9.75       7.75       19.5
      10.75        7.5      19.75
      11.25       7.75       20.5
      10.75       8.25      19.25
      11.75       9.25         19
         11       9.75      18.75
       10.5       9.25      19.25
       11.5       8.75      20.25
      11.25          7      21.25
      12.25       6.25      20.75
       13.5       7.25         21
      13.25       8.25      21.75
      13.25         10      21.75
```

Here the filter implemented is $4y_n = x_n + x_{n-1} + x_{n-2} + x_{n-3}$ or, equivalently, $y_n = (x_n + x_{n-1} + x_{n-2} + x_{n-3})/4$. That is, each column of `temps` is passed through a moving average filter of length 4. Any realizable filter structure can be applied by specifying different coefficients for the input and output coefficient vectors. The function `y=filter(b,a,x)` implements the following general tapped delay-line algorithm:

$$\sum_{k=0}^{N} a_{k+1} y_{n-k} = \sum_{k=0}^{M} b_{k+1} x_{n-k}$$

where the vector `a` is the tap weight vector $a_{k+1}$ on the output, and the vector `b` is the tap weight vector $b_{k+1}$ on the input.

Data can also be sorted. For example,

```
» data = rand(10,1) % create some data
data =
      0.61543
      0.79194
      0.92181
      0.73821
      0.17627
      0.40571
      0.93547
      0.91690
      0.41027
      0.89365
» [sdata,sidx] = sort(data) % sort in ascending order
sdata =
      0.17627
      0.40571
      0.41027
      0.61543
      0.73821
      0.79194
      0.89365
      0.91690
      0.92181
      0.93547
sidx =
      5
      6
      9
      1
      4
      2
     10
      8
      3
      7
```

The second output of the `sort` function is the sorted index order. That is, the fifth element in this example has the lowest value, and the seventh has the largest value.

Sometimes it is important to know the ***rank*** of the data (e.g., What is the rank or position of the *i*th data point in the unsorted array with respect to the sorted array?). Using MATLAB array indexing, the rank is found by the single statement

```
» ridx(sidx) = 1:10 % ridx is rank
ridx =
     4     6     9     5     1     2    10     8     3     7
```

That is, the first element of the unsorted data appears fourth in the sorted data and the last element is seventh.

When the array to be sorted is a matrix like `temps`, each column is sorted and each column produces a column in the optional index matrix. As with the other data analysis functions, the dimension to perform analysis on can be specified as a final input argument.

A vector is strictly monotonic if its elements either always increase or always decrease as one proceeds down the array. The function `diff` is useful for determining monotonicity:

```
» A = diff(data) % check random data
A =
     0.17650
     0.12988
    -0.18361
    -0.56194
     0.22944
     0.52976
    -0.01856
    -0.50663
     0.48338
» mono = all(A>0) | all(A<0) % as expected, not monotonic
mono =
     0
» B = diff(sdata) % check sorted data
B =
     0.22944
     0.00456
     0.20516
     0.12277
     0.05373
     0.10171
     0.02325
     0.00490
     0.01365
» mono = all(B>0) | all(B<0) % as expected, monotonic
mono =
     1
```

Furthermore, a monotonic vector is equally spaced if

```
» all( diff( diff(sdata) )==0 ) % random data is not equally spaced
ans =
     0
» all( diff( diff(1:25) )==0 ) % but numbers from 1 to 25 are equally
spaced
ans =
     1
```

## 17.3 DATA ANALYSIS AND STATISTICAL FUNCTIONS

By default, data analysis in MATLAB is performed on column-oriented matrices. Different variables are stored in individual columns and each row represents a different observation of each variable. Many data analysis functions work along any dimension, provided it is specified as the last input argument. MATLAB data analysis and statistical functions include the following:

| Data Analysis Function | Description |
|---|---|
| `corrcoef(x)` | Correlation coefficients. |
| `cov(x)` | Covariance matrix. |
| `cplxpair(x)` | Sort vector into complex conjugate pairs. |
| `cumprod(x)` | Cumulative product. |
| `cumsum(x)` | Cumulative sum. |
| `cumtrapz(x,y)` | Cumulative trapezoidal integration. |
| `del2(A)` | Discrete Laplacian. |
| `diff(x)` | Differences between elements. |
| `gradient(Z,dx,dy)` | Approximate gradient. |
| `histogram(x)` | Histogram or bar chart. |
| `max(x), max(x,y)` | Maximum component. |
| `mean(x)` | Mean or average value. |
| `median(x)` | Median value. |
| `min(x), min(x,y)` | Minimum component. |
| `prod(x)` | Product of elements. |
| `rand(x)` | Uniformly distributed random numbers. |

| Data Analysis Function | Description |
| --- | --- |
| `randn(x)` | Normally distributed random numbers. |
| `sort(x)` | Sort in ascending order. |
| `sortrows(A)` | Sort rows in a ascending order. |
| `std(x)` | Standard deviation of columns normalized by $N - 1$. |
| `subspace(A,B)` | Angle between two subspaces. |
| `sum(x)` | Sum of elements in each column. |
| `trapz(x,y)` | Trapezoidal integration of $y = f(x)$. |

# 18

# Polynomials

MATLAB provides a number of functions for manipulating polynomials. Polynomials are easily differentiated and integrated, and it is straightforward to find polynomial roots. However, higher-order polynomials pose numerical difficulties in a number of situations and therefore should be used with caution.

## 18.1 ROOTS

Finding the roots of a polynomial (i.e., the values for which the polynomial is zero) is a problem common to many disciplines. MATLAB solves this problem and provides other polynomial manipulation tools as well. In MATLAB a polynomial is represented by a row vector of its coefficients in descending order. For example, the polynomial $x^4 - 12x^3 + 0x^2 + 25x + 116$ is entered as

```
» p = [1 -12 0 25 116]
p =
     1   -12     0    25   116
```

Note that terms with zero coefficients must be included. MATLAB has no way of knowing which terms are zero unless you specifically identify them. Given this form, the roots of a polynomial are found by using the function `roots`:

```
» r = roots(p)
r =
  11.7473
   2.7028
  -1.2251 + 1.4672i
  -1.2251 - 1.4672i
```

> Since both a polynomial and its roots are vectors in MATLAB, MATLAB adopts the convention that polynomials are row vectors and roots are column vectors.

Given the roots of a polynomial, it is also possible to construct the associated polynomial. In MATLAB, the command `poly` performs this task:

```
» pp = poly(r)
pp =
            1          -12    -1.7764e-14           25          116

» pp(abs(pp)<1e-12) = 0  % change small element to zero!
pp =
            1          -12              0           25          116
```

Because of truncation errors, it is not uncommon for the results of `poly` to have near zero components or have components with small imaginary parts. As shown previously, near zero components can be corrected by array manipulation. Similarly, eliminating spurious imaginary parts is simply a matter of using the function `real` to extract the real part of the result.

## 18.2 MULTIPLICATION

Polynomial multiplication is supported by the function `conv` (which performs the convolution of two arrays). Consider the product of the two polynomials $a(x) = x^3 + 2x^2 + 3x + 4$ and $b(x) = x^3 + 4x^2 + 9x + 16$:

```
» a = [1 2 3 4];  b = [1 4 9 16];

» c = conv(a,b)
c =
     1     6    20    50    75    84    64
```

This result is $c(x) = x^6 + 6x^5 + 20x^4 + 50x^3 + 75x^2 + 84x + 64$. Multiplication of more than two polynomials requires repeated use of `conv`.

## 18.3 ADDITION

MATLAB does not provide a direct function for adding polynomials. Standard array addition works if both polynomial vectors are the same size. For example,

```
» d = a + b
d =
      2      6     12     20
```

which is $d(x) = 2x^3 + 6x^2 + 12x + 20$. When two polynomials are of different orders, the one having lower order must be padded with leading zeros to make it have the same effective order as the higher-order polynomial. Consider the addition of polynomials `c` and `d`:

```
» e = c + [0 0 0 d]
e =
      1      6     20     52     81     96     84
```

which is $e(x) = x^6 + 6x^5 + 20x^4 + 52x^3 + 81x^2 + 96x + 84$. Leading zeros are required rather than trailing zeros because coefficients associated with like powers of $x$ must line up.

The *Mastering MATLAB Toolbox* contains the function `mmpadd`, which performs polynomial addition:

```
function p=mmpadd(a,b)
%MMPADD Polynomial Addition. (MM)
% MMPADD(A,B) adds the polynomials A and B.
%
% See also MMPOLY, MMPSIM, MMPSCALE, MMPSHIFT, MMP2STR.

if nargin<2
   error('Not Enough Input Arguments.')
end
a=a(:).';       % make sure inputs are polynomial row vectors
b=b(:).';
na=length(a);  % find lengths of a and b
nb=length(b);
p=[zeros(1,nb-na) a]+[zeros(1,na-nb) b];  % pad with zeros as necessary
```

To illustrate the use of `mmpadd`, reconsider the preceding example:

```
» f = mmpadd(c,d)
f =
      1      6     20     52     81     96     84
```

which is the same as e. Of course, mmpadd can also be used for subtraction:

```
» g = mmpadd(c,-d)
g =
     1     6    20    48    69    72    44
```

which is $g(x) = x^6 + 6x^5 + 20x^4 + 48x^3 + 69x^2 + 72x + 44$.

## 18.4 DIVISION

In some special cases it is necessary to divide one polynomial into another. In MATLAB, this is accomplished with the function deconv. Using the polynomials b and c from before,

```
» [q,r] = deconv(c,b)
q =
     1     2     3     4
r =
     0     0     0     0     0     0     0
```

This result says that b divided into c gives the quotient polynomial q and the remainder r, which is zero in this case since the product of b and q is exactly c.

## 18.5 DERIVATIVES

Because differentiation of a polynomial is simple to express, MATLAB offers the function polyder for polynomial differentiation:

```
» g
g =
     1     6    20    48    69    72    44
» h = polyder(g)
h =
     6    30    80   144   138    72
```

## 18.6 EVALUATION

Given that you can add, subtract, multiply, divide, and differentiate polynomials based on row vectors of their coefficients, you should be able to evaluate them also. In MATLAB this is accomplished with the function polyval:

```
» x = linspace(-1,3);
```

This chooses 100 data points between -1 and 3.

```
» p = [1 4 -7 -10];
```

uses the polynomial $p(x) = x^3 + 4x^2 - 7x - 10$.

```
» v = polyval(p,x);
```

evaluates $p(x)$ at the values in `x` and stores the result in `v`. The result is then plotted using

```
» plot(x,v), title('Figure 18.1: x{^3} + 4x{^2} - 7x -10'), xlabel('x')
```

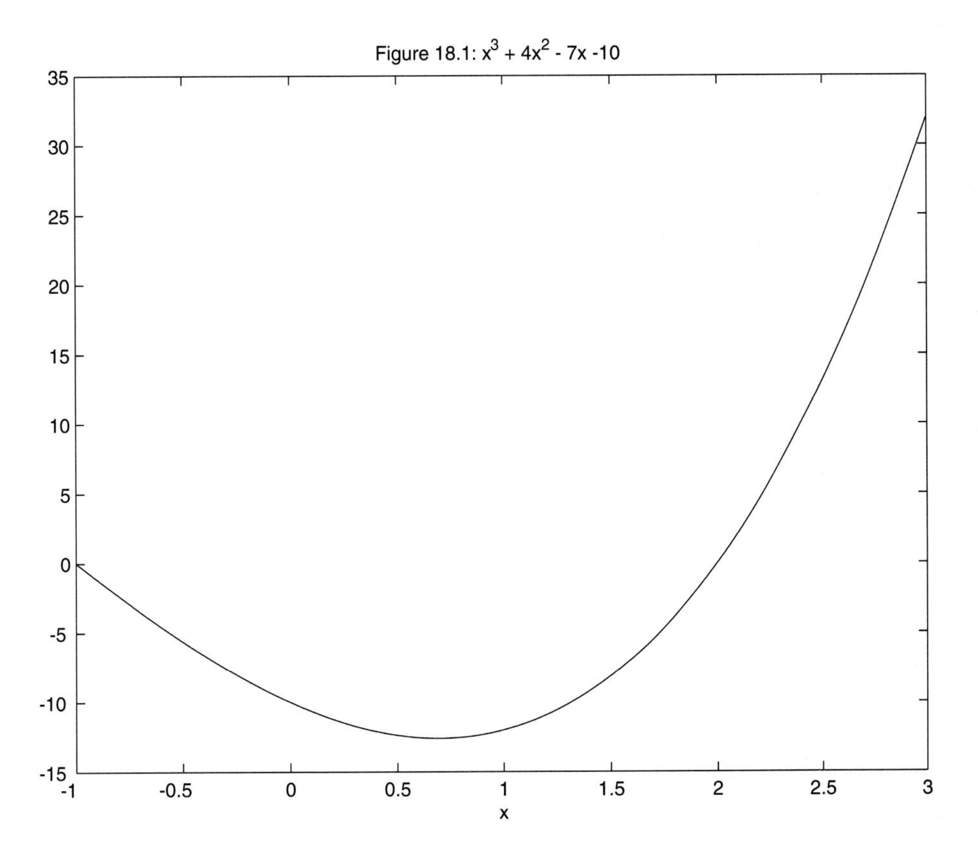


## 18.7 RATIONAL POLYNOMIALS

Sometimes one encounters ratios of polynomials (e.g., transfer functions and Pade approximations to functions) having the form

$$\frac{n(x)}{d(x)} = \frac{N_1x^m + N_2x^{m-1} + \ldots + N_{m+1}}{D_1x^n + D_2x^{n-1} + \ldots + D_{n+1}}$$

In MATLAB, these are manipulated by considering the two polynomials separately. For example,

```
» n = [1 -10 100]  % a numerator
n =
     1   -10   100
» d = [1 10 100 0]  % a denominator
d =
     1    10   100     0
» z = roots(n)  % the zeros of n(x)/d(x)
z =
          5 +      8.6603i
          5 -      8.6603i
» p = roots(d)  % the poles of n(x)/d(x)
p =
          0
         -5 +      8.6603i
         -5 -      8.6603i
```

The derivative of this rational polynomial with respect to $x$ is found using `polyder`:

```
» [nd,dd] = polyder(n,d)
nd =
         -1          20        -100       -2000      -10000
dd =
  Columns 1 through 6
          1          20         300        2000       10000           0
  Column 7
          0
```

Here `nd` and `dd` are the numerator and denominator polynomials of the derivative.

Another common operation is to find the partial fraction expansion of a rational polynomial. For example,

```
» [r,p,k] = residue(n,d)
r =
  9.7954e-17 +      1.1547i
  9.7954e-17 -      1.1547i
           1
p =
         -5 +      8.6603i
         -5 -      8.6603i
          0
k =
     []
```

In this case, the `residue` function returns the residues or partial fraction expansion coefficients `r`, their associated poles `p`, and direct term polynomial `k`. Since the order of the numerator is less than that of the denominator, there are no direct terms. For this example, the partial fraction expansion of the rational polynomial is

$$\frac{n(x)}{d(x)} = \frac{1.1547i}{x + 5 - 8.6603i} + \frac{-1.1547i}{x + 5 + 8.8803i} + \frac{1}{x}$$

Given this information, the original rational polynomial found by using `residue` yet again is

```
» [nn,dd] = residue(r,p,k)
nn =
             1          -10          100
dd =
             1           10          100           0
```

So, in this case, the function `residue` performs two operations that are inverses of one another based on how many input and output arguments are used.

## 18.8 CURVE FITTING

In numerous application areas, one is faced with the task of fitting a curve to measured data. Sometimes the chosen curve passes through the data points, but in others, the curve comes close to, but does not necessarily pass through, the data points. In the most common situation, the curve is chosen so that the sum of the squared errors at the data points is minimized. This choice results in a ***least squares*** curve fit. While least squares curve fitting can be done using any set of basis functions, it is straightforward and common to use a truncated power series (i.e., a polynomial).

In MATLAB, the function `polyfit` solves the least squares polynomial curve fitting problem. To illustrate the use of this function, let's start with the data

```
» x = [0 .1 .2 .3 .4 .5 .6 .7 .8 .9 1];
» y = [-.447 1.978 3.28 6.16 7.08 7.34 7.66 9.56 9.48 9.30 11.2];
```

To use `polyfit`, we must give it this data and the order or degree of the polynomial we wish to best fit to the data. If we choose $n = 1$ as the order, the best straight-line approximation will be found. This is often called ***linear regression.*** On the other hand, if we choose $n = 2$ as the order, a quadratic polynomial will be found. For now, let's choose a quadratic polynomial:

```
» n = 2;
» p = polyfit(x,y,n)
p =
   -9.8108   20.1293   -0.0317
```

The output of `polyfit` is a row vector of the polynomial coefficients. Here the solution is $y = -9.8108x^2 + 20.1293x - 0.0317$. To compare the curve fit solution to the data points, let's plot both:

```
» xi = linspace(0,1,100);
» yi = polyval(p,xi);

» plot(x,y,'-o',xi,yi,'--')
» xlabel('x'), ylabel('y=f(x)')
» title('Figure 18.2: Second Order Curve Fitting')
```

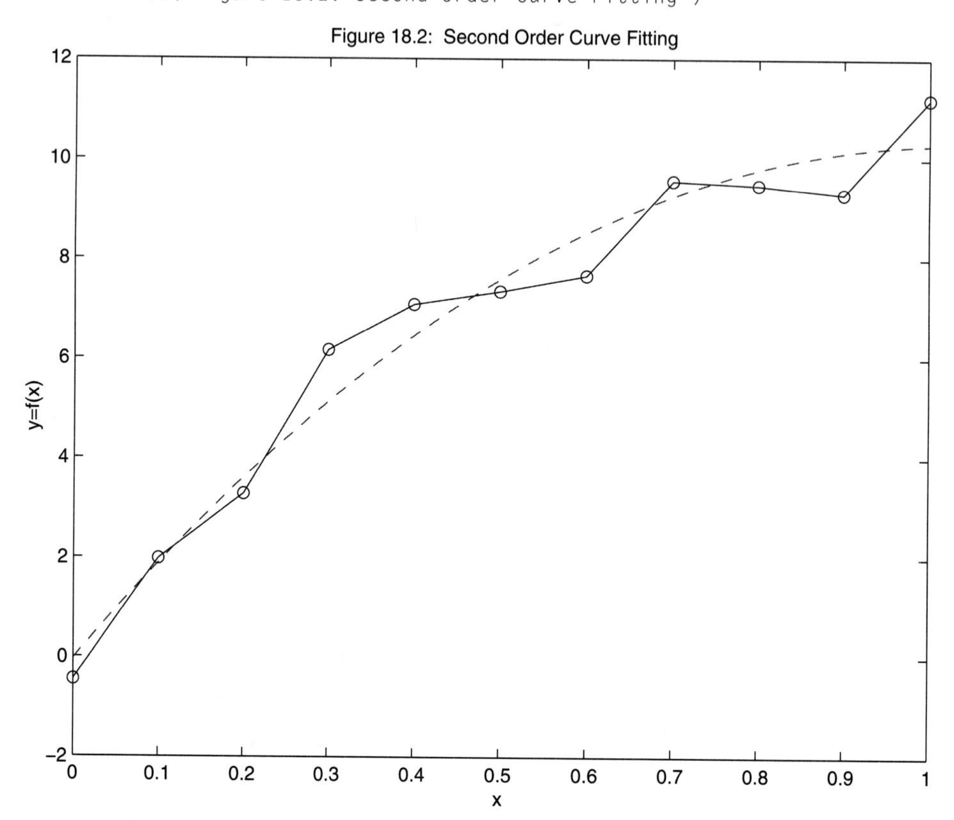


This plots the original data `x` and `y` marking the data points with `'o'` and connecting them with straight lines. In addition, it plots the polynomial data `xi` and `yi` using a dashed line `'--'`.

The choice of polynomial order is somewhat arbitrary. It takes two points to define a straight line or first-order polynomial. (If this is not clear to you, mark two points and draw a straight line between them.) It takes three points to define a quadratic or second-order polynomial. Following this progression, it takes $n + 1$ data points to specify uniquely an $n$th-order polynomial. Thus, in the preceding case, where there are 11 data points, we could choose up to a tenth-order polynomial. However, given the poor numerical properties of higher-order polynomials, one should not choose a polynomial order any higher than necessary. In addition, as the polynomial order increases, the approximation becomes less smooth since higher-order polynomials can be differentiated more times before they become zero. For example, choosing a tenth-order polynomial,

```
» pp = polyfit(x,y,10);

» format short e  % change display format
» pp.'  % display polynomial coefficients as a column
ans =
  -4.6436e+05
   2.2965e+06
  -4.8773e+06
   5.8233e+06
  -4.2948e+06
   2.0211e+06
  -6.0322e+05
   1.0896e+05
  -1.0626e+04
   4.3599e+02
  -4.4700e-01
```

Note the size of the polynomial coefficients in this case compared to those of the earlier quadratic fit. Note also the seven orders of magnitude difference between the smallest (`-4.4700e-01`) and largest (`5.8233e+06`) coefficients and the alternating signs on the coefficients. How about plotting this solution and comparing it to the original data and quadratic curve fit?

```
» y10 = polyval(pp,xi);  % evaluate 10th order polynomial

» plot(x,y,'o',xi,yi,'--',xi,y10) % plot data
» xlabel('x'), ylabel('y=f(x)')
» title('Figure 18.3: 2nd and 10th Order Curve Fitting')
```

In this plot, the original data is marked with `'o'`, the quadratic curve fit is dashed, and the tenth-order fit is solid. Note the wavelike ripples that appear between the data points at the left and right extremes in the tenth-order fit. This example clearly demonstrates the difficulties with higher-order polynomials.

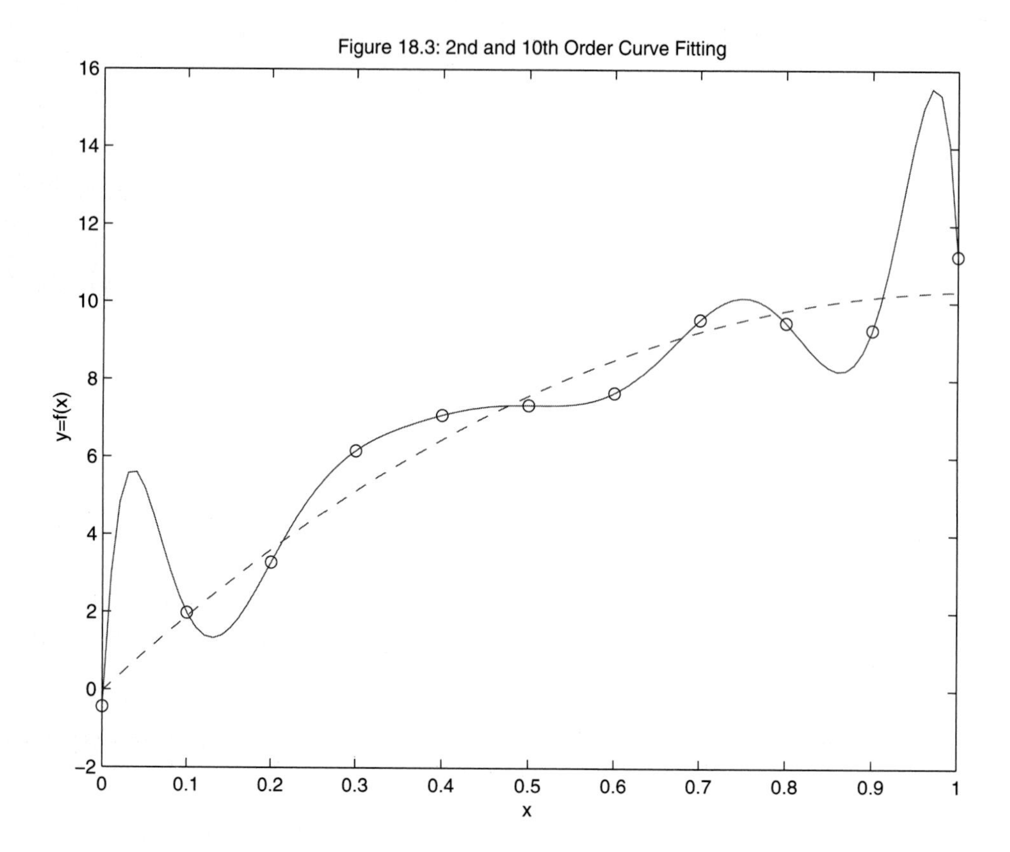
Figure 18.3: 2nd and 10th Order Curve Fitting
y=f(x)
x
16
14
12
10
8
6
4
2
0
−2
0
0.1
0.2
0.3
0.4
0.5
0.6
0.7
0.8
0.9
1

19

# Interpolation

Interpolation is a way of estimating values of a function between those given by some set of data points. In particular, interpolation is a valuable tool when one cannot quickly evaluate the function at the desired intermediate points. For example, this is true when the data points are the result of some experimental measurements or lengthy computational procedure. MATLAB provides tools for interpolating in any number of dimensions by using multidimensional arrays. To illustrate interpolation, only one- and two-dimensional interpolation will be considered in depth here. However, the functions used for higher dimensions are briefly discussed.

## 19.1 ONE-DIMENSIONAL INTERPOLATION

Perhaps the simplest example of interpolation is a MATLAB plot. By default, MATLAB draws straight lines connecting the data points used to make a plot. This linear interpolation guesses that intermediate values fall on a straight line between the entered points. Certainly, as the number of data points increases and the distance between them decreases, linear interpolation becomes more accurate. For example,

```
» x1 = linspace(0,2*pi,60);
» x2 = linspace(0,2*pi,6);

» plot(x1,sin(x1),x2,sin(x2), '--')
» xlabel('x'),ylabel('sin(x)')
» title('Figure 19.1: Linear Interpolation')
```

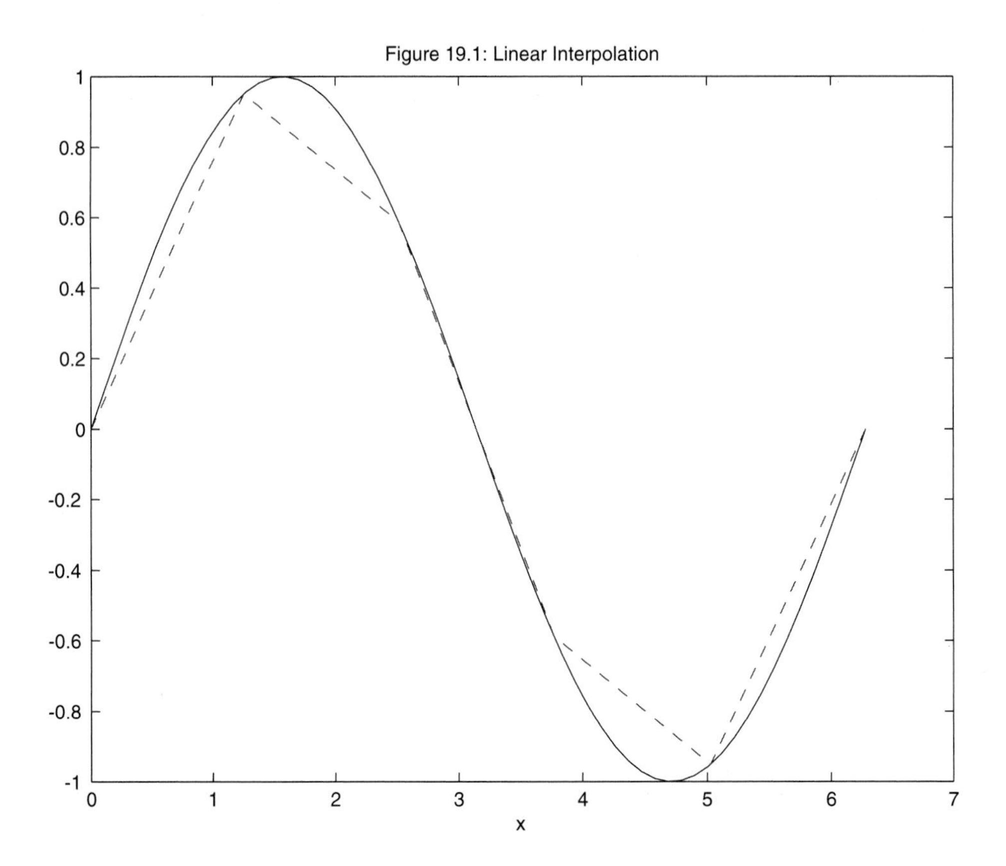


Of the two plots of the sine function, the one using 60 points is much more accurate between the data points than the one using only 6 points.

To illustrate one-dimensional interpolation, consider the following example: The threshold of audibility (i.e., the lowest perceptible sound level) of the human ear varies with frequency. Typical data are as follows:

```
» Hz=[20:10:100 200:100:1000 1500 2000:1000:10000];  % frequencies in Hertz

» spl=[76 66 59 54   49  46   43 40 38 22 ...  % sound pressure level in dB
   14  9   6 3.5 2.5 1.4 0.7  0 -1 -3 ...
   -8 -7 -2   2   7   9  11 12];
```

The sound pressure levels are normalized so that 0 dB appears at 1000 Hz. Since the frequencies span such a large range, plot the data using a logarithmic *x*-axis:

```
» semilogx(Hz,spl,'-o')
» xlabel('Frequency, Hz')
» ylabel('Relative Sound Pressure Level, dB')
» title('Figure 19.2: Threshold of Human Hearing')

» grid on
```

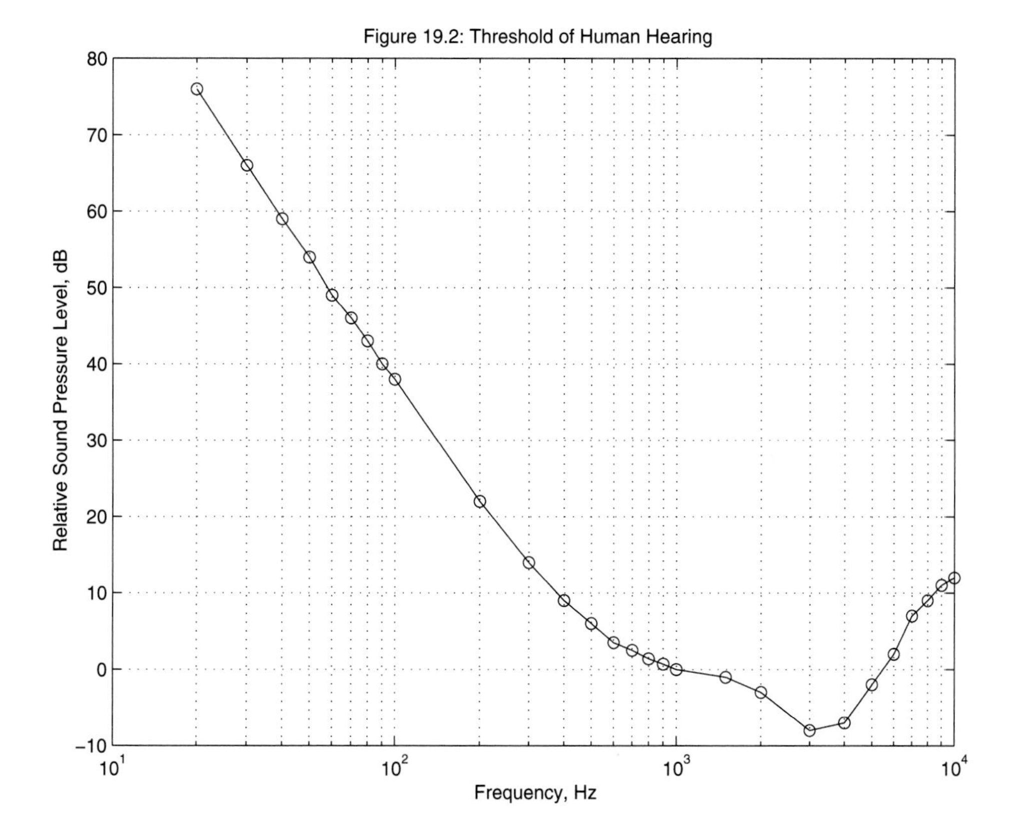


Based on this plot, the human ear is most sensitive to tones around 3kHz. Given these data, let's use the function `interp1` to estimate the sound pressure level in several different ways at a frequency of 2.5 kHz.

```
» s = interp1(Hz,spl,2.5e3)  % linear interpolation
s =
        -5.5
```

```
» s = interp1(Hz,spl,2.5e3,'linear')  % linear interpolation again
s =
        -5.5
» s = interp1(Hz,spl,2.5e3,'cubic')  % cubic interpolation
s =
      -5.6875
» s = interp1(Hz,spl,2.5e3,'spline')  % spline interpolation
s =
      -5.6641
» s = interp1(Hz,spl,2.5e3,'nearest')  % nearest neighbor interpolation
s =
    -8
```

Note the differences in these results. The first two results return exactly what is shown in the figure at 2.5 kHz since MATLAB linearly interpolates between data points on plots. Cubic and spline interpolation fit cubic (i.e., third order) polynomials to each data interval using different constraints. As a result, these results are close to each, but different from the linear interpolation solution. The poorest interpolation in this case is nearest neighbor, which returns the input data point nearest to the given value.

So how do you choose an interpolation method for a given problem? In many cases linear interpolation is sufficient. In fact, that is why it is the default method. While nearest neighbor produced poor results here, it is often used when speed is important or the data set is large. The most time-consuming method is spline, but it often produces the most desirable results.

While the preceding case considered only a single interpolation point, `interp1` can handle any arbitrary number of points. In fact, one of the most common uses of cubic or spline interpolation is to smooth data. That is, given a set of data, use interpolation to evaluate the data at a finer interval. For example,

```
» Hzi = linspace(2e3,5e3);  % look closely near minimum
» spli = interp1(Hz,spl,Hzi,'cubic');  % interpolate near minimum
» i = find(Hz>=2e3 & Hz<=5e3);  % find original data indices near minimum

» semilogx(Hz(i),spl(i),'--o',Hzi,spli) % plot old and new data
» xlabel('Frequency, Hz')
» ylabel('Relative Sound Pressure Level, dB')
» title('Figure 19.3: Threshold of Human Hearing')

» grid on
```

In this plot, the dashed line is linear interpolation, the solid line is the cubic interpolation, and the original data is marked with `'o'`. By asking for a finer resolution on the frequency axis and using cubic interpolation, we have a smoother estimate of the sound pressure level. In particular, note how the slope of the cubic solution does not change abruptly at the data points.

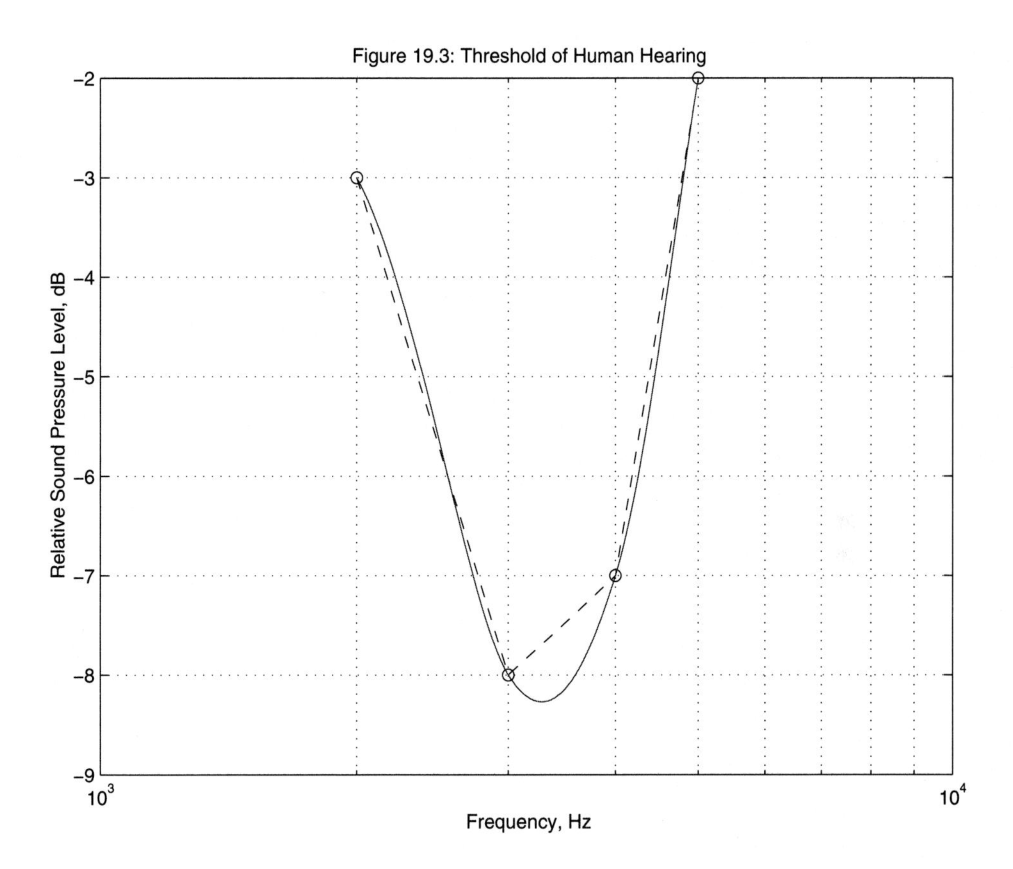


With this data we can make a better estimate of the frequency of greatest sensitivity:

```
» [spl_min,i] = min(spli)  % minimum and index of minimum
spl_min =
      -8.2682
i =
    43
» Hz_min=Hzi(i)  % frequency at minimum
Hz_min =
       3272.7
```

According to this analysis, the human ear is most sensitive to tones near 3.3 kHz.

It is important to recognize the two major restrictions enforced by `interp1`. First, asking for results outside the range of the independent variable—for example, `interp1(Hz,spl,1e5)`—produces `NaN` results. Second, the independent variable must be monotonic. That is, the independent variable must always increase or must always decrease. In our example, `Hz` is monotonic.

Because interpolation is used so often and because it often appears buried in critical areas of larger problems, there are two variations to interp1 that produce results more quickly. When the independent variable data x is equally spaced, prepending an asterisk * before the method choice—for example, yi=interp1(x,y,xi,'*linear') or yi=interp1(x,y,xi,'*cubic')—tells interp1 to assume equally spaced data and immediately perform the interpolation. All testing for nonequally spaced data is avoided. When the independent variable data x is not equally spaced (as is true in the preceding audibility example), the function interp1q(x,y,xi) provides the same results as interp1(x,y,xi) but is faster because interp1q performs no input argument checking. In this case x must be a column vector and y must be either a column vector or an array having length(x) rows.

Finally, it is possible to interpolate more than one data set at a time if y is a column-oriented data array. That is, if x is a vector, y can be either a vector as shown previously, or it can be an array having length(x) rows and any number of columns. For example,

```
» x = linspace(0,2*pi,11)'; % example data
» y = [sin(x) cos(x) tan(x)];
» size(y) % three columns
ans =
    11     3
» xi = linspace(0,2*pi); % interpolate on a finer scale
» yi = interp1(x,y,xi,'cubic');
» size(yi) % result is all three columns interpolated
ans =
   100     3
```

Here sin(x), cos(x), and tan(x) are all interpolated at the points in xi.

## 19.2 TWO-DIMENSIONAL INTERPOLATION

Two-dimensional interpolation is based on the same underlying ideas as one-dimensional interpolation. However, as the name implies, two-dimensional interpolation interpolates functions of two variables, $z = f(x,y)$. To illustrate this added dimension, consider the following example: An exploration company is using sonar to map the ocean floor. At points every 0.5 km on a rectangular grid, the ocean depth in meters is recorded for later analysis. A portion of the data collected is entered into MATLAB in the script M-file ocean.m:

```
% ocean.m, example test data
% ocean depth data
x=0:.5:4;  % x-axis (varies across the rows of z)
y=0:.5:6;  % y-axis (varies down the columns of z)

z=[100   99  100   99  100   99   99   99  100
   100   99   99   99  100   99  100   99   99
    99   99   98   98  100   99  100  100  100
   100   98   97   97   99  100  100  100   99
   101  100   98   98  100  102  103  100  100
   102  103  101  100  102  106  104  101  100
    99  102  100  100  103  108  106  101   99
    97   99  100  100  102  105  103  101  100
   100  102  103  101  102  103  102  100   99
   100  102  103  102  101  101  100   99   99
   100  100  101  101  100  100  100   99   99
   100  100  100  100  100   99   99   99   99
   100  100  100   99   99  100   99  100   99];
```

A plot of these data can be displayed by

```
» mesh(x,y,z)
» xlabel('X-axis, km')
» ylabel('Y-axis, km')
» zlabel('Ocean Depth, m')
» title('Figure 19.4: Ocean Depth Measurements')
```

Using these data, the depth at arbitrary points within the rectangle can be found by using the function `interp2`. For example,

```
» zi = interp2(x,y,z,2.2,3.3)
zi =
       103.92
» zi = interp2(x,y,z,2.2,3.3,'linear')
zi =
       103.92
» zi = interp2(x,y,z,2.2,3.3,'cubic')
zi =
       104.19
» zi = interp2(x,y,z,2.2,3.3,'nearest')
zi =
   102
```

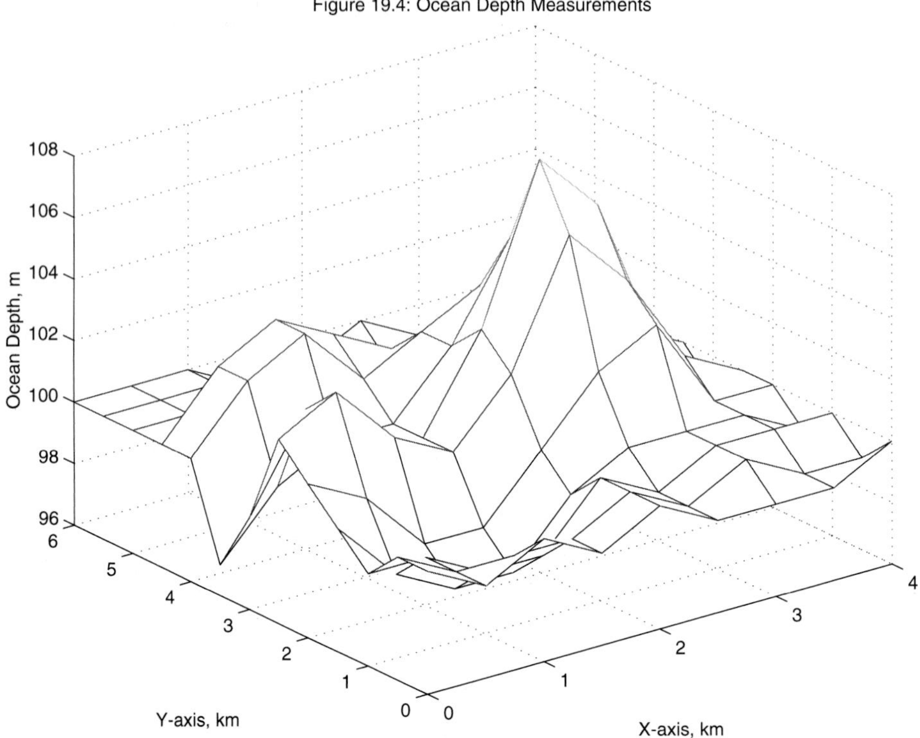


As was the case with one-dimensional interpolation, several interpolation methods are available, with the default method being linear. Once again, we can interpolate on a finer scale or mesh to smooth the plot:

```
» xi = linspace(0,4,30);  % finer x-axis
» yi = linspace(0,6,40);  % finer y-axis
```

For each value in `xi`, we wish to interpolate at all values in `yi`. That is, we wish to create a grid of all combinations of the values of `xi` and `yi`, and then interpolate at all those points. The function `meshgrid` accepts two vectors and produces two arrays containing duplicates of its inputs so that all combinations of the inputs are considered. For example,

```
» xtest = 1:5
xtest =
     1     2     3     4     5
» ytest = 6:9
ytest =
     6     7     8     9
```

```
» [xx,yy] = meshgrid(xtest,ytest)
xx =
     1     2     3     4     5
     1     2     3     4     5
     1     2     3     4     5
     1     2     3     4     5
yy =
     6     6     6     6     6
     7     7     7     7     7
     8     8     8     8     8
     9     9     9     9     9
```

As shown, `xx` contains `length(ytest)` rows, each containing `xtest`. `yy` contains `length(xtest)` columns, each containing `ytest`. With this structure, `xx(i,j)` and `yy(i,j)` for all `i` and `j` covers all combinations of the original vectors `xtest` and `ytest`. Applying `meshgrid` to our ocean depth example produces

```
» [xxi,yyi] = meshgrid(xi,yi);  % grid of all combinations of xi and yi
» size(xxi) % xxi has 40 rows each containing xi
ans =
    40    30
» size(yyi) % yyi has 30 columns each containing yi
ans =
    40    30
```

Given `xxi` and `yyi`, the ocean depth can now be interpolated on the finer scale:

```
» zzi = interp2(x,y,z,xxi,yyi,'cubic');  % interpolate
» size(zzi) % zzi is the same size as xxi and yyi
ans =
    40    30

» mesh(xxi,yyi,zzi)  % plot smoothed data
» hold on
» [xx,yy] = meshgrid(x,y);   % grid original data
» plot3(xx,yy,z+0.1,'ok')  % plot original data up a bit to show nodes
» hold off
» title('Figure 19.5: 2-D Smoothing')
```

Using these data, we can now estimate the peak and its location:

```
» zmax = max(max(zzi))
zmax =
      108.05
» [i,j] = find(zmax==zzi);
» xmax=xi(j)
xmax =
      2.6207
```

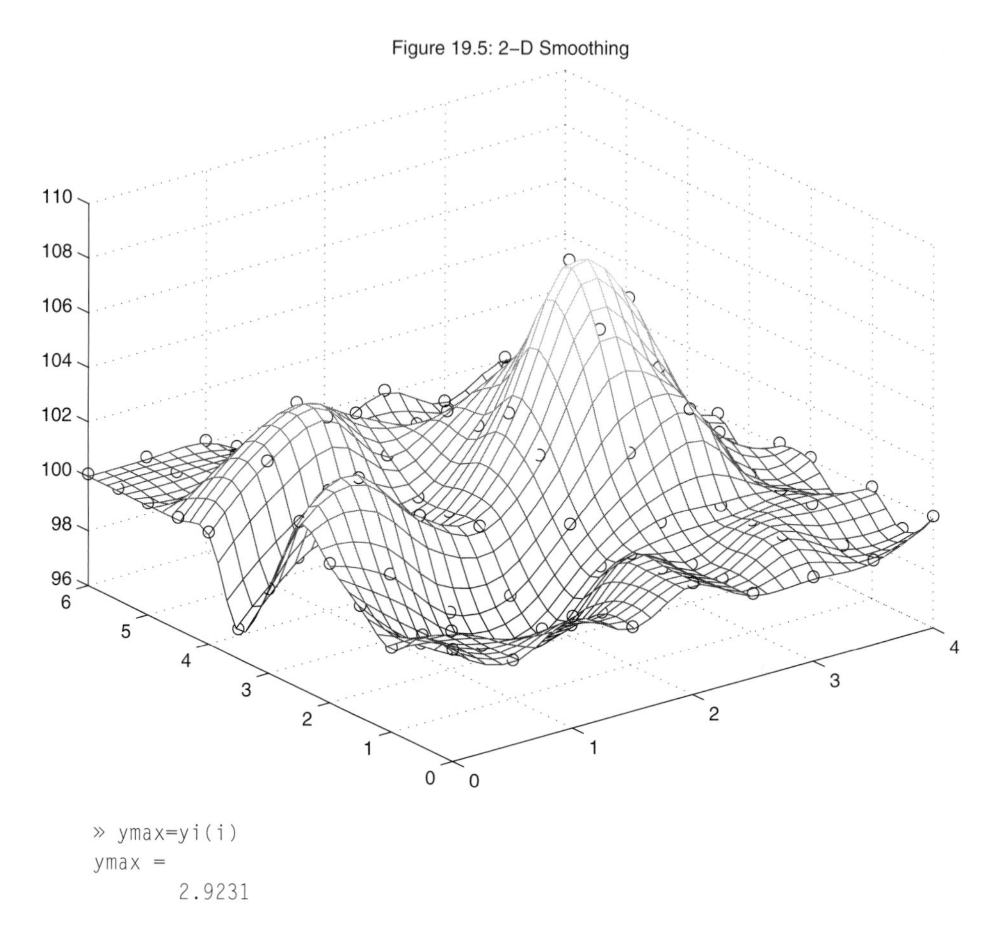


```
» ymax=yi(i)
ymax =
        2.9231
```

The concepts discussed in these first two sections extend naturally to higher dimensions where `ndgrid`, `interp3`, and `interpn` apply. `ndgrid` is the multidimensional equivalent of `meshgrid`. Multidimensional interpolation uses multidimensional arrays in a straightforward way to organize the data and perform the interpolation. `interp3` performs interpolation in three-dimensional space, and `interpn` performs interpolation in higher-order dimensions. Both `interp3` and `interpn` offer method choices of `'linear'`, `'cubic'`, and `'nearest'`. For more information regarding these functions, see MATLAB documentation and on-line help.

## 19.3 TRIANGULATION AND SCATTERED DATA

In a number of applications, such as those that involve geometric analysis, data points are often scattered rather than appearing on a rectangular grid (such as the ocean data in the example considered in the last section). For example, consider the two-dimensional random data:

```
» x = randn(1,12);
» y = randn(1,12);
» z = zeros(1,12); % no z component for now

» plot(x,y,'o')
» title('Figure 19.6: Random Data')
```

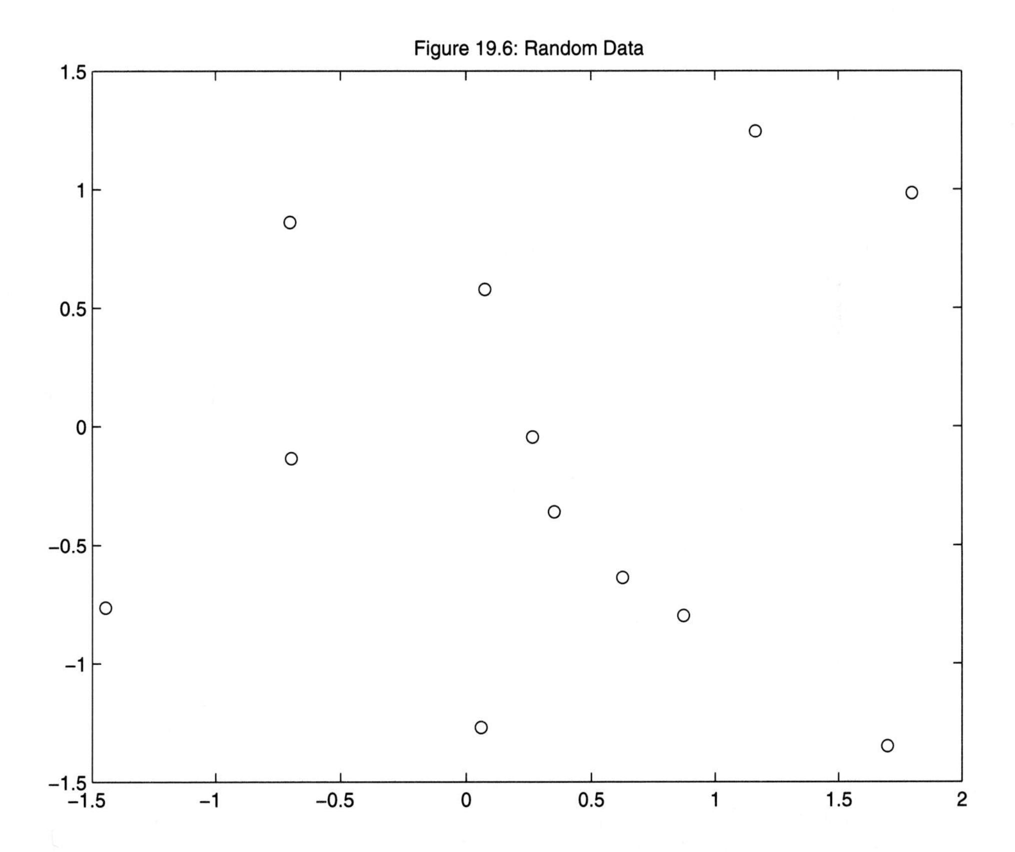


Given scattered data such as that shown, it is common to apply Delaunay triangulation, which returns a set of triangles connecting the data points such that no data points are contained within any triangle. The MATLAB function `delaunay` accepts data points and returns a list of indices into the data that identify the triangle vertices. For the preceding data `delaunay` returns the following:

```
» tri = delaunay(x,y)
     7    10     6
     7     2    10
     7     4     2
     5     7    11
```

```
     5     4     7
     5     9     4
     5     3     9
     9     2     4
    12     3     5
     8     6    10
     2     8    10
     9     8     2
     9     1     8
     3     1     9
    11    12     5
    12     1     3
```

Each row contains indices into x and y that identify triangle vertices. For example, the first triangle is described by the data points in `x([7 10 6])` and `y([7 10 6])`. The triangles can be plotted using the function `trimesh`:

```
» hold on, trimesh(tri,x,y,z), hold off
» hidden off
» title('Figure 19.7: Delaunay Triangulation')
```

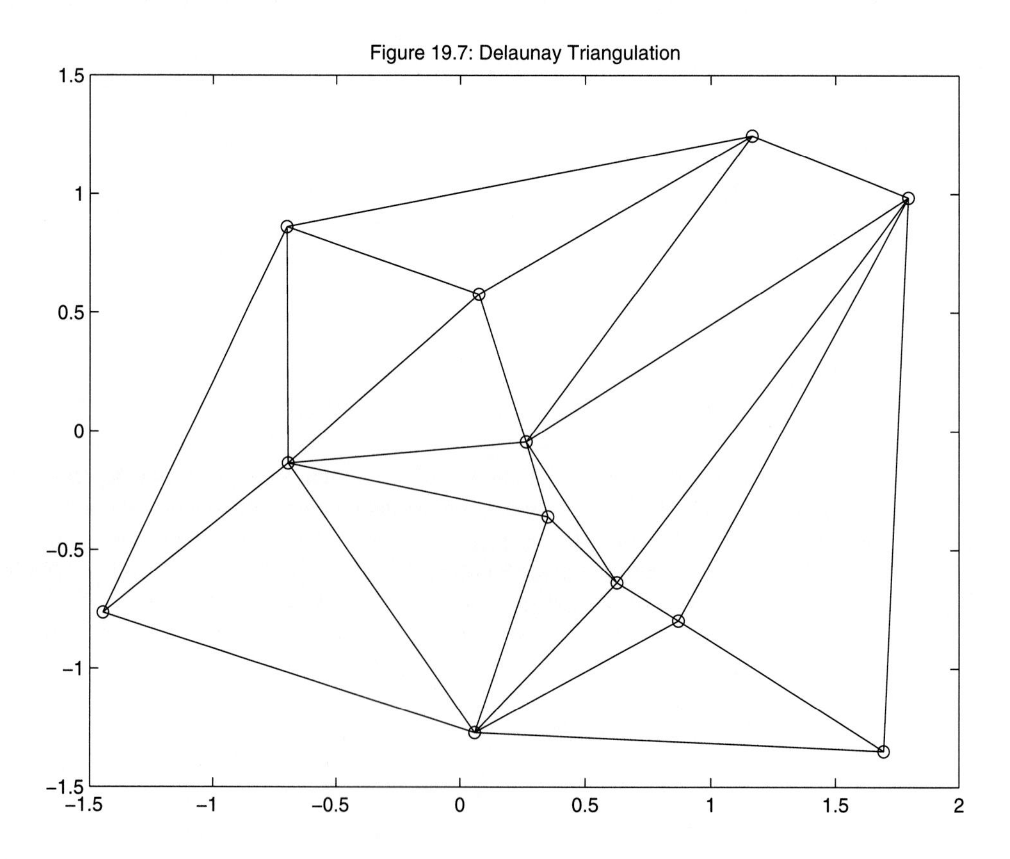

Once the Delaunay triangulation is known, the functions `tsearch` and `dsearch` can be used to interpolate it. For example, the triangle enclosing the origin is

```
» tsearch(x,y,tri,0,0) % find row of tri closest to (0,0)
ans =
     7
» tri(ans,:) % vertices of triangle closest to (0,0)
ans =
     5     3     9
```

Naturally, `tsearch` accepts multiple values. For example,

```
» tsearch(x,y,tri,[-.1 .1],[.5 .9])
ans =
     9    16
```

Here the ninth triangle encloses the point $(-.1, .5)$ and the sixteenth triangle encloses the point $(.1, .9)$.

Rather than returning the triangle enclosing one or more data points, the function `dsearch` returns indices into `x` and `y` that are closest to the desired points. For example,

```
» dsearch(x,y,tri,[-.1 .1],[.5 .9])
ans =
     3     3
```

Here the point (x(3),y(3)) is closest to both points $(-.1, .5)$ and $(.1, .9)$.

In addition to interpolating the data, it is often useful to know which points form the outer boundary or convex hull for the set. The function `convhull` returns indices into `x` and `y` that describe the convex hull. For example,

```
» k = convhull(x,y,tri)
k =
    11     7     6     8     1    12    11
```

Note that `convhull` returns the indices of a closed curve since the first and last index values are the same. Using the data returned by `convhull`, the boundary can be drawn by `plot(x(k),y(k))`.

Finally, it is possible to interpolate a Delaunay triangulation to produce interpolated points on a rectangular grid. In particular, this step is required to use functions such as `contour` and other standard plotting routines. Since this function produces grid data, it is called `griddata`. Use of `griddata` is demonstrated next.

```
» z = rand(1,12); % now use some random z axis data
» xi = linspace(min(x),max(x),30); % desired x points
» yi = linspace(min(y),max(y),30); % desired y points
```

```
» [Xi,Yi] = meshgrid(xi,yi); % create mesh grid
» Zi = griddata(x,y,z,Xi,Yi); % grid the scattered data at Xi, Yi points
```

Now `Zi` contains a 30-by-30 array of points linearly interpolated from the triangulation of the data in `x`, `y`, and `z`. Just as `interp1` and `interp2` support other interpolations, `griddata` also supports others. For example,

```
» Zi = griddata(x,y,z,Xi,Yi,'linear') % same as above (default)
» Zi = griddata(x,y,z,Xi,Yi,'cubic') % triangle based cubic interpolation
» Zi = griddata(x,y,z,Xi,Yi,'nearest') % triangle based nearest neighbor
» Zi = griddata(x,y,z,Xi,Yi,'invdist') % inverse distance method
```

20

# Cubic Splines

It is well known that interpolation using high-order polynomials often produces ill-behaved results. There are numerous approaches to eliminating this poor behavior. Of these approaches, cubic splines are very popular. In MATLAB, basic cubic splines interpolation is accomplished by the functions `spline`, `ppval`, `mkpp`, and `unmkpp`. Of these, only `spline` is documented in MATLAB documentation. In the following sections, the basic features of cubic splines as implemented in these M-file functions will be demonstrated.

## 20.1 BASIC FEATURES

In cubic splines, cubic polynomials are found to approximate the curve between each pair of data points. In the language of splines, these data points are called the breakpoints. Since a straight line is uniquely defined by two points, an infinite number of cubic polynomials can be used to approximate a curve between two points. Therefore, in cubic splines, additional constraints are placed on the cubic polynomials to make the result unique. By constraining the first and second derivatives of each cubic polynomial to match at the breakpoints, all internal cubic polynomials are well defined. Moreover, both the slope and curvature of the approximating polynomials are continuous across the breakpoints. However, the first and last cubic polynomials do not have adjoining cubic polynomials beyond the first and last breakpoints. As a result, the remaining constraints must be determined by some other means. The most common approach, which is adopted by the function `spline`, is to adopt a *not-a-knot* condition. This condition forces the third derivative of the first and second cubic polynomials to be identical, and likewise for the last and second to the last cubic polynomials.

Based on the preceding description, one could guess that finding cubic spline polynomials requires solving a large set of linear equations. In fact, given $N$ breakpoints, there are $N - 1$ cubic polynomials to be found, each having four unknown coefficients. Thus, the set of equations to be solved involves $4(N - 1)$ unknowns. By writing each cubic polynomial in a special form and by applying the contraints, the cubic polynomials can be found by solving a reduced set of $N$ equations in $N$ unknowns. Thus if there are 50 breakpoints, there are 50 equations in 50 unknowns. Luckily, the equations can be concisely written and solved using sparse matrices, which is what the function `spline` uses to compute the unknown coefficients.

## 20.2 PIECEWISE POLYNOMIALS

In its most simple use, `spline` takes data `x` and `y` and desired values `xi`, finds the cubic spline interpolation polynomials that fit `x` and `y`, and then evaluates the polynomials to find the corresponding `yi` values for each `xi` value. This matches the use of `yi = interp1(x,y,yi,'spline')`. For example,

```
» x = 0:12;
» y = tan(pi*x/25);
» xi = linspace(0,12);
» yi = spline(x,y,xi);
» plot(x,y,'o',xi,yi), title('Figure 20.1: Spline Fit')
```

This approach is appropriate if only one set of interpolated values are required. However, if another set of interpolated values are needed from the same set of data, it does not make sense to recompute the same set of cubic spline coefficients a second time. In this situation, one can call `spline` with only the first two arguments:

```
» pp = spline(x,y)
pp =
  Columns 1 through 7
   10.0000     1.0000    12.0000          0     1.0000     2.0000     3.0000
```

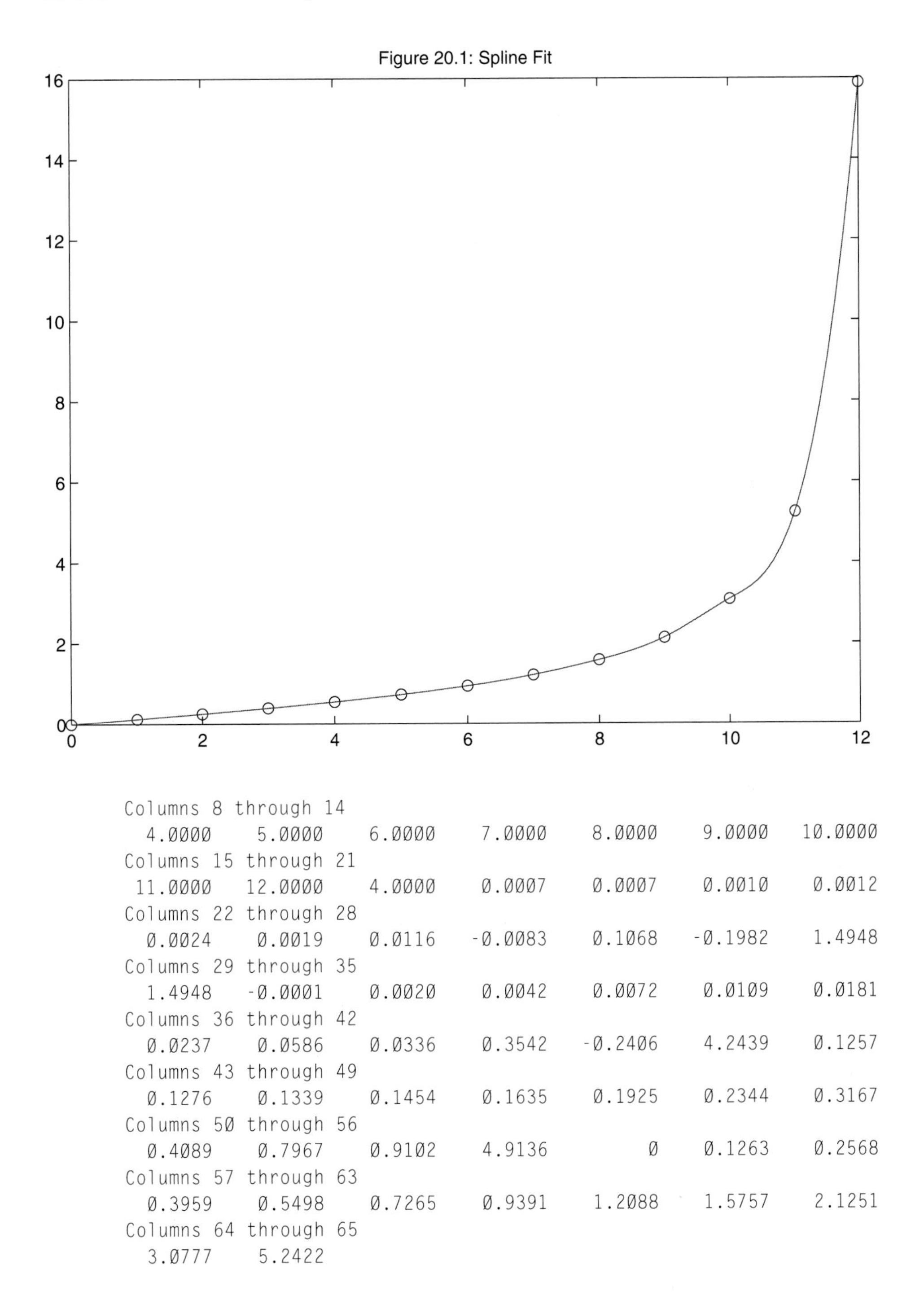


Figure 20.1: Spline Fit

```
Columns 8 through 14
  4.0000    5.0000    6.0000    7.0000    8.0000    9.0000   10.0000
Columns 15 through 21
 11.0000   12.0000    4.0000    0.0007    0.0007    0.0010    0.0012
Columns 22 through 28
  0.0024    0.0019    0.0116   -0.0083    0.1068   -0.1982    1.4948
Columns 29 through 35
  1.4948   -0.0001    0.0020    0.0042    0.0072    0.0109    0.0181
Columns 36 through 42
  0.0237    0.0586    0.0336    0.3542   -0.2406    4.2439    0.1257
Columns 43 through 49
  0.1276    0.1339    0.1454    0.1635    0.1925    0.2344    0.3167
Columns 50 through 56
  0.4089    0.7967    0.9102    4.9136         0    0.1263    0.2568
Columns 57 through 63
  0.3959    0.5498    0.7265    0.9391    1.2088    1.5757    2.1251
Columns 64 through 65
  3.0777    5.2422
```

When called in this way, `spline` returns an array called the ***pp-form*** or piecewise polynomial form of the cubic splines. This array contains all the information necessary to evaluate the cubic splines for any set of desired interpolation values. Given the pp-form, the function `ppval` evaluates the cubic splines. For example,

```
» yi = ppval(pp,xi);
```

computes the same `yi` values computed earlier. Similarly,

```
» xi2 = linspace(10,12);
» yi2 = ppval(pp,xi2);
```

uses the pp-form again to evaluate the cubic splines over a finer spacing restricted to the region between 10 and 12.

```
» xi3 = 10:15;
» yi3 = ppval(pp,xi3)
yi3 =
    3.0777    5.2422   15.8945   44.0038   98.5389  188.4689
```

shows that cubic splines can be evaluated outside the region over which the cubic polynomials were computed. When data appear beyond the last or before the first breakpoint, the last and first cubic polynomial are used, respectively, to find interpolated values.

The cubic splines pp-form just given stores the breakpoints, polynomial coefficients, as well as other information regarding the cubic splines representation. This form is a convenient data structure in MATLAB, since all information is stored in a single vector. When a cubic spline representation is evaluated, the pp-form must be broken into its representative pieces. In MATLAB this process is performed by the function `unmkpp`. Using this function on the preceding pp-form gives

```
» [breaks,coefs,npolys,ncoefs] = unmkpp(pp)
breaks =
  Columns 1 through 12
     0     1     2     3     4     5     6     7     8     9    10    11
  Column 13
    12
coefs =
    0.0007   -0.0001    0.1257         0
    0.0007    0.0020    0.1276    0.1263
    0.0010    0.0042    0.1339    0.2568
    0.0012    0.0072    0.1454    0.3959
    0.0024    0.0109    0.1635    0.5498
    0.0019    0.0181    0.1925    0.7265
```

```
    0.0116    0.0237    0.2344    0.9391
   -0.0083    0.0586    0.3167    1.2088
    0.1068    0.0336    0.4089    1.5757
   -0.1982    0.3542    0.7967    2.1251
    1.4948   -0.2406    0.9102    3.0777
    1.4948    4.2439    4.9136    5.2422
npolys =
    12
ncoefs =
     4
```

Here breaks is the breakpoints, coefs is a matrix whose $i$th row is the $i$th cubic polynomial, npolys is the number of polynomials, and ncoefs is the number of coefficients per polynomial. Note that this form is sufficiently general so that the spline polynomials need not be cubic. This fact is useful when the spline is integrated or differentiated.

Given the preceding broken apart form, the function mkpp restores the pp-form:

```
» pp = mkpp(breaks,coefs)
pp =
  Columns 1 through 7
   10.0000    1.0000   12.0000         0    1.0000    2.0000    3.0000
  Columns 8 through 14
    4.0000    5.0000    6.0000    7.0000    8.0000    9.0000   10.0000
  Columns 15 through 21
   11.0000   12.0000    4.0000    0.0007    0.0007    0.0010    0.0012
  Columns 22 through 28
    0.0024    0.0019    0.0116   -0.0083    0.1068   -0.1982    1.4948
  Columns 29 through 35
    1.4948   -0.0001    0.0020    0.0042    0.0072    0.0109    0.0181
  Columns 36 through 42
    0.0237    0.0586    0.0336    0.3542   -0.2406    4.2439    0.1257
  Columns 43 through 49
    0.1276    0.1339    0.1454    0.1635    0.1925    0.2344    0.3167
  Columns 50 through 56
    0.4089    0.7967    0.9102    4.9136         0    0.1263    0.2568
  Columns 57 through 63
    0.3959    0.5498    0.7265    0.9391    1.2088    1.5757    2.1251
  Columns 64 through 65
    3.0777    5.2422
```

Since the size of the matrix coefs determines npolys and ncoefs, they are not needed by mkpp to reconstruct the pp-form. The pp-form data structure is given simply in mkpp as `pp=[10 1 npolys breaks(:)' ncoefs coefs(:)']`. The first two elements appear in all pp-forms as a means to identify a vector as a pp-form.

## 20.3 INTEGRATION

In many situations it is desirable to know the area under a function described by cubic splines as a function of the independent variable $x$. That is, if the splines are denoted $y = s(x)$, we are interested in computing

$$S(x) = \int_{x_1}^{x} s(x)dx \quad \text{with } S(x_1) = 0$$

where $x_1$ is the first spline breakpoint. Since $s(x)$ is composed of connected cubic polynomials, with the $k$th cubic polynomial being

$$s_k(x) = a_k(x - x_k)^3 + b_k(x - x_k)^2 + c_k(x - x_k) + d_k, \quad x_k \leq x \leq x_{k+1}$$

and whose area over the region $x_k \leq x \leq x_{k+1}$ is

$$S_k(x) = \int_{x_k}^{x} s_k(x)dx = \frac{a_k}{4}(x - x_k)^4 + \frac{b_k}{3}(x - x_k)^3 + \frac{c_k}{2}(x - x_k)^2 + d_k(x - x_k)$$

The area under a cubic splines is easily computed as

$$S(x) = \sum_{i=1}^{k-1} \int_{x_i}^{x_{i+1}} s_i(x)dx + \int_{x_k}^{x} s_k(x)dx$$

where $x_k \leq x \leq x_{k+1}$, or

$$S(x) = \sum_{i=1}^{k-1} S_i(x_{i+1}) + S_k(x)$$

The summation term is the cumulative sum of the areas under all preceding cubic polynomials. As such, it is readily computed and forms the constant term in the polynomial describing $S(x)$ since $S_k(x)$ is a polynomial. With this understanding, the integral itself can be written as a spline. In this case, it is a quartic spline since the individual polynomials are of order 4.

Because the pp-form used in MATLAB can support splines of any order, the preceding splines integration is embodied in the *Mastering MATLAB Toolbox* function `mmspint`. The body of this function is given by the following:

```
function z=mmspint(x,y,xi)
%MMSPINT Cubic Spline Integral Interpolation. (MM)
% YI=MMSPINT(X,Y,XI) uses cubic spline interpolation to fit the
% data in X and Y, integrates the spline, and returns
% values of the integral evaluated at the points in XI.
%
% PPI=MMSPINT(PP) returns the piecewise polynomial vector PPI
% describing the integral of the cubic spline described by
% the piecewise polynomial in PP. PP is returned by the functions
% SPLINE and MMSPLINE and is a data vector containing all information
% to evaluate and manipulate a spline.
%
% YI=MMSPINT(PP,XI) integrates the cubic spline given by
% the piecewise polynomial PP, and returns the values of the
% integral evaluated at the points in XI.
%
% See also MMSPLINE,MMSPDER,MMSPAREA,MMSPCUT,MMSPII,SPLINE,MKPP,UNMKPP,PPVAL.

if nargin==3
   pp=spline(x,y);
else
   pp=x;
end
if pp(1)~=10
   error('Spline Data Does Not Have PP Form.')
end
[br,co,npy,nco]=unmkpp(pp);% take apart pp
sf=nco:-1:1;              % scale factors for integration
ico=[co./sf(ones(npy,1),:) zeros(npy,1)];% integral coefficients
nco=nco+1;                % integral spline has higher order
for k=2:npy               % find constant terms in polynomials
   ico(k,nco)=polyval(ico(k-1,:),br(k)-br(k-1));
end
ppi=mkpp(br,ico);         % build pp form for integral
if nargin==1
   z=ppi;
elseif nargin==2
   z=ppval(ppi,y);
else
   z=ppval(ppi,xi);
end
```

Consider the following example using `mmspint`:

```
» x = (0:.1:1)*2*pi;
» y = sin(x);        % create rough data
```

```
» pp = spline(x,y);  % pp-form fitting rough data
» ppi = mmspint(pp); % pp-form of integral
» xi = linspace(0,2*pi);% finer points for interpolation
» yi = ppval(pp,xi);     % evaluate curve
» yyi = ppval(ppi,xi);   % evaluate integral
```

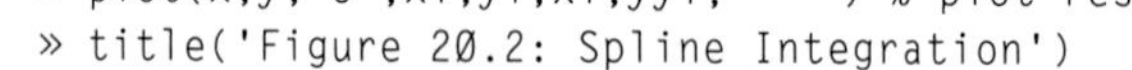

```
» plot(x,y,'o',xi,yi,xi,yyi,'--') % plot results
» title('Figure 20.2: Spline Integration')
```

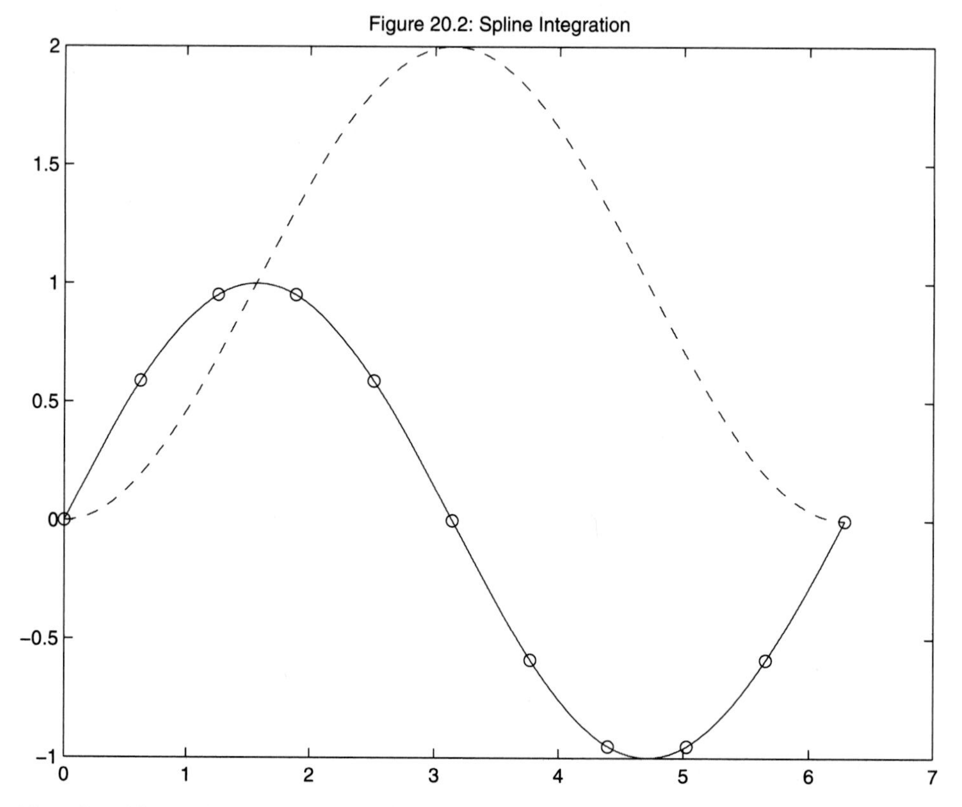


Note that this qualitatively shows the identity

$$\int_0^x \sin(x)dx = 1 - \cos(x)$$

## 20.4 DIFFERENTIATION

Just as one may be interested in spline integration, the derivative or slope of a function described by splines is also useful. Given the $k$th cubic polynomial being

$$s_k(x) = a_k(x - x_k)^3 + b_k(x - x_k)^2 + c_k(x - x_k) + d_k\,, \quad x_k \le x \le x_{k+1}$$

the derivative of $s_k(x)$ is easily written as

$$\frac{ds_k(x)}{dx} = 3a_k(x - x_k)^2 + 2b_k(x - x_k) + c_k$$

where $x_k \le x \le x_{k+1}$. As with integration, the derivative is also a spline. However, in this case it is a quadratic spline, since the order of the polynomial is 2.

Based on the preceding expression, the *Mastering MATLAB Toolboox* function mmspder performs spline differentiation. The body of mmspder is as follows:

```
function z=mmspder(x,y,xi)
%MMSPDER Cubic Spline Derivative Interpolation. (MM)
% YI=MMSPDER(X,Y,XI) uses cubic spline interpolation to fit the
% data in X and Y, differentiates the spline, and returns values
% of the spline derivatives evaluated at the points in XI.
%
% PPD=MMSPDER(PP) returns the piecewise polynomial vector PPD
% describing the cubic spline derivative of the curve described by
% the piecewise polynomial in PP. PP is returned by the function
% SPLINE and MMSPLINE and is a data vector containing all information
% to evaluate and manipulate a spline.
%
% YI=MMSPDER(PP,XI) differentiates the cubic spline given by
% the piecewise polynomial PP, and returns the values of the
% spline derivatives evaluated at the points in XI.
%
% See also MMSPINT,MMSPLINE,MMSPAREA,MMSPCUT,MMSPII,SPLINE,MKPP,UNMKPP,PPVAL.

if nargin==3
   pp=spline(x,y);
else
   pp=x;
end
[br,co,npy,nco]=unmkpp(pp);% take apart pp
if pp(1)~=10
   error('Spline Data Does Not Have PP Form.')
end
sf=nco-1:-1:1;            % scale factors for differentiation
dco=sf(ones(npy,1),:).*co(:,1:nco-1);% derivative coefficients
ppd=mkpp(br,dco);         % build pp form for derivative
if nargin==1
   z=ppd;
elseif nargin==2
   z=ppval(ppd,y);
else
   z=ppval(ppd,xi);
end
```

To demonstrate the use of mmspder, consider the following example:

```
» x = (0:.1:1)*2*pi;      % same data as earlier
» y = sin(x);

» pp = spline(x,y);       % pp-form fitting rough data
» ppd = mmspder(pp);      % pp-form of derivative
» xi = linspace(0,2*pi); % finer points for interpolation
» yi = ppval(pp,xi);      % evaluate curve
» yyd = ppval(ppd,xi);    % evaluate derivative

» plot(x,y,'o',xi,yi,xi,yyd,'--') % plot results
» title('Figure 20.3: Spline Differentiation')
```

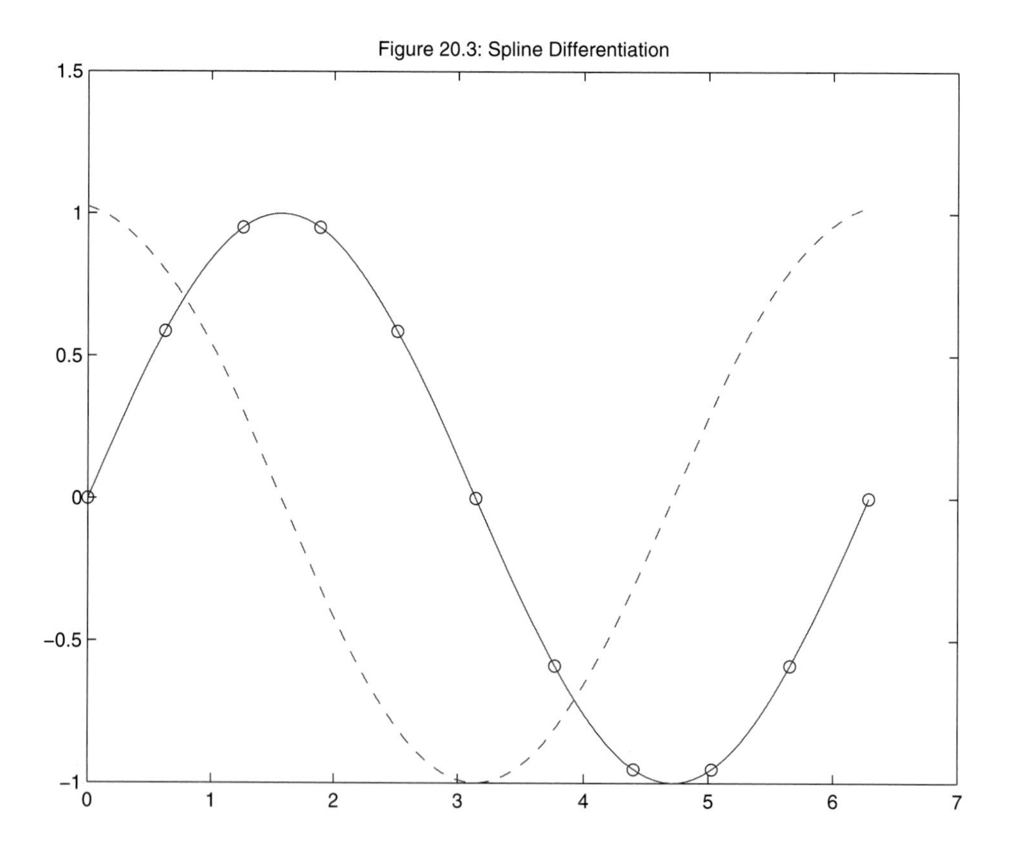


Note that this qualitatively shows the identity

$$\frac{d}{dx}\sin(x) = \cos(x)$$

## 20.5 SPLINE INTERPOLATION ON A PLANE

Spline interpolation as implemented by the function `spline` assumes that the independent variable is monotonic. That is, the spline $y = s(x)$ describes a continuous function. When it is not continuous, there is no one-to-one relationship between $x$ and $y$ and function `ppval` has no way of knowing what $y$ value to return for a given $x$. A common situation where this occurs is when a curve is defined on a plane. For example,

```
» t = linspace(0,3*pi,15);
» x = sqrt(t).*cos(t);
» y = sqrt(t).*sin(t);
» plot(x,y)
» ylabel('Y'), xlabel('X')
» title('Figure 20.4: Spiral Y=f(X)')
```

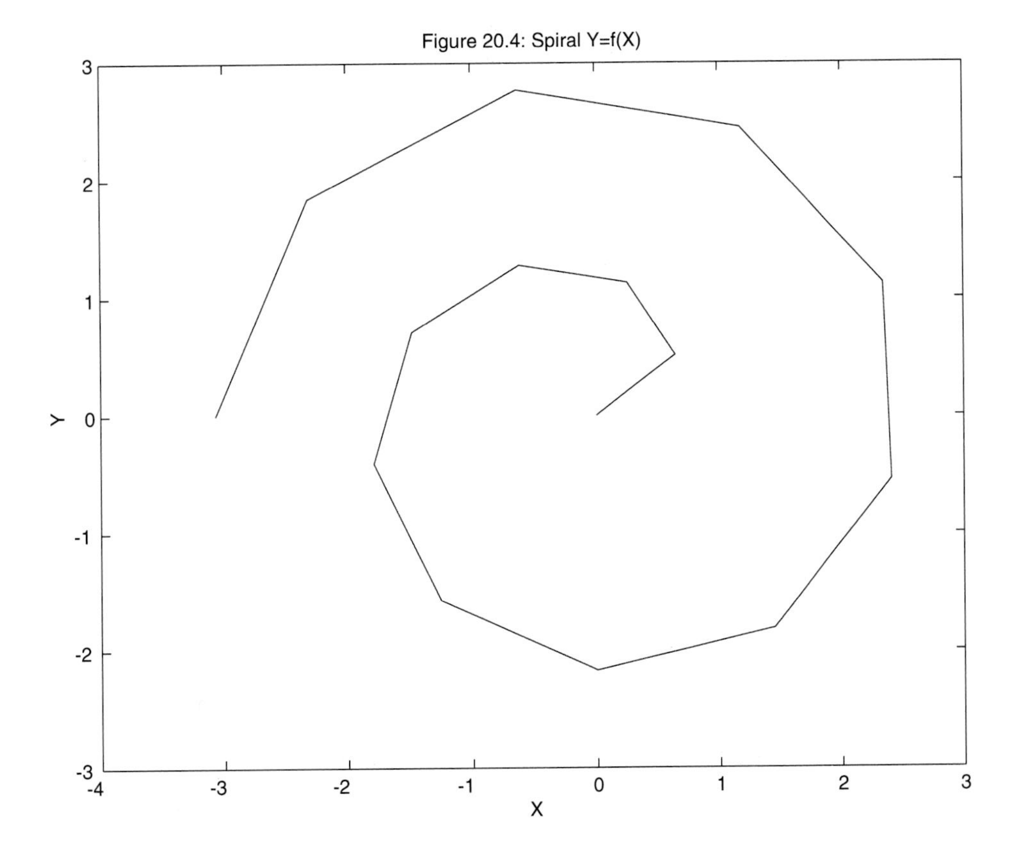


It is not possible to compute a cubic spline for the spiral as a function of $x$ since there are multiple $y$ values for each $x$ near the origin. However, it is possible to compute a spline for each axis with respect to the variable or parameter `t`. For example,

```
» px = spline(t,x); % x=Sx(t)
» py = spline(t,y); % y=Sy(t)
```

Given these two piecewise polynomial vectors, the original data can be interpolated by specifying a desired range of t:

```
» ti = linspace(0,3*pi); % total range, 100 points
» xi = ppval(px,ti);
» yi = ppval(py,ti);

» plot(x,y,'d',xi,yi)
» xlabel('X')
» ylabel('Y')
» title('Figure 20.5: Interpolated Spiral Y=f(X)')
```

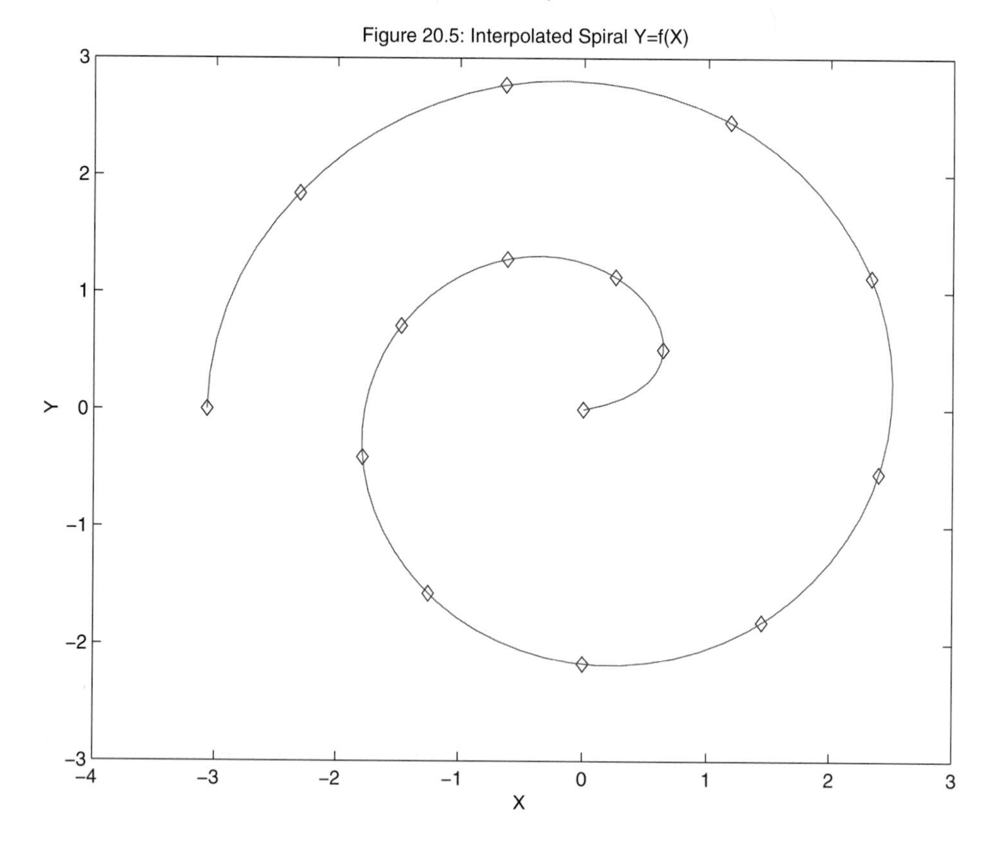


In the preceding case, the parameter t was known because the example was simply conceived. However, in the case where there is no underlying parameterization like t, it can just be created. For example,

```
» x = 1 + rand(1,10);        % example x variable
» y = 2.*(x.^2) - 3*x + 2; % example y variable
```

```
» t = linspace(0,1,length(x)); % artificial parameter

» px = spline(t,x);
» py = spline(t,y);
```

In this example, a parameter t is created to match the length of the x and y data. In particular, t was chosen to vary from 0 to 1.

Finally, the preceding approach is not limited to two dimensions. Any number of dimensions can be handled, with each one having its own spline representation with respect to a monotonic parameter.

# 21

# Fourier Analysis

Frequency domain tools such as Fourier series, Fourier transform, and their discrete time counterparts form a cornerstone in signal processing. These transforms decompose a signal into a sequence or continuum of sinusoidal components that identify the frequency domain content of the signal. MATLAB provides the functions `fft`, `ifft`, `fft2`, `ifft2`, and `fftshift` for Fourier analysis. This collection of functions performs the discrete Fourier transform and its inverse in one or more dimensions. More extensive signal processing tools are available in the optional *Signal Processing Toolbox*.

Because signal processing encompasses such a diverse area, it is beyond the scope of this text to illustrate even a small sample of the type of problems that can be solved using the discrete Fourier transform functions in MATLAB. As a result, one example using the function `fft` to approximate the Fourier transform of a continuous time signal will be illustrated. In addition, use of the `fft` to approximate the Fourier series of a periodic continuous time signal will be demonstrated.

## 21.1 DISCRETE FOURIER TRANSFORM

In MATLAB, the function `fft` computes the discrete Fourier transform of a signal. In those cases where the length of the data is a power of 2 or a product of prime factors, fast Fourier transform (FFT) algorithms are utilized to compute the discrete Fourier transform.

> Because of the substantial increase in computational speed that occurs when data length is a power of 2, whenever possible it is important to choose data lengths equal to a power of 2, or to pad data with zeros to give it a length equal to a power of 2.

The fast Fourier transform implemented in MATLAB follows that commonly used in engineering texts:

$$F(k) = \mathrm{FFT}\{f(n)\}$$

$$F(k) = \sum_{n=0}^{N-1} f(n)e^{-j2\pi nk/N} \quad k = 0, 1, \ldots, N-1$$

Since MATLAB does not allow zero indices, the values are shifted by one index value:

$$F(k) = \sum_{n=1}^{N} f(n)e^{-j2\pi(k-1)\left(\frac{n-1}{N}\right)} \quad k = 1, 2, \ldots, N$$

The inverse transform follows accordingly:

$$f(n) = \mathrm{FFT}^{-1}\{F(k)\}$$

$$f(n) = \frac{1}{N}\sum_{k=1}^{N} F(k)e^{j2\pi(k-1)\left(\frac{n-1}{N}\right)} \quad n = 1, 2, \ldots, N$$

Specific details on the use of MATLAB's `fft` function are described in its help text:

```
» help fft
FFT Discrete Fourier transform.
  FFT(X) is the discrete Fourier transform (DFT) of vector X. If the
  length of X is a power of two, a fast radix-2 fast-Fourier
  transform algorithm is used. If the length of X is not a
  power of two, a slower non-power-of-two algorithm is employed.
  For matrices, the FFT operation is applied to each column.
  For N-D arrays, the FFT operation operates on the first
  non-singleton dimension.

  FFT(X,N) is the N-point FFT, padded with zeros if X has less
  than N points and truncated if it has more.
```

```
FFT(X,[],DIM) or FFT(X,N,DIM) applies the FFT operation across the
dimension DIM.

For length N input vector x, the DFT is a length N vector X,
with elements
                  N
   X(k) =        sum  x(n)*exp(-j*2*pi*(k-1)*(n-1)/N), 1 <= k <= N.
                  n=1
The inverse DFT (computed by IFFT) is given by
                  N
   x(n) = (1/N) sum  X(k)*exp( j*2*pi*(k-1)*(n-1)/N), 1 <= n <= N.
                  k=1
The relationship between the DFT and the Fourier coefficients a and b in
              N/2
x(n) = a0 + sum a(k)*cos(2*pi*k*t(n)/(N*dt))+b(k)*sin(2*pi*k*t(n)/(N*dt))
              k=1
is
   a0 = 2*X(1)/N, a(k) = 2*real(X(k+1))/N, b(k) = 2*imag(X(k+1))/N,
where x is a length N discrete signal sampled at times t with spacing dt.

See also IFFT, FFT2, IFFT2, FFTSHIFT.
```

To illustrate use of the FFT, consider the problem of estimating the continuous Fourier transform of the signal

$$f(t) = \begin{cases} 2e^{-3t} & t \geq 0 \\ 0 & t < 0 \end{cases}$$

Analytically, this Fourier transform is given by

$$F(\omega) = \frac{2}{3 + j\omega}$$

Although using the FFT has little real value in this case since the analytic solution is known, this example illustrates an approach to estimating the Fourier transform of less common signals, especially those whose Fourier transform is not readily found analytically. The following MATLAB statements estimate $|F(\omega)|$ using the FFT and graphically compare it to the preceding analytic expression:

```
» N = 128;  % choose a power of 2 for speed
» t = linspace(0,3,N);  % time points for function evaluation
» f = 2*exp(-3*t);  % evaluate the function and minimize aliasing: f(3) ~ 0
» Ts = t(2) - t(1);  % the sampling period
» Ws = 2*pi/Ts;  % the sampling frequency in rad/sec
» F = fft(f);  % compute the fft
» Fp = F(1:N/2+1)*Ts;
```

This last statement extracts only the positive frequency components from F and multiplies them by the sampling period to estimate $F(\omega)$.

```
» W = Ws*(0:N/2)/N;
```

creates the continuous frequency axis, which starts at zero and ends at the Nyquist frequency Ws/2:

```
» Fa = 2./(3+j*W);                    % evaluate analytical Fourier transform
» plot(W,abs(Fa),W,abs(Fp),'o')  % generate plot, 'o' mark fft results
» xlabel('Frequency, Rad/s')
» ylabel('|F(\omega)|')
» title('Figure 21.1: Fourier Transform Approximation')
```

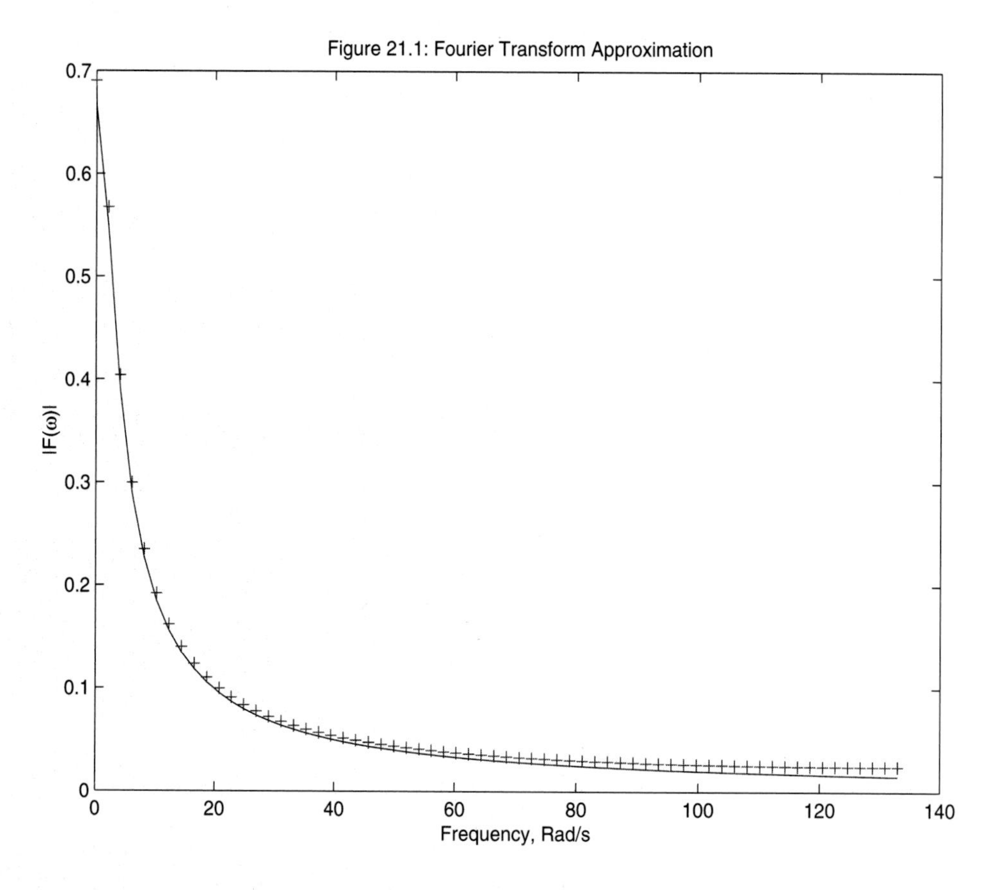


Based on the above plot, it is clear that there is some minor aliasing especially as one nears the Nyquist frequency.

In addition to the function fft and its companions, MATLAB offers a number of functions, shown in the following table, that implement common signal processing tasks.

| Function | Description |
|---|---|
| conv | Convolution. |
| conv2 | 2-D convolution. |
| fft | Discrete Fourier transform. |
| fft2 | 2-D discrete Fourier transform. |
| fftn | *N*-dimensional discrete Fourier transform. |
| ifft | Inverse discrete Fourier transform. |
| ifft2 | 2-D Inverse discrete Fourier transform. |
| ifftn | *N*-dimensional inverse discrete Fourier transform. |
| filter | Discrete time filter. |
| filter2 | 2-D discrete time filter. |
| abs | Magnitude. |
| angle | Four-quadrant phase angle. |
| unwrap | Remove phase angle jumps at 360-degree boundaries. |
| cplxpair | Sort array into complex conjugate pairs. |
| fftshift | Shift FFT results so negative frequencies appear first. |
| nextpow2 | Next higher power of 2. |

## 21.2 FOURIER SERIES

MATLAB itself offers no functions specifically tailored to Fourier series analysis and manipulation. However, they are easily added when one understands the relationship between the discrete Fourier transform of a periodic signal and its Fourier series.

The Fourier series representation of a real-valued periodic signal $f(t)$ can be written in complex exponential form as

$$f(t) = \sum_{n=-\infty}^{\infty} F_n e^{jn\omega_o t}$$

where the Fourier series coefficients are

$$F_n = \frac{1}{T_o} \int_t^{t+T_o} f(t) e^{-jn\omega_o t}\, dt$$

and the fundamental frequency is $\omega_o = 2\pi/T_o$, where $T_o$ is the period—that is, $f(t + T_o) = f(t)$.

The trigonometric form of the Fourier series is given by

$$f(t) = A_0 + \sum_{n=1}^{\infty}\{A_n \cos(n\omega_o t) + B_n \sin(n\omega_o t)\}$$

where the Fourier series coefficients are

$$A_0 = \frac{1}{T_o}\int_t^{t+T_o} f(t)\, dt$$

$$A_n = \frac{2}{T_o}\int_t^{t+T_o} f(t)\cos(n\omega_o t)\, dt$$

$$B_n = \frac{2}{T_o}\int_t^{t+T_o} f(t)\sin(n\omega_o t)\, dt$$

Of these two forms, the complex exponential Fourier series is generally easier to manipulate analytically, whereas the trigonometric form provides a more intuitive understanding because it is easier to visualize sine and cosine waveforms. The relationships between the coefficients of the two forms are

$$A_0 = F_0$$

$$A_n = 2\,\mathrm{Re}\{F_n\}$$

$$B_n = -2\,\mathrm{Im}\{F_n\}$$

$$F_n = F^*_{-n} = \frac{1}{2}(A_n - jB_n)$$

Using these relationships, one can use the complex exponential form analytically and then convert results to the trigonometric form for display.

The discrete Fourier transform can be used to compute the Fourier series coefficients provided the time samples are appropriately chosen and the transform output is scaled. For example, consider computing the Fourier series coefficients of the sawtooth waveform shown next.

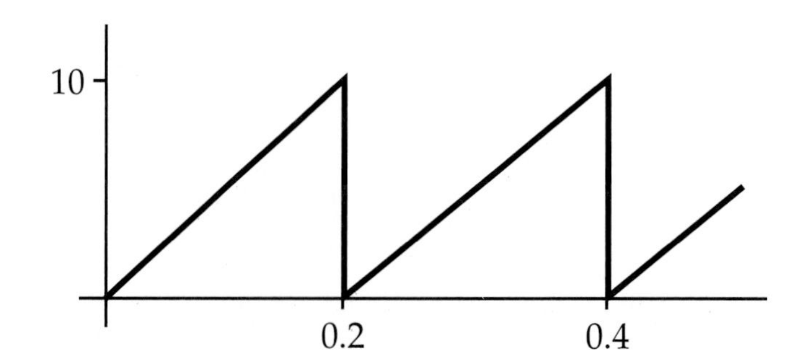

First, one must create a function to evaluate the sawtooth at arbitrary points. For example,

```
function f=sawtooth(t,To)
%SAWTOOTH Sawtooth Waveform Generation.
% SAWTOOTH(t,To) computes values of a sawtooth having
% a period To at the points defined in the vector t.

f = 10*rem(t,To)/To;
f(f==0 | f==10) = 5; % must average value at discontinuity!
```

To minimize aliasing, it is necessary to compute enough harmonics so that the highest harmonic amplitude is negligible. In this case choose

```
» N = 25; % number of harmonics
» To = 0.2; % choose period
```

The number of terms to consider in the discrete Fourier transform is twice the number of harmonics since the discrete Fourier transform computes both positive and negative harmonics:

```
» n = 2*N;
```

The function must be evaluated at `n` points over one period in such a manner that the (`n+1`)th point is one period away from the first point; that is,

```
» t = linspace(0,To,n+1); % (n+1)th point is one period away
» t(end) = []; % throw away undesired last point
» f = sawtooth(t,To); % compute sawtooth
```

We are now ready to compute the transform, rearrange components, and scale the results:

```
» Fn = fft(f);
» Fn = [conj(Fn(N+1)) Fn(N+2:end) Fn(1:N+1)]; % rearrange frequency points
» Fn = Fn/n; % scale results by fft length
```

The vector `Fn` now contains the complex exponential Fourier series coefficients in ascending order. That is, `Fn(1)` is $F_{-25}$, `Fn(26)` is $F_0$, the DC component, and `Fn(51)` is $F_{25}$, the twenty-fifth harmonic.

From these data the trigonometric Fourier series coefficients are as follows:

```
» A0 = Fn(N+1)
A0 =
     5
» An = 2*real(Fn(N+2:end))
An =
Columns 1 through 6
 2.1548e-16  -4.8928e-16  -7.9562e-17  -3.486e-16  -4.0318e-17  -6.9969e-16
```

```
Columns 7 through 12
 1.9072e-17  -4.5554e-16   6.4843e-16 -1.9759e-15  -1.2055e-15  -9.7793e-16
Columns 13 through 18
-5.7089e-16  -8.2137e-16  -3.6439e-16 -5.4984e-16  -1.0076e-16   -3.221e-16
Columns 19 through 24
 3.7984e-16  -4.4959e-15  -9.9383e-16 -9.1789e-16  -3.3734e-16  -3.8442e-16
Column 25
-3.0198e-16
» Bn = -2*imag(Fn(N+2:end))
Bn =
Columns 1 through 6
    -3.1789      -1.5832      -1.0484     -0.77895      -0.61554      -0.50514
Columns 7 through 12
   -0.42502      -0.3638     -0.31515     -0.27528      -0.24176      -0.21298
Columns 13 through 18
   -0.18781     -0.16545     -0.14531     -0.12692      -0.10995     -0.094113
Columns 19 through 24
  -0.079186    -0.064984    -0.051351    -0.038152     -0.025266     -0.012583
Column 25
-2.6886e-17
```

Since the sawtooth waveform has odd symmetry except for its DC component, it makes sense that the cosine coefficients An are negligible. For comparison, the actual Fourier series coefficients for this sawtooth waveform are as follows:

```
» idx = -N:N;
» Fna = 5j./(idx*pi); % actual Fourier series coefficients
» Fna(N+1) = 5;
» Bna = -2*imag(Fna(N+2:end)) % sine terms
» Bna
Bna =
  Columns 1 through 6
    -3.1831      -1.5915       -1.061     -0.79577      -0.63662      -0.53052
  Columns 7 through 12
   -0.45473     -0.39789     -0.35368     -0.31831      -0.28937      -0.26526
Columns 13 through 18
   -0.24485     -0.22736     -0.21221     -0.19894      -0.18724      -0.17684
Columns 19 through 24
   -0.16753     -0.15915     -0.15158     -0.14469       -0.1384      -0.13263
Column 25
   -0.12732
```

The relative error between the fft and analytic results are as follows:

```
» Berr = (Bn-Bna)./Bna
Berr =
  Columns 1 through 6
 -0.0013163   -0.0052693    -0.011872    -0.021144    -0.033117     -0.047829
```

```
Columns 7 through 12
-0.065329    -0.085674    -0.10894    -0.13519    -0.16454    -0.19709
Columns 13 through 18
 -0.23296     -0.27229    -0.31525    -0.36201    -0.41278     -0.4678
Columns 19 through 24
 -0.52734     -0.59169    -0.66122    -0.73631    -0.81744    -0.90513
Column 25
       -1
```

As with the earlier Fourier transform example, aliasing causes errors that increase with increasing frequency. Since all practical signals are not band limited, aliasing is inevitable and a decision must be made about the degree of aliasing that can be tolerated in a given application. As the number of requested harmonics increases, the degree of aliasing decreases. Therefore, to minimize aliasing one can request a larger number of harmonics and then choose a subset of them for viewing and further manipulation.

Finally, the line spectra of the complex exponential Fourier series can be plotted using the `stem` function:

```
» stem(idx,abs(Fn))
» xlabel('Harmonic Index')
» title('Figure 21.2: Sawtooth Harmonic Content')
```

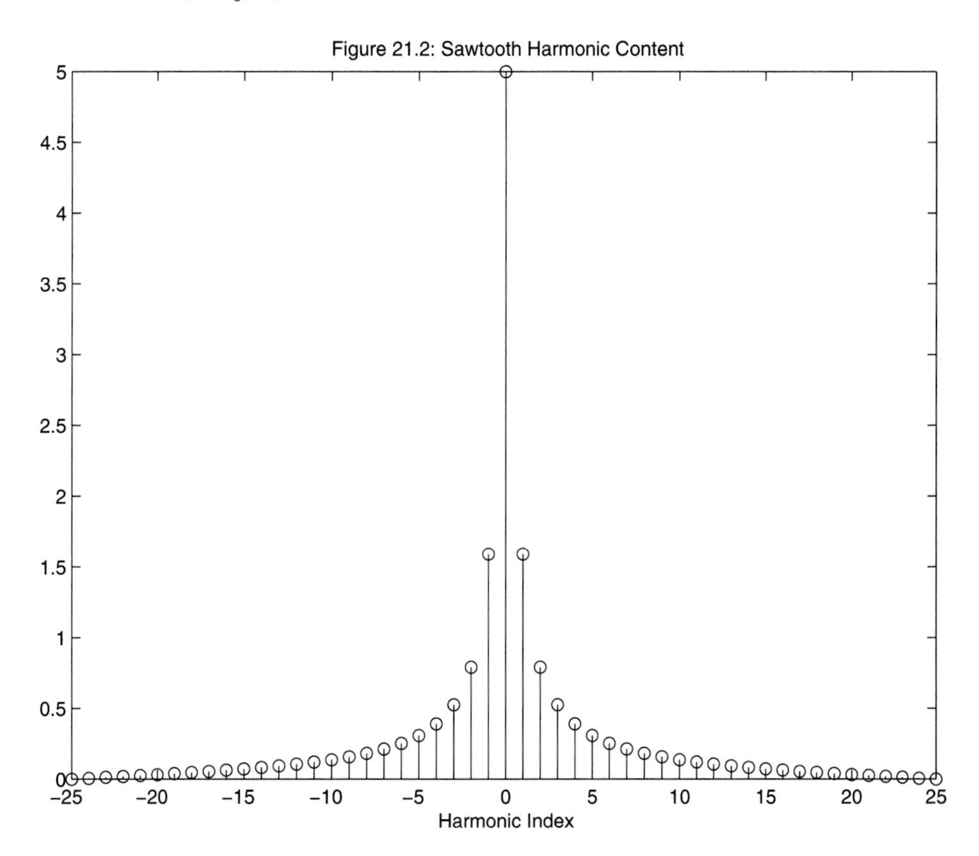

22

# Optimization

Optimization in the context of this chapter refers to the process of determining where a function $y = g(x)$ takes on either specific or extreme values. When a function is defined simply, the corresponding inverse function $x = g^{-1}(y)$ can often be found, in which case one can determine what $x$ values produce a given $y$ by evaluating the inverse function. On the other hand, many functions as well as many simple ones have no inverse. When this is the case, one must estimate the $x$ that produces a known $y$ by some iterative procedure. In practice, this iterative procedure is called ***zero finding,*** because finding $x$ such that $y = g(x)$ for some $y$ is equivalent to finding $x$ such that $f(x) = y - g(x) = 0$.

In addition to knowing where a function takes on specific values, it is also common to know its extreme values (i.e., where it achieves maximum or minimum values). As before, there are numerous times when these extreme values must be estimated by some iterative procedure. Since a function maximum is the minimum of its negative—that is, $\max f(x) = \min\{-f(x)\}$—iterative procedures for finding extreme values typically find only minimum values and the procedures are called ***minimization*** algorithms.

In this chapter, the optimization functions available in basic MATLAB are covered. Many more functions are available in the optional *Optimization Toolbox.*

## 22.1 ZERO FINDING

Finding a zero of a function can be interpreted in a number of ways depending on the function. When the function is one-dimensional the MATLAB function `fzero` can be used to find a zero. The algorithm used by this function is a combination of bisection and inverse quadratic interpolation. When the function is multidimensional (i.e., the function definition consists of multiple scalar functions of a vector variable), one must look beyond basic MATLAB for a solution. Both the *Optimization Toolbox* and *Mastering MATLAB Toolbox* provide functions for solving the multidimensional case.

To illustrate the use of the function `fzero`, consider the function `humps`:

```
» x = linspace(0,2);
» y = humps(x);
» plot(x,y)
» grid on
» Title('Figure 22.1: Humps Function')
```

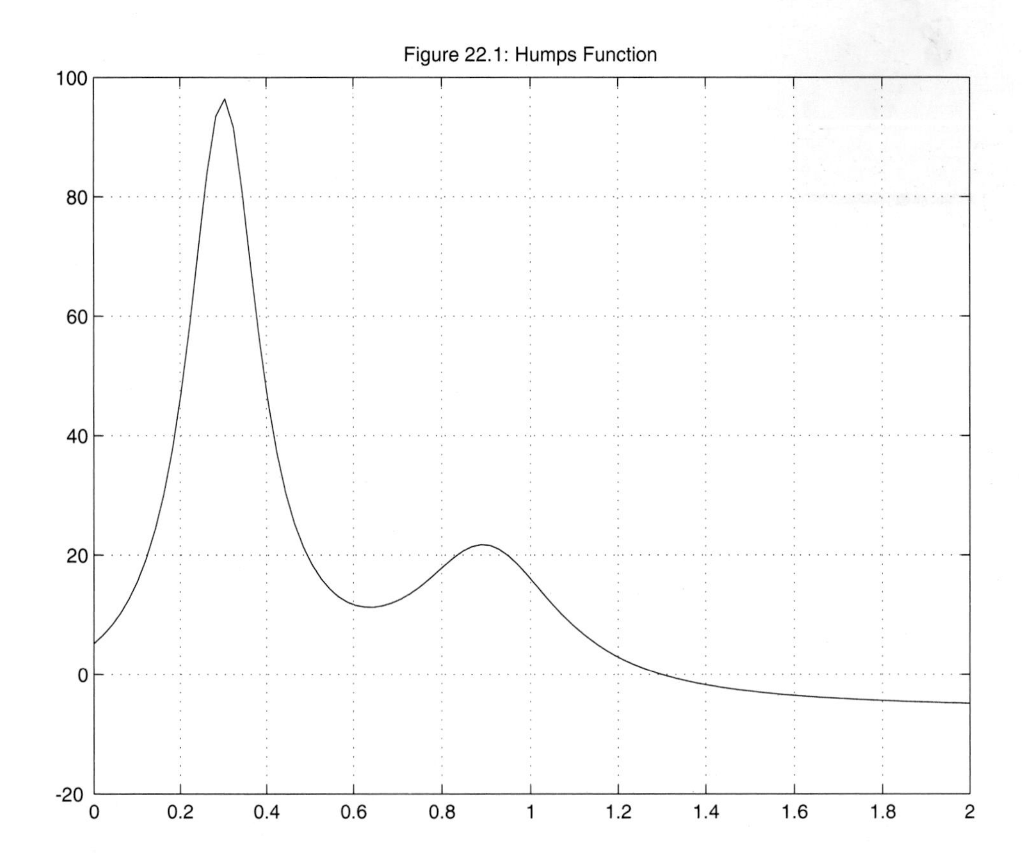


The `humps` M-file evaluates the function

$$\text{humps}(x) = \frac{1}{(x-0.3)^2 + 0.01} + \frac{1}{(x-0.9)^2 + 0.04} - 6$$

This function crosses zero near $x = 1.3$. The function `fzero` provides a way to find a better approximation to the zero crossing:

```
» format long % display more precision
» x = fzero('humps',1.3) % call fzero(Fname,initial_guess)
x =
   1.29954968258482
» humps(x) % evaluate result at zero
ans =
     0
```

So the zero occurs very close to 1.3 and `fzero` finds the zero with maximum precision. More information can be obtained from `fzero` by providing a nonzero fourth input argument (the third input argument, if provided and nonempty, sets the desired solution tolerance):

```
» x = fzero('humps',1.3,[],1)
 Func evals         x               f(x)        Procedure
     1            1.3      -0.00990099          initial
     2        1.26323         0.882416          search

   Looking for a zero in the interval [1.2632, 1.3]

     3        1.29959     -0.00093168          interpolation
     4        1.29955     1.23235e-07          interpolation
     5        1.29955    -1.37597e-11          interpolation
     6        1.29955               0          interpolation
x =
   1.29954968258482
```

The first two function evaluations are used to find a bracket around the solution. In other words, since `humps(1.3)` has a different sign than `humps(1.26323)`, humps($x$) must cross through zero somewhere in the interval `[1.2632, 1.3]`. The last set of function evaluations shows the iterations used to find the solution.

In the preceding example, the function to be iterated is defined by a function M-file. One can also specify the function as a character string with $x$ being the argument. In this case, MATLAB converts the character string into an in-line function and proceeds. For example, consider finding the points where humps($x$) = 60:

```
» x = fzero('humps(x)-60',[0.2 0.3]) % fzero(string,bracket)
x =
   0.22497151787552
» humps(x)
ans =
  60.00000000000001
```

```
» x = fzero('humps(x)-60',[0.3 0.4])
x =
   0.37694647486211
» humps(x)
ans =
  60.00000000000004
```

In this example, `'humps(x)-60'` defines the function to be iterated. Internally, MATLAB searches the string, finds `x` as the argument, and converts the string to an in-line function. In addition, the preceding example shows that one can optionally enter a bracket instead of an initial guess. In the first case, `humps(0.2)` is less than 60 and `humps(0.3)` is greater than 60. Likewise, `humps(0.4)` is less than 60. When called in this way, the two-element vector must form a bracket or an error is returned. For example,

```
» x = fzero('humps(x)-60',[0.5 0.4])
??? Error using ==> fzero
The function values at the interval endpoints must differ in sign.
```

## 22.2 MINIMIZATION IN ONE DIMENSION

In addition to the visual information provided by plotting, it is often necessary to determine other more specific attributes of a function. Of particular interest in many applications are function extremes—that is, its maxima (peaks) and its minima (valleys). Mathematically, these extremes are found analytically by determining where the derivative (slope) of a function is zero. This fact can be readily understood by inspecting the slope of the humps plot at its peaks and valleys. Clearly, when a function is simply defined, this process often works. However, even for many simple functions that can be differentiated readily, it is often not possible to find where the derivative is zero. In these cases and in cases where it is difficult or impossible to find the derivative analytically, it is necessary to search for function extremes numerically. MATLAB provides two functions that perform this task, `fmin` and `fmins`. These two functions find minima of one-dimensional and $n$-dimensional functions, respectively. `fmin` utilizes a combination of golden section search and parabolic interpolation. Since a maximum of $f(x)$ is equal to a minimum of $-f(x)$, `fmin` and `fmins` can be used to find both minima and maxima. If this fact is not clear, visualize the preceding plot flipped upside down. In the upside-down state, peaks become valleys and valleys become peaks.

To illustrate one-dimensional minimization and maximization, consider the preceding humps($x$) example once again. From the figure, there is a maximum near $x_{max} = 0.3$ and a minimum near $x_{min} = 0.6$. Using `fmin`, these extremes can be found with more accuracy:

```
» xmin = fmin('humps',0.5,0.8) % fmin(Fname,lower_limit,upper_limit)
xmin =
   0.63700821196362
```

```
» humps(xmin)
ans =
  11.25275412587769

» xmax = fmin('-humps(x)',0.2,0.4) % creates in-line function
xmax =
   0.30036413790024
» humps(xmax)
ans =
  96.50140724387050
```

As with `fzero`, `fmin` handles both M-file functions and character string functions that are converted to in-line functions. In fact, anywhere character string functions can be used, explicit in-line functions can also be used. For example,

```
» ilhumps = inline('-humps(x)')
ilhumps =
     Inline function:
     ilhumps(x) = -humps(x)

» xmax = min(ilhumps,0.2,0.4)
xmax =
   0.30036413790024

» feval(ilhumps,xmax)
ans =
 -96.50140724387050
```

In addition to the default properties shown previously, `fmin` accepts a fourth input argument containing an `options` vector. If `options(1)` is nonzero, intermediate steps in the iterative solution process are displayed. If `options(2)` exists, it determines the relative tolerance on `x` acceptable for the solution (`1e-4` is default). If `options(14)` exists, it determines the maximum number of allowable function evaluations allowed (`500` is default). This options vector is shared among `fmin`, `fmins`, and all the functions in the *Optimization Toolbox*. At some point in the future, it is anticipated that this vector will become a MATLAB structure with appropriately named fields, thereby making it easier to identify each option visually. Perhaps this change will come with version 2.0 of the *Optimization Toolbox*.

An example, showing use of the `options` vector, follows:

```
» options(1) = 1; % turn on display of intermediate results
» options(2) = 1e-8; % tighten termination criteria
» xmin = min('humps',0.5,0.8,options)
Func evals        x             f(x)          Procedure
     1         0.61459        11.4103         initial
     2         0.68541        11.9288         golden
```

```
       3        0.57082        12.7389        golden
       4        0.638866       11.2538        parabolic
       5        0.637626       11.2529        parabolic
       6        0.637046       11.2528        parabolic
       7        0.637008       11.2528        parabolic
       8        0.637009       11.2528        parabolic
       9        0.637009       11.2528        parabolic
      10        0.637009       11.2528        parabolic
xmin =
   0.63700898473489
» humps(xmin) % very little change in results!
ans =
  11.25275412569616
```

## 22.3 MINIMIZATION IN HIGHER DIMENSIONS

As described previously, the function fmins provides a simple algorithm for minimizing a function of several variables. That is, fmins attempts to solve

$$\min_{x} f(\mathbf{x})$$

where $f(\mathbf{x})$ is a scalar function of a vector argument $\mathbf{x}$. fmins implements the Nelder-Mead simplex search algorithm. This algorithm is not very efficient on smooth functions, but on the other hand, it does not require gradient information, which is often expensive to compute. It also tends to be more robust on functions that are not smooth where gradient information is less valuable.

To illustrate usage of fmins, consider the banana function, also called Rosenbrock's function:

$$f(\mathbf{x}) = 100(x_2 - x_1^2)^2 + (1 - x_1)^2$$

This function can be visualized by creating a 3-D mesh plot with $x_1$ being the $x$-dimension and $x_2$ being the $y$-dimension:

```
x = [-1.5:0.125:1.5];
y = [-.6:0.125:2.8];
[X,Y] = meshgrid(x,y) ;
Z = 100.*(Y-X.*X).^2 + (1-X).^2;
mesh(X,Y,Z)
hidden off
xlabel('x(1)'), ylabel('x(2)')
title('Figure 22.2: Banana Function')
hold on
plot3(1,1,1,'k.','markersize',30)
hold off
```

Figure 22.2: Banana Function

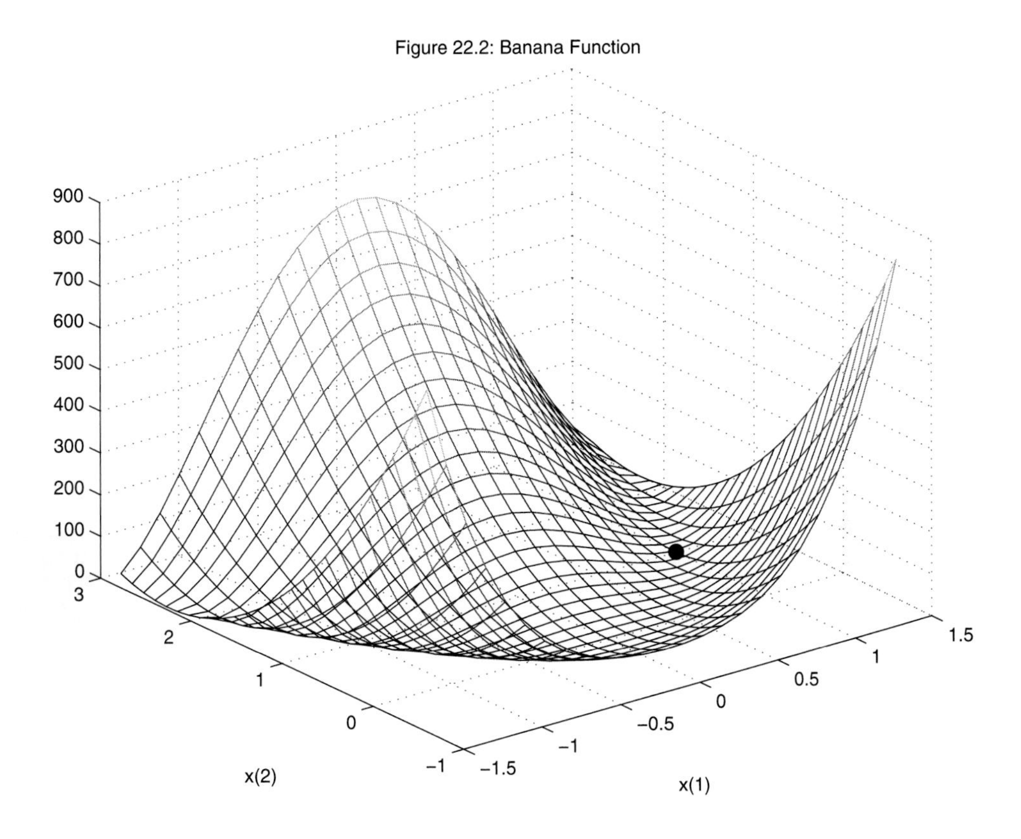


As shown in the plot, the banana function has a unique minimum of 0 at $\mathbf{x} = [1; 1]$. To find the minimum of this function, it can be entered as the M-file:

```
function f=banana(x)
% Rosenbrock's banana function

f =100*(x(2)-x(1)^2)^2 + (1-x(1))^2;
```

or it can be entered as a character string

```
» fstr = '100*(x(2)-x(1)^2)^2 + (1-x(1))^2';
```

or as an in-line function:

```
» ilfstr = inline(fstr)
```

```
ilfstr =
     Inline function:
     ilfstr(x) = 100*(x(2)-x(1)^2)^2 + (1-x(1))^2
```

Just like fzero and fmin, the function fmins accepts all three representations.

Using these representations with the function fmins produces

```
» xmin = fmins('banana',[-1.9,2]) % fmins(Fname,initial_guess_vector)
xmin =
   1.00002562931386   1.00005319450299
» xmin = fmins(fstr,[-1.9,2])
xmin =
   1.00002562931386   1.00005319450299
» xmin = fmins(ilfstr,[-1.9,2])
xmin =
   1.00002562931386   1.00005319450299
» feval(ilfstr,xmin)
ans =
   1.0314e-09
```

Of these representations, the M-file form is much faster than either the character string or in-line function forms. In-line functions add a great deal of flexibility, but with that flexibility comes complexity that is hidden from the user. For simple functions, this complexity and associated computational burden is hardly noticeable. However, if the function is complicated it is beneficial and perhaps easier to use the M-file form.

As with fmin, fmins accepts an options vector. If options(1) is nonzero, intermediate steps in the iterative solution process are displayed. If options(2) exists, it determines the relative tolerance on x acceptable for the solution (1e-4 is default). If options(3) exists, it determines the relative tolerance on the function acceptable for the solution (1e-4 is default). If options(14) exists, it determines the maximum number of allowable function evaluations allowed (200*length(x) is default). For example,

```
» xmin = fmins('banana',[1.1,0.9],1) % just options(1)=1
func_evals =
     3
initial
v =
   1.1000e+00   1.1000e+00   1.1550e+00
   9.4500e-01   9.0000e-01   9.0000e-01
f =
   7.0325e+00
func_evals =
     5
expand
```

```
% (intermediate iteration information discarded for brevity)
func_evals =
    50
shrink

v =
   9.9000e-01   9.9000e-01   9.9005e-01
   9.8009e-01   9.8011e-01   9.8008e-01
fv =
   1.0001e-04   1.0002e-04   1.0054e-04

xmin =

    0.9900    0.9801
```

## 22.4 PRACTICAL ISSUES

Iterative solutions such as those found by `fzero`, `fmin`, and `fmins` all make some assumptions about the function to be iterated. Since there are essentially no limits to the function provided, it makes sense that these ***function functions*** may not converge or may take many iterations to converge. At worst, these functions may produce a MATLAB error that terminates the iteration without producing a result. And even if they do terminate promptly, there is no guarantee that they stopped at the desired result. To make the most efficient use of these function functions, consider the following points:

1. Start with a good initial guess. This is the most important consideration. A good guess keeps the problem in the neighborhood of the solution where its numerical properties should be stable.
2. If components of the solution (e.g., in `fmins`) are separated by several orders of magnitude or more, consider scaling them to improve iteration efficiency and accuracy. For example, if `x(1)` is known to be near `1` and `x(2)` is known to be near `1e6`, scale `x(2)` by `1e-6` in the function definition, and then scale the returned result by `1e6`.
3. If the problem is complicated, look for ways to simplify the problem into a sequence of simpler problems that have fewer variables.
4. Make sure your function cannot return complex numbers, `Inf`, or `NaN`. These results usually result in convergence failure. The functions `isreal`, `isfinite`, and `isnan` can be used to test results before returning them.
5. Avoid functions that are discontinuous. Functions such as `abs`, `min`, and `max` all produce discontinuities that can lead to divergence.
6. Constraints on the allowable range of $x$ can be included by adding a penalty term to the function to be iterated such that the algorithm is persuaded to avoid out-of-range values.

# 23

# Integration and Differentiation

Integration and differentiation are fundamental tools in calculus. Integration computes the area under a function, and differentiation describes the slope or gradient of a function. MATLAB provides functions for numerically approximating the integral and slope of a function. Functions are provided for making approximations when functions exists as M-files or in-line functions as well as when functions are tabulated at uniformly spaced points over the region of interest.

## 23.1 INTEGRATION

MATLAB provides three functions for computing integrals of functions described by M-files or in-line functions: `quad`, `quad8`, and `dblquad`. In the MATLAB version 5 hard copy documentation set, `dblquad` is referred to as `quad2`, reflecting the fact that the function name was changed prior to the release of MATLAB 5.0. Moreover, while other function functions in MATLAB such as `fzero` handle M-file functions, character string functions, and in-line functions transparently, these integration functions do not accept character string functions. It is expected that this feature may be added in a future release.

To illustrate integration, consider the function humps*(x)* as shown in the following figure. As is apparent, the sum of the trapezoidal areas approximates the integral of the function. Clearly, as the number of trapezoids increases, the fit between the function and the trapezoids gets better, leading to a better integral or area approximation.

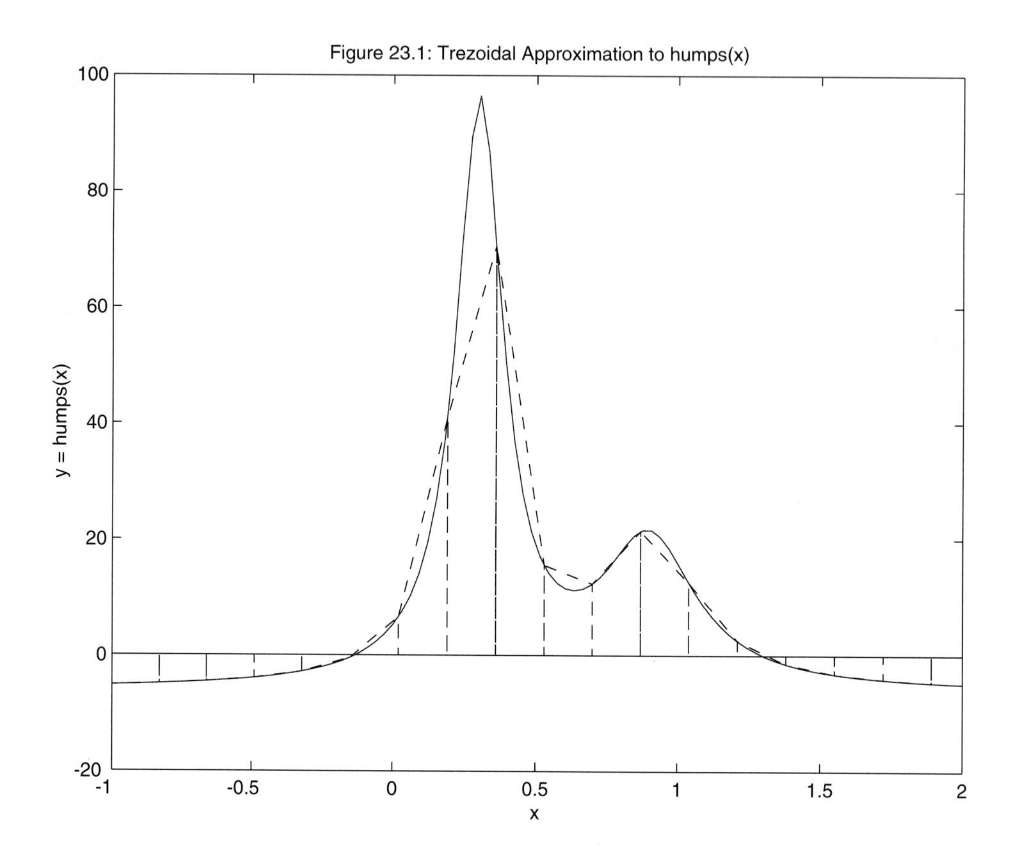


Using tabulated values from the `humps` function, the MATLAB function `trapz` approximates the area using the trapezoidal approximation. Duplicating the trapezoids shown in the preceding figure produces

```
» x = -1:.17:2;
» y = humps(x);
» area = trapz(x,y)
area =
        25.917
```

Based on the figure, this is probably not a very accurate estimate of the area. However, if a finer discretization is used, more accuracy is achieved:

```
» x = linspace(-1,2,100); % linear spacing required!
» y = humps(x);
» format long
» area = trapz(x,y)
area =
       26.34473119524597
```

This area agrees with the analytic integral through five significant digits.

Sometimes one is interested in the integral as a function of $x$; that is,

$$\int_{x_1}^{x} f(x)dx$$

where $x_1$ is a known lower limit of integration. The definite integral from $x_1$ to any point $x$ is then found by evaluating the function at $x$. Using the trapezoidal rule, tabulated values of the cumulative integral are computed using the function `cumtrapz`. For example,

```
» x = linspace(-1,2,100);
» y = humps(x);
» z = cumtrapz(x,y);
» size(z)
ans =
     1    100

» plotyy(x,y,x,z)
» grid on
» xlabel('x')
» ylabel('humps(x) and integral of humps(x)')
» title('Figure 23.2: Cumulative Integral of humps(x)')
```

Depending on the properties of the function at hand, it may be difficult to determine an optimum trapezoidal width. Clearly, if one could somehow vary the individual trapezoid widths to match the characteristics of the function, much greater accuracy could be achieved.

The MATLAB functions `quad` and `quad8`, which are based on the mathematical concept called quadrature, take this approach. These integration functions operate in the same way. Both evaluate the function to be integrated at whatever intervals are necessary to achieve accurate results. Moreover, both functions make higher-order approximations than a simple trapezoid, with `quad8` being more rigorous than `quad`. As an example, consider computing the integral of the `humps` function again:

```
» quad('humps',-1,2)
ans =
  26.34497558341242
» quad8('humps',-1,2)
ans =
  26.34496024631924
```

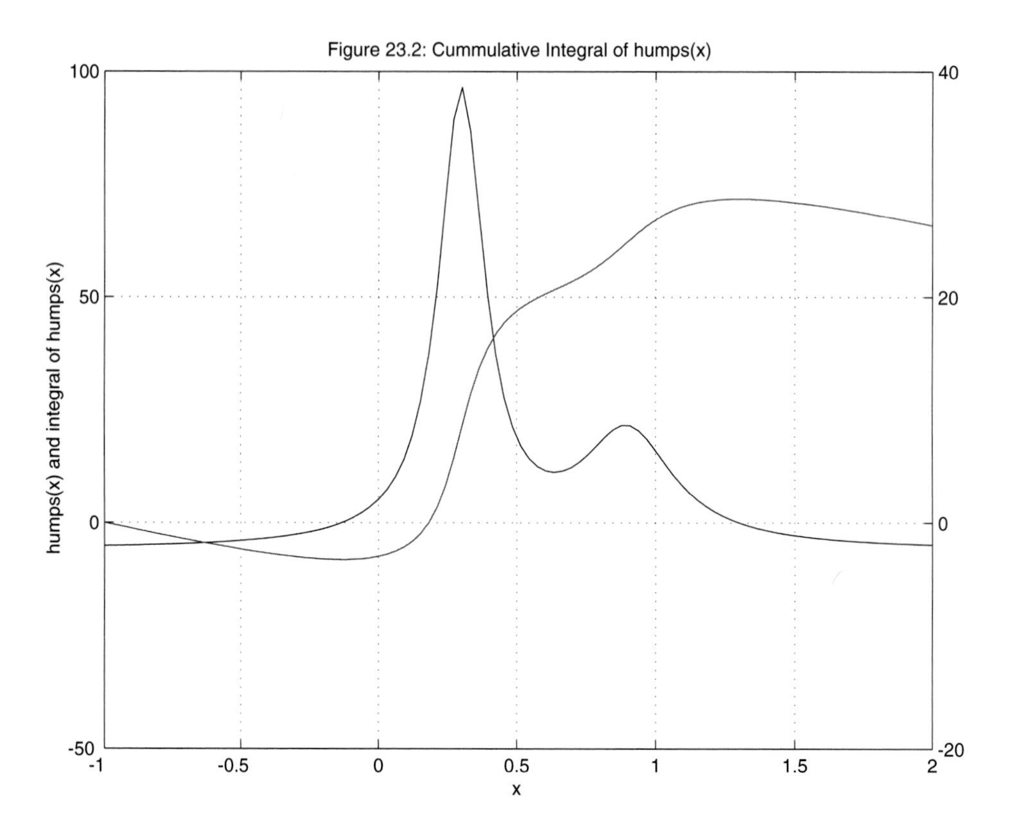


Here the quad solution achieves six significant digit accuracy and the quad8 solution achieves eight significant digit accuracy. As stated earlier, quad and quad8 cannot accept character string function arguments in MATLAB version 5.0. However, they do accept in-line function arguments:

```
» hstr = '1./((x-.3).^2 + .01) + 1./((x-.9).^2 + .04) - 6'
hstr =
1./((x-.3).^2 + .01) + 1./((x-.9).^2 + .04) - 6
» hinline = inline(hstr)
hinline =
     Inline function:
     hinline(x) = 1./((x-.3).^2 + .01) + 1./((x-.9).^2 + .04) - 6
» quad(hinline,-1,2)
ans =
  26.34497558341242
```

quad and quad8 also allow one to specify relative and absolute error tolerances as a two-element vector [rel_tol abs_tol] as a fourth input argument. The default values are [1.e-3 0]. For example,

```
» quad('humps',-1,2,[1e-3 0]) % default tolerances
ans =
  26.34497558341242
» quad('humps',-1,2[1e-4 0]) % tighter relative tolerance

Warning: Recursion level limit reached in quad. Singularity likely.
> In HD:Applications:MATLAB:Toolbox:matlab:funfun:quad.m (quadstp) at line 104
  In HD:Applications:MATLAB:Toolbox:matlab:funfun:quad.m (quadstp) at line 142
  In HD:Applications:MATLAB:Toolbox:matlab:funfun:quad.m (quadstp) at line 142
  In HD:Applications:MATLAB:Toolbox:matlab:funfun:quad.m (quadstp) at line 142
  In HD:Applications:MATLAB:Toolbox:matlab:funfun:quad.m (quadstp) at line 142
  In HD:Applications:MATLAB:Toolbox:matlab:funfun:quad.m (quadstp) at line 143
  In HD:Applications:MATLAB:Toolbox:matlab:funfun:quad.m (quadstp) at line 142
  In HD:Applications:MATLAB:Toolbox:matlab:funfun:quad.m (quadstp) at line 142
  In HD:Applications:MATLAB:Toolbox:matlab:funfun:quad.m (quadstp) at line 142
  In HD:Applications:MATLAB:Toolbox:matlab:funfun:quad.m (quadstp) at line 143
  In HD:Applications:MATLAB:Toolbox:matlab:funfun:quad.m (quadstp) at line 143
  In HD:Applications:MATLAB:Toolbox:matlab:funfun:quad.m at line 74

Warning: Recursion level limit reached 2 times.
> In HD:Applications:MATLAB:Toolbox:matlab:funfun:quad.m at line 82
ans =
  26.34496101516815
```

In addition to one-dimensional integration, MATLAB supports two-dimensional integration with the function `dblquad`, which supports only M-file functions in version 5.0 of MATLAB. That is, `dblquad` approximates the integral

$$\int_{y_{min}}^{y_{max}} \int_{x_{min}}^{x_{max}} f(x,y)\, dxdy$$

To use `dblquad`, one must first create a function that evaluates $f(x, y)$. Consider, for example, the function `myfun.m`:

```
function z=myfun(x,y)
%MYFUN(X,Y) computes an example function of two variables

z = sin(x).*cos(y) + 1;
```

This function can be plotted by the commands

```
» x = linspace(0,pi,20);
» y = linspace(-pi,pi,20);
» [xx,yy] = meshgrid(x,y);
» zz = myfun(xx,yy);
» mesh(xx,yy,zz)
```

```
» xlabel('x'), ylabel('y')
» title('Figure 23.3: myfun.m plot')
```

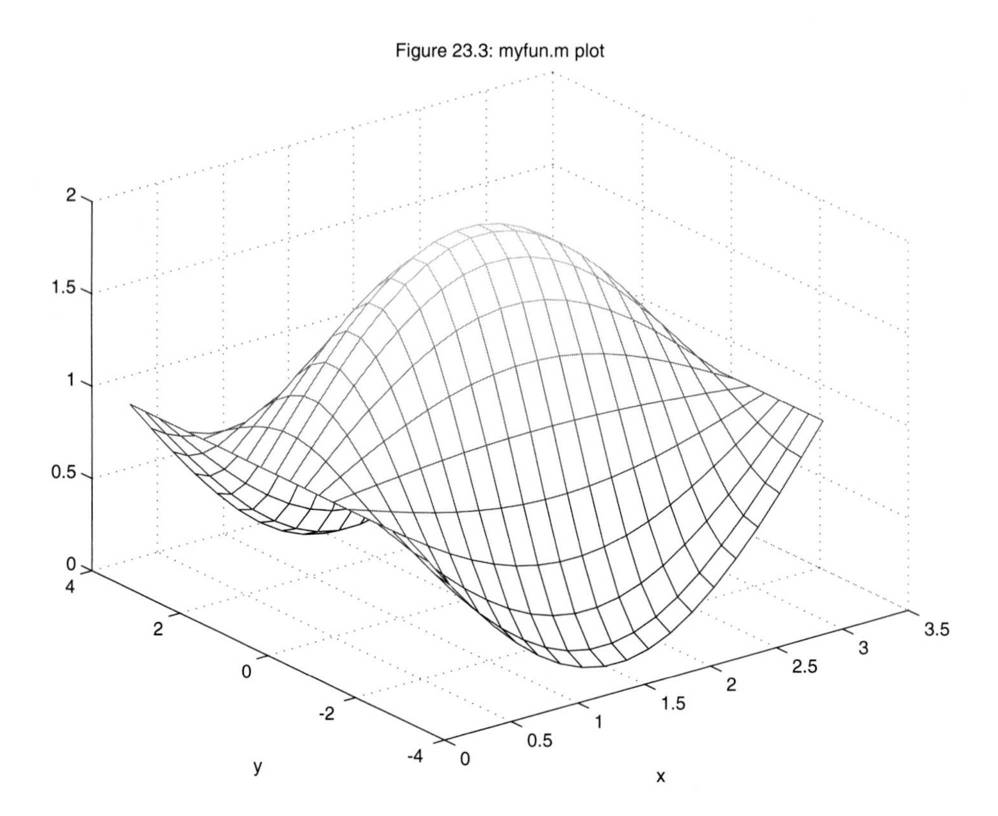


The volume under this function is computed by calling `dblquad` as

```
dblquad('Fname',xmin,xmax,ymin,ymax)
```

which, in this example, becomes

```
» dblquad('myfun',0,pi,-pi,pi)
ans =
  19.73921476256606
```

A vector containing desired relative and absolute error tolerances can also be appended as a final argument to `dblquad`. For example,

```
dblquad('myfun',0,pi,-pi,pi,[rel_tol abs_tol])
```

By default, `dblquad` calls `quad` to do its work, so the same default tolerances apply to `dblquad`.

## 23.2 DIFFERENTIATION

As opposed to integration, numerical differentiation is much more difficult. Integration describes an overall or macroscopic property of a function, whereas differentiation describes the slope of a function at a point, which is a microscopic property of a function. As a result, integration is not sensitive to minor changes in the shape of a function, whereas differentiation is. Any small change in a function can easily create large changes in its slope in the neighborhood of the change.

Because of this inherent sensitivity in differentiation, numerical differentiation is avoided whenever possible, especially if the data to be differentiated are obtained experimentally. In this case, it is best to perform a least squares curve fit to the data and then differentiate the resulting polynomial. Alternatively, one could fit cubic splines to the data and then find the spline representation of the derivative, as discussed in the chapter on cubic splines (Chapter 20). For example, reconsider the example from the chapter on polynomials (Chapter 18):

```
» x = [0 .1 .2 .3 .4 .5 .6 .7 .8 .9 1];
» y = [-.447 1.978 3.28 6.16 7.08 7.34 7.66 9.56 9.48 9.30 11.2]; % data
» n = 2; % order of fit
» p = polyfit(x,y,n)  % find polynomial coefficients
p =
   -9.8108   20.1293   -0.0317

» xi = linspace(0,1,100);
» yi = polyval(p,xi); % evaluate polynomial
» plot(x,y,'-o',xi,yi,'-')
» xlabel('x'), ylabel('y=f(x)')
» title('Figure 23.4: Second Order Curve Fitting')
```

The derivative in this case is found by using the polynomial derivative function `polyder`:

```
» pd = polyder(p)
pd =
   -19.6217   20.1293
```

The derivative of $y = -9.8108x^2 + 20.1293x - 0.0317$ is $dy/dx = -19.6217x + 20.1293$. Since the derivative of a polynomial is yet another polynomial of the next lowest order, the derivative can also be evaluated at any point. In this case, the polynomial fit was second order, making the resulting derivative first order. As a result, the derivative is a straight line, meaning that it changes linearly with $x$.

MATLAB provides a function for computing an approximate derivative given tabulated data describing some function. This function, named `diff`, computes the difference between elements in an array. Since differentiation is defined as

$$\frac{dy}{dx} = \lim_{h\to 0} \frac{f(x+h) - f(x)}{h}$$

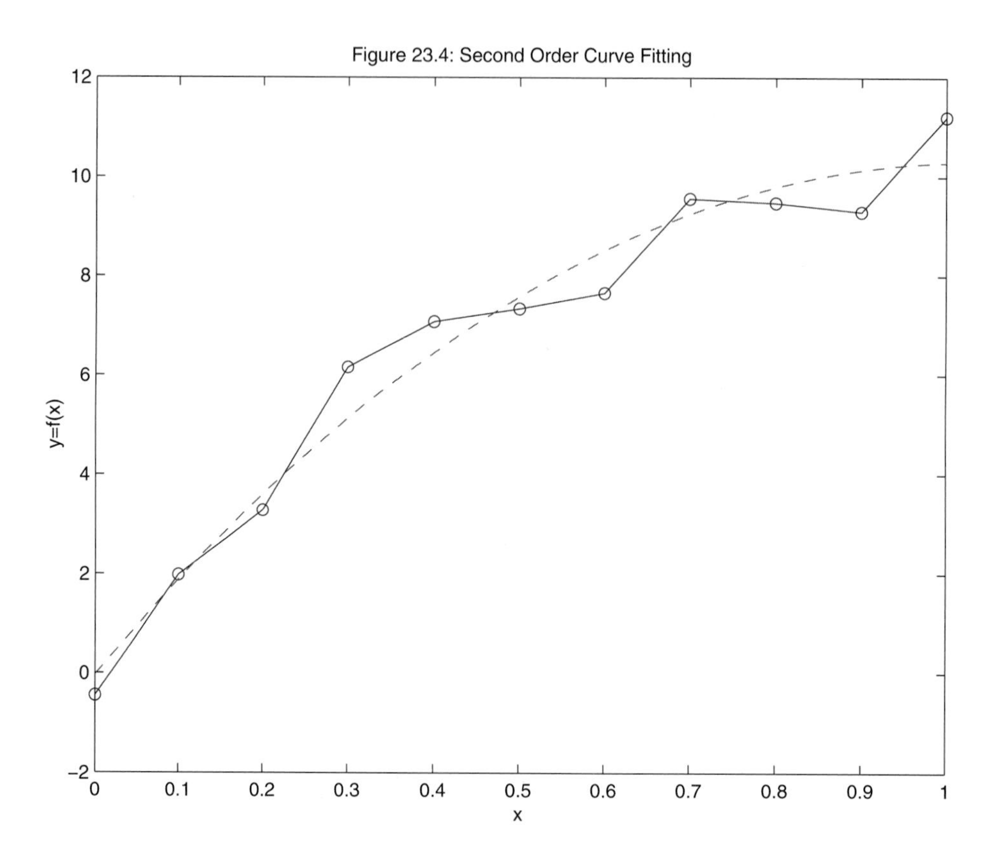


the derivative of $y = f(x)$ can be approximated by

$$\frac{dy}{dx} \approx \frac{\Delta y}{\Delta x} = \frac{f(x+h) - f(x)}{h}$$

which is the forward finite difference of $y$ divided by the finite difference in $x$. Since `diff` computes differences between array elements, differentiation can be approximated in MATLAB. Continuing with the prior example,

```
» dy = diff(y)./diff(x);  % compute differences and use array division
» xd = x(1:end-1);  % create new x axis array since dy is shorter than y
» plot(xd,dy)
» ylabel('dy/dx'), xlabel('x')
» title('Figure 23.5: Forward Difference Derivative Approximation')
```

Since `diff` computes the difference between elements of an array, the resulting output contains one less element than the original array. Thus, to plot the derivative, one element of the `x` array must be thrown out. When the first element of `x` is thrown out, the foregoing procedure gives a backward difference approximation, which uses information

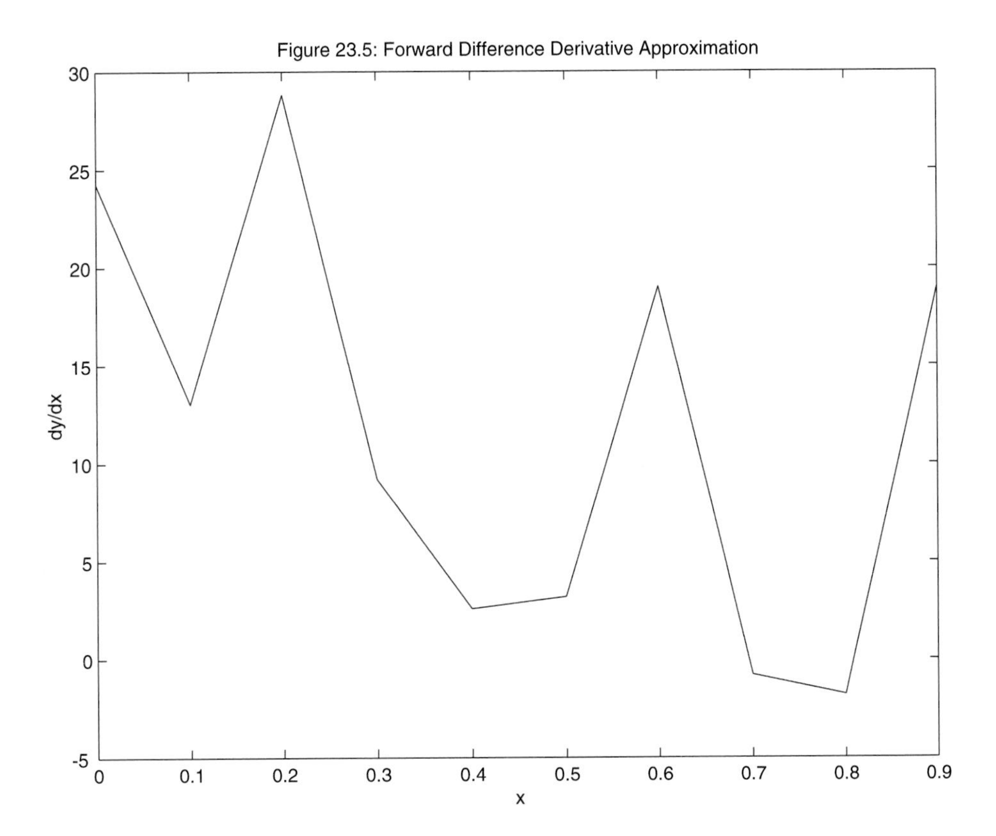


at `x(n-1)` and `x(n)` to approximate the derivative at `x(n)`. On the other hand, throwing out the last element gives a forward difference approximation, which uses `x(n+1)` and `x(n)` to compute results at `x(n)`. Comparing the last two plots, it is overwhelmingly apparent that approximating the derivative by finite differences can lead to poor results, especially if the data originate from experimental or noisy measurements. When the data used do not have uncertainty, the results of using `diff` can be acceptable, especially for visualization purposes. For example,

```
» x = linspace(0,2*pi);
» y = sin(x);
» dy = diff(y)/(x(2)-x(1));
» xd = x(2:end);
» plot(x,y,xd,dy)
» xlabel('x'), ylabel('sin(x) and cos(x)')
» title('Figure 23.6: Backward Difference Derivative Approximation')
```

In this example, `x` was linearly spaced so dividing by `x(2)-x(1)` gives the same answer as `diff(x)`, which is required if `x` is not linearly spaced. In addition, the first element in

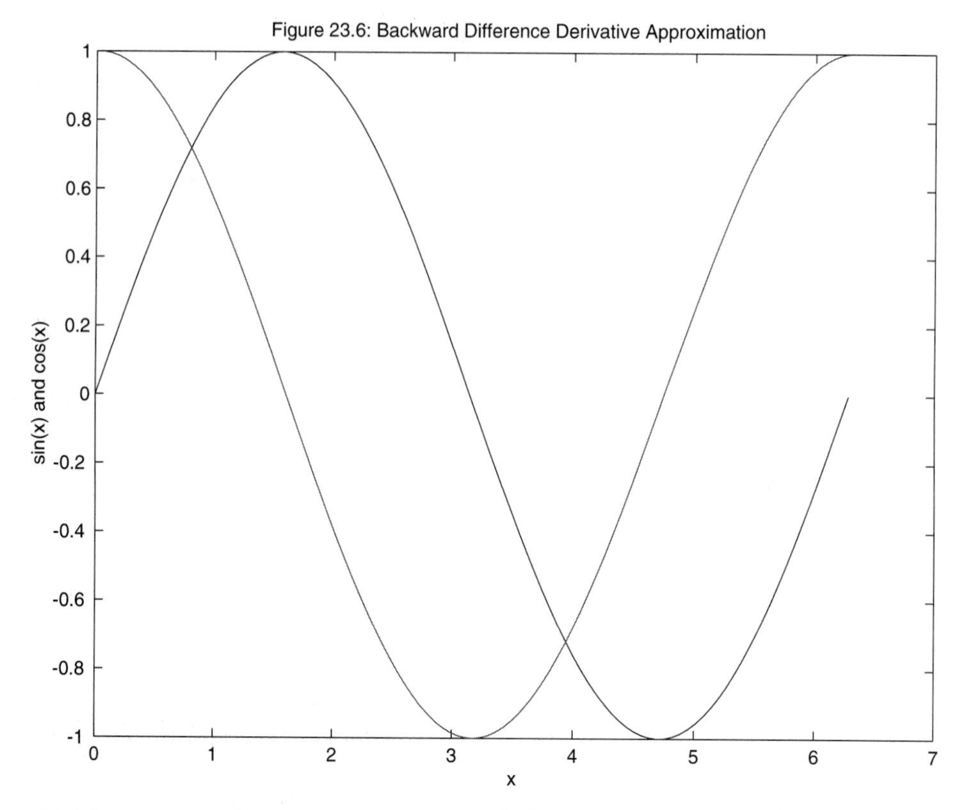


x is thrown out, making the result a backward difference derivative approximation. Visually, the preceding derivative is quite accurate. In fact, the maximum error is

```
» max(abs(cos(xd)-dy))
ans =
    0.0317
```

When forward or backward difference approximations are not sufficient, central differences can be computed by performing the required array operations directly. The first central difference for linearly spaced data is given by

$$\left.\frac{dy}{dx}\right|_{x_n} \approx \frac{f(x_{n+1}) - f(x_{n-1})}{x_{n+1} - x_{n-1}}$$

Therefore, the slope at $x_n$ is a function of its neighboring data points. Repeating the previous example gives

```
» dy = (y(3:end)-y(1:end-2)) / (x(3)-x(1));
» xd = x(2:end-1);
» max(abs(cos(xd)-dy))
ans =
    0.00067086
```

In this case, the first and last data points do not have a central difference approximation because there is no data at `n = 0` and `n = 101`, respectively. However, at all intermediate points, the central difference approximation is nearly two orders of magnitude more accurate than the forward or backward difference approximations.

When dealing with two-dimensional data, the function `gradient` uses central differences to compute the slope in each direction at each tabulated point. Forward differences are used at the initial points and backward differences are used at the final points so that the output has the same number of data points as the input. The function `gradient` is used primarily for graphical data visualization. For example,

```
» [x,y,z] = peaks(20);    % simple 2-D function
» dx = x(1,2) - x(1,1);   % spacing in x direction
» dy = y(2,1) - y(1,1);   % spacing in y direction

» [dzdx,dzdy] = gradient(z,dx,dy);

» contour(x,y,z)
» hold on
» quiver(x,y,dzdx,dzdy)
» hold off
» title('Figure 23.7: Gradient Arrow Plot')
```

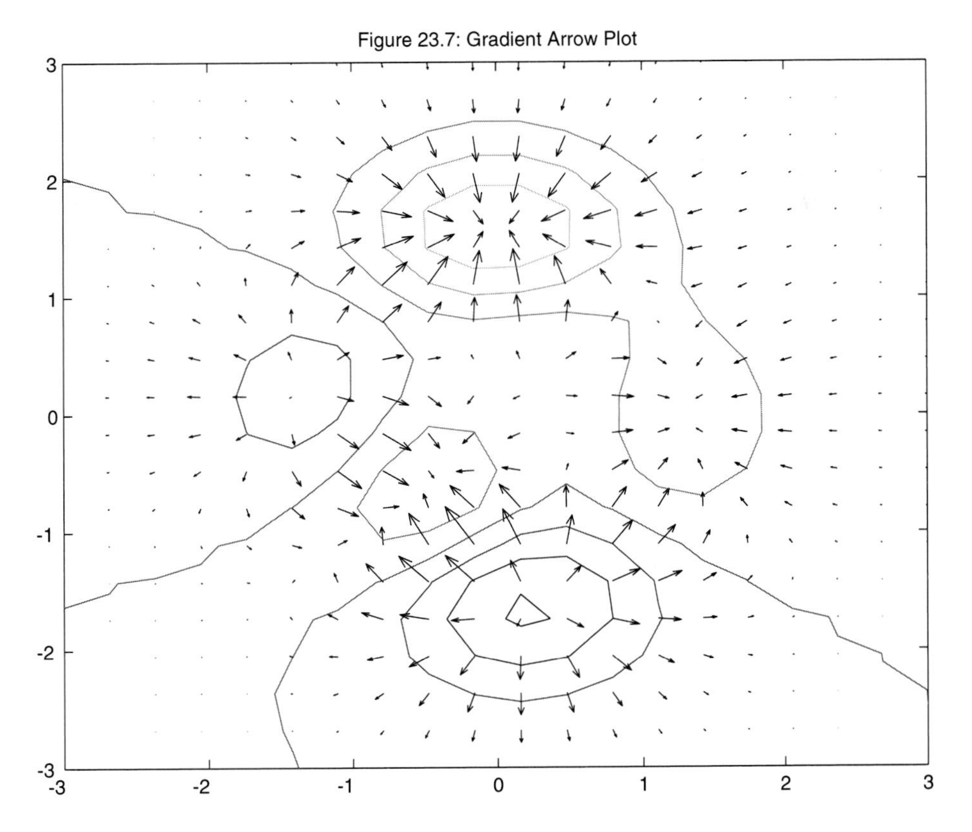

In this example, `gradient` computes *dz/dx* and *dz/dy* from the tabulated data output of `peaks`. These data are supplied to the function `quiver`, which draws arrows normal to the underlying surface with lengths scaled by the slope at each point.

In addition to the gradient, it is sometimes useful to know the curvature of a surface. The curvature or change in slope at each point is computed by the function `del2`, which computes the discrete approximation to the Laplacian

$$\nabla^2 z = \frac{d^2 z}{dx^2} + \frac{d^2 z}{dy^2}$$

In its simplest form, this is computed by taking each surface element and subtracting from it the average of its four neighbors. If the surface is flat in each direction at a given point, the element does not change. In MATLAB version 5, central second differences are used at interior points to produce more accurate results. Consider the following example, where surface curvature influences the color of the surface:

```
» [x,y,z] = peaks; % default output of peaks

» dx = x(1,2) - x(1,1); % spacing in x direction
» dy = y(2,1) - y(1,1); % spacing in y direction

» L = del2(z,dx,dy);

» surf(x,y,z,L)
» shading interp
» title('Figure 23.8: Discrete Laplacian Color')
```

Figure 23.8: Discrete Laplacian Color

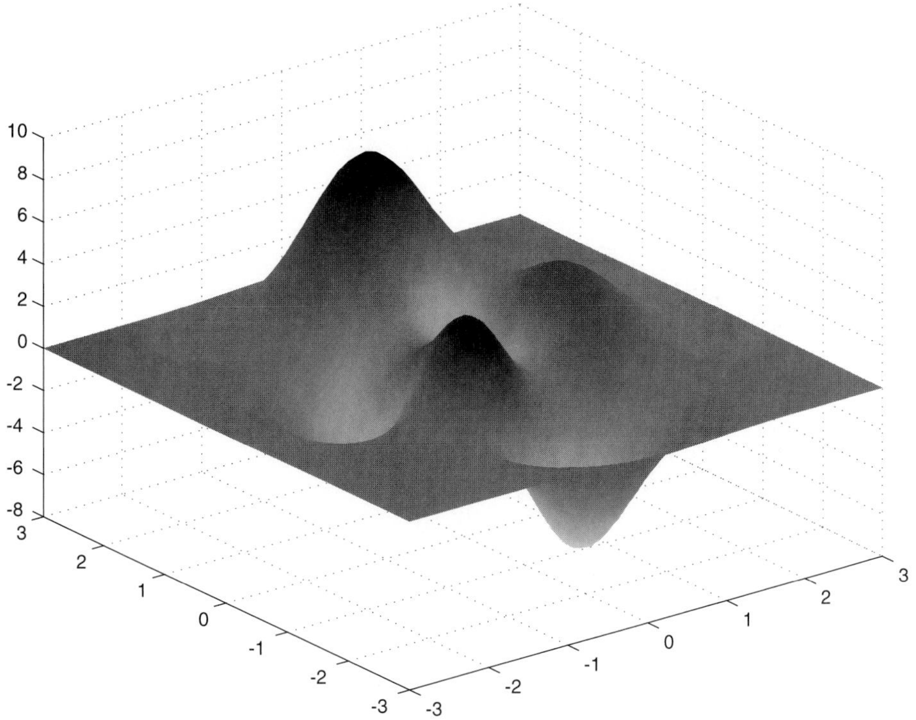

# 24

# Ordinary Differential Equations

In 1995, MATLAB introduced a collection of M-files, called The MATLAB ODE suite, for solving ordinary differential equations (ODEs). These files superseded the original `ode23` and `ode45` solver functions with five new solvers and a host of associated utility functions. With the introduction of MATLAB 5, the MATLAB ODE suite became a standard part of MATLAB. The new suite of functions offers solvers suitable for just about any problem type. Before discussing ODE solver details, you can explore the MATLAB ODE suite by using the GUI-based ODE demo, which is accessed by typing `odedemo` at the MATLAB prompt.

## 24.1 INITIAL VALUE PROBLEM FORMAT

The MATLAB ODE suite computes the time history of a set of coupled, first-order differential equations with known initial conditions. In mathematical terms, these problems are called initial value problems and have the form

$$\dot{y} = f(t, y) \quad y(t_0) = y_0$$

which is vector notation for the set of differential equations

$$\dot{y}_1 = f_1(t, y_1, y_2, \ldots, y_n) \qquad y_1(t_0) = y_{10}$$
$$\dot{y}_2 = f_2(t, y_1, y_2, \ldots, y_n) \qquad y_2(t_0) = y_{20}$$
$$\vdots \qquad\qquad\qquad \vdots$$
$$\dot{y}_n = f_n(t, y_1, y_2, \ldots, y_n) \qquad y_n(t_0) = y_{n0}$$

where $\dot{y}_i = dy_i/dt$, $n$ is the number of first-order differential equations, and $y_{i0}$ is the initial condition associated with the $i$th equation. When an initial value problem is not specified as a set of first-order differential equations, it must be rewritten as one. For example, consider the classic van der Pol equation:

$$\ddot{x} - \mu(1 - x^2)\dot{x} + x = 0$$

where $\mu$ is a parameter greater than zero. If we choose $y_1 = x$ and $y_2 = dx/dt$, then the van der Pol equation becomes

$$\dot{y}_1 = y_2$$
$$\dot{y}_2 = \mu(1 - y_1^2)y_2 - y_1$$

This initial value problem will be used throughout this chapter to demonstrate aspects of the MATLAB ODE suite.

## 24.2 ODE SUITE SOLVERS

The MATLAB ODE suite offers five initial value problem solvers. Each has characteristics appropriate for different initial value problems. The calling syntax for each solver is identical, making it relatively easy to change solvers for a given problem. A description of each solver is given in the following table.

| Solver | Description |
|---|---|
| `ode23` | An explicit, one-step Runge-Kutta low-order (2–3) solver. Suitable for problems that exhibit mild stiffness, problems where lower accuracy is acceptable, or problems where $f(t, \mathbf{y})$ is not smooth (e.g., discontinuous). |
| `ode45` | An explicit, one-step Runge-Kutta medium-order (4–5) solver. Suitable for nonstiff problems that require moderate accuracy. *This is typically the first solver to try on a new problem.* |
| `ode113` | A multistep Adams-Bashforth-Moulton PECE solver of varying order (1–13). Suitable for nonstiff problems that require moderate to high accuracy involving problems where $f(t, \mathbf{y})$ is expensive to compute. Not suitable for problems where $f(t, \mathbf{y})$ is not smooth (i.e., where it is discontinuous or has discontinuous lower-order derivatives). |
| `ode23s` | An implicit, one-step modified Rosenbrock solver of order 2. Suitable for stiff problems where lower accuracy is acceptable, or where $f(t, \mathbf{y})$ is discontinuous. *Stiff problems are generally described as problems where the underlying time constants vary by several orders of magnitude or more.* |
| `ode15s` | An implicit, multistep numerical differentiation solver of varying order (1–5). Suitable for stiff problems that require moderate accuracy. *This is typically the solver to try if* `ode45` *fails or is too inefficient.* |

This table uses terminology (e.g., explicit, implicit, stiff, etc.) that requires a substantial theoretical background to understand. If you understand the terminology, the preceding table describes the basic properties of each solver. If you do not understand the terminology, just follow the guidelines presented in the table where `ode45` and `ode15s` are the first and second solvers to try on a given problem.

It is important to note that the MATLAB ODE suite is provided as a set of M-files that can be viewed. In addition, these same solvers are included internally in SIMULINK for the simulation of dynamic systems.

## 24.3 BASIC USE

Before a set of differential equations can be solved, they must be coded in a function M-file as `ydot=odefile(t,y)`. That is, the file must accept a time `t` and solution `y` and return values for the derivatives. For the van der Pol equation, this ODE file can be written as follows:

```
function ydot=vdpol(t,y)
%VDPOL van der Pol equation.
% Ydot=VDPOL(t,Y)
% Ydot(1) = Y(2)
% Ydot(2) = mu*(1-Y(1)^2)*Y(2)-Y(1)
% mu = 2

mu = 2;
ydot = [y(2); mu*(1-y(1)^2)*y(2)-y(1)];
```

Note that the input arguments are `t` and `y`, but the function does not use `t`. Note also that the output `ydot` must be a column vector.

Given the preceding ODE file, this set of ODEs is solved using the following commands:

```
» tspan = [0 20]; % time span to integrate over
» yo = [2; 0];     % initial conditions (must be a column)
» [t,y] = ode45('vdpol',tspan,yo);
» size(t)          % number of time points
ans =
   333     1
» size(y)          % (i)th column is y(i) at t(i)
ans =
   333     2
» plot(t,y(:,1),t,y(:,2),'--')
» xlabel ('time')
» title('Figure 24.1: van der Pol Solution')
```

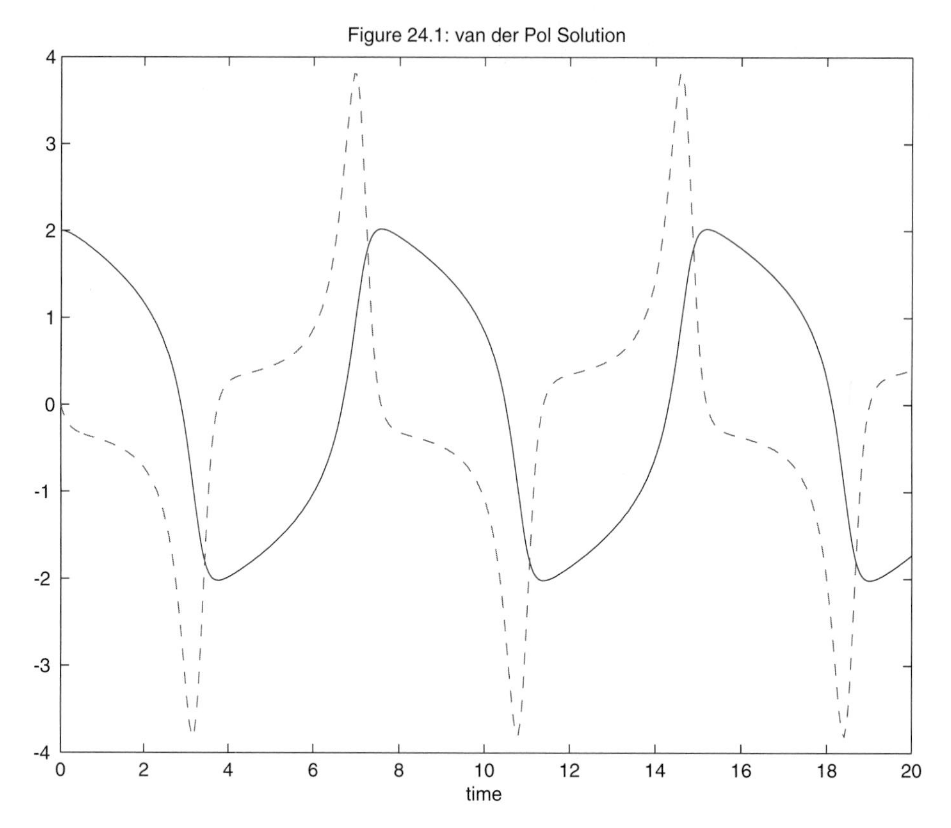


By default, if a solver is called with no output arguments—for example,

```
ode45('vdpol',tspan,yo)
```

the solver generates no output variables, but generates a time plot similar to that shown in the above figure.

In addition to specifying the initial and final time points in `tspan`, one can identify the specific solution time points desired by simply adding them to `tspan`. For example,

```
» tspan = linspace(0,20,100);
» [t,y] = ode45('vdpol',tspan,yo);
» size(t)
ans =
   100     1
» size(y)
ans =
   100     2
```

Here 100 points are gathered over the same time interval as the earlier example. When called in this way, the solver still uses automatic step size control to maintain accuracy. ***It does not***

***used fixed step integration.*** To find the solution at the desired points, the solver interpolates its own solution in an efficient way that does not deteriorate the solution's accuracy.

## 24.4 ODE FILE OPTIONS

In addition to the simple ODE file and calling syntax illustrated previously, the MATLAB ODE suite provides a variety of optional features and approaches to specifying an ODE file for solution. Some of these features are described in the help text for `odefile` (i.e., `helpwin odefile`).

Rather than passing the time span and initial conditions as input arguments to a solver, it is possible to store the time span `tspan` and initial conditions `yo` in the ODE file itself. To do so, one must modify the ODE file to accept three inputs and three outputs. For the van der Pol example the ODE file becomes as follows:

```
function [out1,out2,out3]=vdpol(t,y,flag)
%VDPOL van der Pol equation.
% Ydot=VDPOL(t,Y)
% Ydot(1) = Y(2)
% Ydot(2) = mu*(1-Y(1)^2)*Y(2)-Y(1)
% mu = 2

if nargin<3 | isempty(flag) % return ydot
   mu = 2;
   out1 = [y(2); mu*(1-y(1)^2)*y(2)-y(1)];

elseif strcmp(flag,'init') % set initial parameters
   out1 = [0 20]; % tspan
   out2 = [2; 0]; % initial conditions
   out3 = [];     % solver options

else
   error(['Unknown flag: ' flag])
end
```

According to the above ODE file, the solver calls the ODE file as

```
[tspan,yo,options] = vdpol([],[],'init')
```

to get the time span, initial conditions, and solver options (which will be discussed later in this chapter). When the ODE file is coded in this way, the differential equations can be solved using the syntax

```
» ode45('vdpol')
```

or

```
» [t,y] = ode45('vdpol');
```

Often a set of ordinary differential equations has one or more parameters that can be chosen. In the van der Pol example, $\mu$ is a parameter that changes the fundamental characteristics of the solution. Rather than open the ODE file and change the parameter as desired, it is possible to supply all extra parameters as additional input arguments to the solver. That is, if a solver is called as

```
solver('odefile',tspan,yo,options,P1,P2,...,Pn)
```

it calls the ODE file as

```
odefile(t,y,flag,P1,P2,...,Pn)
```

In doing so, the parameters `P1,P2,...,Pn` are made available to the ODE file. The input argument `options` in the preceding solver calling syntax is an optional structure for setting solver features. This structure will be discussed later in this chapter. For now, it is sufficient to set `options` equal to an empty array.

Modifying the van der Pol example to accommodate $\mu$ as a parameter gives the following:

```
function [out1,out2,out3]=vdpol(t,y,flag,mu)
%VDPOL van der Pol equation.
% Ydot=VDPOL(t,y)
% Ydot(1) = Y(2)
% Ydot(2) = mu*(1-Y(1)^2)*Y(2)-Y(1)
% mu = 2 is default

if nargin< 4 | isempty(mu) % set default mu
   mu = 2;
end
if nargin<3 | isempty(flag) % return ydot
   out1 = [y(2); mu*(1-y(1)^2)*y(2)-y(1)];

elseif strcmp(flag,'init') % set initial parameters
   out1 = [0 20]; % tspan
   out2 = [2; 0]; % initial conditions
   out3 = [];     % default solver options

else
   error(['Unknown flag: ' flag])
end
```

Now the differential equations can be solved using the syntax

```
» mu = 10;
» ode45('vdpol',[],[],[],mu)
```

where empty array inputs for `tspan`, `yo`, and `options` tell the solver to call the ODE file to get them as described earlier and a value of $\mu = 10$ is passed to `vdpol`.

Given the ability to change $\mu$, let's compare results for various $\mu$ values:

```
» tic, [t,y] = ode45('vdpol',[],[],[],2); toc
elapsed_time =
     0.58239
» tic, [t,y] = ode45('vdpol',[],[],[],20); toc
elapsed_time =
      1.5772
» tic, [t,y] = ode45('vdpol',[],[],[],200); toc
elapsed_time =
      24.128
```

As $\mu$ increases, it becomes increasingly difficult for `ode45` to compute the solution. The reason for this is that the van der Pol equation becomes increasingly ***stiff*** as $\mu$ increases. To increase the solution efficiency when $\mu$ is large, a different solver should be tried. For example,

```
» tic, [t,y] = ode23('vdpol',[],[],[],200); toc
elapsed_time =
      13.293
» tic, [t,y] = ode15s('vdpol',[],[],[],200); toc
elapsed_time =
     0.18255
```

In the first case, `ode23`, a lower-order Runge-Kutta algorithm, is used. Being lower order, it handles moderate stiffness better than `ode45` and is approximately twice as fast as `ode45`. Better yet is the stiff solver `ode15s`. It is more than two orders of magnitude more efficient than `ode45`.

If the stiff solver `ode15s` works better for stiff problems, how well does it work for nonstiff problems? Testing it on our van der Pol example with $\mu = 2$ produces

```
» tic, [t,y] = ode15s('vdpol',[],[],[],2); toc
elapsed_time =
      1.3517
» tic, [t,y] = ode45('vdpol',[],[],[],2); toc
elapsed_time =
     0.58041
» tic, [t,y] = ode23('vdpol',[],[],[],2); toc
elapsed_time =
     0.63304
```

As expected, `ode45` works better for nonstiff problems. This example clearly demonstrates that it may be beneficial to test more than one solver on a given problem and that the same problem under different circumstances may benefit from a different solver.

## 24.5 SOLVER OPTIONS

Up to this point we have just accepted all default tolerances and options. When these are not sufficient, an options structure can be passed to a solver as a fourth input argument, or included in the ODE file when it is called to return initialization information. The MATLAB ODE suite contains the functions `odeset` and `odeget` to manage this options structure. `odeset` works similar to the Handle Graphics `set` function in that parameters are specified in name/value pairs. For example,

```
options = odeset('Name1',Value1,'Name2',...);
```

The available parameter names and values are described in the following table.

| Name: Values {default} | Description |
|---|---|
| `RelTol:` `[positive scalar,{1e-3}]` | Relative error tolerance for all solution components. The estimated error at each step satisfies `e(i) <= max(RelTol*abs(y(i)),AbsTol(i))` |
| `AbsTol:` `[positive scalar or vector, {1e-6}]` | Absolute error tolerance for each solution component. If a scalar, it applies to all components. Controls error when solution components are near zero. |
| `Refine:` `[positive integer, {1 for all solvers except ode45, where it is 4}]` | Output refinement factor. Increases the number of output points to produce smoother output by interpolating the internal solution by the specified factor using continuous extension formulas. For example, a value of 4 produces four equally spaced output points for each successful integration step. Does not apply when `length(tspan)>2`. |
| `OutputFcn:` `[string name of user's output function]` | Output function called by the solver after each time step. When solver is called with output arguments, i.e., `[t,y]=solver(...),OutputFcn = ''` (empty string). When solver is called with no output arguments, `OutputFcn` defaults to `'odeplot'`. The output function, e.g., `'myfun'`, used is called as `myfun(tspan,yo,'init')` before integration begins. After each integration step it is called as `stop=myfun(t,y)`, where stop is a logical output that stops the integration when returned as True. Output functions in the MATLAB ODE suite include `odeplot` (times series plot), `odephase2` (2-D phase plane plot), `odephase3` (3-D phase plane plot), and `odeprint` (print solution to *Command* window.) |
| `OutputSel:` `[vector of integer indices, {1:length(y)}]` | Output selection indices. Indices of solution components that are passed to `OutputFcn`. For example, `odeset('OutputSel',[1 3 4])` passes the first, third, and fourth solution components to the output function specified by `'OutputFcn'`. |

| Name: Values {default} | Description |
|---|---|
| `Stats: ['on' {'off'}]` | Display cost statistics at job completion. |
| `Events: ['on' {'off'}]` | Event location desired. If `'on'`, `odefile(t,y,'events')` returns event function results. |
| `MaxStep: [positive scalar]` | Maximum integration step size. Default is one tenth of total time span. Not to be used to produce more output points; adjust `'Refine'` instead. Not to be used to increase accuracy; adjust `'RelTol'` and `'AbsTol'` instead. `'MaxStep'` is useful for oscillatory solutions to guarantee that the solver does not jump over periods of the solution. That is, it could be set to some fraction; e.g., one fifth of the period of the solution. |
| `InitialStep: [positive scalar]` | Suggested initial step size. If it fails, the solver automatically chooses a better one. |
| `NormControl: ['on' {'off'}]` | Control error relative to norm of solution. If `'on'`, error satisfies the weaker criteria `norm(e) <= max(RelTol*norm(y),AbsTol)`. |
| `Jacobian: ['on' {'off'}]` | Analytic Jacobian availability from ODE file. Applies only to stiff solvers `ode23s` and `ode15s`. `odefile(t,y,'jacobian')` returns Jacobian. |
| `JConstant: ['on' {'off'}]` | Constant Jacobian matrix. Set to `'on'` if Jacobian is not a function of `t` or `y`. Applies only to stiff solvers `ode23s` and `ode15s`. |
| `JPattern: ['on' {'off'}]` | Jacobian sparsity pattern available from ODE file. Applies only to stiff solvers `ode23s` and `ode15s`. `odefile(t,y,'jpattern')` returns sparse matrix with ones showing the nonzero pattern of the Jacobian. |
| `Vectorized: ['on' {'off'}]` | Vectorized ODE file for fast numeric Jacobian computation. If `'on'`,`odefile(t,[y1 y2 ...])` returns multiple columns with *i*th column being `odefile(t,yi)`. |
| `Mass: ['on' {'off'}]` | Mass matrix available from ODE file. If `'on'`, `odefile(t,[],'mass')` returns mass matrix $\boldsymbol{M}(t)$. Applies only to stiff solvers `ode23s` and `ode15s`. ODEs with a mass matrix have the form $\boldsymbol{M}(t)\dot{\boldsymbol{y}} = f(t, y)$, where $\boldsymbol{M}(t)$ is nonsingular and usually sparse. |
| `MassConstant: ['on' {'off'}]` | Constant mass matrix. If `'on'`, mass matrix is not a function of time. Applies only to stiff solvers `ode23s` and `ode15s`. `'on'` is required and default in `ode23s`, `'off'` is default in `ode15s`. |
| `MaxOrder: [positive integer less than 6, {5}]` | Maximum order used in `ode15s`. |

The first ten options in the preceding table apply to all solvers, whereas the remainder apply only to the stiff solvers ode23s and ode15s. Before considering the first ten options, consider the general use of the remaining options. The stiff solvers ode23s and ode15s must solve a set of nonlinear equations at each step that involves the Jacobian of the ODEs, which is a matrix of the form.

$$\begin{bmatrix} \frac{\partial f_1}{\partial y_1} & \frac{\partial f_1}{\partial y_2} & \cdots & \frac{\partial f_1}{\partial y_n} \\ \frac{\partial f_2}{\partial y_1} & \frac{\partial f_2}{\partial y_2} & \cdots & \frac{\partial f_2}{\partial y_n} \\ \vdots & \vdots & \ddots & \cdots \\ \frac{\partial f_n}{\partial y_1} & \frac{\partial f_n}{\partial y_2} & \cdots & \frac{\partial f_n}{\partial y_n} \end{bmatrix}$$

By default, the solvers numerically compute the Jacobian by finite differences that require calling the ODE file repeatedly. By vectorizing the ODE file as described in the table, the effort required to compute the Jacobian is dramatically reduced. Vectorizing the ODE file usually means replacing `y(i)` with `y(i,:)` and using array operators (e.g., `.*` and `./`), as shown in the following modified vdpol:

```
function [out1,out2,out3]=vdpol(t,y,flag,mu)
%VDPOL van der Pol equation.
% Ydot=VDPOL(t,Y)
% Ydot(1)  = Y(2)
% Ydot(2)  = mu*(1-Y(1)^2)*Y(2)-Y(1)
% mu = 2

if nargin< 4 | is empty(mu) % set default mu
    mu = 2;
end
if nargin<3 | isempty(flag) % return ydot (vectorized)
    out1 = [y(2,:); mu*(1-y(1,:)^2).*y(2,:)-y(1,:)];

elseif strcmp(flag,'init') % set initial parameters
    out1 = [0 20]; % tspan
    out2 = [2; 0]; % initial conditions
    out3 = odeset('Vectorized','on');     % turn on Vectorized option

else
    error(['Unknown flag: ' flag])
end
```

The use of out3 when `flag = 'init'` is now apparent. out3 is the options structure returned by odeset, which in this example simply tells the solver that the ODEs have been vectorized.

For simple ODEs, the Jacobian can often be written analytically. When this is possible, the ODE file can be modified to include it as shown in the following van der Pol example.

```
function [out1,out2,out3]=vdpol(t,y,flag,mu)
%VDPOL van der Pol equation.
% Ydot=VDPOL(t,y)
% Ydot(1) = Y(2)
% Ydot(2) = mu*(1-Y(1)^2)*Y(2)-Y(1)
% mu = 2

if nargin< 4 | isempty(mu) % set default mu
   mu = 2;
end
if nargin<3 | isempty(flag) % return ydot (vectorized)
   out1 = [y(2,:); mu*(1-y(1,:)^2).*y(2,:)-y(1,:)];

elseif strcmp(flag,'init') % set initial parameters
   out1 = [0 20]; % tspan
   out2 = [2; 0]; % initial conditions
   out3 = odeset('jacobian','on'); % enable analytic Jacobian

elseif strcmp(flag,'jacobian') % return analytic Jacobian
   out1 = [          0                     1
           (-2*mu*y(1)*y(2)-1) (mu*(1-y(1)^2))];

else
   error(['Unknown flag: ' flag])
end
```

Obviously, there is no need to vectorize the set of equations and provide an analytic Jacobian. However, doing so in this case allows one to test both options.

As described previously, `odeset` returns a MATLAB structure with user-chosen parameters set. Once set, they are passed to the solver within the ODE file, as shown in the last two of the preceding examples, or are passed to the solver as its fourth input argument. For example,

```
» options = odeset('RelTol',1e-4,'AbsTol',1e-8)
options =
          AbsTol: 1e-08
             BDF: []
          Events: []
     InitialStep: []
        Jacobian: []
       JConstant: []
        JPattern: []
            Mass: []
```

```
    MassConstant: []
        MaxOrder: []
         MaxStep: []
     NormControl: []
       OutputFcn: []
       OutputSel: []
          Refine: []
          RelTol: 0.0001
           Stats: []
      Vectorized: []
» [t,y] = ode45('vdpol'),[],[],options);
» size(t)
ans =
   497     1
» [t,y] = ode45('vdpol',[0 20],[2;0],options); % same as above
» size(t)
ans =
   497     1
```

By decreasing the relative and absolute error tolerances, it now takes 497 steps to compute results over 20 seconds as opposed to the 333 steps required for default tolerances. Note that unspecified options are empty. Whenever a solver encounters an empty option, it replaces it with the corresponding default value. For example, when a solver is called with no output arguments, `'OutputFcn'` is set to `'odeplot'`, which is a utility plotting function in the MATLAB ODE suite.

The `'Refine'` property determines how much output data to generate. It does not affect the step sizes chosen by the solvers or the solution accuracy. It merely dictates how many intermediate points to interpolate the solution at within each integration step. For example,

```
» newopts = odeset(options,'Refine',1);
» odeget(newopts,'Refine')
ans =
     1
» [t,y] = ode45('vdpol',[0 20],[2;0],newopts);
» size(t)
ans =
   125     1
```

Now only 125 time points are generated over the same time span. As shown previously, `odeset` allows one to append new options to an old options structure by simply providing the old options structure as the first input argument to `odeset`. In addition, the function `odeget` allows one to extract data from an options structure.

For further information regarding solver options and `odeset`, see MATLAB documentation.

## 24.6 FINDING EVENTS

The `'Events'` property allows one to flag one or more events that occur as an ODE solutions evolves in time. For example, a simple event could be when some solution component reaches a maximum, a minimum, or crosses through zero. Optionally, the occurrence of an event can force a solver to stop integrating.

To make use of this feature one must enable it—for example, `options=odeset('Events','on')`. In addition, one must modify the ODE file so that event location information is returned when the ODE file is called as

```
[values,halt,direction] = odefile(t,y,'events');
```

Here `values` is a vector containing the evaluation of the event functions given `t` and `y`. `halt` is a logical vector the same length as `values` with True `(1)` for those events that are to halt the integration. An event occurs when an element of `values` crosses through zero in a direction dictated by the vector `direction`. `direction` is a vector the same length as `values` with `-1, +1, 0` at indices corresponding to negative, positive, and don't care zero crossings in `values`.

To demonstrate finding events, let's use the van der Pol example to find those points where $\dot{y} = 0$ and $\ddot{y} = 0$, which corresponds to the inflection points in $y_1$ and $y_2$. Implementing this in the ODE file gives the following:

```
function [out1,out2,out3]=vdpol(t,y,flag,mu)
%VDPOL van der Pol equation.
% Ydot=VDPOL(t,Y)
% Ydot(1) = Y(2)
% Ydot(2) = mu*(1-Y(1)^2)*Y(2)-Y(1)
% mu = 2

if nargin< 4 | isempty(mu) % set default mu
   mu = 2;
end
if nargin<3 | isempty(flag) % return ydot (vectorized)
   out1 = [y(2,:); mu*(1-y(1,:)^2).*y(2,:)-y(1,:)];

elseif strcmp(flag,'init') % set initial parameters
   out1 = [0 20] % tspan
   out2 = [2; 0]; % initial conditions
   out3 = odeset('events','on'); % enable event location !!!!!!!

elseif strcmp(flag,'jacobian') % return analytic Jacobian
   out1 = [        0                    1
           (-2*mu*y(1)*y(2)-1) (mu*(1-y(1)^2))];

elseif strcmp(flag,'events') % event location
   out1 = [y(2); mu*(1-y(1,:)^2).*y(2,:)-y(1,:)]; % values
   out2 = [0; 0]; % halt: not at this time
   out3 = [0; 0]; % direction: don't care
else
   error(['Unknown flag: ' flag])
end
```

To find the events, the solver is called as

```
» [t,y,te,ye,ie] = ode45('vdpol');
```

where `te` is a vector of times when the events occurred, `ye` is the ODE solution at the events, and `ie` is a vector of indices identifying which event occurred. This information can be plotted as

```
» e1 = (ie==1); % True where event 1 occurred
» e2 = (ie==2); % True where event 2 occurred
» plot(t,y,te(e1),ye(e1,1),'ok',te(e2),ye(e2,2),'xk')
» xlabel('time')
» title('Figure 24.2: van der Pol Event Location')
```

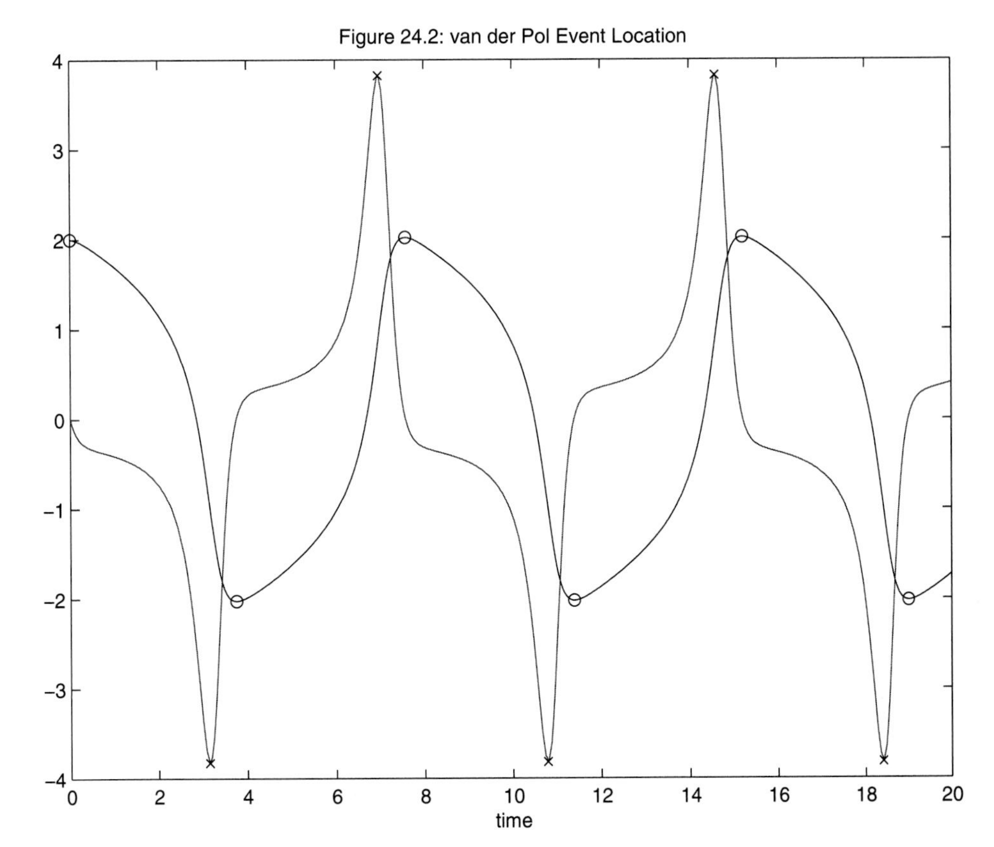


Since the expressions used to compute `values` are arbitrary, and given the options provided by `halt` and `direction`, events can be defined as just about anything.

25

# Object-oriented Programming

Object-oriented programming refers to the practice of defining new data types and a set of operations that can be performed on them while hiding the implementation details from the user. ***Class*** describes the structure of the new data type and the set of operations or functions defined for this data type. An ***object*** is a variable having a specific class. MATLAB has a number of built-in classes, as shown in the following table:

| Class | Description |
|---|---|
| `double` | Double-precision floating-point numeric array. |
| `char` | Character array. |
| `cell` | Cell array. |
| `struct` | Structure array. |
| `unint8` | Unsigned 8-bit integer array. |
| `sparse` | Two-dimensional sparse matrix. |

Other toolboxes supply additional classes to MATLAB. Examples include the `inline` class in the *MATLAB Toolbox,* the `sym` class in the *Symbolic Math Toolbox,* and the `lti` class in the *Control System Toolbox.*

## 25.1 OBJECT IDENTIFICATION

MATLAB supplies a number of functions to assist in the identification of variables. The `class` function identifies the class of a variable. The form `p=class(v)` returns a string in `p` containing the name of the class of the variable or value `v`. For example,

```
» v = 'Hi there.';
» class(v)
ans =
char
```

A related function is the `isa` function. This is a logical function where `tf=isa(v,'class')` returns True (`1`) if `v` is a member of class `'class'`.

```
» isa(v, 'char')
ans =
    1
```

The logical function `isobject` can be used to determine if a variable is an object. The definition of object for *this* purpose is a variable of a class that is ***not*** one of MATLAB's built-in classes, as listed previously. For example, `isobject(A)` returns True (`1`) if `A` is of class `inline` or `sym`, but returns False (`0`) if `A` is of class `double`, `struct`, `cell`, `sparse`, or `char`.

Finally, the `whos` function identifies the class and labels each variable as an array or an object:

```
» whos
Name      Size          Bytes  Class

c         1x17             34  char array
f         1x1            1002  inline object
v         1x1               8  double array
```

## 25.2 CREATING A CLASS

You can add your own classes to MATLAB by defining a structure for each class, a *constructor* function that creates an instance of each class, and *methods* for each class. Methods are the set of functions that implement the operations you define for the class. These methods are implemented as function M-files that are collected in their own subdirectory. This directory must be a subdirectory on the MATLAB search path, but is itself not listed on the path. The directory must have the same name as the constructor function name with an `'@'` prepended to the name, such as `'@myclass'`. This way, MATLAB knows where to look for the methods of a class.

To illustrate the development of a class, consider a simple implementation of a last-in-first-out (LIFO) `stack`. The first step is to create a subdirectory in any directory on the

MATLAB path. A good choice is a subdirectory of your own working directory that has been added to the MATLAB path. Under Unix this might be ~/matlab/@stack. Another possibility might be C:\MATLAB\TOOLBOX\LOCAL\@STACK.

The next step is to define the data structure for a stack object. MATLAB implements classes as structures (data structures of type struct). This example implements the stack datatype as an array of structures containing a variable field v and a name field n. The variable field contains the element pushed onto the stack, while the name field contains a string representation of the original name of the variable element in the *Command* window. This is not the only possible implementation of a stack class; other data structures are equally valid and are only limited by the imagination of the creator.

## 25.3 THE CONSTRUCTOR

To create a new stack class, we need a function to create a stack *object,* or an instance of the stack *class.* This function is known as a ***constructor.*** The name of the function **must** be the name of the class (i.e., the name of the directory containing the M-files for this class without the preceding '@'). This stack constructor @stack/stack.m is as follows:

```
function r = stack(p)
%STACK Constructor for a stack object
%  R=STACK creates an empty stack object R.
%  R=STACK(P) creates a stack object R containing the variable P.

superiorto('double','sparse','struct','cell','char','inline');

if nargin > 1
  error('Too many arguments.');
end

if nargin == 0        % initial call; create an empty stack.
  s.v = [];
  s.n = '';
  s = class(s, 'stack');

elseif isa(p,'stack') % a stack object was supplied; return it.
  s = p;

else              % create a stack containing the input variable.
  s.v = p;
  s.n = inputname(1);
  if isempty(s.n)
    s.n = ['(' class(p) ')'];
  end
  s = class(s,'stack');
end

r = s;
```

This function creates an instance of a `stack` object. If no argument is given, it creates an empty `stack` object. If the argument is already a `stack`, it is simply passed to the output. If any other argument is given, a new `stack` object is created containing the name and contents of the argument variable as its single element. If the argument is a value rather than a named variable (such as a numeric argument or a string literal), the name field of the new `stack` element contains the name of the object class of which it is a member. For example, if the argument is the numeric variable `pi`, then the value field contains `3.1416` and the name field contains the string `'(double)'`.

As shown earlier, when called from the command line or from another function, `c=class(v)` returns a string containing the name of the class of the variable `v`. However, the `class` function ***when called from within a constructor*** is used to assign a tag that identifies the new object type. For example, in `stack.m`, `s=class(s,'stack')` tags `s` as being a `'stack'` object. The form `st=class(s,'`*`cname`*`')` is only valid within a constructor. Additional arguments can be used to inherit methods and fields of parent objects when creating subclasses. These possibilities are discussed later in this chapter.

## 25.4 OBJECT PRECEDENCE

When MATLAB evaluates an expression containing objects of different classes or a function with arguments of different classes, MATLAB makes the assumption that all objects have equal precedence and uses the *method* associated with the leftmost object in the called function input argument list. This is similar to the way MATLAB evaluates expressions containing mathematical operators. For example, given the expression `d op s` or `op(d,s)` where `d` is of class double, `s` is a `stack`, and `op` is an operator or a function, MATLAB calls the method associated with the `double` class to implement the operator or function. If a precedence relationship has been established, MATLAB calls the method for the class with the highest precedence instead.

In this example, we wish to use only `stack` methods when `stack` objects are involved, since most of the standard operators make no sense when applied to stacks. The `superiorto` and `inferiorto` functions are called within a class constructor M-file to establish the desired class precedence. This example establishes the precedence of the `stack` class as superior to all other listed MATLAB classes. As a result, when any expression contains a `stack` object, `stack` methods will *always* be used. This allows any object belonging to any listed class to be pushed onto the stack. If the method is not defined for the class (such as addition involving `stack` objects), MATLAB reports an error.

Note, however, that if you wish to be able to push objects belonging to unlisted classes onto `stack` objects, the object class must be listed. For example, objects of the `sym` class cannot be pushed onto a `stack` unless the `superiorto` function contains the string `'sym'` as one of its arguments. If the *Symbolic Math Toolbox* is not installed on your computer and `'sym'` is listed as an argument to `superiorto`, MATLAB generates an error message.

## 25.5 DISPLAYING OBJECTS

A function named `display` is called whenever an object is returned by a MATLAB statement that is not terminated by a semicolon. Simply put, `display` dictates how the object is displayed in the *Command* window. For example, the `@stack/display.m` function M-file is as follows:

```
function display(s)
%DISPLAY Command window display of stack.

[m,n] = size(s);
var = inputname(1);

if isempty(var)
  fprintf('\nans = \n');
else
  fprintf('\n%s = \n',var);
end

if isempty(s) % is s an empty stack?
  fprintf('\n    [ empty stack ]\n\n')
elseif m*n == 1
  fprintf('\n%s: \n',s.n);
  disp(s.v);
  fprintf('\n');
else
  fprintf('\n    [ %dx%d stack ]\n\n',m,n);
end
```

This `display` function is called if you type the name of a `stack` object in the *Command* window.

> A function named `display` is a required method for any new class in addition to the constructor unless inheritance is used in the class statement. These two functions are the *only* methods required for new classes, although additional methods are usually defined to make use of the new class.

## 25.6 OVERLOADING FUNCTIONS

Notice that the `isempty(s)` function call is used in the preceding `display` M-file. Since MATLAB's built-in `isempty` function returns False for stack objects, we need to ***overload*** the `isempty` function by creating a `@stack/isempty.m` function M-file:

```
function tf = isempty(s)

%ISEMPTY True for empty stack.
%  TF = ISEMPTY(S) returns 1 if S is an empty stack and 0 otherwise.

if nargin ~= 1, error('Too many input arguments.'); end
tf = 0;
if prod(size(s)) == 1
  if isempty(s(1).v) & isempty(s(1).n)
    tf = 1;
  end
end
```

***MATLAB always looks in the appropriate*** `@class` ***directory for a function or method before looking anywhere else on the search path.*** Now, whenever `isempty` is called with a `stack` input argument, the `@stack/isempty` function is used rather than the built-in MATLAB function. The `isempty` function has been *overloaded* for `stack` objects. Note also that object type checking is not necessary here, since `@stack/isempty.m` is only called when the argument is a `stack`.

Note that the `isempty` function is also used within the `@stack/isempty.m` M-file itself. MATLAB chooses which `isempty` function to call based on the class of its argument. Therefore, using `isempty` within the `isempty.m` M-file will not call itself recursively since its arguments are not stack objects.

So far we have defined methods to create an empty `stack` object or a `stack` object containing one element, display the `stack` object, and check for an empty `stack`. Using these basics, consider the following examples:

```
» s1 = stack
s1 =
   [ empty stack ]
» isempty(s1)
ans =
     1
» a = sqrt(5);
» s2 = stack(a)
s2 =
a:
       2.2361
» isempty(s2)
ans =
     0
» class(s1)
ans =
stack
```

```
» whos
  Name        Size           Bytes  Class

  a           1x1                8  double array
  s1          1x1              248  stack object
  s2          1x1              258  stack object
```

As illustrated previously, the `stack` objects are recognized as members of class `stack` and are reported as such by `whos`. The `size` and `length` functions return the correct values for `stacks` as well.

Occasionally, it is desirable to have MATLAB call a built-in MATLAB function rather than an overloaded method. The `builtin` function forces MATLAB to ignore overloaded methods and use MATLAB's built-in version of a function. `builtin` is often used within method M-files to call the original function. The following `@char/cd.m` function M-file is a good example:

```
function wd=cd(str)
%CD Change directory.
%   Implementation for UNIX so that ~ is supported.
%   Copyright (c) 1984-96 by The MathWorks, Inc.
%   $Revision: 1.4 $  $Date: 1996/10/25 21:54:48 $

if (nargin == 0)
  wdd = builtin('cd');
else
  if isunix & ~isempty(str)
    idx = find(str=='~');
    if (~isempty(idx))
      home = getenv('HOME');
      for k = length(idx):-1:1
        str = fullfile(str(1:idx(k)-1),home,str(idx(k)+1:end),'');
      end
    end
  end
  wdd = eval('builtin(''cd'',str)','error(lasterr)');
end

if (nargout > 0)
  wd = wdd;
end
```



The `builtin` function uses the same syntax as the `feval` function discussed earlier. Note also that `builtin` should ***never*** be overloaded.

## 25.7 ADDING STACK ELEMENTS

To be a useful class, methods must be defined to add or *push* an element onto a `stack` and remove or *pop* an element off a stack. The function M-file `@stack/push.m` implements this feature:

```
function s = push(p,r)
%PUSH Pushes a variable onto a stack.
%  PUSH(P,R) pushes the variable P onto stack R.
%  S = PUSH(P,R) copies the resulting stack R into S.

if nargin ~= 2
  error('Push requires two arguments.')
end
if ~isa(r,'stack'), error([inputname(2), ' is not a stack.']); end

% Create a stack object containing P preserving the original variable name.
q = stack(evalin('caller',inputname(1)));

% Push the new object onto the stack.
if ~isempty(r)
  q = [q r];
end

% Copy the new stack to R
assignin('caller',inputname(2),q);

if nargout == 0
 evalin('caller',inputname(2));
else
  s = q;
end
```

This function creates a `stack` object containing one element from the first input argument and places it at the front (or top) of the `stack` specified by the second input argument. The input `stack` is then updated in the calling workspace. Any object except another `stack` may be pushed onto the `stack`, including arrays, structures, and cell arrays. When a `stack` object is pushed onto another `stack`, the two `stacks` are simply concatentated. Since the structure of a `stack` object is visible to M-file functions inside the `@stack` directory, the line

```
q = [q r];
```

where `q` and `r` are both `stacks`, simply pushes the elements of `q` onto the stack `r` and assigns the result to `q`.

## 25.8 COMMUNICATING BETWEEN WORKSPACES

The `push` M-file uses three functions that were discussed in the function M-files chapter (Chapter 14): `assignin`, `evalin`, and `inputname`. These functions allow you to pass information between the workspace of the executing M-file function and the workspace of the *calling* function or the *Command* window workspace.

The `assignin` function assigns a value to a variable in the selected workspace. The `push` function uses the `assignin` command to change the contents of the stack passed as an argument to `push`. If `push` is called from the command line, the `'caller'` workspace is the *Command* window workspace. If `push` is called from another function M-file, the `'caller'` workspace is that of the calling function. Using `'caller'` makes `push` work correctly when called either from the command line or from another function. For example, if `push` is called from the command line as follows

```
» push(stuff,mystack)
```

where `stuff` is a data element or object of some kind and `mystack` is an existing `stack` object, the line

```
assignin('caller',inputname(2),q);
```

in the `push` M-file assigns the contents of `q` to `mystack`, the second argument to `push`, in the *Command* window workspace.

The dual to `assignin` is `evalin`. The `evalin` function evaluates an expression in the selected workspace and returns the result to the *current* workspace. In this example, the line

```
evalin('caller',inputname(2));
```

in the `push` M-file has the same effect as typing

```
» mystack
```

on the command line. It is used to display the result of the `push` operation to the user. The line

```
q = stack(evalin('caller',inputname(1)));
```

creates a `stack` object from the first input argument and passes the name used for the argument *in the calling workspace* to `stack`. This ensures that `stack` will use the correct string in the name field of the `stack` element.

Continuing our example, let us `push` some more elements onto a stack.

```
» clear
» s1 = stack;
```

```
» a = sqrt(5);
» s2 = stack(a);
» b = 'Hi there.';
» c = [1 2 3; 4 5 6];
» d = {c, 'A string', (0:pi/3:pi)};
» e = struct('field1','string','field2',[100 150; -150 100]);
» f = inline('sin(x)*cos(x)');

» push(f,s1) % push f onto stack s1
si =
f:
     Inline function:
     (x) = sin(x)*cos(x)
» push(e,s1) % push e onto stack s1
s1 =
   [ 1x2 stack]

» push(b,s2) % push b onto stack s2
s2 =
   [1x2 stack]
» push(c,s2) % push c onto stack s2
s2 =
   [ 1x3 stack ]
» push(d,s2) % push d onto stack s2
s2 =
    [ 1x4 stack ]
» whos
  Name      Size          Bytes  Class

  a         1x1               8  double array
  b         1x9              18  char array
  c         2x3              48  double array
  d         1x3             372  cell array
  e         1x1             292  struct array
  f         1x1             842  inline object
  s1        1x2            1570  stack object
  s2        1x4            1254  stack object
```

## 25.9 REMOVING STACK ELEMENTS

Now we need a way to remove an element from a `stack`. For that `@stack/pop.m` is useful:

```
function [v,n,r] = pop(s)
%POP Pops a variable from a stack.
%  V = POP(S) pops a variable V from stack S.
%  [V,N] = POP(S) pops a variable V with name N from stack S.
%  [V,N,R] = POP(S) pops a variable V with name N from stack S and
%     returns a copy of the resulting stack in R.

if nargin == 0, error('No stack specified'); end
if nargin > 1, error('Too many input arguments.'); end
if ~isa(s,'stack'), error([inputname(1), ' is not a stack.']); end

if isempty(s)
  q = s;

else
  v1 = s(1).v;
  n1 = s(1).n;
  if prod(size(s)) == 1
    q = stack;
  else
    q = s(2:end);
  end;
end

assignin('caller',inputname(1),q);

if nargout == 0, evalin('caller',inputname(1)), end
if nargout >= 1, v = v1; end
if nargout >= 2, n = n1; end
if nargout == 3, r = q; end
```

Again, `evalin` is used to display the result if no output arguments are specified, and `assignin` is used to update the stack that is passed as the input argument. Continuing our example,

```
» pop(s2) % pop variable d off stack s2
s2 =
   [ 1x3 stack ]
»[v,n] = pop(s2) % pop next variable off stack s2
v =
       1      2      3
       4      5      6

n =
c
```

As expected, the first time `pop` was called, it removed the variable `d` from the stack. The last element pushed is the first element popped. The second time `pop` was called, it returned the value into `v` and the name of the variable as a character string into `n`.

## 25.10 EXAMINING STACK CONTENTS

The `stack` class now contains methods to create and display `stack` objects, `push` and `pop` stack elements, and test for an empty stack. The one thing we cannot do is view the contents of a `stack` without popping the stack. This is a consequence of one of the basic features of object-oriented programming: hiding implementation details such as the `stack` data structure from the user. Therefore, we will create a function `@stack/peek.m` that allows the user to view the contents of a `stack`:

```
function [v,n] = peek(s,m)
%PEEK Display the contents of a stack without altering the stack.
%  PEEK(S) displays the top variable from stack S.
%  PEEK(S,M) where M > 0 displays the top M variables from stack S.
%  PEEK(S,0) or PEEK(S,'all') displays all variables from stack S.
%  V = PEEK(S,...) copies one or more variables from stack S
%    into the cell array V.
%  [V,N] = PEEK(S,...) copies one or more variables from stack S
%    into the cell array V and their names into the cell array N.

if nargin == 0, error('No stack specified'); end
if nargin > 2, error('Too many input arguments.'); end
if ~isa(s,'stack'), error([inputname(1), ' is not a stack.']); end

if nargin == 1
  m = 1;
elseif ischar(m) & strcmp(m,'all')
  m = length(s);
else
  m = fix(m);
  if m < 1, m = length(s); end
  m = min(m,length(s));
end

if nargin == 0
  evalin('caller',inputname(1))
  for i = 1:m
    fprintf('\n%s: \n',s(i).n);
    disp(s(i).v);
    fprintf('\n');
  end
else
  v = cell(1,m);
  v{:} = deal(s.v);
  if nargout == 2
    n = cell(1,m);
    n{:} = deal(s.n);
  end
end
```

Continuing our example, we get

```
» peek(s1)
s1 =
   [ 1x2 stack ]
```

```
e:
    field1: 'string'
    field2: [2x2 double]
» peek(s2,3)
s2 =
   [ 1x2 stack]
b:
Hi there.
a:
        2.2361
```

The first time `peek` was called, it displayed the first element in the stack `s1`. The second time it displayed both elements in the stack `s2`.

```
» push(d,s2)
s2 =
   [ 1x3 stack ]
» peek(s2)
s2 =
   [ 1x3 stack ]
d:
   [2x3 double]    'A string'    [1x4 double]
```

As expected, the first element of `s2`, a cell array, was displayed. This shows the flexibility of the `stack` object. Now we have a reasonably complete set of methods for the `stack` class. One more method will be added to demonstrate the concept of operator overloading.

## 25.11 OVERLOADING OPERATORS

Earlier, the `isempty` function was overloaded for the `stack` class. Operators as well as functions may be overloaded. Each operator in MATLAB has an associated function name that can be used to create overloaded methods. Any operator may be overloaded by creating a function M-file within the `@class` directory with the appropriate name. The following table lists the function names that can be used to override most MATLAB operations.

| Operation | Function | Description |
|---|---|---|
| `a + b` | `plus(a,b)` | Addition. |
| `a - b` | `minus(a,b)` | Subtraction. |
| `-a` | `uminus(a)` | Unary minus. |
| `+a` | `uplus(a)` | Unary plus. |
| `a .* b` | `times(a,b)` | Element-by-element multiplication. |

| Operation | Function | Description |
|---|---|---|
| a * b | mtimes(a,b) | Matrix multiplication. |
| a ./ b | rdivide(a,b) | Right element-by-element division. |
| a .\ b | ldivide(a,b) | Left element-by-element division. |
| a / b | mrdivide(a,b) | Right matrix division. |
| a \ b | mldivide(a,b) | Left matrix division. |
| a .^ b | power(a,b) | Element-by-element power operation. |
| a ^ b | mpower(a,b) | Matrix power operation. |
| a < b | lt(a,b) | Logical less than operation. |
| a > b | gt(a,b) | Logical greater than operation. |
| a <= b | le(a,b) | Logical less than or equal to operation. |
| a > = b | ge(a,b) | Logical greater than or equal to operation. |
| a ~= b | ne(a,b) | Logical not equal to operation. |
| a == b | eq(a,b) | Logical equal to operation. |
| a & b | and(a,b) | Logical AND. |
| a \| b | or(a,b) | Logical OR. |
| ~a | not(a) | Logical NOT. |
| a:b | colon(a,b) | Colon operator. |
| a:c:b | colon(a,c,b) | Colon operator. |
| a' | ctranspose(a) | Complex conjugate transpose. |
| a.' | transpose(a) | Matrix transpose. |
| a | display(a) | Display method. |
| [a b] | horzcat(a,b,...) | Horizontal concatenation. |
| [a; b] | vertcat(a,b,...) | Vertical concatenation. |
| a(s1,s2,...,sn) | subsref(a,s) | Subscripted reference. |
| a(s1,s2,...,sn) = b | subsasgn(a,s,b) | Subscripted assignment. |
| b(a) | subsindex(a,b) | Subscript index. |

On-line help is available for each of the functions shown in the preceding table. Most of these operations make no sense for `stack` objects. One operation that does make sense is the horizontal concatenation operation. The command

```
» stackc = [stacka stackb]
```

should create a new stackc that is a concatenation of stacka and stackb with stacka on top. If stacka is 1-by-4 and stackb is 1-by-4, then stackc should be 1-by-8. It turns out that the standard MATLAB operator will do this for you. The push function takes advantage of this. However, the command

```
» stackc = [stacka; stackb]
```

normally creates a 2-by-4 stack object. As stacks are inherently one dimensional, a two-dimensional stack object makes no sense. There are two options here. The first is to make the operation illegal. This can be done by creating a @stack/vertcat.m function M-file:

```
function c = vertcat(varargin)
%VERTCAT Vertical concatenation of stacks (not allowed).

% This function is called once for EVERY concatenation since
% even horizontal concatenation contains one row.
% A single argument implies a valid row vector such as [s] or [s t].
% Multiple arguments such as [s; t] are illegal.

if nargin == 1
  c = varargin{1};
else
  error('Operation not allowed on stack objects.')
end
```

This function overloads the standard MATLAB vertical concatenation operation for stacks. Note that if @stack/vertcat is called with a single argument, the argument must be a stack with only one row by definition. The argument is returned intact so that horizontal concatenation will work correctly.

The second option is to create a function @stack/vertcat.m that turns vertical concatenation into horizontal concatenation:

```
function c = vertcat(varargin)
%VERTCAT Vertical concatenation of stacks.

%  Stacks must be row vectors. Column vectors are not defined.
c = cat(2,varargin{:});
```

Continuing our example,

```
»[s1 s2]
ans =
   [ 1x5 stack ]
» peek(ans)
ans =
   [ 1x5 stack ]
e:
    field1: 'string'
    field2: [2x2 double]
» [s2;s1]
ans =
   [ 1x5 stack ]
» peek(ans)
ans =
   [ 1x5 stack ]
d:
   [2x3 double]     'A string'    [1x4 double]
» [pi+1,s1]
ans =
   [ 1x3 stack ]
» peek(ans)
ans =
   [ 1x3 stack ]
(double):
       4.1416
```

The first two examples verify that both concatenation functions produced valid stacks. The third example illustrates a feature of the `stack` function. The result of a calculation was pushed onto the stack and the name of the variable became the name of the class of the variable.

The function `methods('`*`class`*`')` or the command line form `methods class` shows all of the methods available for class `'class'`.

```
» methods stack
Methods for class stack:

display isempty peek   pop    push   stack  vertcat
```

In addition, the methods are all available to the `help`, `helpwin`, or `type` commands using the form `class/function` as the argument (e.g., `help stack/push`). Debugging commands are also available for methods (e.g., `dbstop in stack/vertcat`).

## 25.12 CONVERTER FUNCTIONS

A ***converter*** function converts one class into another. The form of the function is `p=classname(q)`, where the argument `q` is not of class `classname`. MATLAB first looks for the `classname` function in the methods directory associated with the object class of which `q` is a member. If `q` is already a member of class `classname`, then the appropriate converter function is the constructor itself, which normally returns the input. For example,

```
» c = 'string';
» double(c)           % convert from char to double
ans =
    115    116    114    105    110    103
» char(ans)           % convert from double to char
ans =
    string
```

In this example, MATLAB first searched for a `@char/double.m` function M-file and used the built-in `double` function when the file was not found. The inline methods directory contains a good example of a char converter function `@inline/char.m`:

```
function str = char(obj)
%CHAR Convert INLINE object to character array.
%   CHAR(FUN) when FUN is an inline function object returns
%   the string that can be used to recreate the object. This
%   is the opposite of the constructor INLINE.
%
%   See also INLINE.

%   Steven L. Eddins, August 1995
%   Copyright (c) 1984-1996 by The MathWorks, Inc.
%   $Revision: 1.2 $  $Date: 1996/10/28 22:27:26 $

str = sprintf('inline(''%s'',',obj.expr);
for k = 1:obj.numArgs
  str = sprintf('%s ''%s'',', str, deblank(obj.args(k,:)));
end
str(end) = [];
str = [str ')'];
```

Reprinted with permission of The MathWorks, Inc., Natick, MA.

This function is called whenever the statement `char(f)` is encountered, where `f` is an `inline` object. It returns the appropriate string for an `inline` object. Standard conversion functions should be included in the methods directory for new classes when such conversions are appropriate. Objects of the `stack` class should not be converted to standard classes since such conversions are of dubious value.

## 25.13 INHERITANCE

One of the key features of object-oriented programming is the concept of *inheritance,* the process by which a *child* class can acquire the structure and methods of one or more *par-*

*ent* classes. MATLAB supports both *simple* inheritance, where the child class inherits from a single parent class, and *multiple* inheritance, where the child inherits from more than one parent. MATLAB also supports *aggregation,* the use of one object as one of the fields of another object.

The `class` statement in a constructor M-file for a child class has the form

```
obj = class(class_structure,'child_class',parent1_obj,parent2_obj,...)
```

where the *parentn_obj* arguments are optional. For example, if we wish to create a new class of object called a `queue` and base it on the `stack` class, the `@queue/queue.m` constructor could contain the following statements:

```
s = stack;    % create an empty stack object to define a parent structure
q.v. = [];    % create an empty queue structure
q.n = '';
q = class(q,'queue',s);  % the queue inherits methods from the stack object
```

If no other methods are defined, a `queue` object inherits the methods of the `stack` class, including `isempty`, `push`, and `display`.

```
» q=queue
q =
    [ empty stack ]
» isempty(q)
ans =
    1
» a = 1;
» push(a,q)
q =
a:
1
```

To create a first-in-first-out (FIFO) `queue`, simply copy the `@stack/pop.m` file to the `@queue` directory and replace this fragment:

```
v1 = s(1).v;
n1 = s(1).n;
if prod(size(s)) == 1
  q = stack;
else
  q = s(2:end);
end
```

with this fragment:

```
v1 = s(end).v;
n1 = s(end).n;
if prod(size(s)) == 1
  q = queue;
else
  q = s(1:end-1);
end
```

to pop from the other end of the row vector. Alternatively, you could use the existing pop function and modify the push function instead. The display function should be modified as well since the word 'stack' in the output is incorrect. The peek function could be used as is or modified as desired depending on your implementation strategy.

As you have seen, MATLAB includes a comprehensive set of tools to support user-defined objects and object-oriented programming. More detailed information is available in MATLAB hard copy documentation and in on-line documentation.

26

# 2-D Graphics

Throughout this text several of MATLAB's graphics features were introduced. In this and the next several chapters, the graphics features in MATLAB will be more rigorously illustrated.

## 26.1 THE plot FUNCTION

As you have seen in earlier examples, the most common function for plotting two-dimensional data is the `plot` function. This versatile function plots sets of data arrays on appropriate axes and connects the points with straight lines. Here is an example you have seen before:

```
» x = linspace(0,2*pi,30);
» y = sin(x);
» plot(x,y), title('Figure 26.1')
```

This example creates 30 data points over $0 \le x \le 2\pi$ to form the horizontal axis of the plot, and creates another vector `y` containing the sine of the data points in `x`. The `plot`

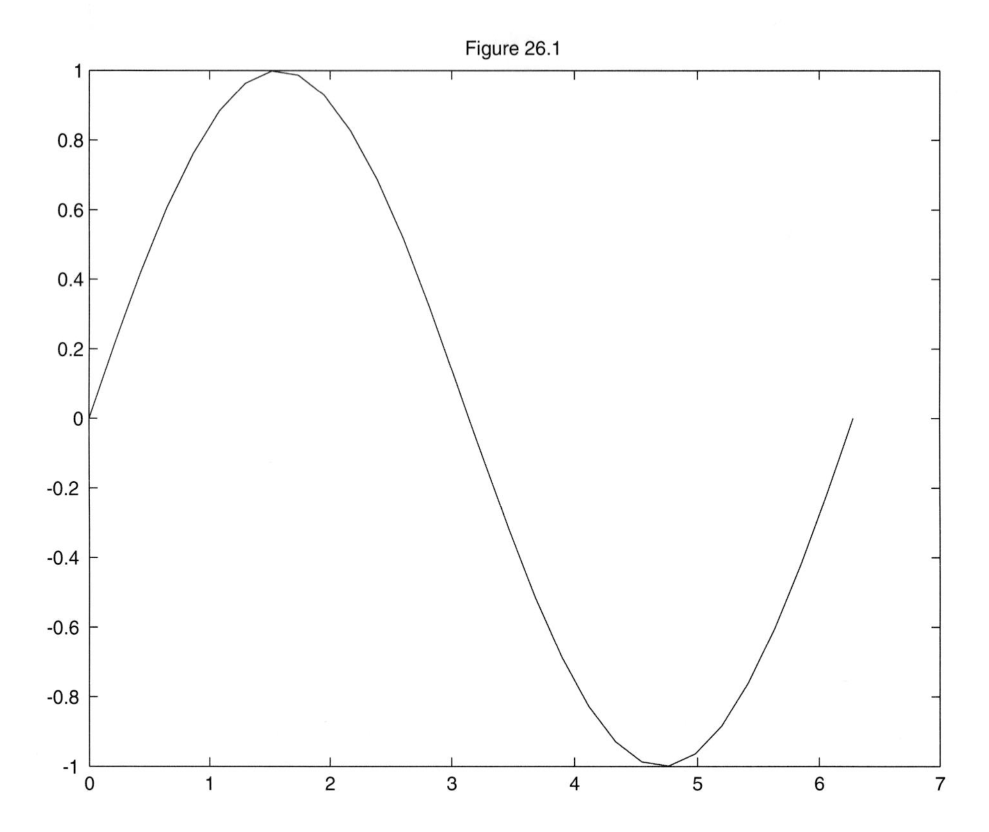


function opens a graphics window, called a *Figure* window, scales the axes to fit the data, plots the points, and then connects the points with straight lines. It also adds numerical scales and tick marks to the axes automatically. If a *Figure* window already exists, `plot` generally clears the current *Figure* window and draws a new plot.

Let's plot a sine and cosine on the same plot:

```
» z = cos(x);
» plot(x,y,x,z), title('Figure 26.2')
```

This example shows that you can plot more than one set of data at the same time, just by giving `plot` another pair of arguments. This time sin($x$) versus $x$ and cos($x$) versus $x$ were plotted on the same plot. `plot` automatically drew the second curve in a different color on the screen. Many curves may be plotted at one time by supplying additional pairs of arguments to plot.

If one of the arguments is a matrix and the other a vector, the `plot` function plots each column of the matrix versus the vector:

```
» W = [y;z]; % create a matrix of the sine and cosine
» plot(x,W)  % plot the columns of W vs. x
» title('Figure 26.3')
```

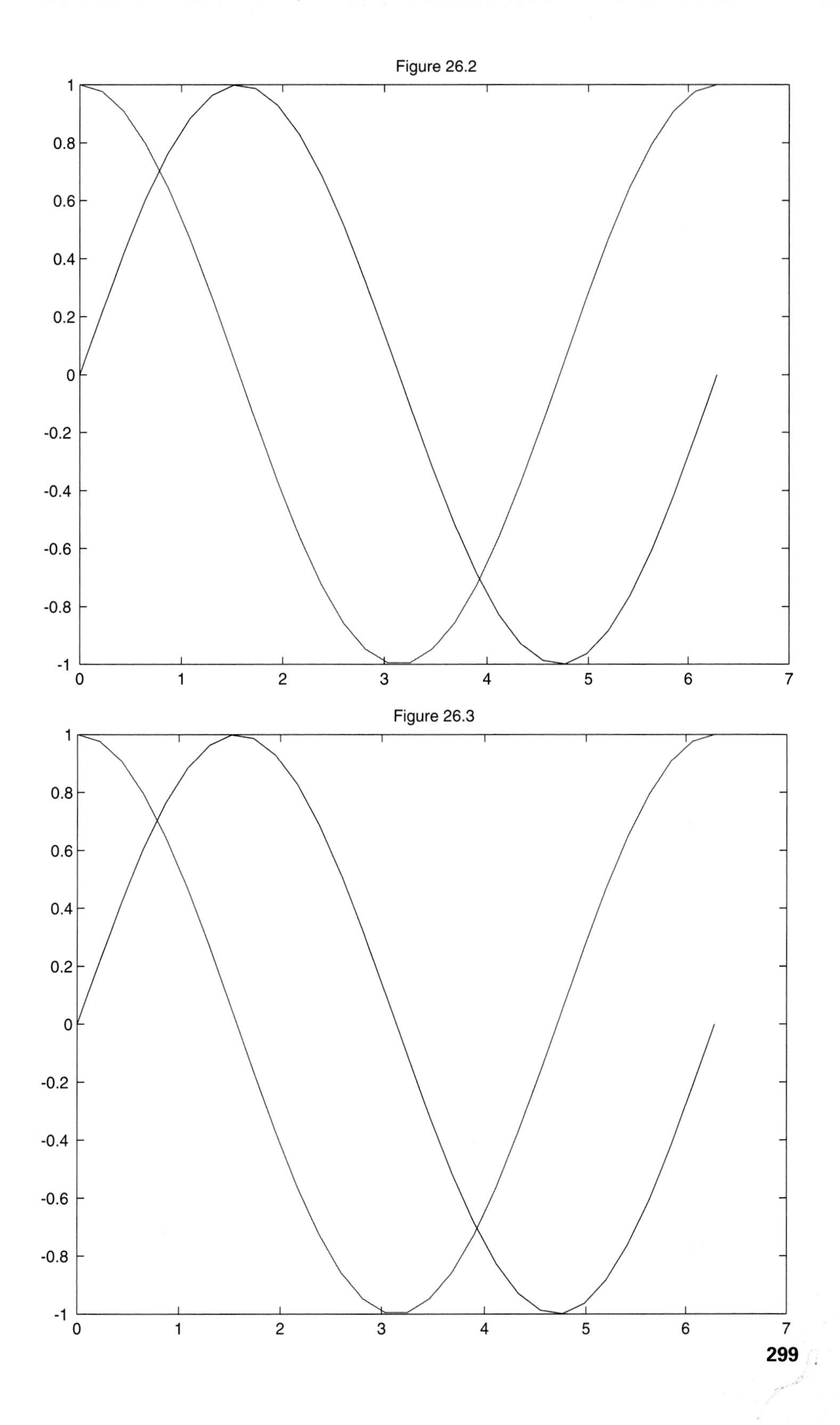


Figure 26.2

Figure 26.3

If you change the order of the arguments, the plot rotates 90 degrees:

```
» plot(W,x) % plot x vs. the columns of W
» title('Figure 26.4')
```

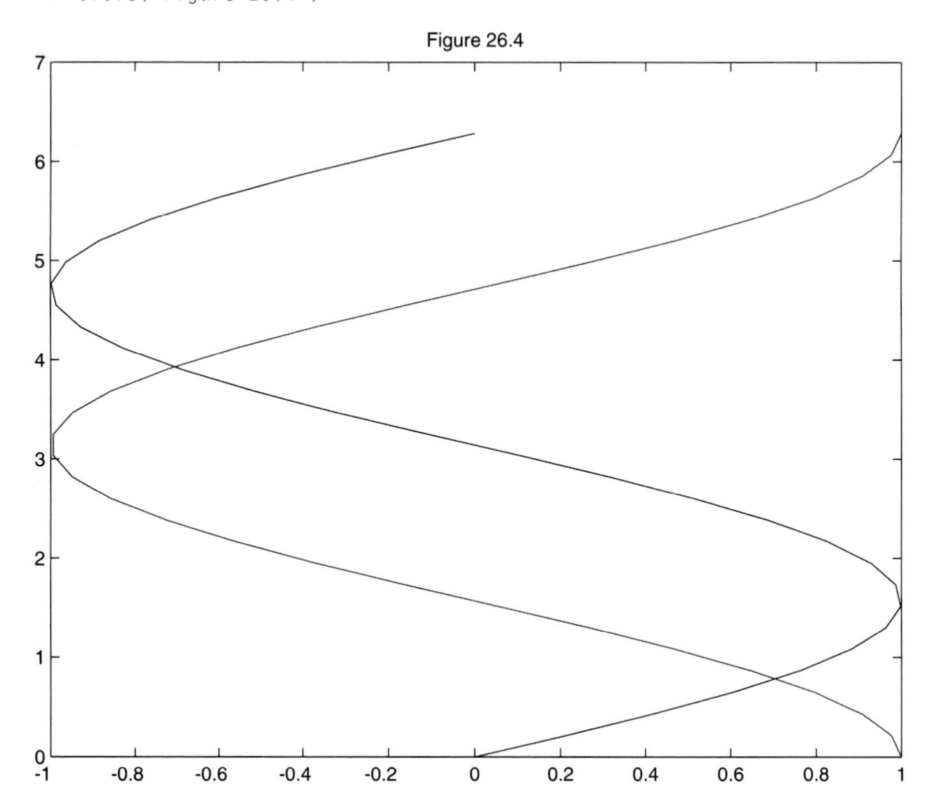


When the `plot` function is called with only one argument—for example, `plot(Y)`—the `plot` function acts differently depending on the data contained in `Y`. If `Y` is a *complex*-valued vector, `plot(Y)` is interpreted as `plot(real(Y),imag(Y))`. In all other cases, the imaginary components of the input vectors are ignored. On the other hand, if `Y` is *real* valued, then `plot(Y)` is interpreted as `plot(1:length(Y),Y)` (i.e., `Y` is plotted versus an index of its values). When `Y` is a matrix, the preceding interpretations are applied to each column of `Y`.

## 26.2 LINESTYLES, MARKERS, AND COLORS

In the previous examples, MATLAB chose the *solid* linestyle and the colors *blue* and *green* for the plots. You can specify your own colors, markers, and linestyles by giving `plot` a third argument after each pair of data arrays. This optional argument is a character string consisting of one or more characters from the following table.

| Symbol | Color | Symbol | Marker | Symbol | Linestyle |
|---|---|---|---|---|---|
| b | blue | . | point | - | solid line |
| g | green | o | circle | : | dotted line |
| r | red | x | cross | -. | dash-dot line |
| c | cyan | + | plus sign | -- | dashed line |
| m | magenta | * | asterisk | | |
| y | yellow | s | square | | |
| k | black | d | diamond | | |
| w | white | v | triangle (down) | | |
| | | ^ | triangle (up) | | |
| | | < | triangle (left) | | |
| | | > | triangle (right) | | |
| | | p | pentagram | | |
| | | h | hexagram | | |

If you do not specify a color and you are using the default color scheme, MATLAB starts with blue and cycles through the first seven colors in the table for each additional line. The default linestyle is the solid line unless you specify a different linestyle. There is no default marker. If no marker is selected, no markers are drawn. Use of any marker places the chosen symbol at each data point, but does not connect the data points with a straight line unless a linestyle is specified as well.

If a color, marker, and linestyle are all included in the string, the color applies to both the marker and the line. To specify a different color for the markers, plot the same data with a different specification string. Here is an example using different linestyles, colors, and point markers:

```
» plot(x,y,'b:p',x,z,'c-',x,z,'m+')
» title('Figure 26.5')
```

As with many of the plots in this section, your computer displays color but the figures printed here do not. If you are following along in MATLAB, just enter the commands listed in the examples to see the effects of color.

## 26.3 PLOTTING STYLES

The `colordef` command selects an overall style for your plots. The default style is `colordef white`. This style uses a white axes background, a light gray figure

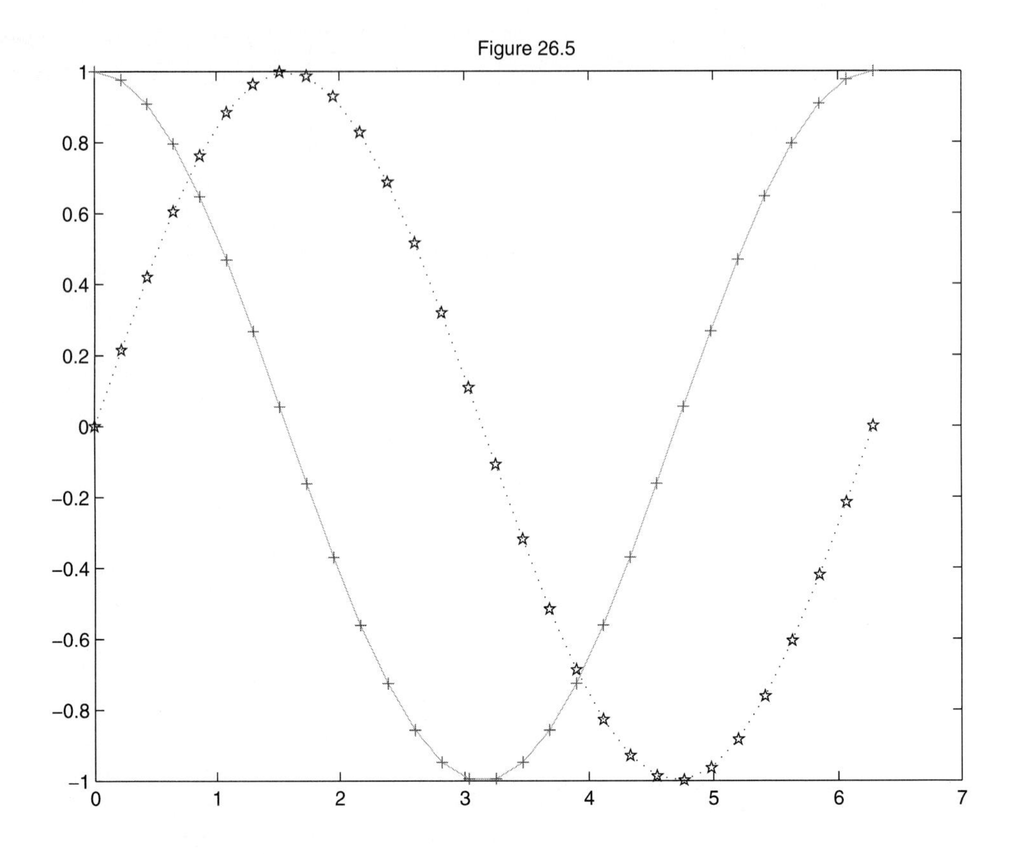


background, black axis labels, the `jet` colormap, and blue, dark green, and red as the first three plot colors. If you like a black background, use `colordef black`. This style uses a black axes background, a dark gray figure background, white axis labels, the `jet` colormap, and yellow, magenta, and cyan as the first three plot colors. If you use `colordef none`, MATLAB will default to the same style that was used in previous versions of MATLAB. This style uses a black axis and figure background color, uses white axis labels, the `hsv` colormap, and yellow, magenta, and cyan as the first three plot colors.

## 26.4 PLOT GRIDS, AXES BOX, AND LABELS

The `grid on` command adds grid lines to the current plot at the tick marks. The `grid off` command removes the grid. `grid` with no arguments alternately turns them on and off (i.e., ***toggles*** them). MATLAB starts with `grid off` for most plots by default. If you like to have grids on all your plots by default, add the following lines to your `startup.m` file:

```
set(0,'DefaultAxesXgrid','on')
set(0,'DefaultAxesYgrid','on')
set(0,'DefaultAxesZgrid','on')
```

The `set` function will be discussed in much more detail in the Handle Graphics chapter (Chapter 31).

Normally, 2-D axes are enclosed by solid lines, called an *axes box*. This box can be turned off with `box off`. `box on` restores the axes box. The `box` command toggles the state of the axes box. Horizontal and vertical axes can be labeled with the `xlabel` and `ylabel` functions, respectively. The `title` function adds a line of text at the top of the plot. Let's use the sine and cosine plot again as an example.

```
>> x = linspace(0,2*pi,30);
>> y = sin(x);
>> z = cos(x);
>> plot(x,y,x,z)
```

Now remove the axes box, and add a plot title and axis labels:

```
>> box off        % turn off the axes box
>> xlabel('Independent Variable X')      % label horizontal axis
>> ylabel('Dependent Variables Y and Z') % label vertical axis
>> title('Figure 26.6: Sine and Cosine Curves')  % title the plot
```

You can add a label or any other text string to any specific location on your plot with the `text` function. The format is `text(x,y,'string')`, where `(x,y)` represents the coordinates of the center left edge of the text string in units taken from the plot axes. To add a label identifying the sine curve at the location `(2.5,0.7)`,

```
>> grid on, box on % turn axes box and grid lines on
>> text(2.5,0.7,'sin(x)')
>> title('Figure 26.7: Sine and Cosine Curves')
```

If you want to add a label but do not want to stop to figure out the coordinates to use, you can place a text string with the mouse. The `gtext` function switches to the current *Figure* window, puts up a crosshair that follows the mouse, and waits for a mouse click or keypress. When either one occurs, the text is placed with the lower left corner of the first character at that location. Try labeling the second curve in the plot:

```
>> gtext('cos(x)')
>> title('Figure 26.8: Sine and Cosine Curves')
```

Note that the general procedure for customizing plots is to issue the `plot` function to generate the default plot and then customize it with ensuing functions.

Figure 26.6: Sine and Cosine Curves

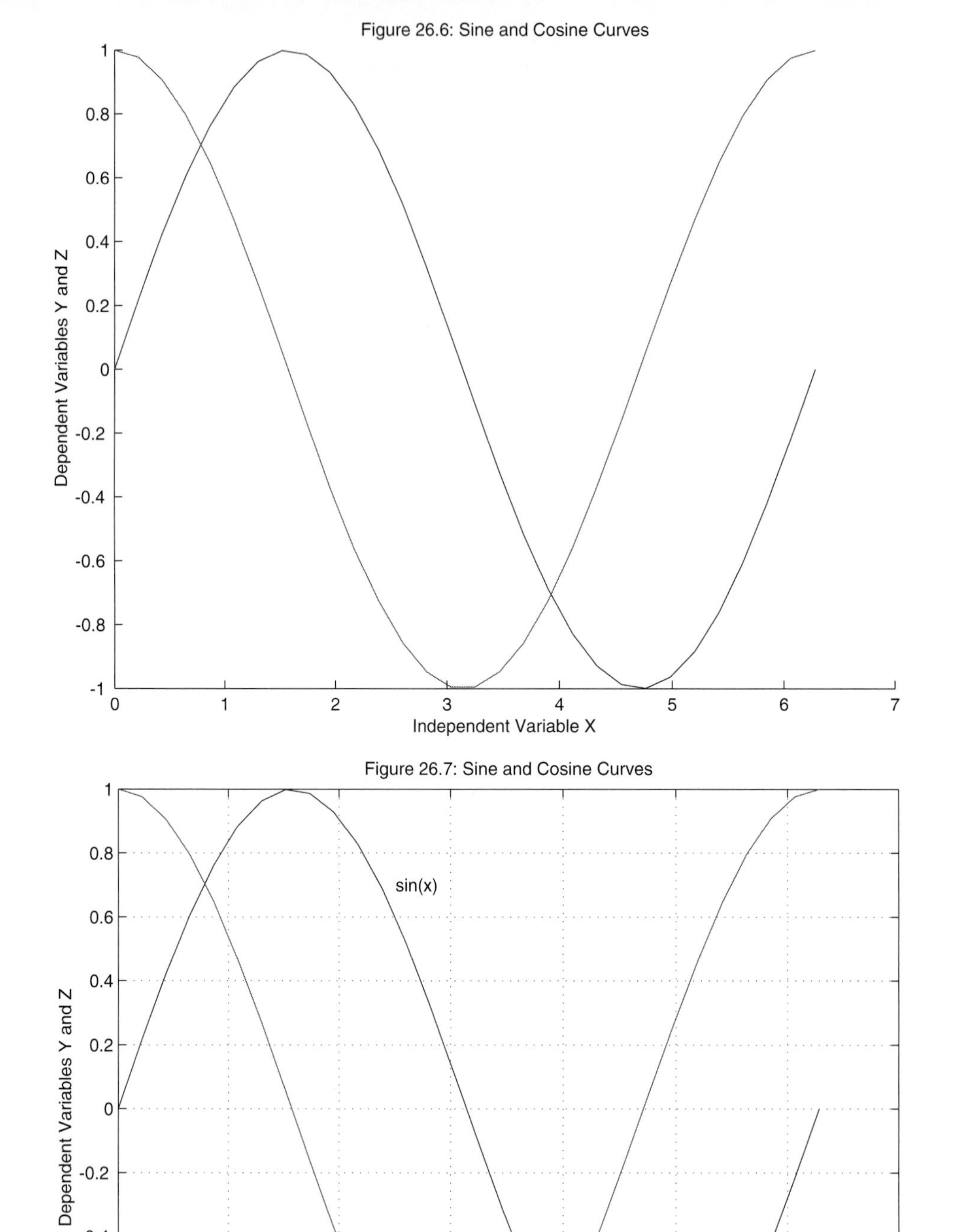


Figure 26.7: Sine and Cosine Curves

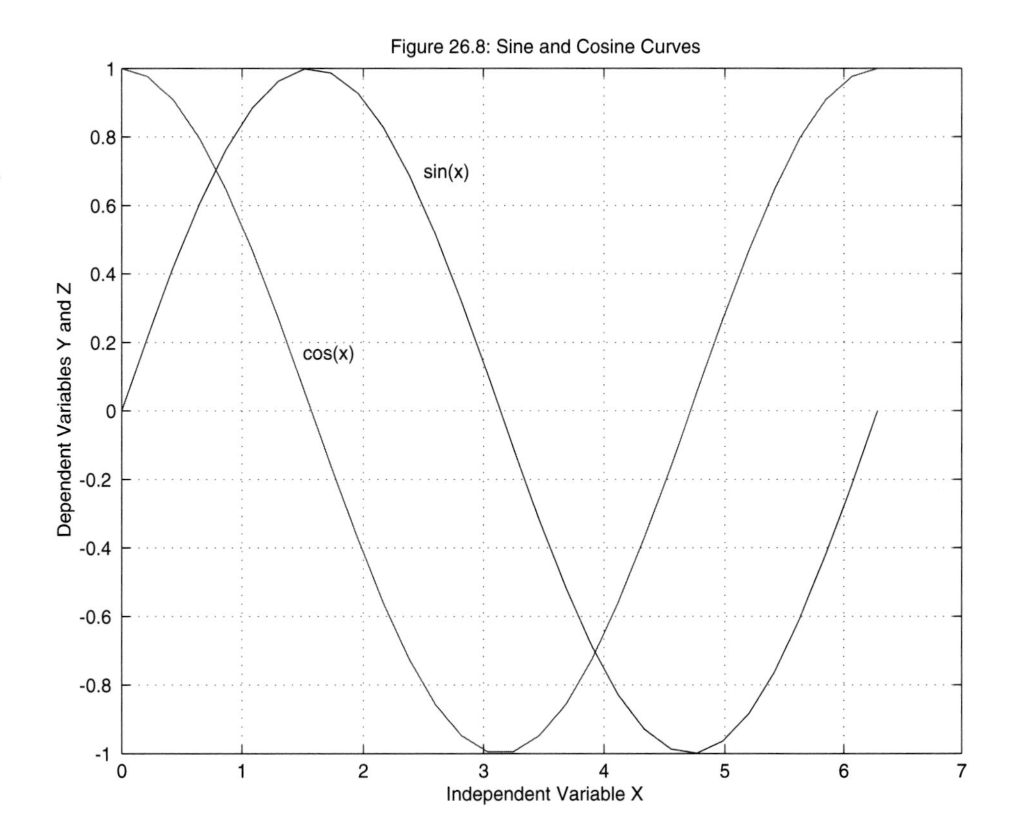


## 26.5 CUSTOMIZING PLOT AXES

MATLAB gives you complete control over the scaling and appearance of both the horizontal and vertical axes of your plot with the `axis` command. Because this command has so many features, only the most useful will be described here. The primary features of the `axis` command are given in the following table.

| Commands | Description |
|---|---|
| `axis([xmin xmax ymin ymax])` | Set the minimum and maximum values of the axes using values given in the row vector. |
| `V = axis` | `V` is a row vector containing the scaling for the current plot: `[xmin xmax ymin ymax]`. |
| `axis auto` | Return the axis scaling to its automatic defaults: `xmin = min(x), xmax = max(x)`, etc. |

| Commands | Description |
|---|---|
| `axis manual` | Freeze scaling at the current limits, so that if `hold` is turned on, subsequent plots use the same axis limits. |
| `axis xy` | Use the (default) cartesian coordinate form, where the *system origin* (the smallest coordinate pair) is at the lower left corner. The horizontal axis increases left to right, and the vertical axis increases bottom to top. |
| `axis ij` | Use the *matrix* coordinate form, where the system origin is at the top left corner. The horizontal axis increases left to right, but the vertical axis increases top to bottom. |
| `axis square` | Set the current plot to be a square rather than the default rectangle. |
| `axis equal` | Set the scaling factors for both axes to be equal. |
| `axis tight` | Set axis limits to the range of the data. |
| `axis vis3d` | Keeps MATLAB from altering the proportions of axes if the view is changed. |
| `axis normal` | Turn off `axis square`, `equal`, `tight`, and `vis3d`. |
| `axis off` | Turn off axis background, labeling, grid, box, and tic marks. Leave the title and any labels placed by the `text` and `gtext` functions. |
| `axis on` | Turn on axis background, labeling, tic marks, and box and grid if they are enabled. |

Multiple commands to `axis` can be given at once. For example, `axis auto on xy` is the default axis scaling. The `axis` command affects the current plot only. Therefore, it is issued after the `plot` command just as `grid`, `xlabel`, `ylabel`, `title`, `text`, etc. are issued after the plot is on the screen. For example,

```
» x = linspace(0,2*pi,30); y = sin(x);
» plot(x,y), title('Figure 26.9')
» axis([0 2*pi -1.5 2])  % change axis limits
```

Note that by specifying the maximum x-axis value to be `2*pi`, the plot axis ends at exactly `2*pi`, rather than rounding the axis limit up to 7.

Try out some of the `axis` commands on your plots. Using the previous example plot produces the following results:

```
» axis off            % turn off the axes
» title('Figure 26.10: Sine and Cosine Curves')
```

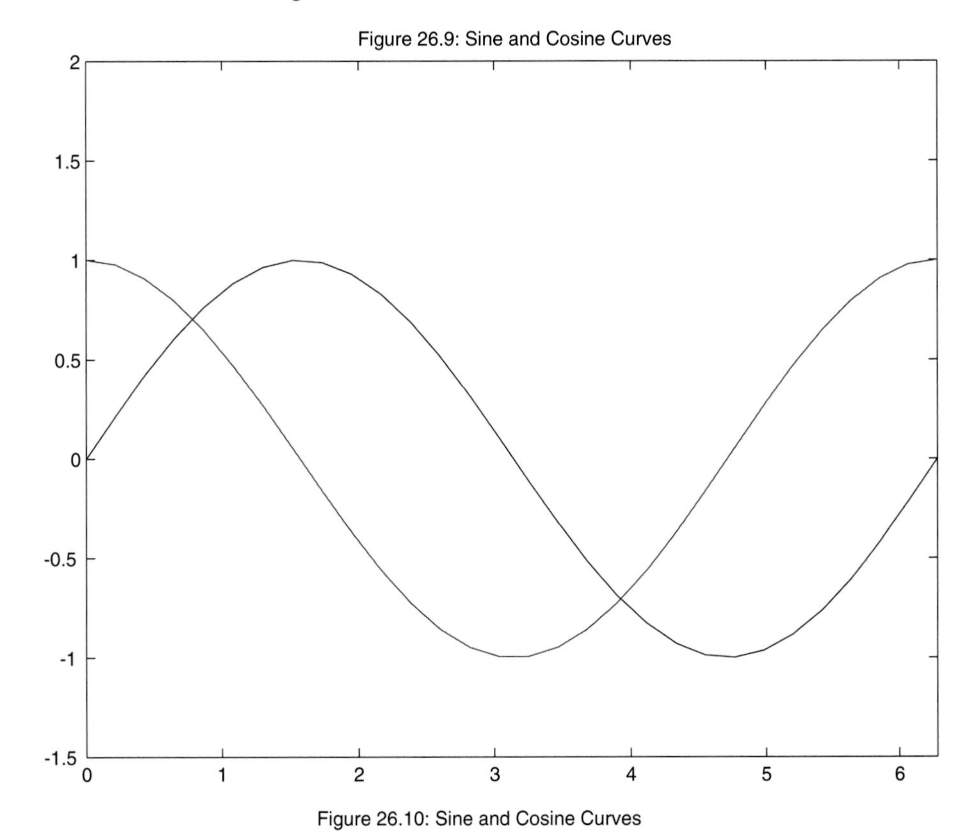


Figure 26.10: Sine and Cosine Curves

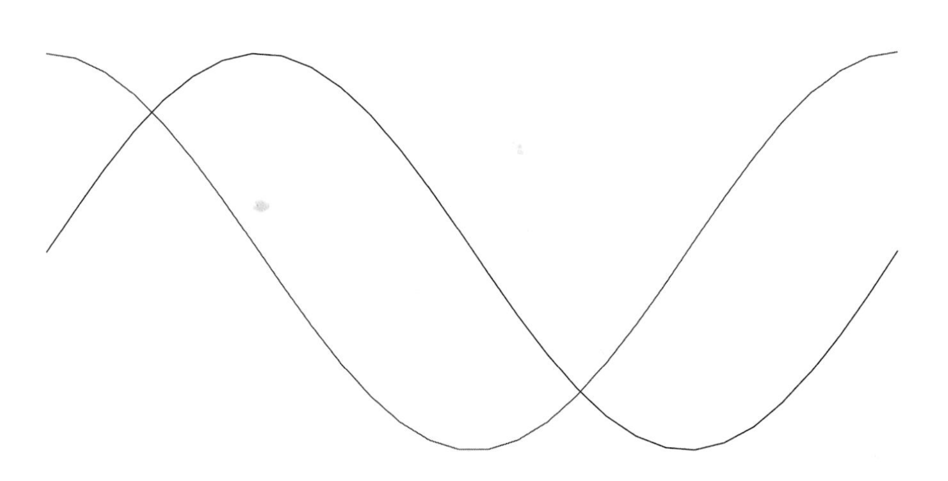

```
» axis on ij        % turn axis back on and flip y-axis
» title('Figure 26.11: Sine and Cosine Curves')
```

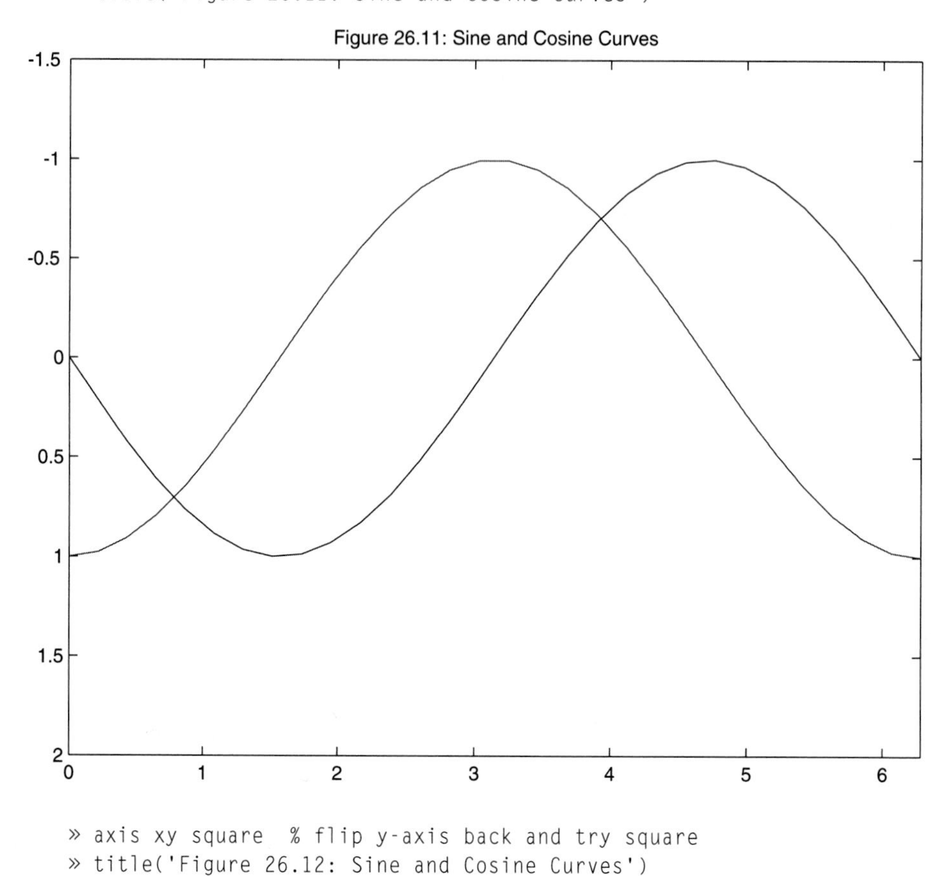


```
» axis xy square  % flip y-axis back and try square
» title('Figure 26.12: Sine and Cosine Curves')
```

## 26.6 MULTIPLE PLOTS

You can add new plots to an existing plot using the `hold` command. When you enter `hold on`, MATLAB does not remove the existing axes when new `plot` functions are issued. Instead, it adds new curves to the current axes, However, if the new data do not fit within the current axes limits, the axes are rescaled. Entering `hold off` releases the current *Figure* window for new plots. The `hold` command without arguments toggles the `hold` setting. Going back to our previous example,

```
» x = linspace(0,2*pi,30);
» y = sin(x); z = cos(x);
» plot(x,y)
```

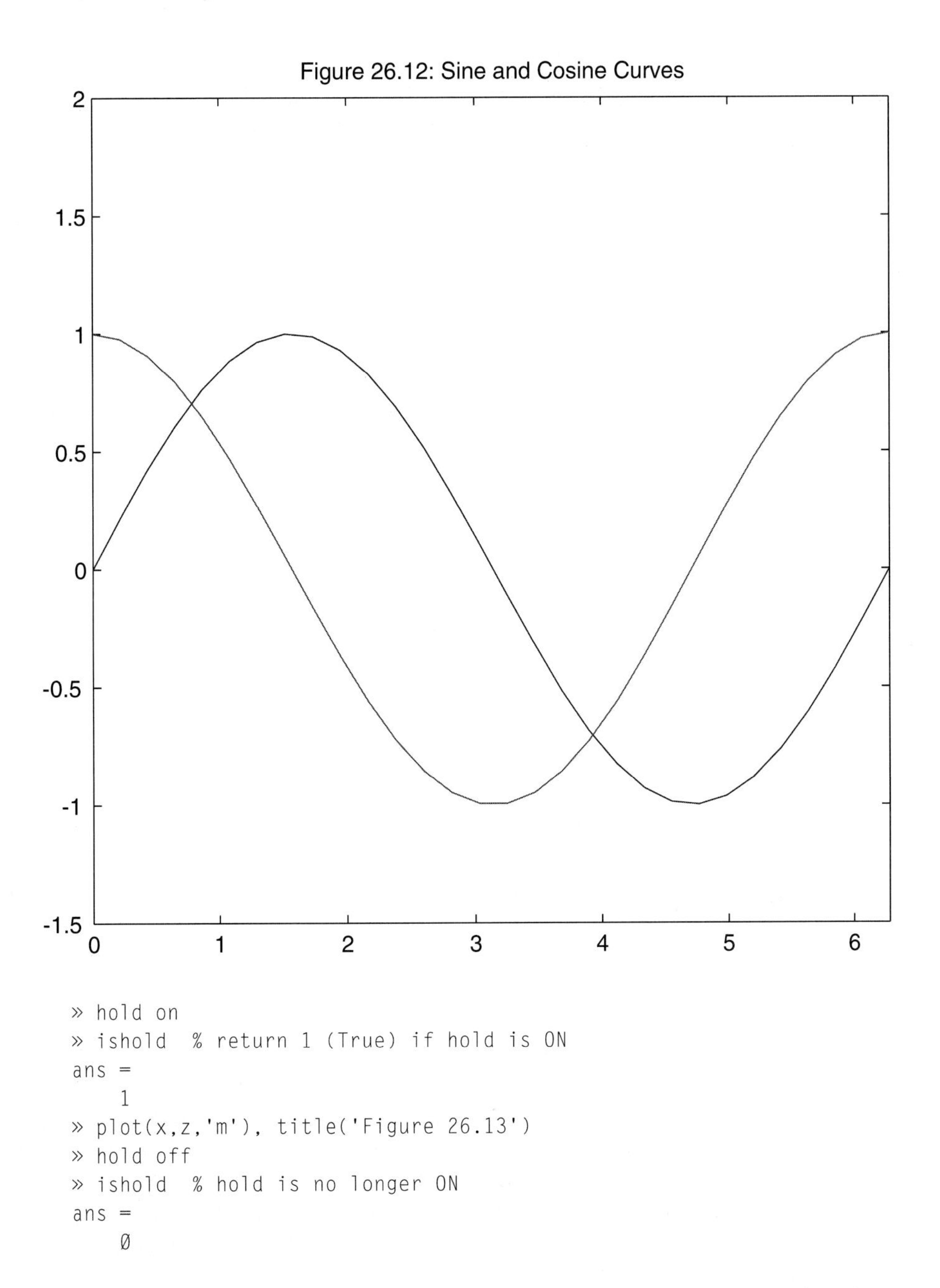


```
» hold on
» ishold  % return 1 (True) if hold is ON
ans =
     1
» plot(x,z,'m'), title('Figure 26.13')
» hold off
» ishold  % hold is no longer ON
ans =
     0
```

Notice that this example specified the color of the second curve. Since there is only one set of data arrays in each `plot` function, the line color for each `plot` function would otherwise default to the first color in the list, resulting in two lines plotted in the same color on the plot.

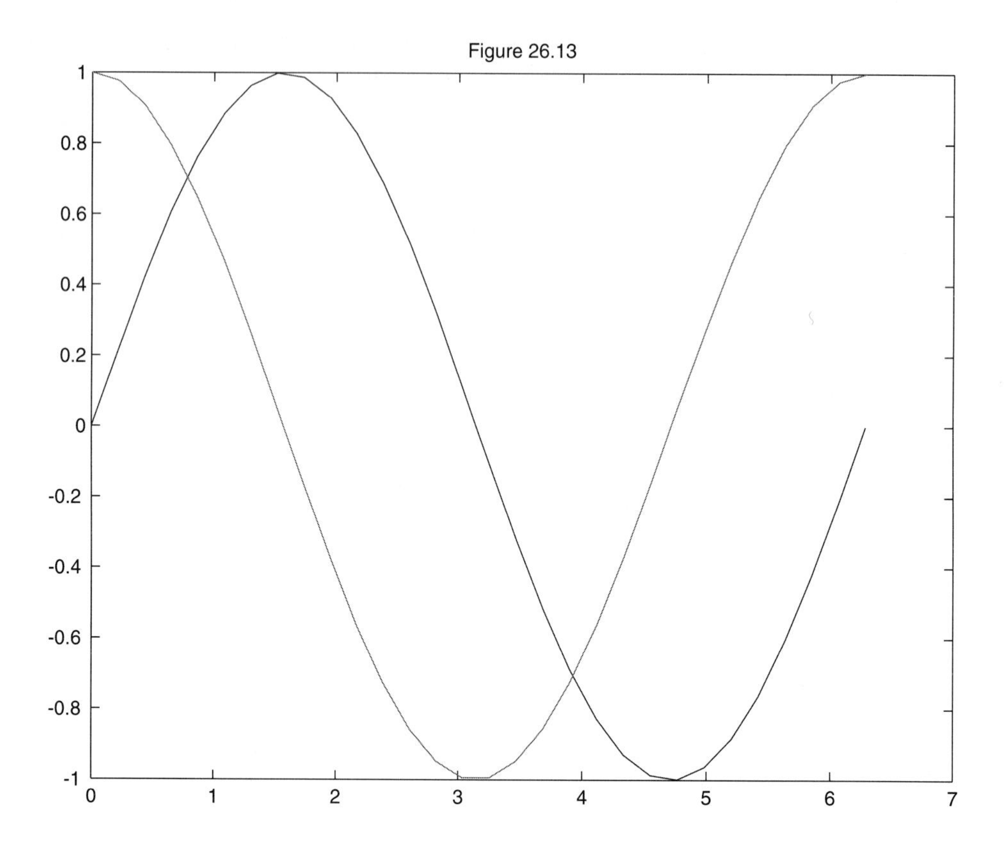


## 26.7 MULTIPLE FIGURES

It is possible to create multiple *Figure* windows and plot different data sets in different ways in each one. To create new *Figure* windows, use the `figure` command in the *Command* window, or the **New Figure** selection from the **File** menu in the *Command* or *Figure* windows. You can choose a specific *Figure* window to be the *active* or *current* figure by selecting it with the mouse, or by using `figure(n)`, where `n` is the number of the window. The current *Figure* window is the window that is active for subsequent plotting functions.

Every time a new *Figure* window is created, a number identifying it (i.e., its ***handle***) is returned and stored for future use. The figure handle is also displayed in the *Figure* window title bar. When a new *Figure* window is created, it is placed in the default figure position on the screen. As a result, when more than one *Figure* window is created, each new window covers all preceding windows. To see the windows simultaneously, simply drag them around using the mouse on the *Figure* window title bar.

To reuse a *Figure* window for a new plot, it must be made the active or *current* figure. Clicking on the figure of choice with the mouse makes it the current figure. From within MATLAB, `figure(h)`, where `h` is the figure handle, makes the corresponding figure active or current. Only the current figure is responsive to `axis`, `hold`, `xlabel`, `ylabel`, `title`, and `grid` commands.

*Figure* windows can be deleted by closing them with a mouse in the way you are familiar with closing windows on your computer, if such a feature exists. Alternatively, the command `close` can be issued:

```
» close
```

closes the current *Figure* window.

```
» close(h)
```

closes the *Figure* window having handle `h`.

```
» close all
```

closes all *Figure* windows.

If you simply want to erase the contents of a *Figure* window without closing it, use the command `clf`:

```
» clf
```

clears the current *Figure* window.

```
» clf reset
```

clears the current *Figure* window and resets all properties, such as `hold`, to their default state.

## 26.8 SUBPLOTS

One *Figure* window, on the other hand, can hold more than one set of axes. The `subplot(m,n,p)` command subdivides the current *Figure* window into an `m`-by-`n` matrix of plotting areas and chooses the `p`th area to be active. The subplots are numbered left to right along the top row, then the second row, etc. For example,

```
» x = linspace(0,2*pi,30);
» y = sin(x);   z = cos(x);

» a = 2*sin(x).*cos(x);
» b = sin(x)./(cos(x)+eps);

» subplot(2,2,1) % pick the upper left of a 2-by-2 grid of subplots
» plot(x,y), axis([0 2*pi -1 1]), title('Figure 26.14a: sin(x)')

» subplot(2,2,2) % pick the upper right of the 4 subplots
» plot(x,z), axis([0 2*pi -1 1]), title('Figure 26.14b: cos(x)')

» subplot(2,2,3) % pick the lower left of the 4 subplots
» plot(x,a), axis([0 2*pi -1 1]), title('Figure 26.14c: 2sin(x)cos(x)')

» subplot(2,2,4) % pick the lower right of the 4 subplots
» plot(x,b), axis([0 2*pi -20 20]), title('Figure 26.14d: sin(x)/cos(x)')
```

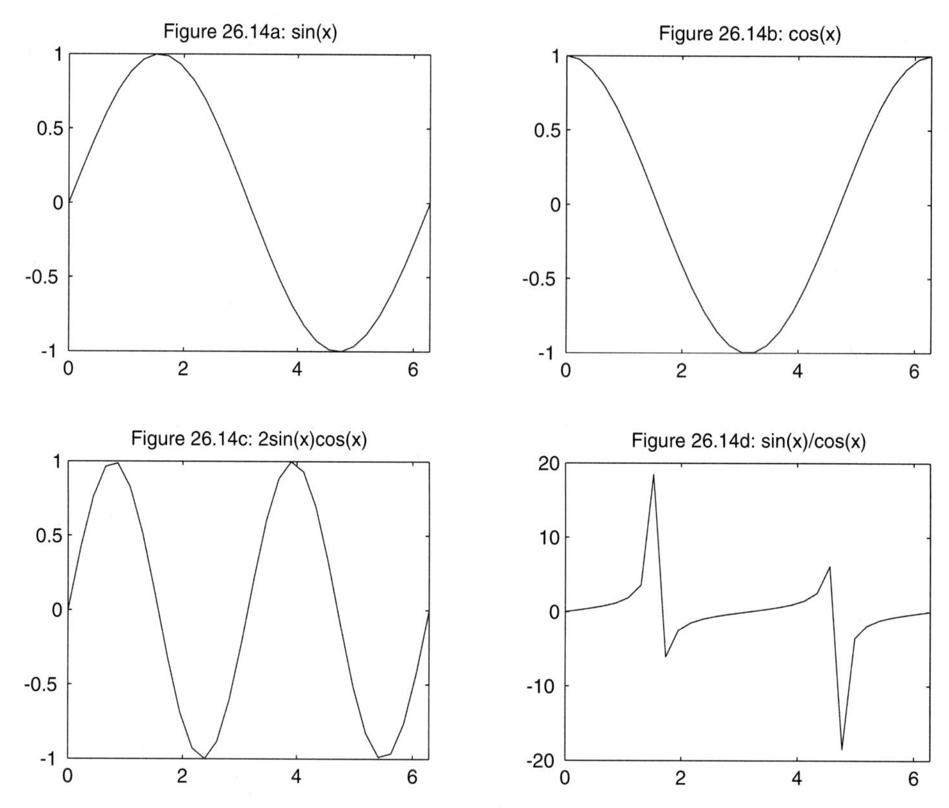


Note that when a particular subplot is active, it is the only subplot or axis that is responsive to `axis`, `hold`, `xlabel`, `ylabel`, `title`, `grid`, and `box` commands. The other subplots are not affected. In addition, the active subplot remains active until another `subplot` or `figure` command is issued. If a new `subplot` command changes the number of subplots in the *Figure* window, previous subplots are erased to make room for the new orientation. To return to the default mode and use the entire *Figure* window for a single set of axes, use the command `subplot(1,1,1)` or the `clf` command. If you print a *Figure* window containing multiple plots, all of them will be printed on the same page. For example, if the current *Figure* window contains four subplots and the orientation is landscape mode, each of the plots will use one quarter of the printed page.

## 26.9 INTERACTIVE PLOTTING TOOLS

Rather than using individual text strings to identify the data sets on your plot, a legend can be used. The `legend` command creates a legend box on the plot keying any text you supply to each line in your plot. If you wish to move the legend, simply click and hold down the mouse button near the edge of the legend and drag the legend to the desired location. `legend off` deletes the legend. Try this example:

```
» clf
» plot(x,y,x,z)
» legend('sin(x)','cos(x)')
```

Try moving the legend around on your plot with the mouse. Then remove the legend with

```
» legend off
```

The `legend` function works with other types of plots as well. In addition, `legend` has an optional argument specifying an initial location for the legend box.

MATLAB provides an interactive tool to expand sections of a 2-D plot to see more detail or to ***zoom in*** on a region of interest. The command `zoom on` turns on zoom mode. Clicking the mouse button on a Macintosh, or left mouse button on most other platforms, within the *Figure* window expands the plot by a factor of 2 centered around the point under the mouse pointer. Each time you click, the plot expands. Click the right mouse button on a PC or shift click on the Macintosh to zoom out by a factor of 2. You can also click and drag a rectangular area to zoom into a specific area. `zoom(n)` zooms in by a factor of `n`. `zoom out` returns the plot to its initial state. `zoom off` turns off zoom mode. `zoom` with no arguments toggles the zoom state of the active *Figure* window.

Try zooming in and out on a plot created using the M-file called `peaks.m`. This is an interesting function that generates a square matrix of data. The data are based on a function of two variables and contain data points for `x` and `y` in the range −3 to 3. The function is defined by the expression

$$z = f(x,y) = 3(1-x)^2\exp\left(-x^2-(y+1)^2\right) - 10\left(\frac{x}{5}-x^3-y^5\right)\exp\left(-x^2-y^2\right) - \tfrac{1}{3}\exp\left(-(x+1)^2-y^2\right)$$

You can specify the size of the square matrix `peaks` generates by passing it an argument. If you omit the argument, it defaults to 49. Try this example:

```
» M = peaks(25)        % create a 25-by-25 matrix of data
» plot(M)             % plot the columns of M
» title('Peaks Plot for ZOOM Practice')
» zoom on
```

The function `plot(M)`, where `M` is a matrix, plots each column of `M` versus its index. The preceding example plotted `25` lines on the plot. Zoom in and out with the mouse to experiment with zooming.

Since the `legend` and `zoom` commands both respond to mouse clicks in the *Figure* window, they interfere with each other. ***Therefore, if* `zoom` *is to be used,* `legend` *must first be turned off.***

In some situations it is convenient to select coordinate points from a plot in a *Figure* window. In MATLAB this feature is embodied in the `ginput` function. The form `[x,y]=ginput(n)` gets `n` points from the current plot or subplot based on mouse click positions within the plot or subplot. If you press the **Return** or **Enter** key before all `n` points are selected, `ginput` terminates with fewer points. The points returned in the vectors `x` and `y` are the respective $x$ and $y$ data coordinate points selected. The returned data are not necessarily points from the data set used to create the plot, but rather the explicit $x$ and $y$ coordinate values where the mouse was clicked. If points are selected outside the plot or subplot axes limits (e.g., outside the plot box), the points returned are extrapolated values.

This function can be somewhat confusing when used in a *Figure* window containing subplots. The data returned are with respect to the current or active subplot. Thus if `ginput` is issued after a `subplot(2,2,3)` command, the data returned are with respect to the axes of the data plotted in `subplot(2,2,3)`. If points are selected from other subplots, the data are still with respect to the axes of the data in `subplot(2,2,3)`.

When an unspecified number of data points are desired, the form `[x,y]=ginput` without an input argument can be used. Here data points are gathered until the **Return** key is pressed. The `gtext` function described earlier in this chapter utilizes the function `ginput` along with the function `text` for placing text with the mouse.

As an example, let's plot a function and then plot a line connecting eight mouse-selected points.

```
» x = linspace(-2*pi,2*pi,60);
» y = sin(x).^2./(x+eps);
» plot(x,y)
» title('Figure 26.15: Plot of sin(x)^2/x')

» [a,b] = ginput(8); % get up to 8 points from the plot

» hold on
» plot(a,b,'mo') % plot the data points just collected
» hold off
```

Since the `legend`, `zoom`, and `ginput` commands respond to mouse clicks in a *Figure* window, they interfere with each other. ***As a result, before using*** `ginput`, `zoom` ***and*** `legend` ***must be turned off.***

## 26.10 SCREEN UPDATES

Because screen rendering is relatively time consuming. MATLAB does not always update the screen after each graphics command. For example, if the following commands are entered at the MATLAB prompt, MATLAB updates the screen after each graphics command (`plot`, `axis`, and `grid`):

```
» x = linspace(0,2*pi); y = sin(x);
» plot(x,y)
» axis([0 2*pi -1.2 1.2])
» grid
```

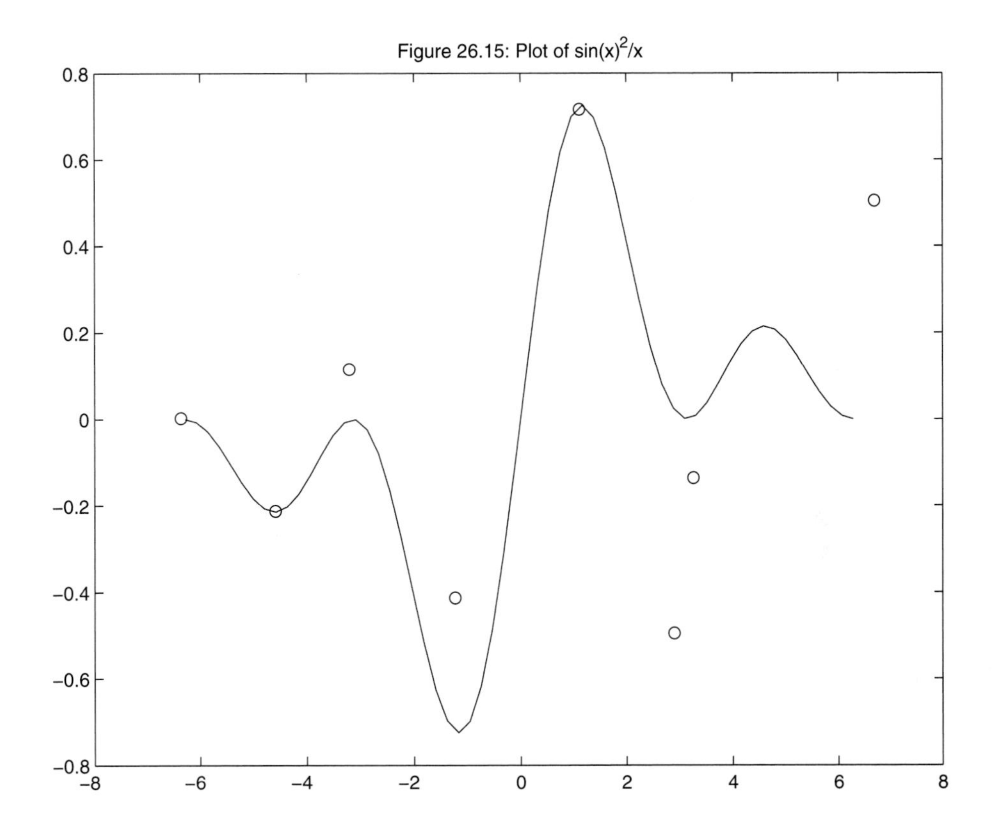


However, if the same graphics commands are entered on a single line—that is,

```
» plot(x,y), axis([0 2*pi -1.2 1.2]), grid
```

MATLAB renders the figure only once—when the MATLAB prompt reappears. A similar procedure occurs when graphics commands appear as part of a script or function M-file. In this case, even if the commands appear on separate lines in the file, the screen is rendered only once—when all commands are completed and the MATLAB prompt reappears.

In general, five events cause MATLAB to render the screen:

1. a return to the MATLAB prompt
2. encountering a function that temporarily stops execution, such as `pause`, `keyboard`, `input`, and `waitforbuttonpress`
3. execution of a `getframe` command
4. execution of a `drawnow` command
5. resizing of a *Figure* window.

Of these, the `drawnow` command specifically allows one to force MATLAB to redraw the screen at arbitrary times.

## 26.11 SPECIALIZED 2-D PLOTS

Up to this point the basic plotting function `plot` has been illustrated. In many situations, plotting lines or points on linearly scaled axes does not convey the desired information. As a result, MATLAB offers other basic 2-D plotting functions, as well as specialized plotting functions that are embodied in function M-files.

In addition to `plot`, MATLAB provides the functions `semilogx` for plotting with a logarithmically scaled *x*-axis, `semilogy` for plotting with a logarithmically scaled *y*-axis, and `loglog` for plotting with both axes logarithmically spaced. All the features discussed previously with respect to the function `plot` apply to these functions as well.

An interesting variation on the standard `plot` function is `comet`. The form `comet(x,y)` generates the same plot as `plot(x,y)` except that the plot is drawn incrementally with a tail like a comet. For example,

```
» t = linspace(-pi,pi,200);
» s = tan(cos(t)) - cos(tan(t));
» comet(t,s), title'Figure 26.16: Comet Plot')
```

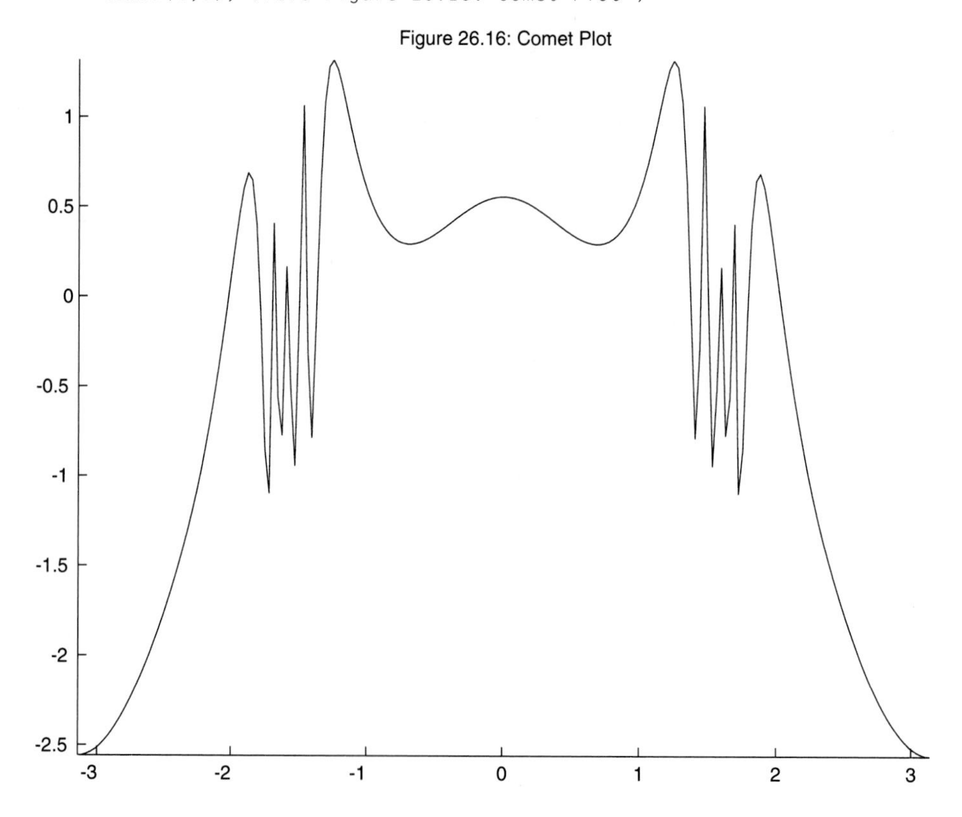

The `area` function is useful for building a stacked area plot. `area(x,y)` is the same as `plot(x,y)` for vectors `x` and `y`, except that the area under the plot is filled in with color. The lower limit for the filled area may be specified but defaults to zero. To stack areas, use the form `area(X,Y)`, where `Y` is a matrix and `X` is a matrix or vector whose length equals the number of rows in `Y`. If `X` is omitted, `area` uses the default value `X=1:size(Y,1)`. For example,

```
» z = -pi:pi/5:pi;
» area([sin(z);cos(z)])
» title('Figure 26.17: Stacked Area Plot')
```

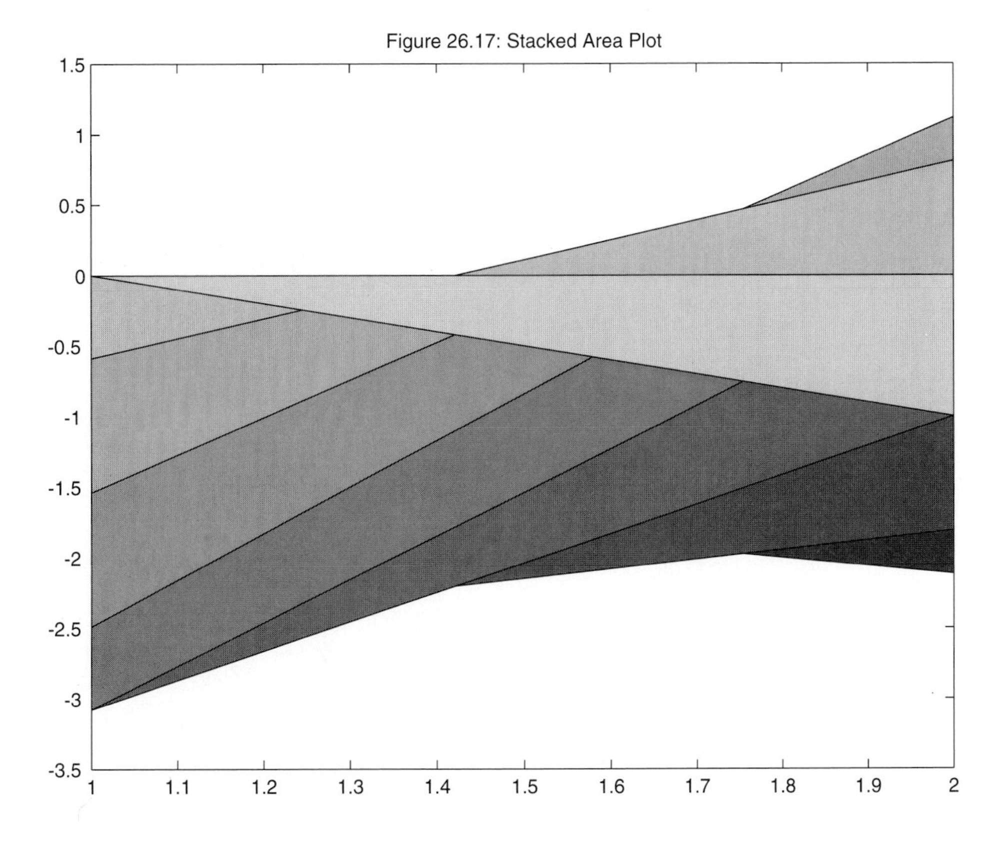


Filled polygons can be drawn using the `fill` function. `fill(x,y,'c')` fills the 2-D polygon defined by the column vectors `x` and `y` with the color specified by `c`. The vertices of the polygon are specified by the pairs $(x_i, y_i)$. If necessary, the polygon is closed by connecting the last vertex to the first. As with the `plot` function, `fill` can have any number of pairs of vertices and associated colors. Moreover, if `x` and `y` are matrices of the same dimension, the columns of `x` and `y` are assumed to describe separate polygons. For example,

```
» t = (1:2:15)'*pi/8; % column vector [π/8;3π8;...;15π/8]
» x = sin(t);
» y = cos(t);
» fill(x,y,'r') % a filled red circle using only 8 data points
» axis square off
» text(0,0,'STOP',...
  'Color',[1 1 1],...
  'FontSize',80,...
  'HorizontalAlignment','center')
» title('Figure 26.18: Red Stop Sign')
```

Figure 26.18: Red Stop Sign

This example uses the `text(x,y,'string')` function with extra arguments. The `Color`, `FontSize`, and `HorizontalAlignment` arguments tell MATLAB to use Handle Graphics to modify the text. Handle Graphics is the name of MATLAB's underlying graphics functions. You can access this rich set of powerful and versatile graphics functions yourself. See the Handle Graphics chapter (Chapter 31) for more information on these features.

Standard pie charts can be created using `pie(a,b)` function, where `a` is a vector of values and `b` is an optional logical vector describing a slice or slices to be pulled out of the pie chart. The `pie3` function renders the pie chart with a three-dimensional appearance. For example,

```
» a = [.5 1 1.6 1.2 .8 2.1];
» pie(a,a==max(a));   % chart a and pull out the biggest slice
» title('Figure 26.19: Example Pie Chart')
```

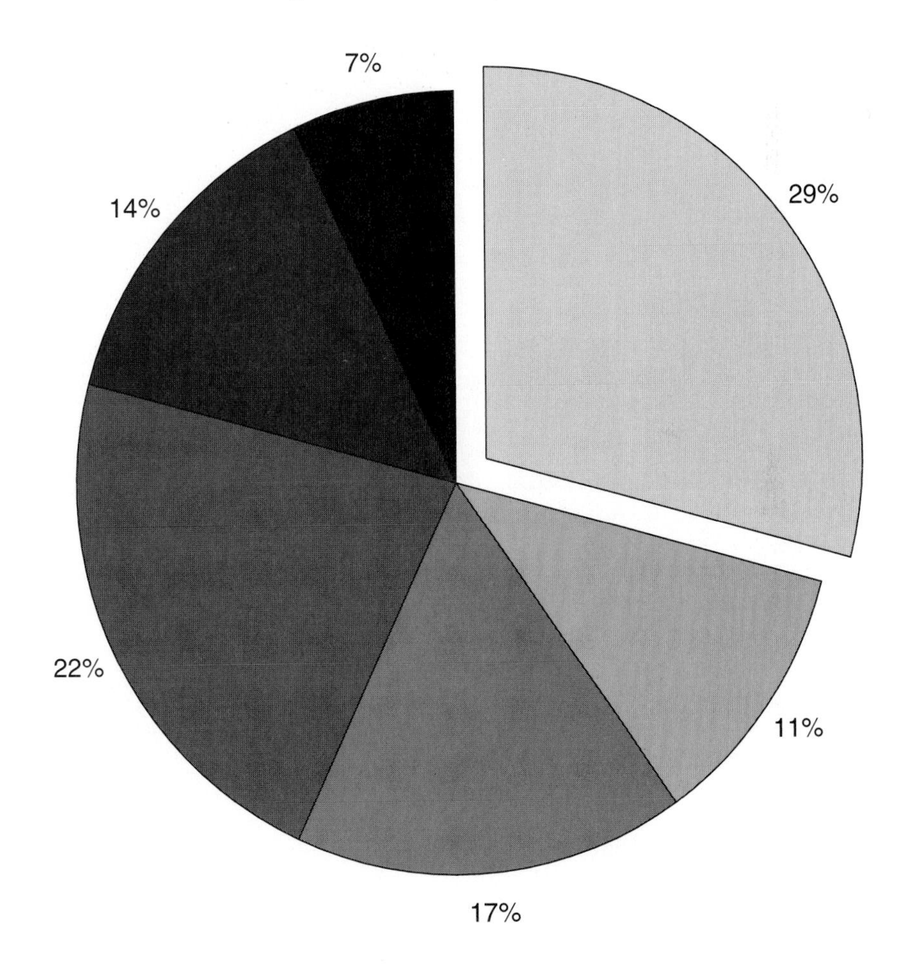


Another way to visualize the same data is with a Pareto chart, where the values in the vector argument are drawn as bars in descending order along with an associated accumulated-value line plot. Using the vector a from the previous example, we get

```
» pareto(a)
» title('Figure 26. 20: Example Pareto Chart')
```

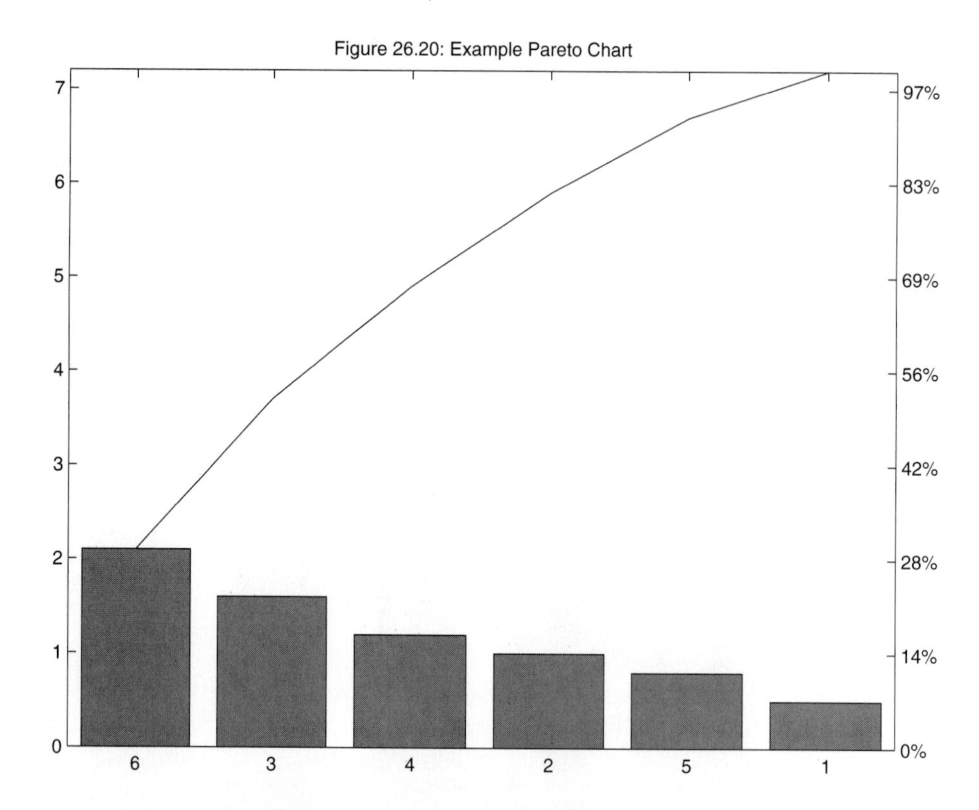


Sometimes you might want to plot two different functions on the same axes using different *y*-axis scales. The function plotyy does just that:

```
» x = -2*pi:pi/10:2*pi;
» y = sin(x);
» z = 2*cos(x);
» subplot(2,1,1), plot(x,y,x,z)
» title('Figure 26.21a: Two plots on the same scale.');
» subplot(2,1,2), plotyy(x,y,x,z)
» title('Figure 26.21b: Two plots on different scales.');
```

Please note that both pareto and plotyy actually create two sets of axes in the same *Figure* window. Any command that affects the current axes (like axis, zoom, or legend) will affect only one set of axes and can lead to unintended results. So, in general, it is best to avoid these commands when using pareto and plotyy functions.

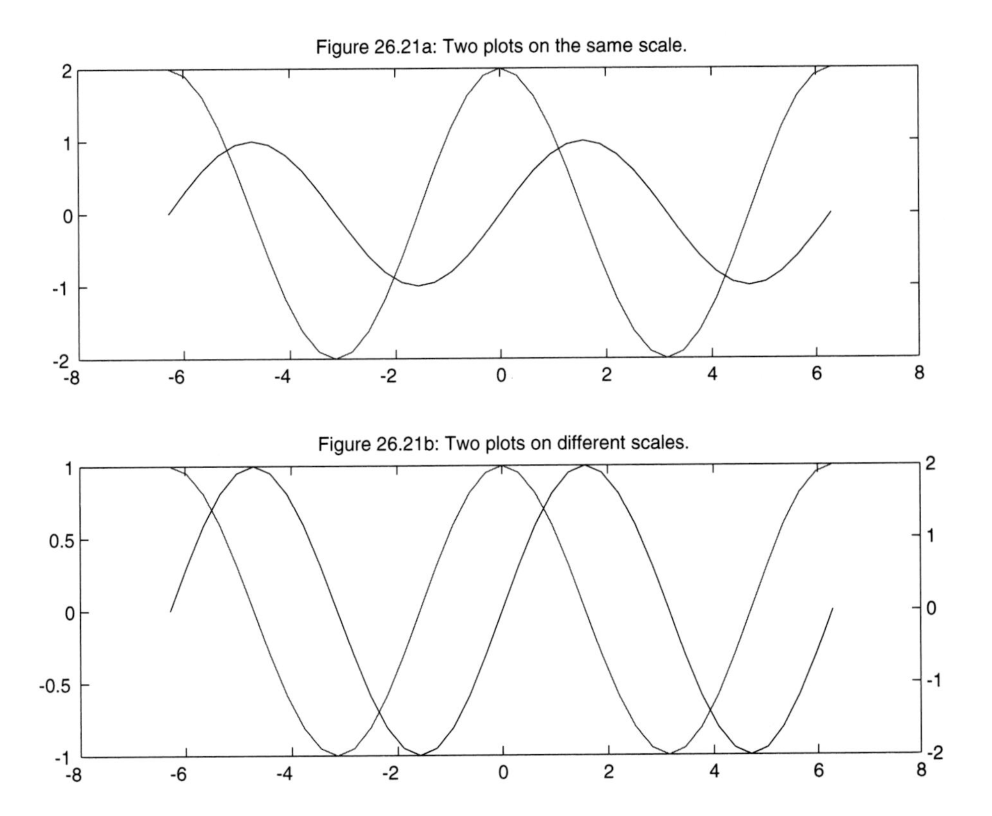


Bar and stair plots can be generated using the `bar`, `barh`, and `stairs` plotting functions. The `bar3` and `bar3h` functions render the bar charts with a three dimensional appearance. Here are examples of a bell curve:

```
» x = -2.9:0.2:2.9;
» y = exp(-x.*x);
» subplot(2,2,1)
» bar(x,y)
» title('Figure 26.22a: 2-D Bar Chart')
» subplot(2,2,2)
» bar3(x,y,'r')
» title('Figure 26.22b: 3-D Bar Chart')
» subplot(2,2,3)
» stairs(x,y)
» title('Figure 26.22c: Stair Chart')
» subplot(2,2,4)
» barh(x,y)
» title ('Figure 26.22d: Horizontal Bar Chart')
```

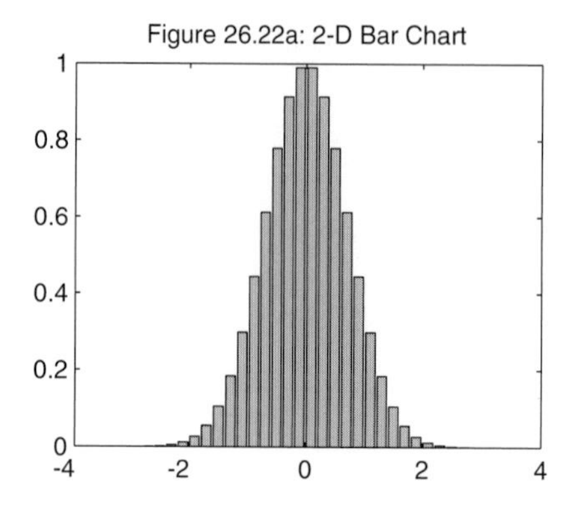


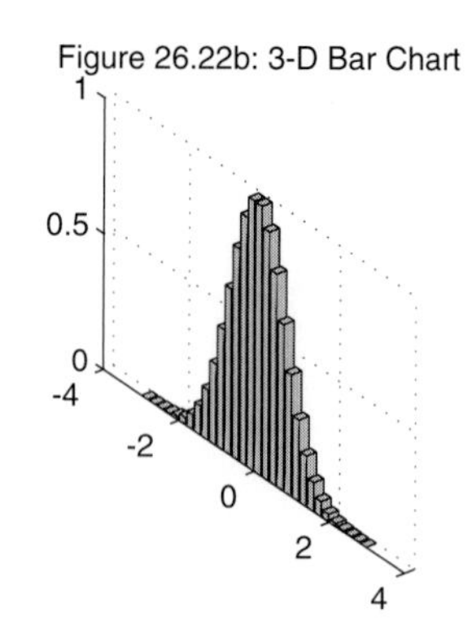


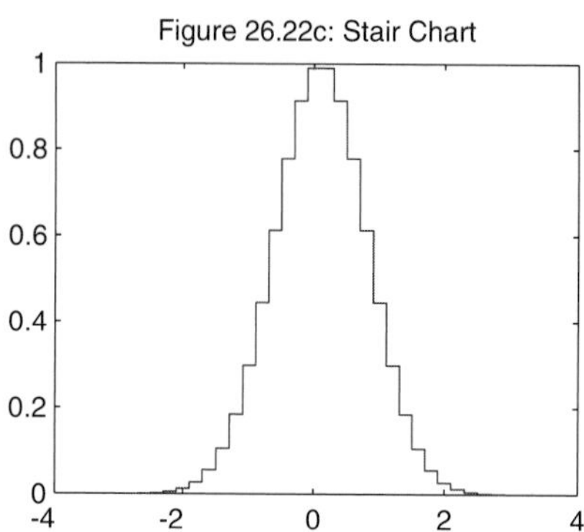


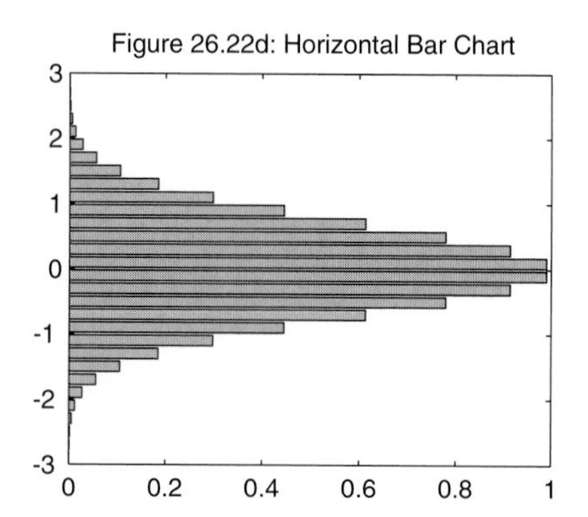


The various `bar` functions accept a single color argument for all bars. Bars may be grouped or stacked as well. The form `bar(x,Y)` for vector `x` and matrix `Y` draws groups of bars corresponding to the columns of `Y`. `bar(x,Y,'stacked')` draws the bars stacked vertically. `barh`, `bar3`, and `bar3h` have similar options.

Histograms illustrate the distribution of values in a vector. `hist(y)` draws a 10-bin histogram for the data in vector `y`. `hist(y,n)`, where `n` is a scalar, draws a histogram with `n` bins. `hist(y,x)`, where `x` is a vector, draws a histogram using the bins specified in `x`. Here is an example of a bell-curve histogram from Gaussian data:

```
» x = -2.9:0.2:2.9; % specify the bins to use
» y = randn(5000,1); % generate 5000 random data points
» hist(y,x) % draw the histogram
» title('Figure 26.23: Histogram of Gaussian Data')
```

Discrete sequence data can be plotted using the `stem` function. `stem(z)` creates a plot of the data points in vector `z` connected to the horizontal axis by a line. An optional character string argument can be used to specify linestyle. `stem(x,z)` plots the data points in `z` at the values specified in `x`.

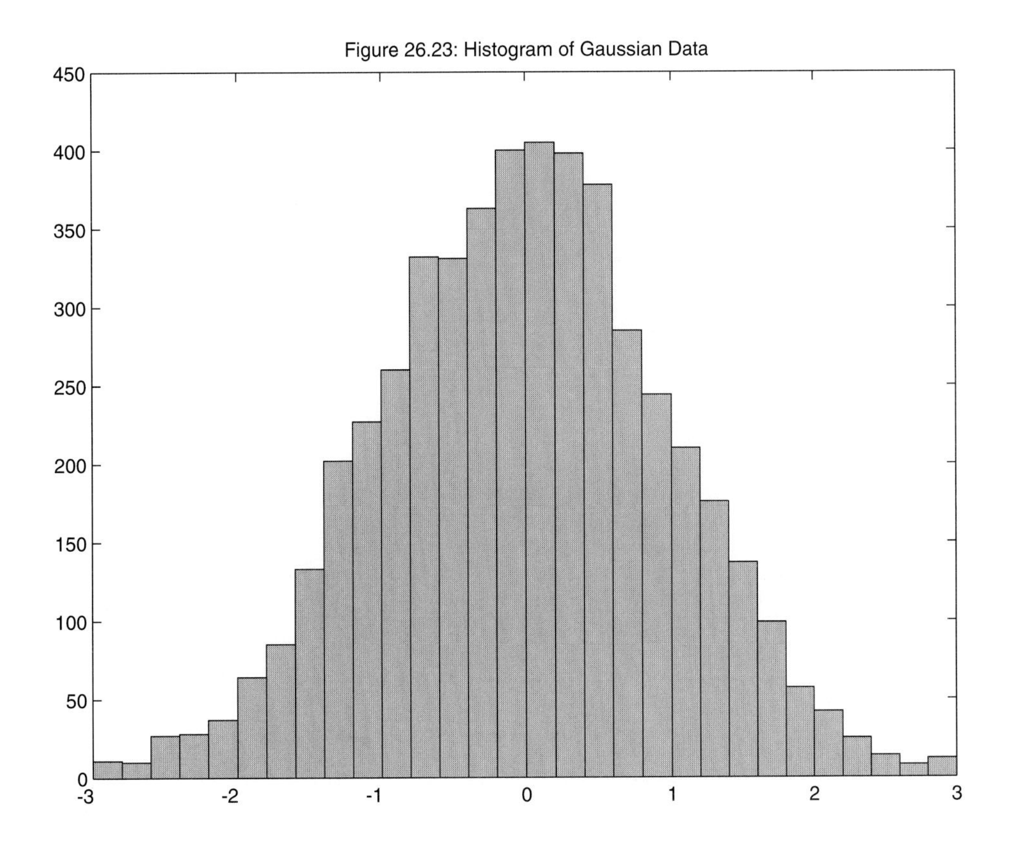


```
» z = randn(50,1); % create some random data
» stem(z,'--') % draw a stem plot using dashed linestyle
» title('Figure 26.24: Stem Plot of Random Data')
```

A plot can include error bars at the data points. `errorbar(x,y,e)` plots the graph of vector `x` versus vector `y` with error bars specified by vector `e`. All vectors must be the same length. For each data point $(x_i, y_i)$, an error bar is drawn a distance $e_i$ above and $e_i$ below the data point. For example,

```
» x = linspace(0,2,21); % create a vector
» y = erf(x); % y is the error function of x
» e = rand(size(x))/10; % e contains random error values
» errorbar(x,y,e) % create the plot
» title('Figure 26.25: Errorbar Plot')
```

Plots in polar coordinates can be created using the `polar(t,r,S)` function, where `t` is the angle vector in radians, `r` is the radius vector, and `S` is an optional character string describing color, marker symbol, and/or linestyle. For example,

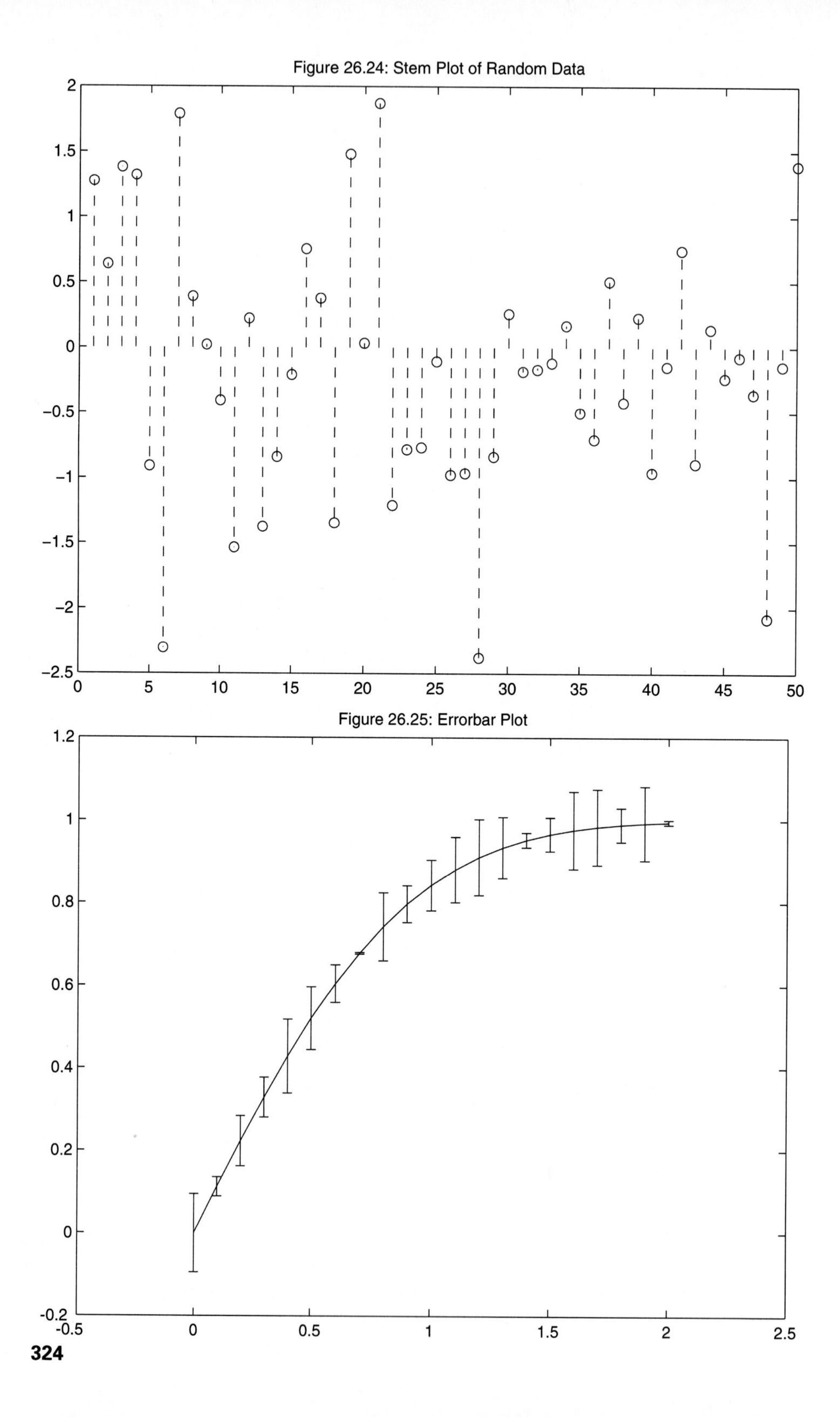

Figure 26.24: Stem Plot of Random Data

Figure 26.25: Errorbar Plot

```
» t = linspace(0,2*pi);
» r = sin(2*t).*cos(2*t);
» subplot(2,2,1)
» polar(t,r), title('Figure 26.26a: Polar Plot')
```

Complex data can be plotted using `compass` and `feather`. `compass(z)` draws a plot that displays the angle and magnitude of the complex elements of `z` as arrows emanating from the origin. `feather(z)` plots the same data using arrows emanating from equally spaced points on a horizontal line. `compass(x,y)` and `feather(x,y)` are equivalent to `compass(x+i*y)` and `feather(x+i*y)`. For example,

```
» z = eig(randn(20));
» subplot(2,2,2)
» compass(z)
» title('Figure 26.26b: Compass Plot')
» subplot(2,2,3)
» feather(z)
» title('Figure 26.26c: Feather Plot')
```

`rose(v)` draws a 20-bin polar histogram for the angles in vector `v`. `rose(v,n)`, where `n` is a scalar, draws a histogram with `n` bins. `rose(v,x)`, where `x` is a vector, draws a histogram using the bins specified in `x`. For example,

```
» subplot(2,2,4)
» v = randn(1000,1)*pi;
» rose(v)
» title('Figure 26.26d: Angle Histogram')
```

Scatter plots can be created with the `plotmatrix` function. `plotmatrix(X,Y)` scatter plots the columns of `X` against the columns of `Y`.

```
» x = randn(50,3);
» y = randn(3);
» plotmatrix(x,x*y)
» title('Figure 26.27: Scatter Plots')
```

`plotmatrix(Y)` is the same as `plotmatrix(Y,Y)` except that the diagonal is replaced by `hist(Y(:,I))`.

## 26.12 QUICK PLOTS

The `fplot` function gives you the ability automatically to plot a function of one variable between specified limits without creating a data set for the variable.

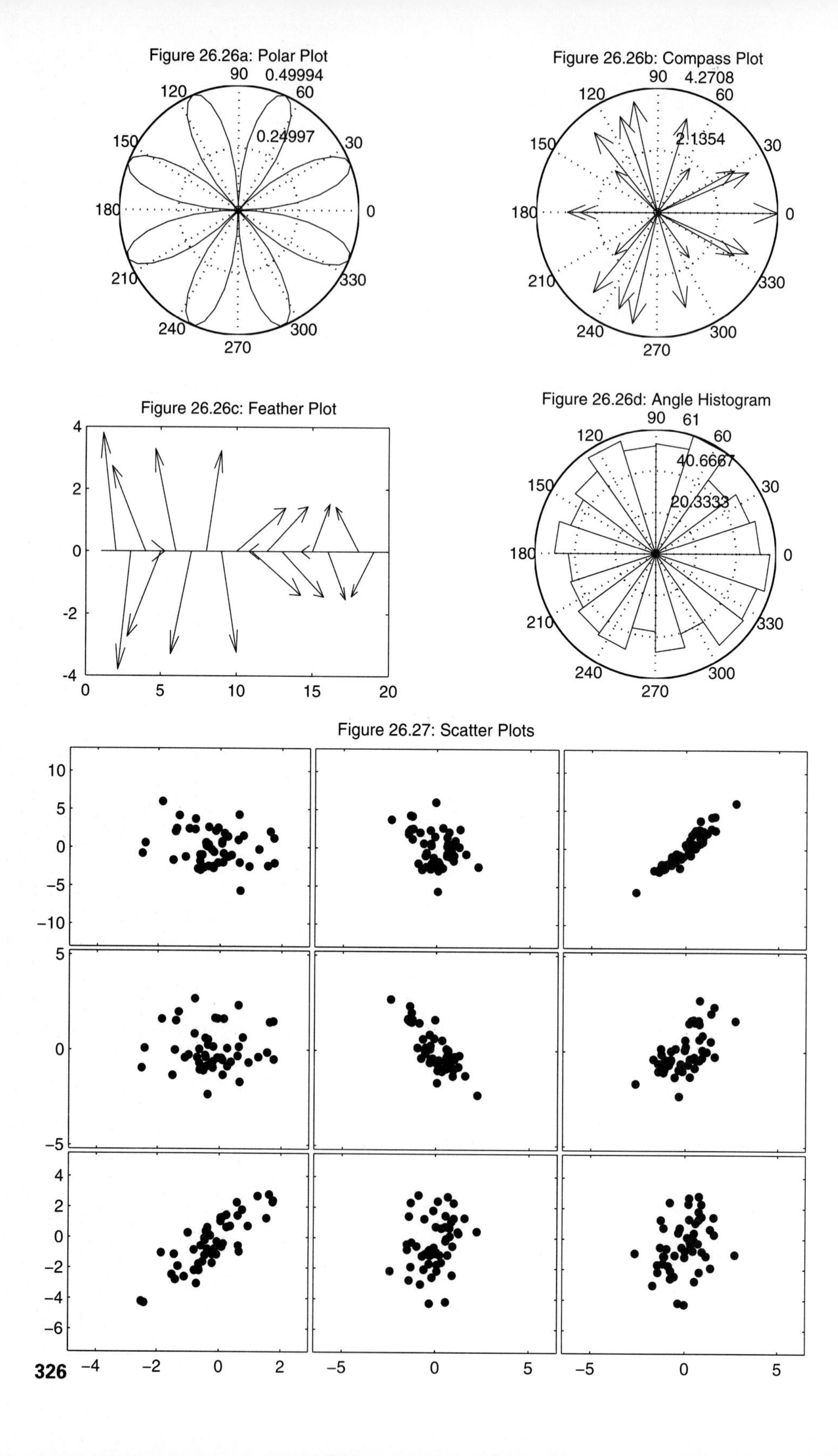
Figure 26.26a: Polar Plot
0.49994
0.24997
Figure 26.26b: Compass Plot
4.2708
2.1354
Figure 26.26c: Feather Plot
Figure 26.26d: Angle Histogram
61
40.6667
20.3333
Figure 26.27: Scatter Plots

`fplot('FUN',[xmin xmax])` plots the function `FUN(x)` over the domain `xmin` $\leq$ `x` $\leq$ `xmax` with automatic scaling of the *y*-axis.
`fplot('FUN',[xmin xmax ymin ymax])` specifies the *y*-axis limits as well. `FUN` must be a string containing a symbolic expression in one variable or the name of an M-file function of the form `y=f(x)` for vector `x`. There are other restrictions on the type of function that can be plotted, and additional arguments can be specified. `fplot` uses adaptive step control to produce a representative graph, concentrating its evaluation in regions where the function's rate of change is the greatest.

```
» fplot('erf',[-2 2 -1.1 1.1])
» title('Figure 26.28: Fplot of the Error Function')
» xlabel('x')
» ylabel('erf(x)')
```

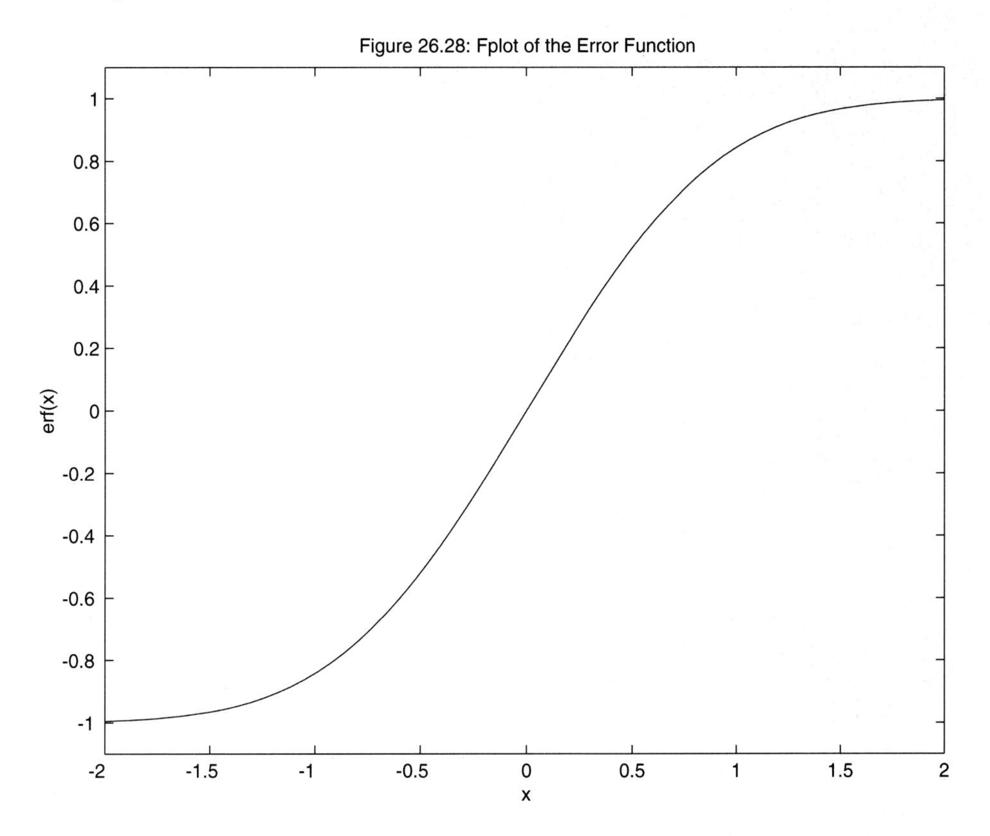


MATLAB includes another tool that will quickly plot a function of a single variable over a range of values and annotate the plot for you as well. `ezplot('FUN')`, where `FUN` is a string containing a function or symbolic expression in one variable, plots the function over the domain `[-2*pi 2*pi]`. The *x*-domain can be specified using the form `ezplot('FUN',[xmin xmax])`. The *x*-axis label is the variable name, and the plot title is the function FUN. ezplot determines the interval of the $x$-axis by sampling the function between -2*pi and 2*pi and then selecting a subinterval where the variation is significant. ezplot omits extreme values associated with singularities when determining the range of the $y$-axis.

```
» ezplot('sin(x^2)/(x+eps)')
```

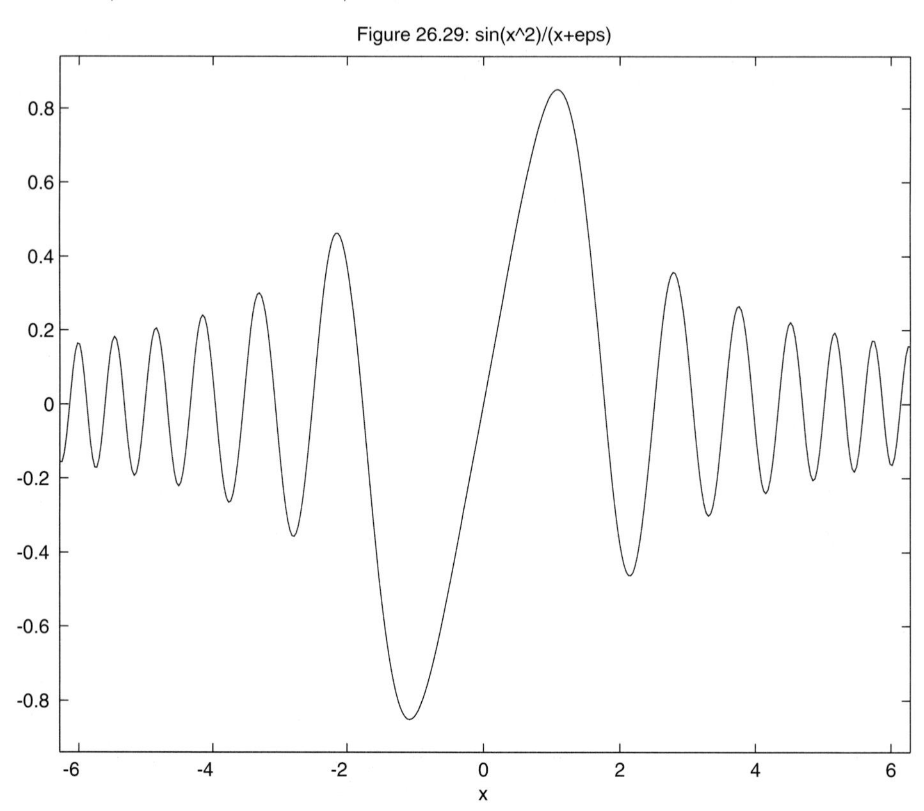


## 26.13 TEXT FORMATTING

Multiline text can be used in any text string, including titles and axis labels as well as the text and gtext functions. Simply use string arrays or cell arrays for multiline text. For example,

```
» xlabel({'This is the first line', ' and this is the second.'});
```

labels the $x$-axis with two lines of text. Note that the string separator can be a space, comma, or semicolon; each style produces the same result. See the chapters on cell arrays (Chapter 12) and character strings (Chapter 10) for more detail.

A selection of over 75 symbols, including Greek letters and other special characters, can be included in MATLAB text strings by embedding a subset of TeX commands within the string. The available symbols and the character strings used to define them are listed in the following table.

| `\alpha` | $\alpha$ | `\upsilon` | $\upsilon$ | `\Re` | $\Re$ | `\neq` | $\neq$ |
|---|---|---|---|---|---|---|---|
| `\beta` | $\beta$ | `\omega` | $\omega$ | `\Im` | $\Im$ | `\approx` | $\approx$ |
| `\chi` | $\chi$ | `\xi` | $\xi$ | `\aleph` | $\aleph$ | `\equiv` | $\equiv$ |
| `\delta` | $\delta$ | `\psi` | $\psi$ | `\wp` | $\wp$ | `\cong` | $\cong$ |
| `\epsilon` | $\varepsilon$ | `\zeta` | $\zeta$ | `\otimes` | $\otimes$ | `\pm` | $\pm$ |
| `\theta` | $\phi$ | `\Delta` | $\varDelta$ | `\oplus` | $\oplus$ | `\propto` | $\propto$ |
| `\gamma` | $\gamma$ | `\Phi` | $\varPhi$ | `\oslash` | $\varnothing$ | `\partial` | $\partial$ |
| `\eta` | $\eta$ | `\Gamma` | $\varGamma$ | `\cap` | $\cap$ | `\bullet` | $\bullet$ |
| `\iota` | $\iota$ | `\Lambda` | $\varLambda$ | `\cup` | $\cup$ | `\circ` | $\circ$ |
| `\phi` | $\varphi$ | `\Pi` | $\Pi$ | `\supset` | $\supset$ | `\div` | $\div$ |
| `\kappa` | $\varkappa$ | `\Theta` | $\varTheta$ | `\supseteq` | $\supseteq$ | `\infty` | $\infty$ |
| `\lambda` | $\lambda$ | `\Sigma` | $\varSigma$ | `\subset` | $\subset$ | `\leftarrow` | $\leftarrow$ |
| `\mu` | $\mu$ | `\Upsilon` | $\varUpsilon$ | `\subseteq` | $\subseteq$ | `\rightarrow` | $\rightarrow$ |
| `\nu` | $\nu$ | `\Omega` | $\varOmega$ | `\ni` | $\backslash\theta$ | `\leftrightarrow` | $\leftrightarrow$ |
| `\o` | $o$ | `\Xi` | $\varXi$ | `\in` | $\in$ | `\uparrow` | $\uparrow$ |
| `\pi` | $\pi$ | `\Psi` | $\varPsi$ | `\o` | $o$ | `\downarrow` | $\downarrow$ |
| `\theta` | $\theta$ | `\varsigma` | $\varsigma$ | `\int` | $\int$ | `\clubsuit` | $\clubsuit$ |
| `\rho` | $\rho$ | `\vartheta` | $\upsilon$ | `\sim` | $\sim$ | `\diamondsuit` | $\blacklozenge$ |
| `\sigma` | $\sigma$ | `\forall` | $\forall$ | `\leq` | $\leq$ | `\heartsuit` | $\heartsuit$ |
| `\tau` | $\tau$ | `\exist` | $\exists$ | `\geq` | $\geq$ | `\spadesuit` | $\spadesuit$ |

A limited subset of TeX formatting commands are also available. Superscripts and subscripts are specified by ^ and _, respectively. The text font and size may be selected using the `\fontname` and `\fontsize` commands, and a style may be specified using the `\bf`, `\it`, `\sl`, or `\rm` commands to select boldface, italic, oblique or slant, and normal Roman fonts, respectively. For example,

```
» gtext('\fontname{courier} \fontsize{16} \it x_{\alpha} + y^{2\pi}')
```

creates the character string

$$x_{\alpha} + y^{2\pi}$$

To print the special characters used to define TeX strings, prefix them with the backslash (\) character. The characters affected are the backslash (\), left and right curly braces { }, underscore (_), and carat (^). Alternatively, the following command will disable interpretation of TeX strings and the string will print exactly as specified.

```
» gtext('Some special characters: ^, \, _, {, and }', 'Interpreter','none');
```

See the Handle Graphics chapter (Chapter 31) for more information.

# 27

# 3-D Graphics

MATLAB provides a variety of functions to display 3-D data. Some functions plot lines in three dimensions, while others draw surfaces and wire frames. In addition, color can be used to represent a fourth dimension. When color is used in this manner it is called ***pseudocolor,*** since color is not an inherent or natural property of the underlying data in the way that color in a photograph is a natural characteristic of the image. To simplify the discussion of 3-D graphics, the use of color is postponed until the next chapter. In this chapter, the fundamental concepts of producing useful 3-D plots are discussed.

## 27.1 LINE PLOTS

The `plot` function from the 2-D world can be extended into three dimensions with `plot3`. The format is the same as the 2-D `plot`, except the data are in triples, rather than pairs. The generalized format of `plot3` is `plot3(`$x_1$`,`$y_1$`,`$z_1$`,`$S_1$`,`$x_2$`,`$y_2$`,`$z_2$`,`$S_2$`,... )`, where $x_n$, $y_n$, and $z_n$ are vectors or matrices, and $S_n$ are optional character strings specifying color, marker symbol, and/or linestyle. `plot3` is commonly used to plot a three-dimensional function of a single variable. For example,

```
» t = linspace(0,10*pi);
» plot3(sin(t),cos(t),t)
» xlabel('sin(t)'), ylabel('cos(t)'), zlabel('t')
» text(0,0,0,'Origin')
» grid on
» title('Figure 27.1: Helix')
» v = axis
v =
    -1      1     -1      1      0     35
```

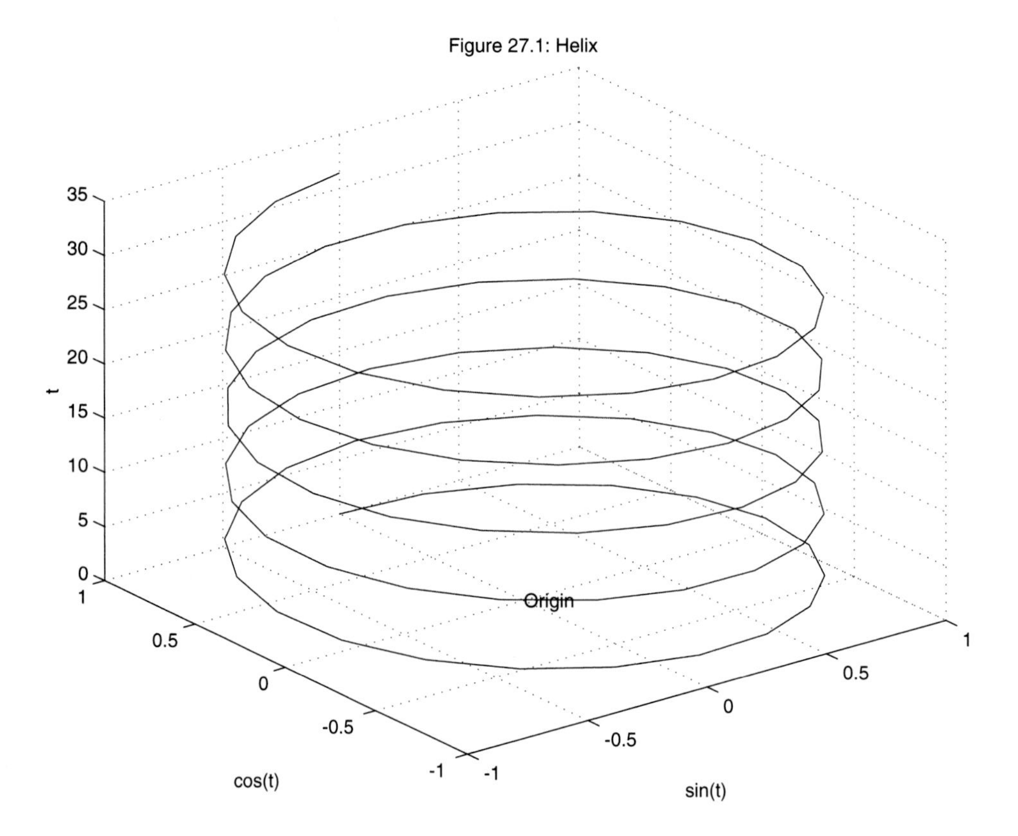


From this simple example it is apparent that all the basic features of 2-D graphics exist in 3-D also. The `axis` command extends to 3-D by returning the *z*-axis limits (0 and 35) as two additional elements in the axis vector. There is a `zlabel` function for labeling the *z*-axis. The `grid` command toggles a 3-D grid underneath the plot and the `box` command creates a 3-D box around the plot. The defaults for `plot3` are `grid off` and `box off`. The function `text(x,y,z,'string')` places a character string at the position identified by the triplet x, y, z. In addition, subplots and multiple *Figure* windows apply directly to 3-D graphics functions.

In the last chapter, multiple lines or curves were plotted on top of one another by specifying multiple arguments to the plot function, or by using the hold command. plot3 and the other 3-D graphics functions offer the same capabilities. For example, the added dimension of plot3 allows multiple 2-D plots to be stacked next to one another along one dimension, rather than directly on top of one another.

```
» x = linspace(0,3*pi);  % x-axis data
» z1 = sin(x);           % plot in x-z plane
» z2 = sin(2*x);
» z3 = sin(3*x);
» y1 = zeros(size(x));   % spread out along y-axes
» y3 = ones(size(x));    % by giving each curve different y-axis values
» y2 = y3/2;

» plot3(x,y1,z1,x,y2,z2,x,y3,z3)
» grid on
» xlabel('X-axis'), ylabel('Y-axis'), zlabel('Z-axis')
» title('Figure 27.2: sin(x), sin(2x), sin(3x)')
```

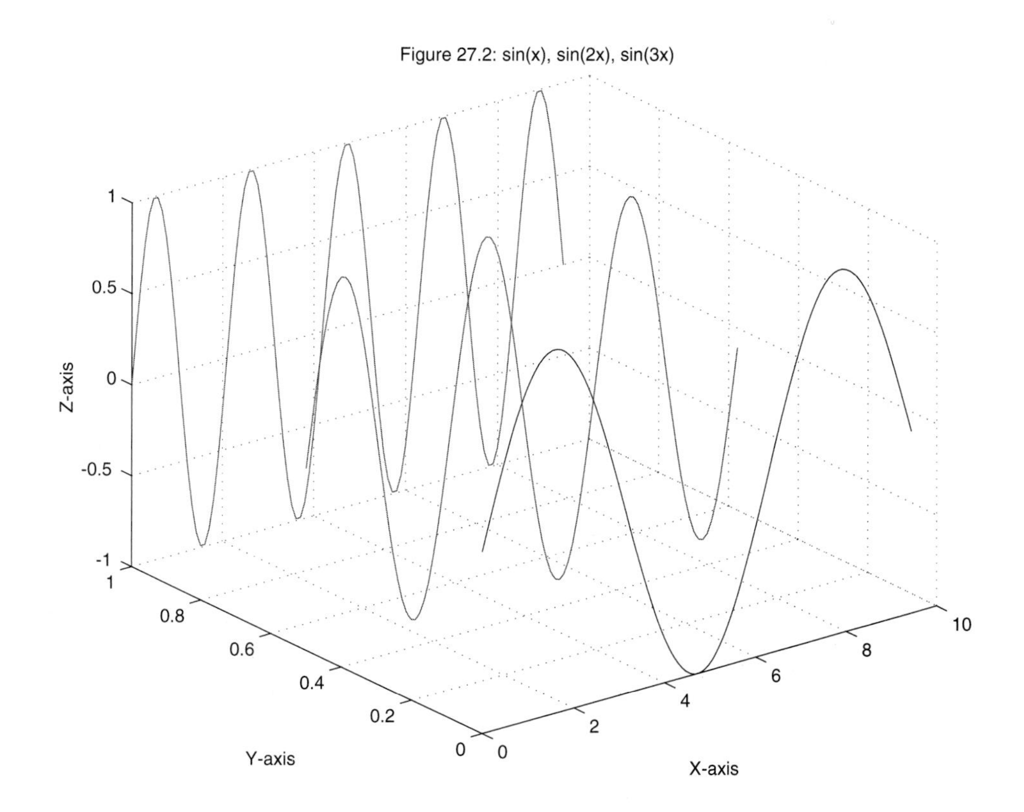

The preceding can also be stacked in other directions. For example,

```
» plot3(x,z1,y1,x,z2,y2,x,z3,y3)
» grid on
» xlabel('X-axis'), ylabel('Y-axis'), zlabel('Z-axis')
» title('Figure 27.3: sin(x), sin(2x), sin(3x)')
```

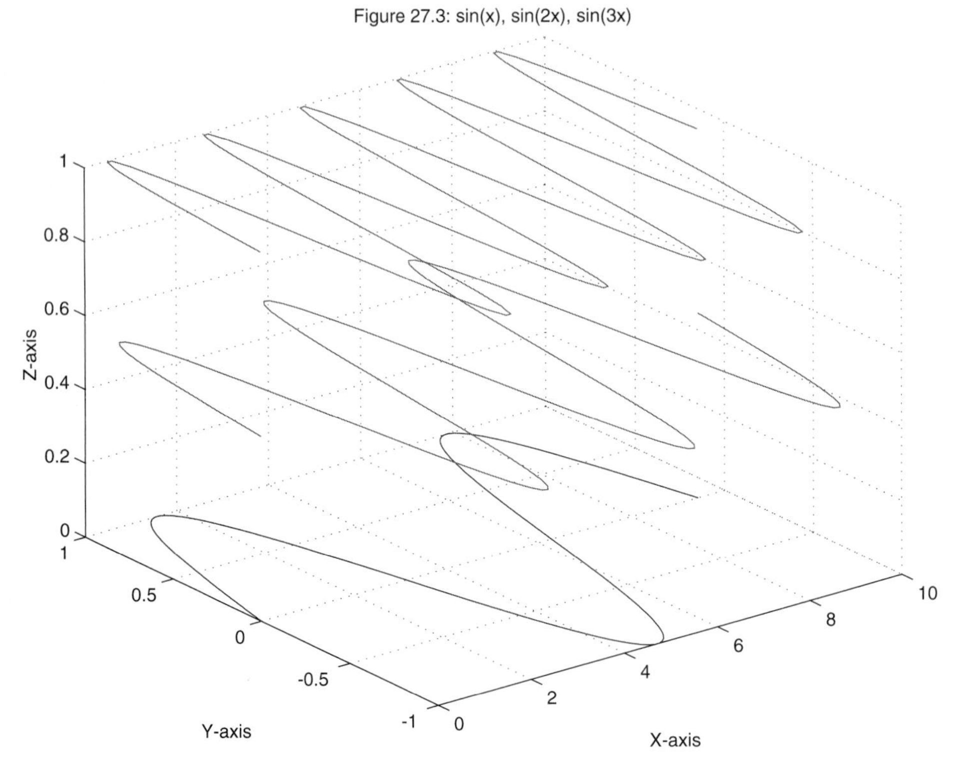


## 27.2 SCALAR FUNCTIONS OF TWO VARIABLES

As opposed to line plots generated with `plot3`, it is often desirable to visualize a scalar function of two variables; that is,

$$z = f(x,y)$$

Here each pair of values for $x$ and $y$ produces a value for $z$. A plot of $z$ as a function of $x$ and $y$ is a surface in three dimensions. To plot this surface in MATLAB, the values for $z$ are stored in a matrix. As described in the section on two-dimensional interpolation (in Chapter 19), given that `x` and `y` are the independent variables, `z` is a matrix of the dependent variable and the association of `x` and `y` to `z` is

```
z(i,:) = f(x,y(i))          and          z(:,j) = f(x(j),y)
```

That is, the $i$th row of z is associated with $i$th element of y and the $j$th column of z is associated with $j$th element of x. In other words, y varies down the columns of z and x varies across the rows of z.

When z=f(x,y) can be expressed simply, it is convenient to use array operations to compute all the values of z in a single statement. To do so requires that we create matrices of all x and y values in the proper orientation. This orientation is sometimes called ***plaid*** by *The Mathworks Inc.* MATLAB provides the function meshgrid to perform this step:

```
» x = -3:3;   % choose x-axis values
» y = 1:5;    % y-axis values
» [X,Y] = meshgrid(x,y)
X =
    -3    -2    -1     0     1     2     3
    -3    -2    -1     0     1     2     3
    -3    -2    -1     0     1     2     3
    -3    -2    -1     0     1     2     3
    -3    -2    -1     0     1     2     3
Y =
     1     1     1     1     1     1     1
     2     2     2     2     2     2     2
     3     3     3     3     3     3     3
     4     4     4     4     4     4     4
     5     5     5     5     5     5     5
```

As you can see, meshgrid duplicated x for each of the five rows in y. Similarly, it duplicated y as a column for each of the seven columns in x. This orientation agrees with the earlier statement where y varies down its columns and x varies across its rows. Given X and Y, if $z = f(x,y) = (x + y)^2$, then the matrix of data values defining the three-dimensional surface is

```
» Z = (X+Y).^2
Z =
     4     1     0     1     4     9    16
     1     0     1     4     9    16    25
     0     1     4     9    16    25    36
     1     4     9    16    25    36    49
     4     9    16    25    36    49    64
```

When a function cannot be expressed simply as shown previously, one must use For Loops or While Loops to compute the elements of Z. In many cases it may be possible to compute the elements of Z row-wise or column-wise. For example, if it is possible to compute Z row-wise, the following script file fragment can be helpful:

```
x= ???  % statement defining vector of x axis values
y= ???  % statement defining vector of y axis values

nx=length(x);  % length of x is no. of rows in Z
ny=length(y);  % length of y is no. of columns in Z

Z=zeros(nx,ny);  % initialize Z matrix for speed

for r=1:nx
  (preliminary commands)
  Z(r,:)= {a function of y and x(r) defining r-th row of Z}
end
```

On the other hand, if Z can be computed column-wise, the following script file fragment can be helpful:

```
x= ???  % statement defining vector of x axis values
y= ???  % statement defining vector of y axis values

nx=length(x);  % length of x is no. of rows in Z
ny=length(y);  % length of y is no. of columns in Z

Z=zeros(nx,ny);  % initialize Z matrix for speed

for c=1:ny
  (preliminary commands)
  Z(:,c)= {a function of y(c) and x defining c-th column of Z}
end
```

Only when the elements of Z must be computed element by element does the computation usually require a nested For Loop, such as the following script file fragment:

```
x= ???  % statement defining vector of x axis values
y= ???  % statement defining vector of y axis value

nx=length(x);  % length of x is no. of rows in Z
ny=length(y);  % length of y is no. of columns in Z

Z=zeros(nx,ny);  % initialize Z matrix for speed

for r=1:nx
  for c=1:ny
    (preliminary command)
    Z(r,c)= {a function of y(c) and x(r) defining (r,c)-th element}
  end
end
```

## 27.3 MESH PLOTS

MATLAB defines a ***mesh*** surface by the $z$-coordinates of points above a rectangular grid in the $x$-$y$ plane. It forms the mesh plot by joining adjacent points with straight lines. The result looks like a fishing net with knots at the data points. For example, consider the `peaks` function:

```
» [X,Y,Z] = peaks(30);
» mesh(X,Y,Z)
» xlabel('X-axis'), ylabel('Y-axis'), zlabel('Z-axis')
» title('Figure 27.4: Mesh Plot of Peaks')
```

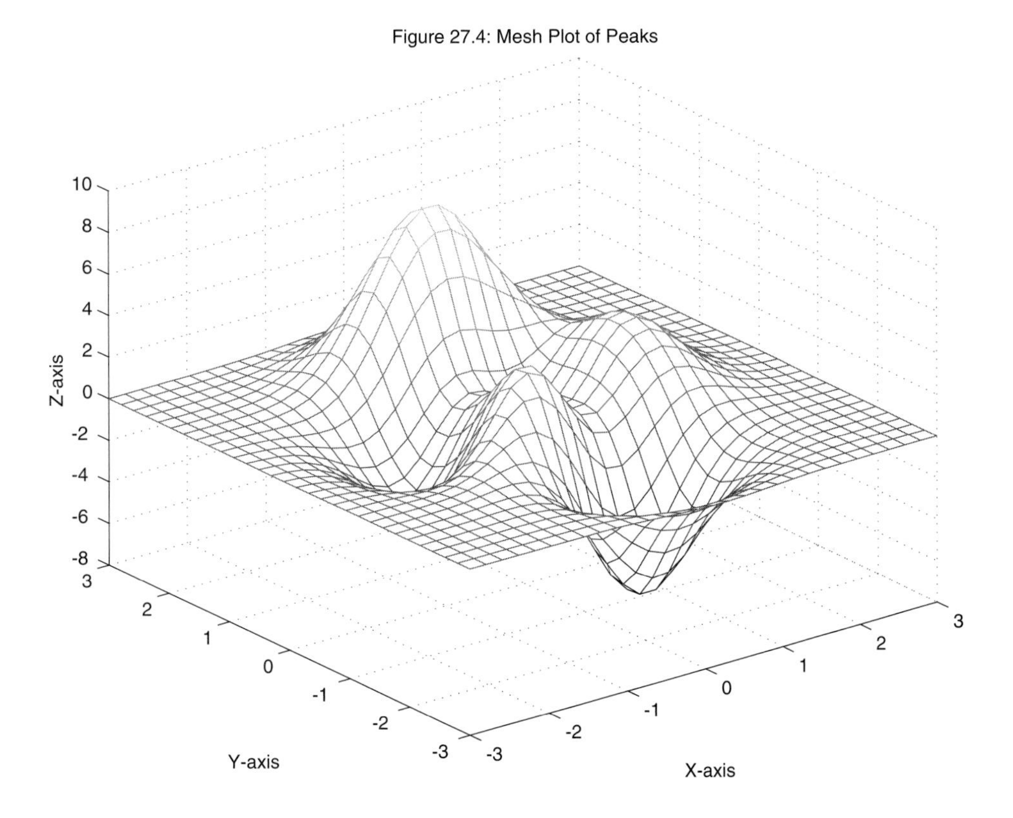


Note on your monitor how the line colors are related to the height of the mesh. In general, `mesh` will accept optional arguments to control color use in the plot. This ability to change how MATLAB uses color will be discussed in the next chapter. In any case, the use of color is called pseudocolor since color is used to add a fourth effective dimension to the graph. Note also that the plot was drawn with a grid. ***Most 3-D plots other than*** `plot3` ***and a few other exceptions default to*** `grid on`.

In addition to the preceding input arguments, `mesh` and most 3-D plot functions can also be called with a variety of input arguments. The syntax used here is the most specific in that information is supplied for all three axes. `mesh(Z)` plots the matrix `Z` versus its row and column indices. The most common variation is to use the vectors that were passed to `meshgrid` for the $x$- and $y$-axes—for example, `mesh(x,y,Z)`.

As shown in Figure 27.4, the areas between the mesh lines are opaque rather than transparent. The MATLAB command `hidden` controls this aspect of mesh plots. For example, using the MATLAB `sphere` function to generate two spheres gives

```
» [X,Y,Z]=sphere(12);
» subplot(1,2,1)
» mesh(X,Y,Z), title('Figure 27.5a: Opaque')
» hidden on
» axis off

» subplot(1,2,2)
» mesh(X,Y,Z), title('Figure 27.5b: Transparent')
» hidden off
» axis off
```

Figure 27.5a: Opaque

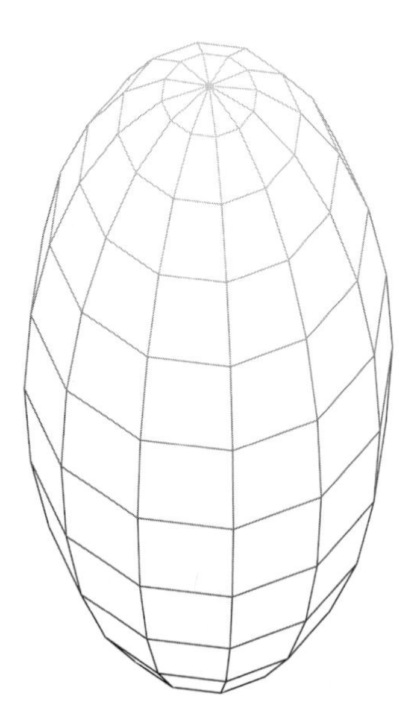

Figure 27.5b: Transparent

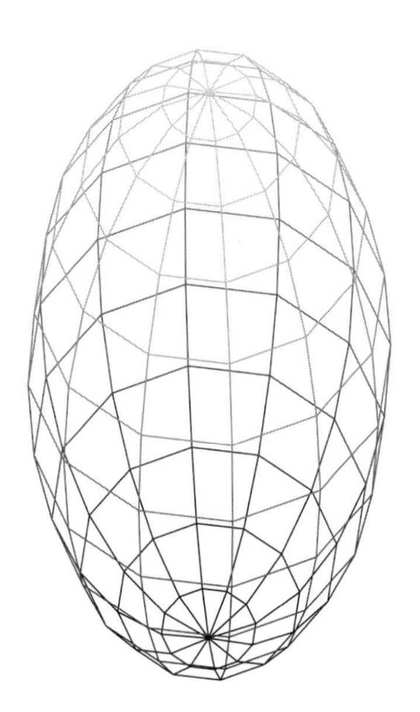

The sphere on the left is opaque (lines are hidden) whereas the one on the right is transparent (lines are not hidden).

The MATLAB `mesh` function has two siblings: `meshc`, which is a mesh plot and underlying contour plot, and `meshz`, which is a mesh plot including a zero plane.

```
» [X,Y,Z] = peaks(30);
» meshc(X,Y,Z)   % mesh plot with underlying contour plot
» title('Figure 27.6: Mesh Plot with Contours')
» meshz(X,Y,Z)  % mesh plot with zero plane
» title('Figure 27.7: Mesh Plot with Zero Plane')
```

The function `waterfall` is identical to `mesh` except that the mesh lines appear only in the *x*-direction. For example,

```
» waterfall(X,Y,Z)
» xlabel('X-axis'), ylabel('Y-axis'), zlabel('Z-axis')
» title('Figure 27.8: Waterfall Plot')
```

## 27.4 SURFACE PLOTS

A ***surface*** plot looks like a mesh plot, except that the spaces between the lines, called ***patches,*** are filled in. Plots of this type are generated using the `surf` function:

```
» [X,Y,Z] = peaks(30);
» surf(X,Y,Z)
» xlabel('X-axis'), ylabel('Y-axis'), zlabel('Z-axis')
» title('Figure 27.9: Surface Plot of Peaks')
```

Note how this plot type is a *dual* of sorts to a `mesh` plot. Here the lines are black and the patches have color, whereas in `mesh`, the patches are the color of the axes and the lines have color. As with `mesh`, color varies along the *z*-axis with each patch or line having constant color. Surface plots default to `grid on` also.

In a surface plot one does not think about hidden line removal as in a `mesh` plot, but rather one thinks about different ways to shade the surface. In the preceding `surf` plot, the shading is ***faceted*** like a stained-glass window or object, where the black lines are the joints between the constant-color patches. In addition to faceted, MATLAB provides ***flat*** shading and ***interpolated*** shading. These are applied by using the function `shading`:

```
» [X,Y,Z] = peaks(30);
» surf(X,Y,Z)  % same plot as above
» shading flat
» xlabel('X-axis'), ylabel('Y-axis'), zlabel('Z-axis')
» title('Figure 27.10: Surface Plot with Flat Shading')
```

In the flat shading shown, the black lines are removed and each patch retains its single color.

```
» shading interp
» title('Figure 27.11: Surface Plot with Interpolated Shading')
```

Figure 27.6: Mesh Plot with Contours

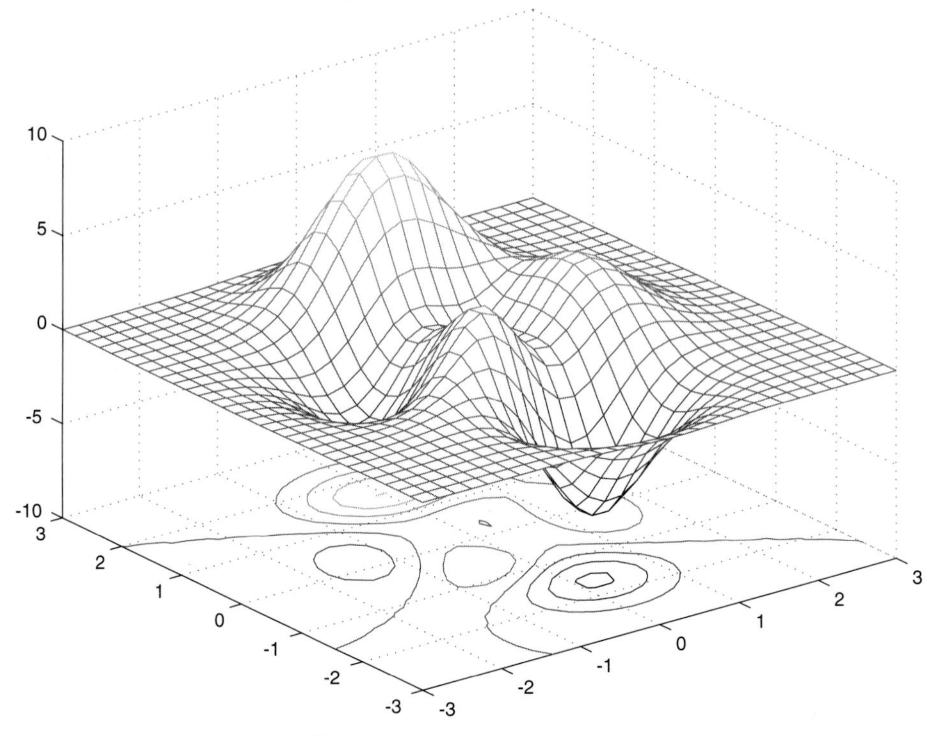


Figure 27.7: Mesh Plot with Zero Plane

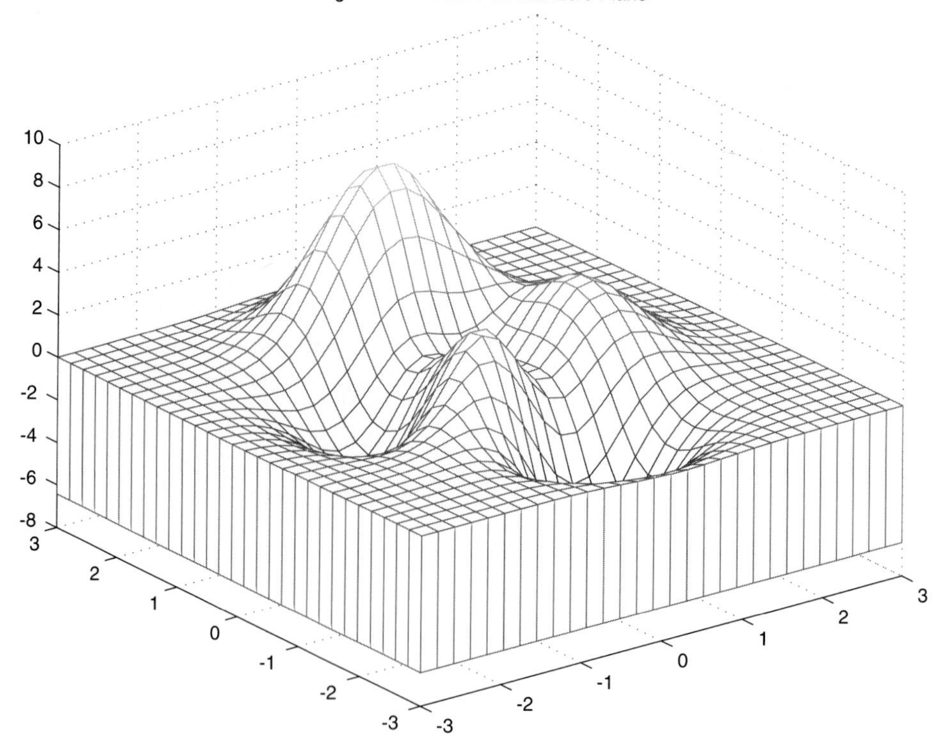

Figure 27.8: Waterfall Plot

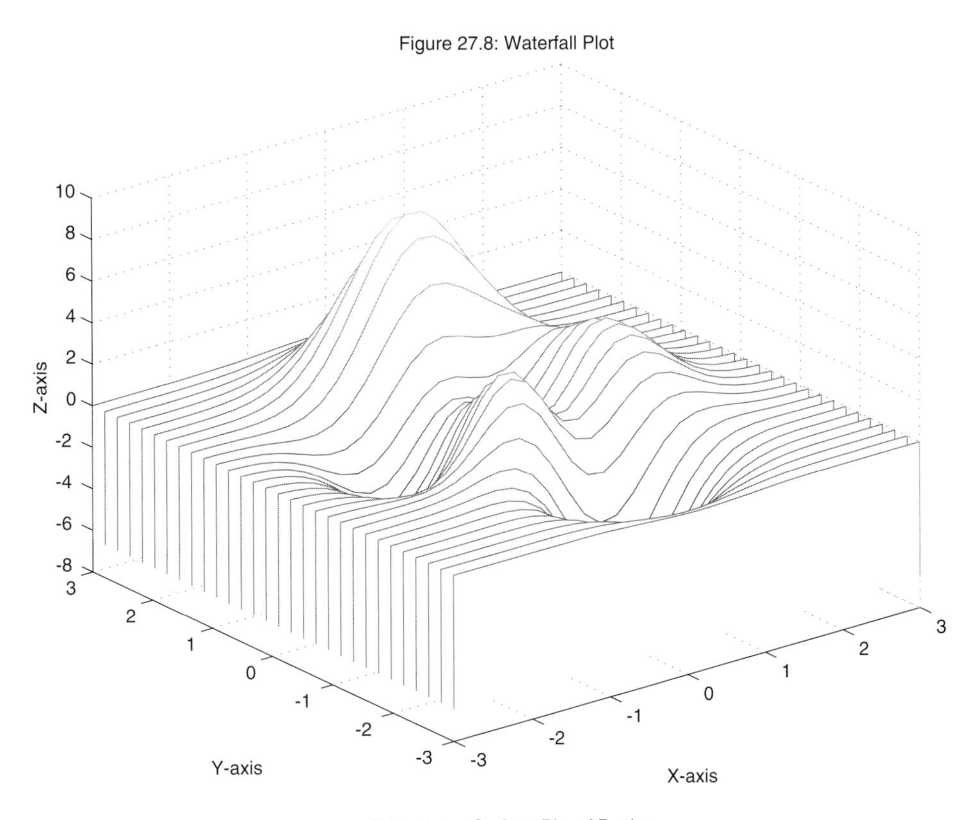


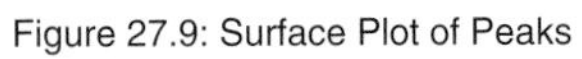
Figure 27.9: Surface Plot of Peaks

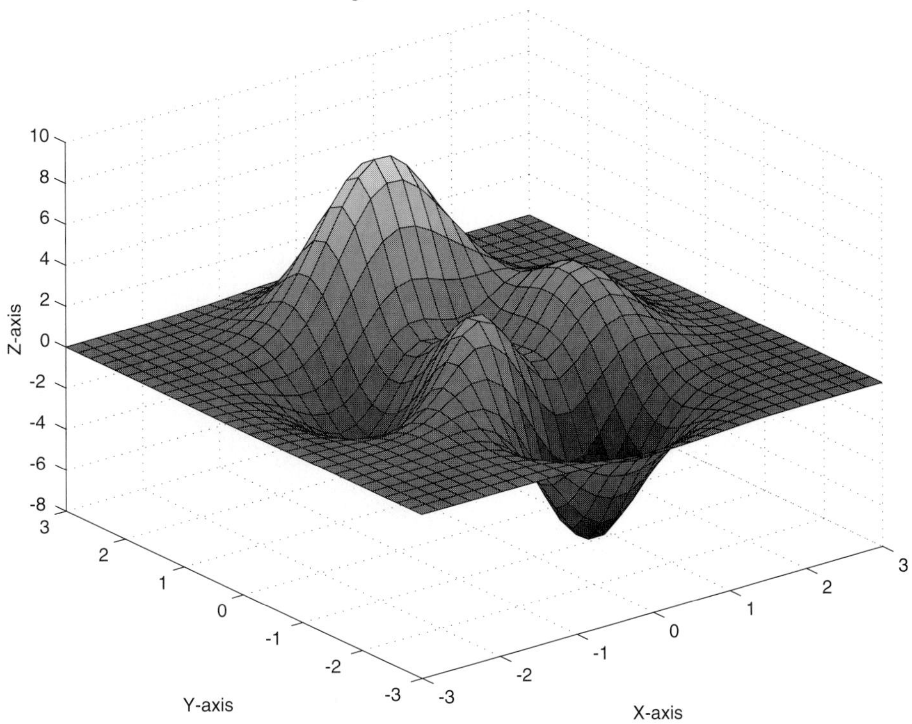

Figure 27.10: Surface Plot with Flat Shading

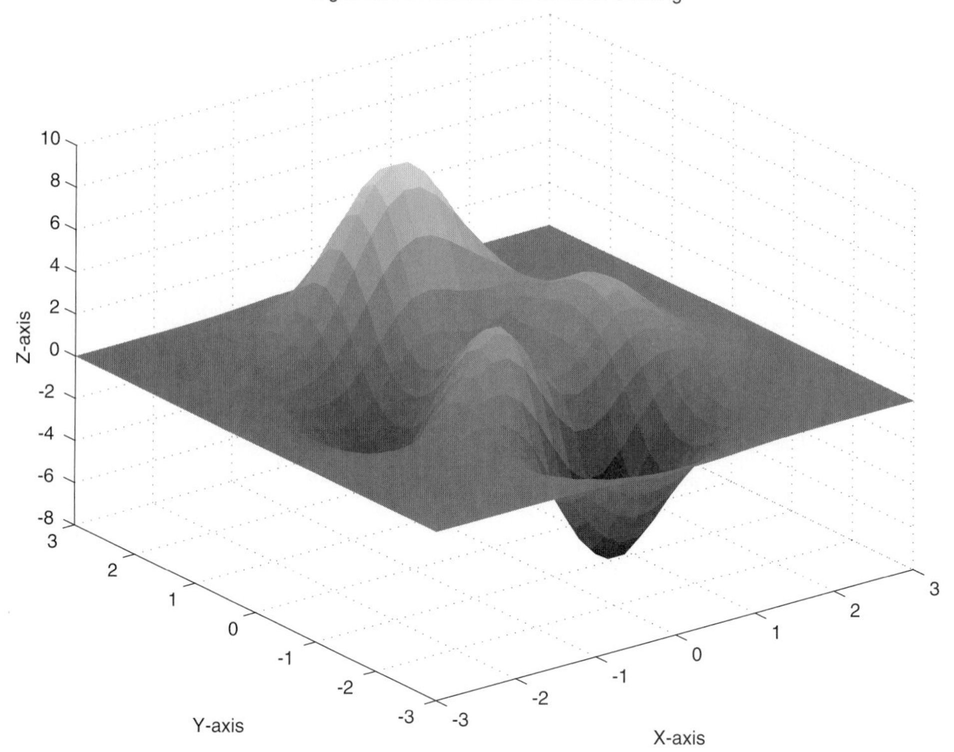


Figure 27.11: Surface Plot with Interpolated Shading

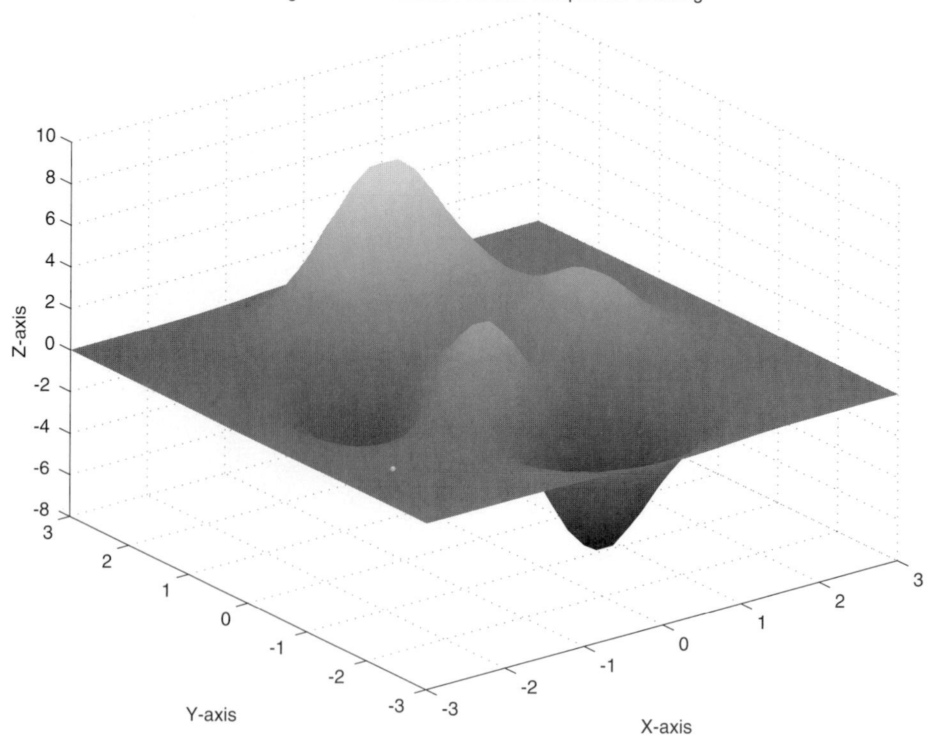

In the interpolated shading shown, the lines are also removed but each patch is given interpolated shading. That is, the color of each patch is interpolated over its area based on the color values assigned to each of its vertices. Needless to say, interpolated shading requires much more computation than faceted and flat shading.

On some computer systems, interpolated shading creates extremely long printing delays or at worst printing errors. These problems are not due to the size of the PostScript data file, but rather due to the enormous amount of computation required in the printer to generate shading that continually changes over the surface of the plot. Often the easiest solution to this problem is to use flat shading for printouts.

While shading has a significant visual impact on `surf` plots, it also applies to `mesh` plots, although in this case the visual impact is relatively minor since only the lines have color. Shading also affects `pcolor` and `fill` plots.

Since surface plots cannot be made transparent, in some situations it may be convenient to remove part of a surface so that underlying parts of the surface can be seen. In MATLAB this is accomplished by setting the data values where holes are desired to the special value `NaN`. ***Since*** `NaNs` ***have no value, all MATLAB plotting functions simply ignore*** `NaN` ***data points, leaving a hole in the plot where they appear.*** For example,

```
» [X,Y,Z] = peaks(30);
» x = X(1,:);   % vector of x axis
» y = Y(:,1);   % vector of y axis

» i = find(y>.8 & y<1.2);   % find y axis indices of hole
» j = find(x>-.6 & x<.5);   % find x axis indices of hole
» Z(i,j) = nan*Z(i,j);   % set values at hole indices to NaNs

» surf(X,Y,Z)
» xlabel('X-axis'), ylabel('Y-axis'), zlabel('Z-axis')
» title('Figure 27.12: Surface Plot with a Hole')
```

The MATLAB `surf` function also has two siblings: `surfc`, which is a surface plot and underlying contour plot, and `surfl`, which is a surface plot with lighting. For example,

```
» [X,Y,Z] = peaks(30);
» surfc(X,Y,Z)   % surf plot with contour plot
» xlabel('X-axis'), ylabel('Y-axis'), zlabel('Z-axis')
» title('Figure 27.13: Surface Plot with Contours')
```

Figure 27.12: Surface Plot with a Hole

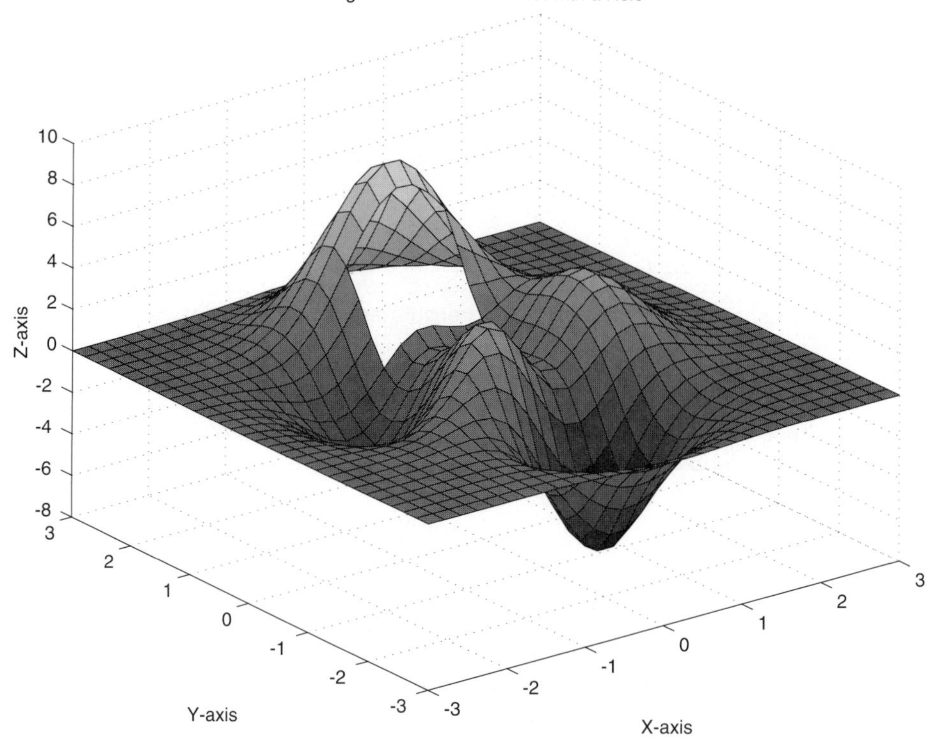


Figure 27.13: Surface Plot with Contours

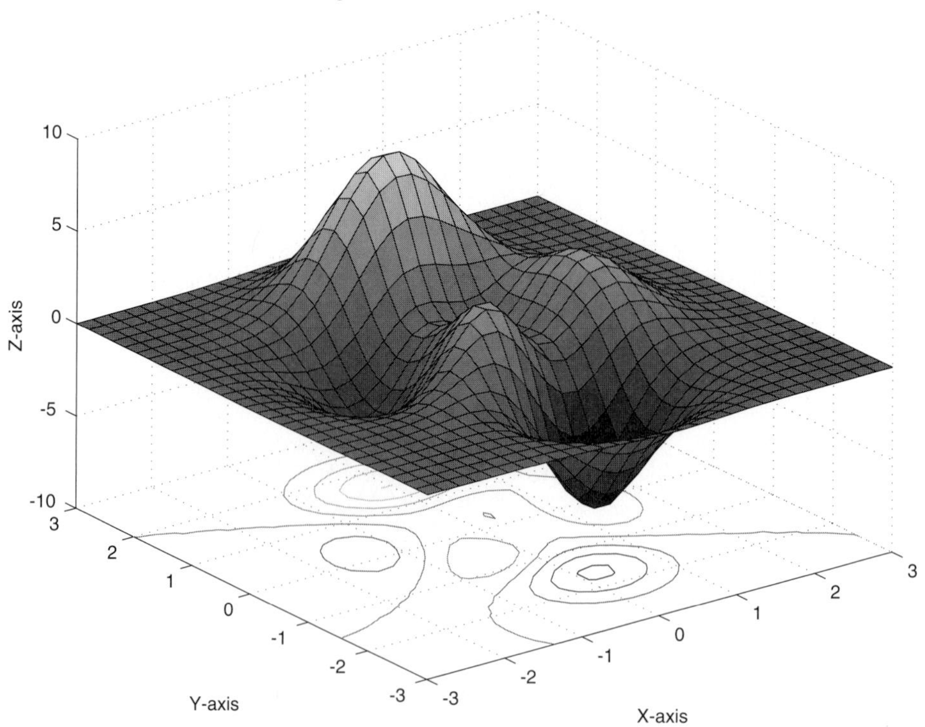

```
» [X,Y,Z] = peaks(30);
» surfl(X,Y,Z)     % surf plot with lighting
» shading interp   % surfl plots look best with interp shading
» colormap pink    % they also look better in a single color
» xlabel('X-axis'), ylabel('Y-axis'), zlabel('Z-axis')
» title('Figure 27.14: Surface Plot with Lighting')
```

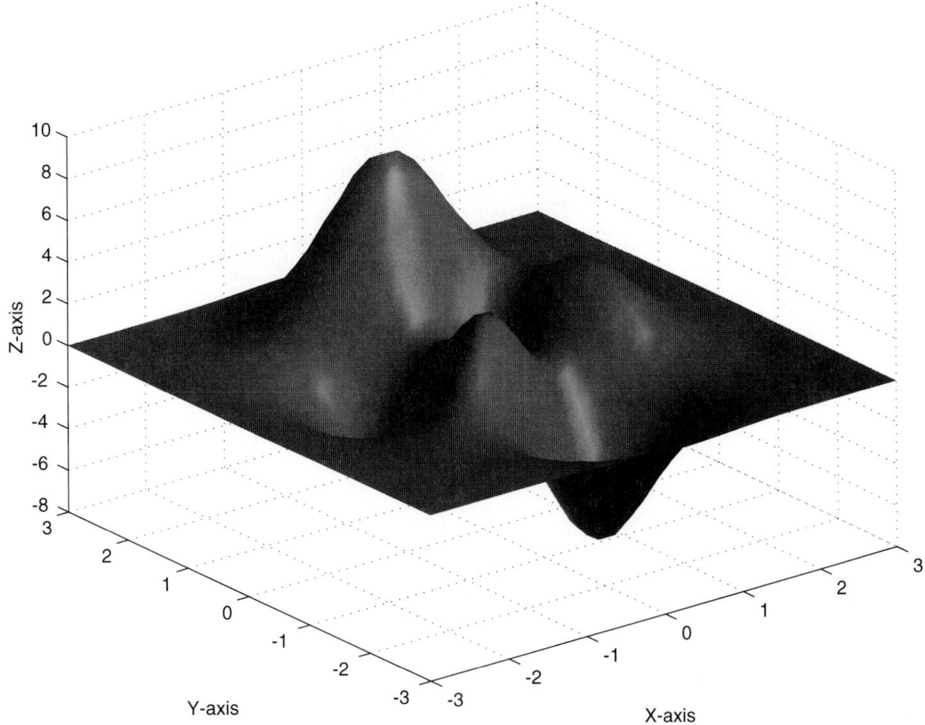


The function `surfl` makes a number of assumptions about the light applied to the surface. The next chapter provides more information regarding light properties. Also, in the preceding executed commands, `colormap` is a MATLAB function for applying a different set of colors to a figure. This function is discussed in the next chapter as well.

The `surfnorm(X,Y,Z)` function computes surface normals for the surface defined by `X`, `Y`, and `Z`, plots the surface, and plots vectors normal to the surface at the data points. The surface normals are unnormalized and valid at each vertex. The form `[Nx,Ny,Nz]=surfnorm(X,Y,Z)` computes the three-dimensional surface normals and returns their components, but does not plot the surface.

```
» [X,Y,Z] = peaks(28);
» surfnorm(X,Y,Z)
» xlabel('X-axis'), ylabel('Y-axis'), zlabel('Z-axis')
» title('Figure 27.15: Surface Plot with Normals')
```

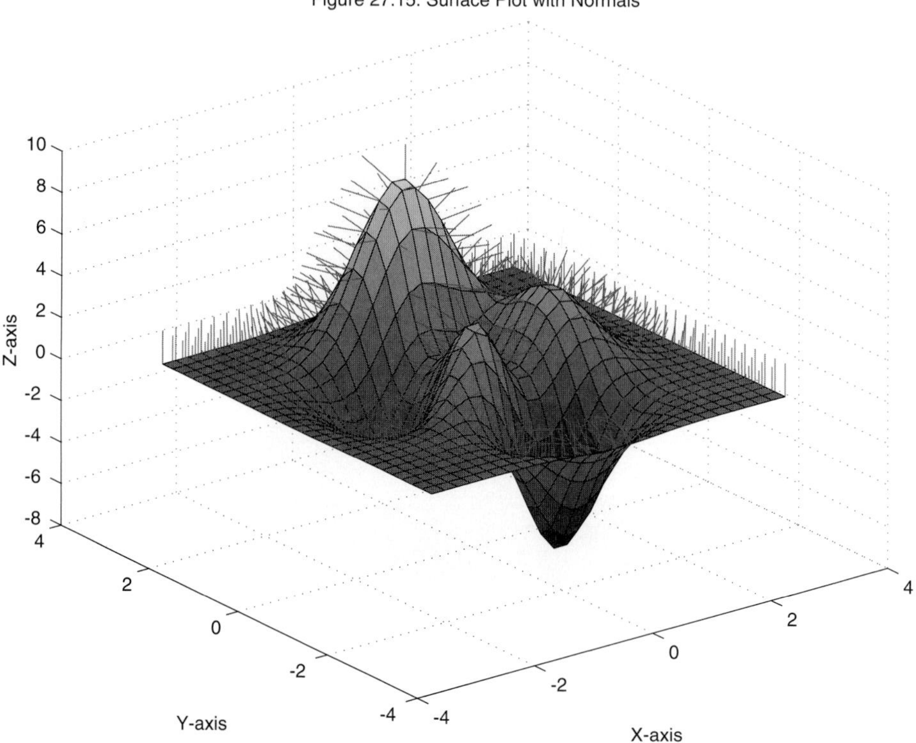


## 27.5 MESH AND SURFACE PLOTS OF IRREGULAR DATA

Irregular or nonuniformly spaced data can be visualized using the functions `trimesh` and `trisurf`:

```
» x = rand(1,50);
» y = rand(1,50);
» z = peaks(x,y*pi);
» t = delaunay(x,y);

» trimesh(t,x,y,z)
» hidden off
» title('Figure 27.16: Triangular Mesh Plot')
```

The `delaunay` function returns a set of triangles connecting the data points with their *natural* neighbors. `trisurf` is the surface equivalent to `trimesh`:

```
» trisurf(t,x,y,z)
» title('Figure 27.17: Triangular Surface Plot')
```

Figure 27.16: Triangular Mesh Plot

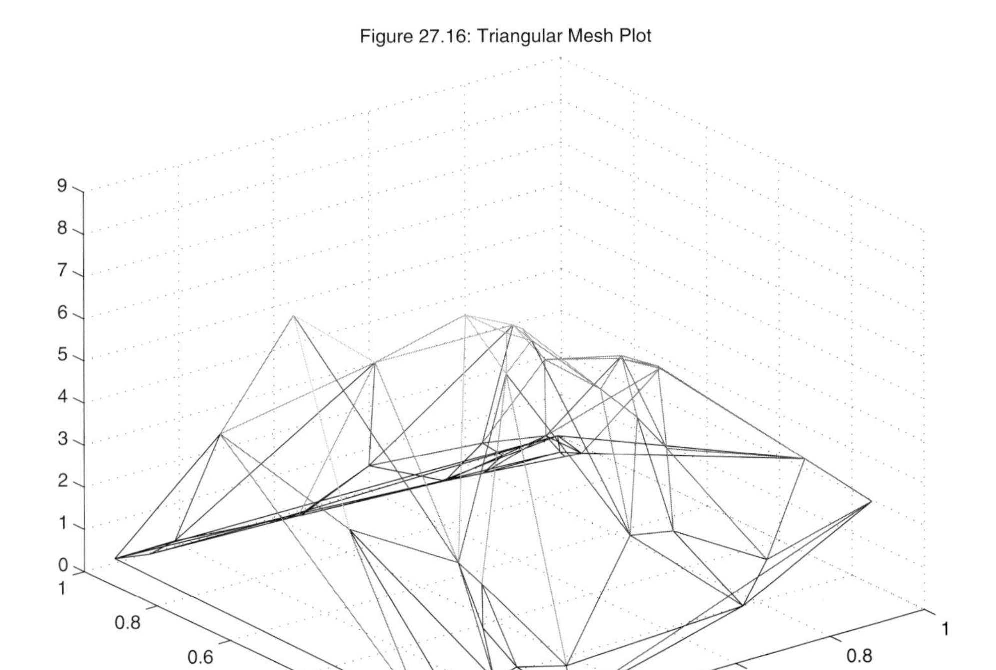

Figure 27.17: Triangular Surface Plot

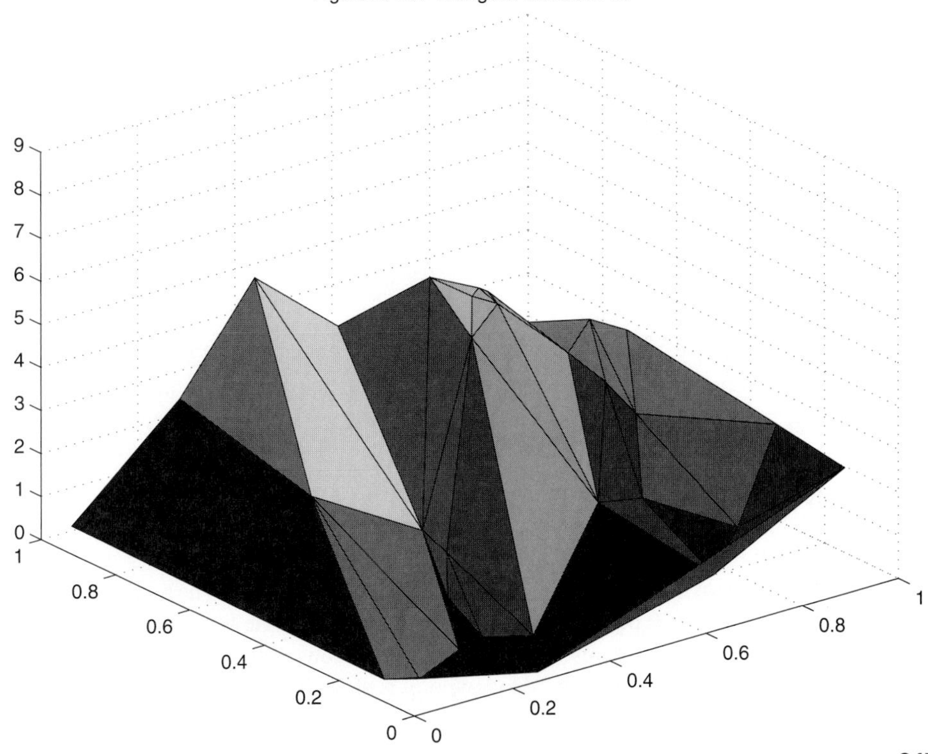

Voronoi diagrams are a closest-point plotting technique related to Delaunay triangulation. Each polygon encloses all points that are closer to the enclosed data point than to any other point in the data set. A Voronoi diagram of the $(x,y)$ data points can be generated with the `voronoi` function:

```
» voronoi(x,y)
» hold on
» plot(x,y,'rx')  % emphasize the data points
» hold off
» title('Figure 27.18: Voronoi Plot')
```

Figure 27.18: Voronoi Plot



The interpolation chapter (Chapter 19) discusses Delaunay triangulation and Voronoi diagrams in greater detail.

## 27.6 CHANGING VIEWPOINTS

Note that the default viewpoint of 3-D plots is looking down at the $z = 0$ plane at an angle of 30° and looking up at the $x = 0$ plane at an angle of 37.5°. The angle of orientation with respect to the $z = 0$ plane is called ***elevation*** and the angle with respect to the $x = 0$ plane is called ***azimuth.*** Thus the default 3-D viewpoint is elevation = 30° and azimuth = −37.5°. The default 2-D viewpoint is elevation = 90° and azimuth = 0°. The concepts of azimuth and elevation are described visually in the following figure.

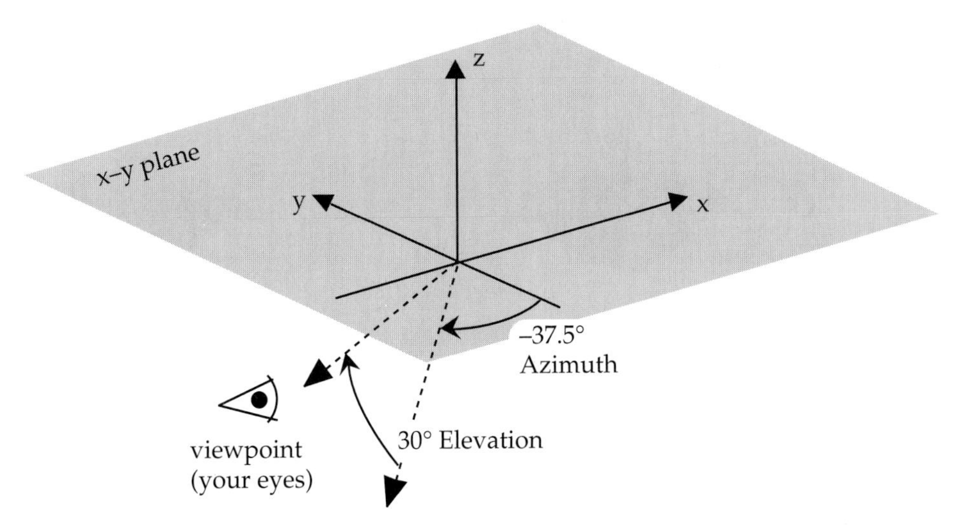


In MATLAB, the function `view` changes the graphical viewpoint for all types of 2-D and 3-D plots. `view(az,el)` and `view([az,el])` change the viewpoint to the specified azimuth `az` and elevation `el`. Consider the following examples:

```
» x = -7.5:.5:7.5; y=x;         % create a data set
» [X,Y] = meshgrid(x,y);
» R = sqrt(X.^2+Y.^2)+eps;
» Z = sin(R)./R;

» subplot(2,2,1)
» surf(X,Y,Z)
» xlabel('X-axis'), ylabel('Y-axis'), zlabel('Z-axis')
» title('Figure 27.19a: Default Az = -37.5, El = 30')
» view(-37.5,30)

» subplot(2,2,2)
» surf(X,Y,Z)
» xlabel('X-axis'), ylabel('Y-axis'), zlabel('Z-axis')
» title('Figure 27.19b: Az Rotated to 52.5')
» view(-37.5+90,30)

» subplot(2,2,3)
» surf(X,Y,Z)
» xlabel('X-axis'), ylabel('Y-axis'), zlabel('Z-axis')
» title('Figure 27.19c: El Increased to 60')
» view(-37.5,60)

» subplot(2,2,4)
» surf(X,Y,Z)
» xlabel('X-axis'), ylabel('Y-axis')
» title('Figure 27.19d: Az = 0, El = 90')
» view(0,90)
```

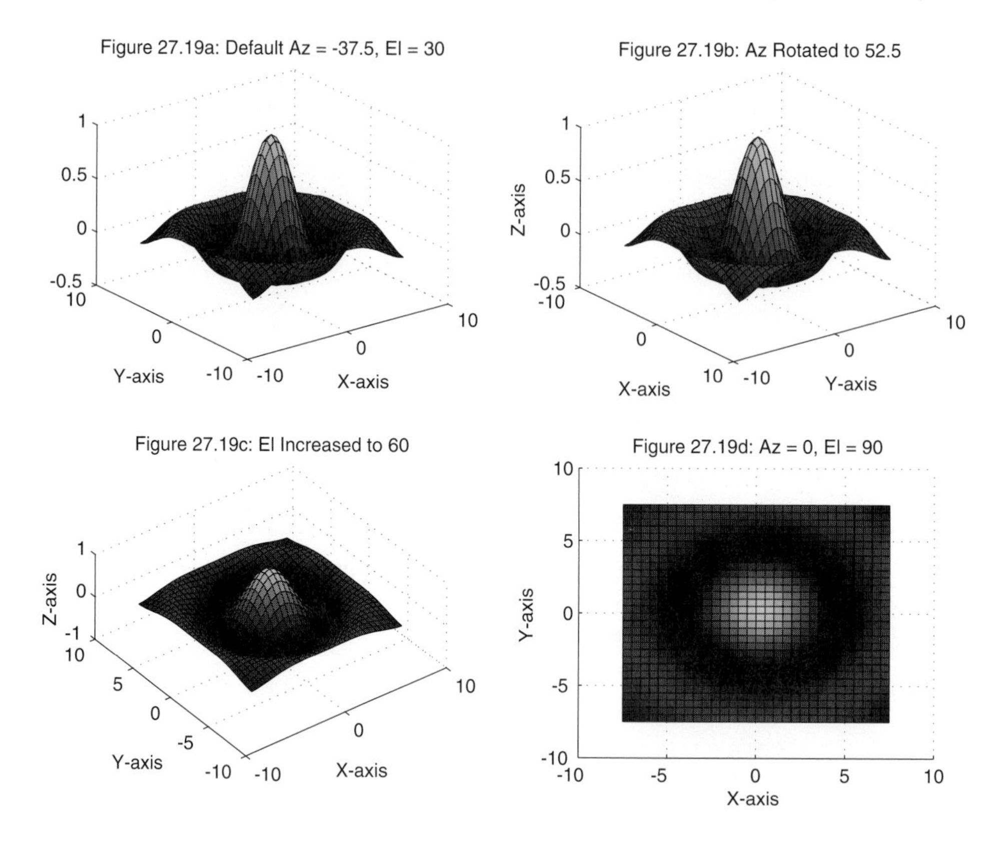


In addition to the preceding form, `view` also offers additional features summarized in the following table.

| `view` syntax | Description |
|---|---|
| `view(az,el)`<br>`view([az el])` | Sets view to azimuth `az` and elevation `el`. |
| `view([x,y,z]` | Sets view looking toward the origin along the vector `[x,y,z]` in cartesian coordinates, e.g., `view([0 0 1])=view(0,90)`. |
| `view(2)` | Sets the default 2-D view, `az` = 0, `el` = 90. |
| `view(3)` | Sets the default 3-D view, `az` = −37.5, `el` = 30. |
| `[az,el]=view` | Returns the current azimuth `az` and elevation `el`. |
| `view(T)` | Uses the 4 by 4 transformation matrix `T` to set the view. |
| `T=view` | Returns the current 4 by 4 transformation matrix. |

The `view` function has another form that may be more useful in some instances. `view([x y z])` places your viewpoint on a vector connecting the origin and the cartesian coordinate `(x,y,z)` in 3-D space. The distance you are from the origin is not affected. For example, `view([0 -10 0])`, `view([0 -1 0])`, and `view(0,0)` all produce the same view. In addition, the azimuth and elevation of the current view can be obtained using `[az,el]=view`. For example,

```
» view([-7 -9 7]) % on a line from the origin through (-7,-9,7)
» [az,el] = view % find the azimuth and elevation of this viewpoint
az =
   -37.8750
el =
    31.5475
```

In addition to the `view` function, the viewpoint can be set interactively with the mouse by using the function `rotate3d`. `rotate3d on` turns on mouse-based view rotation, `rotate3d off` turns it off, and `rotate3d` with no arguments toggles the state. Try `rotate3d` with a mesh or surface plot to see how it works.

## 27.7 CONTOUR PLOTS

Contour plots shows lines of constant elevation or height. If you have ever seen a topographical map, you know what a contour plot looks like. In MATLAB contour plots in 2-D and 3-D are generated using the `contour` and `contour3` functions, respectively. For example,

```
» [X,Y,Z] = peaks;
» contour(X,Y,Z,20)          % generate 20 2-D contour lines
» xlabel('X-axis'), ylabel('Y-axis')
» title('Figure 27.20: 2-D Contour Plot')
```

Note that the color of each line follows the color order of the current colormap in the same way that `mesh` and `surf` plots do. Plotting the same data in 3-D gives

```
» contour3(X,Y,Z,20)         % the same contour plot in 3-D
» xlabel('X-axis'), ylabel('Y-axis'), zlabel('Z-axis')
» title('Figure 27.21: 3-D Contour Plot')
```

The `pcolor` function maps height to a set of colors and presents the same information as the contour plot at the same scale. For example,

```
» pcolor(X,Y,Z)
» shading interp  % remove the grid lines
» title('Figure 27.22: Pseudocolor Plot')
```

Figure 27.20: 2-D Contour Plot

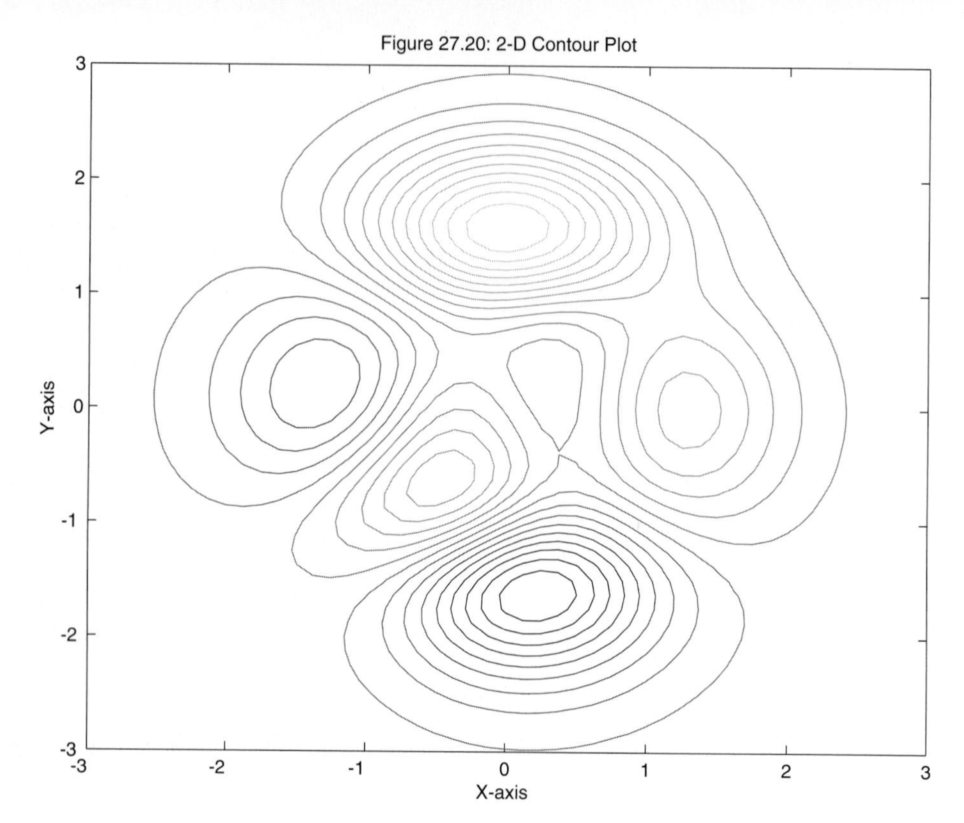


Figure 27.21: 3-D Contour Plot

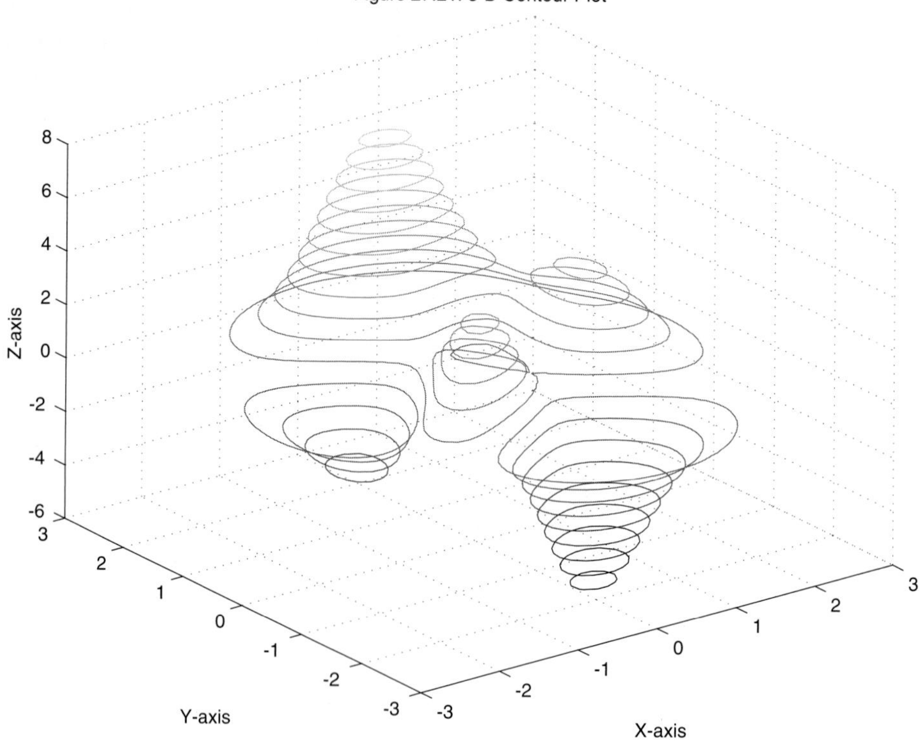

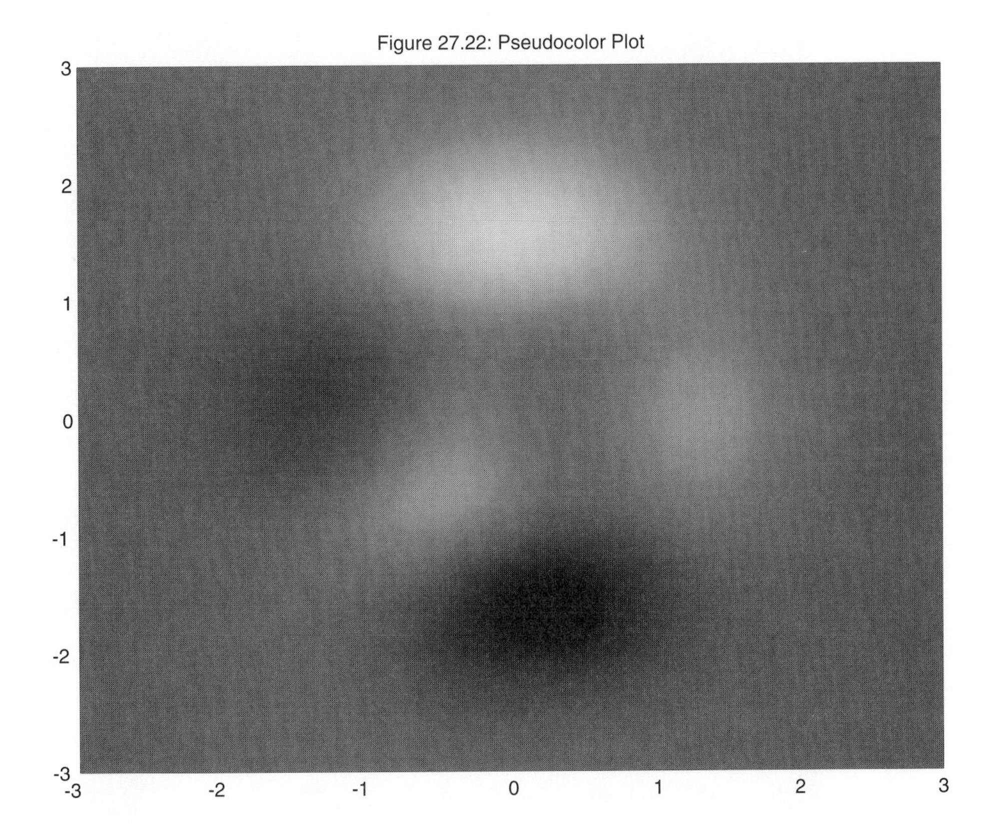


Figure 27.22: Pseudocolor Plot

Since both `pcolor` and `contour` show the same information at the same scale, it is often useful to superimpose the two:

```
» pcolor(X,Y,Z)        % generate the pseudocolor plot
» shading interp       % remove the grid lines
» xlabel('X-axis'), ylabel('Y-axis')
» hold on
» contour(X,Y,Z,12,'k') % plot twelve contour lines in black
» hold off
» title('Figure 27.23: Pseudocolor Plot with Contours')
```

As a matter of fact, these plots are so useful that MATLAB provides the `contourf` function to generate filled contour plots:

```
» contourf(X,Y,Z,12)          % filled contour plot with 12 contours
» xlabel('X-axis'), ylabel('Y-axis')
» title('Figure 27.24: Filled Contour Plot')
```

Figure 27.23: Pseudocolor Plot with Contours

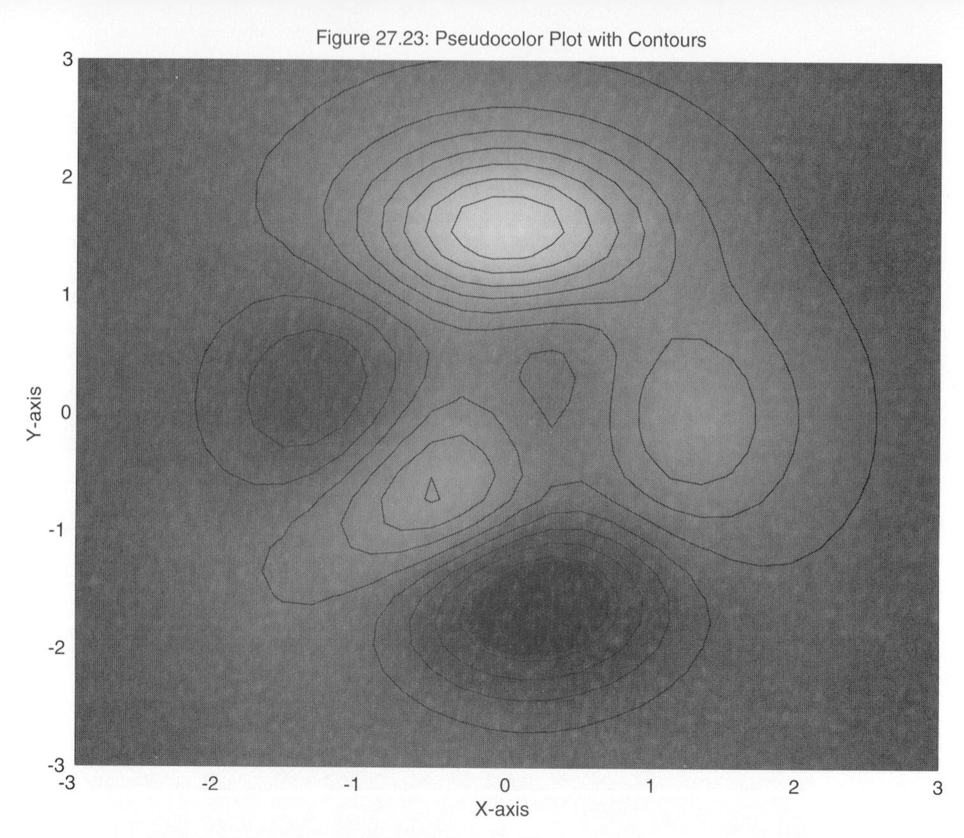


Figure 27.24: Filled Contour Plot

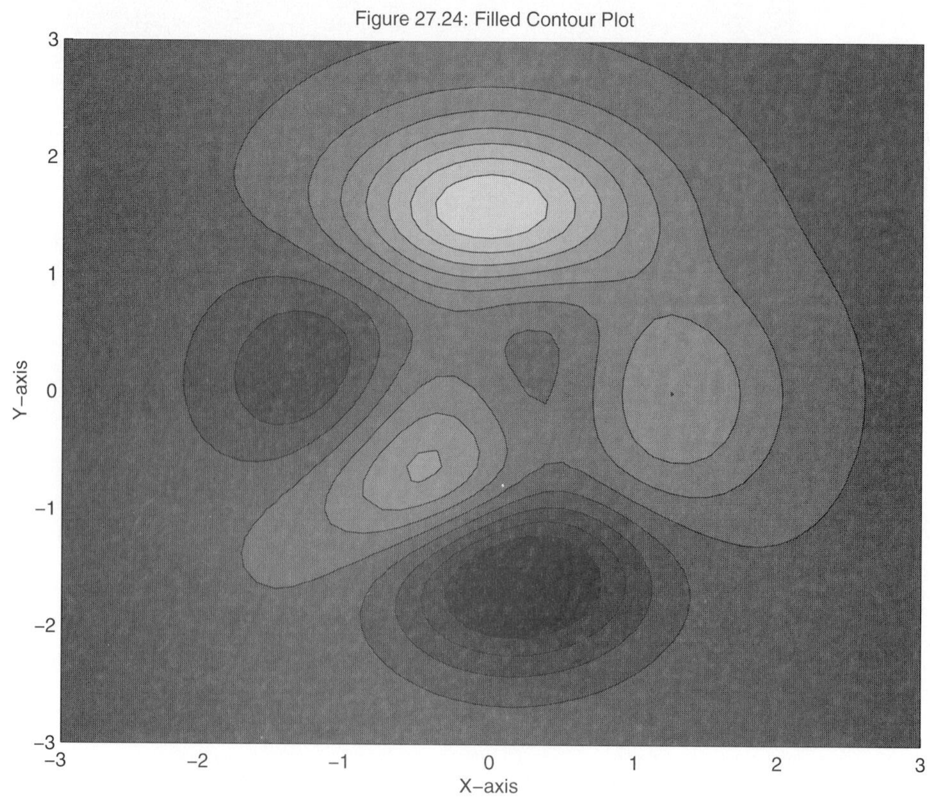

Contour lines can be labeled using the `clabel` function. `clabel` requires a matrix of lines and optional text strings that are returned by `contour`, `contourf`, and `contour3`. For example,

```
» C = contour(X,Y,Z,12);
» clabel(C)
» title('Figure 27.25: Contour Plot With Labels')
```

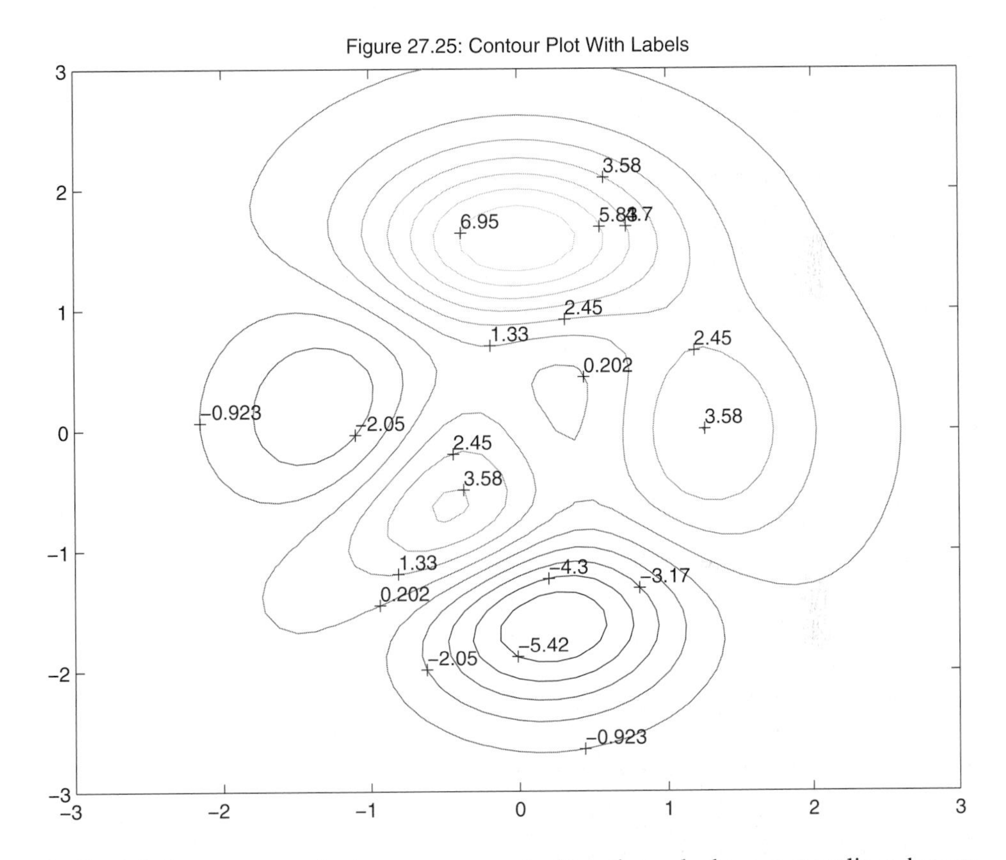


In-line labels are generated using two arguments. Note that only those contour lines that are large enough to accommodate an in-line label are labeled.

```
» [C,h] = contour(X,Y,Z,8);
» clabel(C,h)
» title('Figure 27.26: Contour Plot With In-line Labels')
```

Finally, you can select which contours to label using the mouse and the `'manual'` option. In-line or horizontal labels are used depending on the presence of the second argument `h`, as illustrated previously.

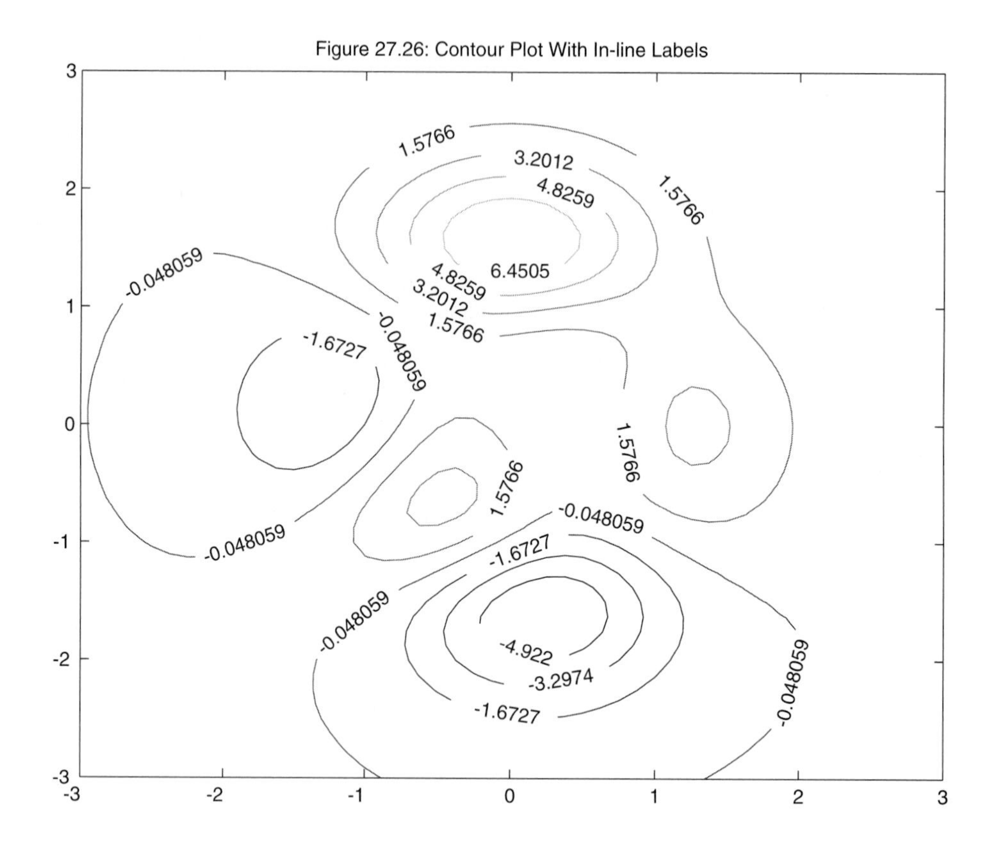

Figure 27.26: Contour Plot With In-line Labels

## 27.8 SPECIALIZED 3-D PLOTS

MATLAB provides a number of specialized plotting functions in addition to those discussed previously.

The function `ribbon(Y)` plots the columns of `Y` as separate ribbons. `ribbon(x,Y)` plots `x` versus the columns of `Y`. The width of the ribbons may be specified as well using the syntax `ribbon(x,Y,width)`, where the default width is 0.75. For example,

```
» Z = peaks
» ribbon(Z)
» title('Figure 27.27: Ribbon Plot of Peaks')
```

The function `quiver(x,y,dx,dy)` draws directional or velocity vectors `(dx,dy)` at the points `(x,y)`. For example,

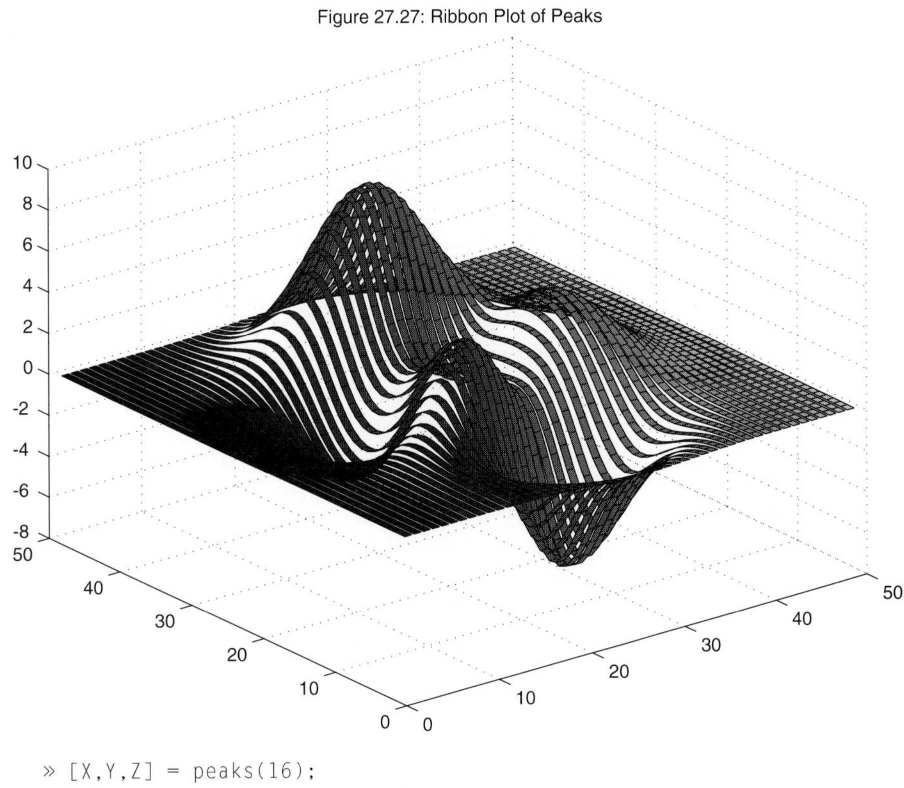


```
» [X,Y,Z] = peaks(16);
» [DX,DY] = gradient(Z,.5,.5);
» contour(X,Y,Z,10)
» hold on
» quiver(X,Y,DX,DY)
» hold off
» title('Figure 27.28: 2-D Quiver Plot')
```

Three-dimensional quiver plots of the form `quiver3(x,y,z,Nx,Ny,Nz)` display the vectors (`Nx`, `Ny`, `Nz`) at the points (`x`, `y`, `z`). For example,

```
» [X,Y,Z] = peaks(20);
» [Nx,Ny,Nz] = surfnorm(X,Y,Z);
» surf(X,Y,Z)
» hold on
» quiver3(X,Y,Z,Nx,Ny,Nz)
» hold off
» title('Figure 27.29: 3-D Quiver Plot')
```

`fill3`, being the 3-D equivalent of `fill`, draws filled polygons in 3-D space. `fill3(X,Y,Z,C)` uses the arrays `X`, `Y`, and `Z` as the vertices of the polygon and `C`

Figure 27.28: 2-D Quiver Plot

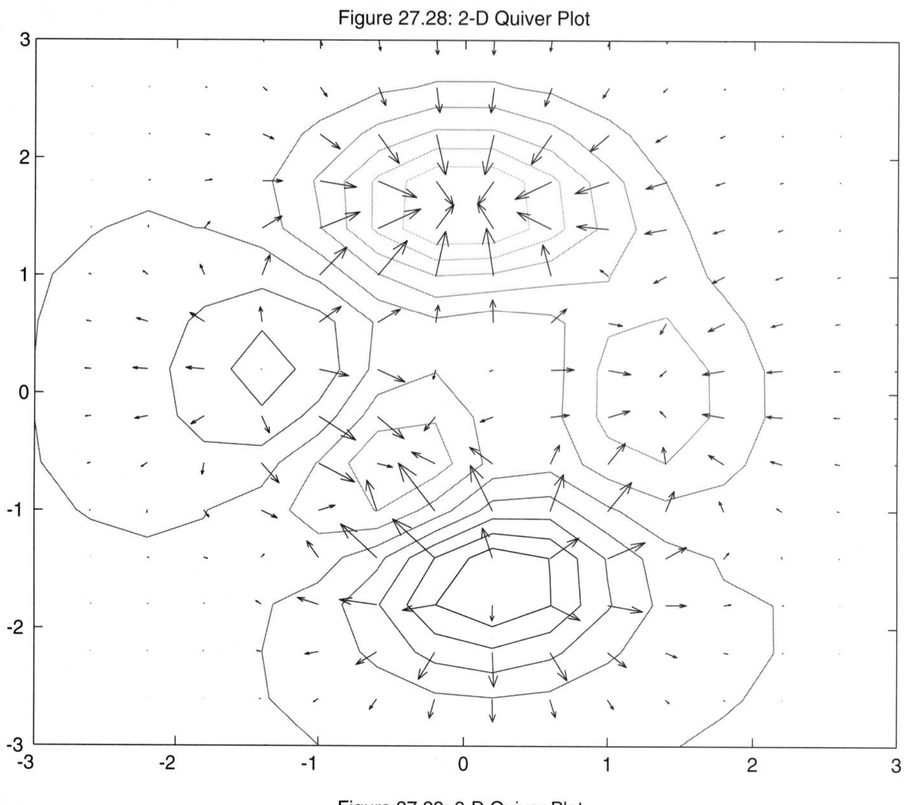


Figure 27.29: 3-D Quiver Plot

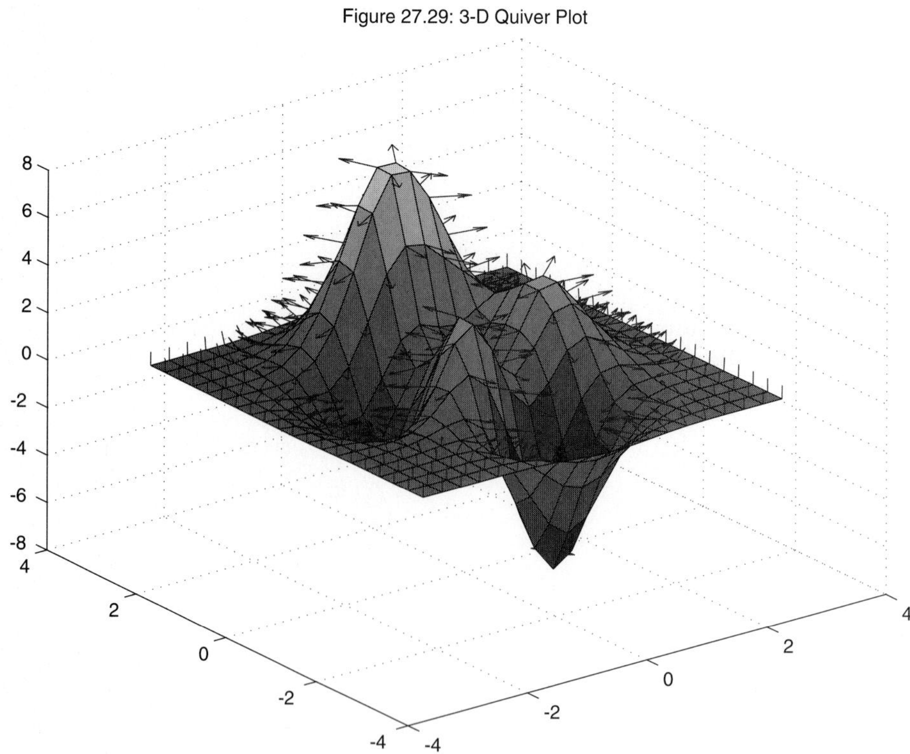

specifies the fill color. The following example draws five triangles with random vertices and fills them with colors from the current colormap:

```
» fill3(rand(3,5),rand(3,5),rand(3,5),rand(3,5))
» grid on
» title('Figure 27.30: Five Filled Triangles')
```

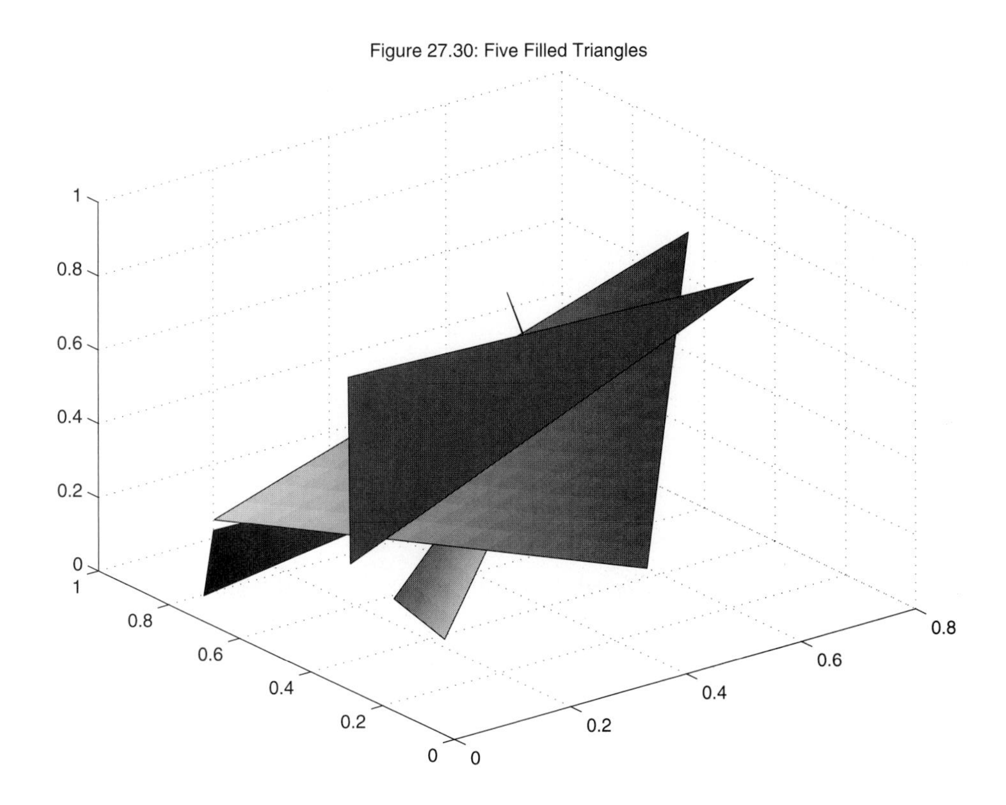


The 3-D equivalent of `stem` plots discrete sequence data in 3-D space. `stem3(X,Y,Z,C,'filled')` plots the data points in `(X,Y,Z)` with lines extending to the `X-Y` plane. The optional argument `C` specifies the marker style and/or color, and the optional `'filled'` argument causes the marker to be filled in. `stem3(Z)` plots the points in `Z` and automatically generates `X` and `Y` values. For example,

```
» Z = rand(5);
» stem3(Z,'ro','filled');
» grid on
» title('Figure 27.31: Stem Plot of Random Data')
```

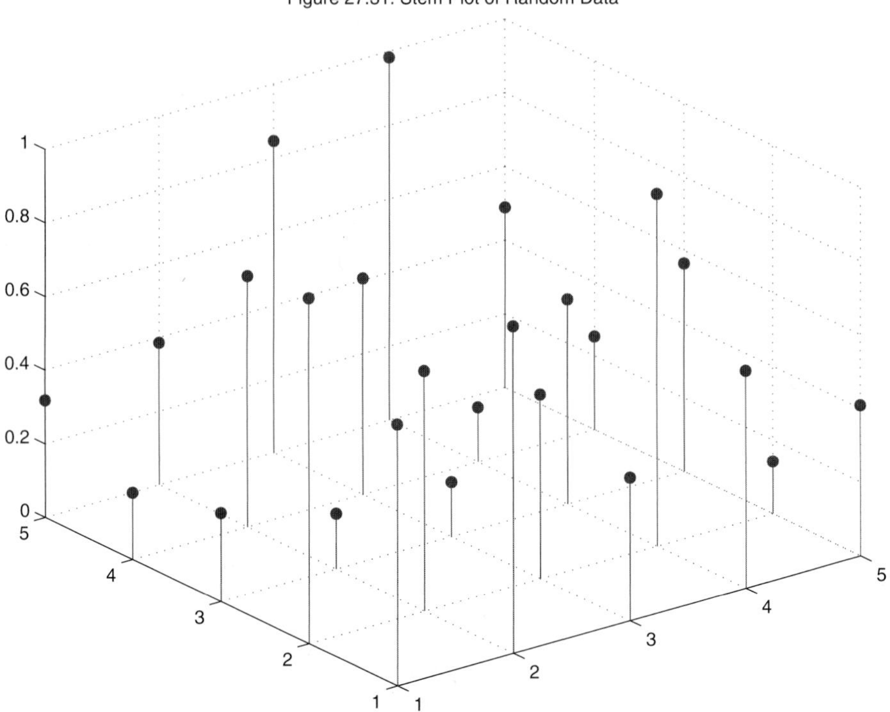


The three-dimensional version of the `comet` plot is `comet3`. `comet(X)` creates an animated plot of the vector `X`. `comet3(X,Y,Z)` plots the points in (`X`, `Y`, `Z`). The length of the comet tail may be specified with an optional argument. For example,

```
» x = -5*pi:pi/100:5*pi;
» comet3(cos(2*t),sin(2*t),t);
» title('Figure 27.32: 3-D Comet Plot')
```

MATLAB provides the function `slice` for visualizing three-dimensional volumes. The best way to explain how `slice` works is to use a few examples. First, generate 3-D plaid data and a function `V` within the volume.

```
» [X,Y,Z] = meshgrid(-10:10, -10:2:10, -10:1.5:10);
» V = sqrt(X,^2 + Y.^2 + Z.^2);
```

Now slice the volume through the `X=0`, `Y=0`, and `Z=0` planes.

```
» slice(X,Y,Z,V,0,0,0)
» xlabel('X-axis'), ylabel('Y-axis'), zlabel('Z-axis')
» title('Figure 27.33: Slices through the X=0, Y=0, Z=0 planes')
```

Figure 27.32: 3-D Comet Plot

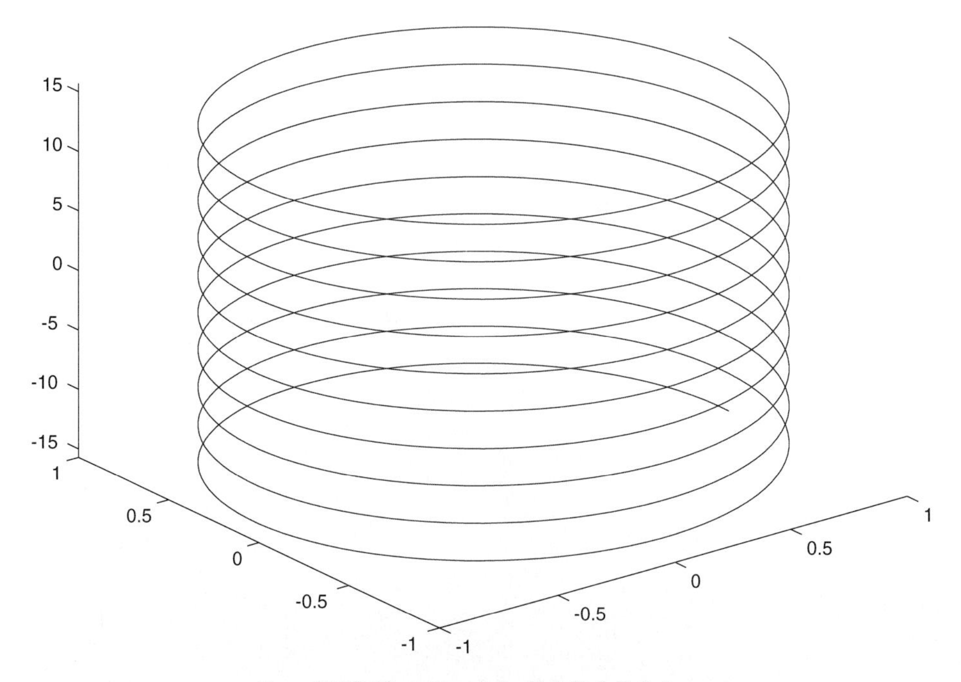


Figure 27.33: Slices through the X=0, Y=0, Z=0 planes

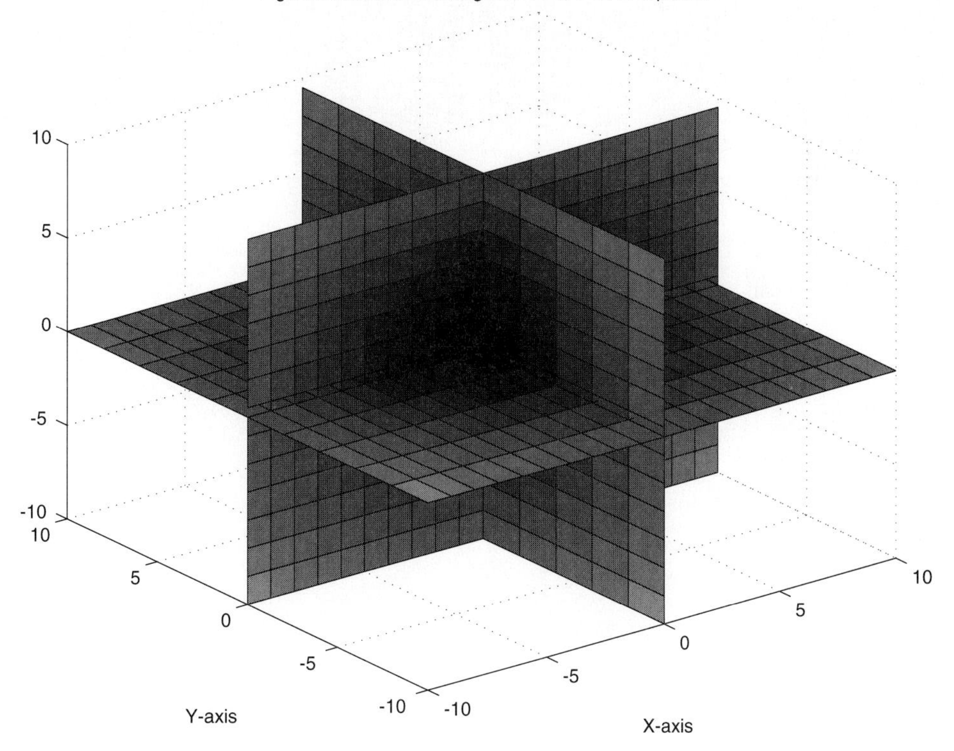

The color at each point is determined by 3-D interpolation into the volume V. The next example illustrates slices through the X=-4, X=7, Y=5, and Z=-8 planes:

```
» slice(X,Y,Z,V,[-4 7],5,-8)
» xlabel('X-axis'), ylabel('Y-axis'), zlabel('Z-axis')
» title('Figure 27.34: Slices through the X=[-4 7], Y=5, Z=-8 planes')
```

Figure 27.34: Slices through the X=[-4 7], Y=5, Z=-8 planes

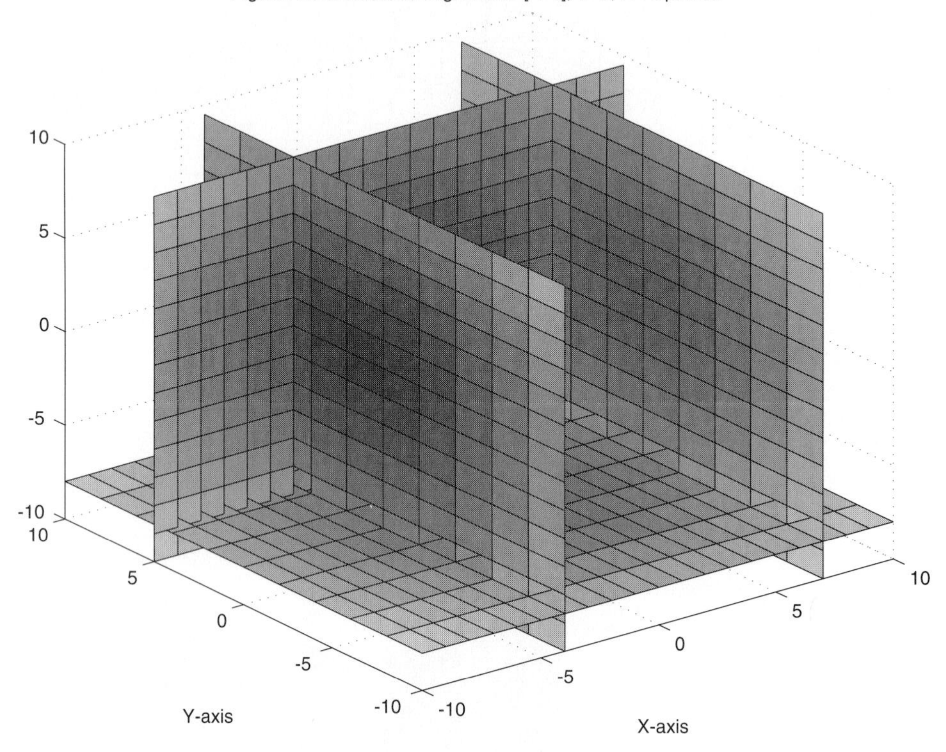


Slices can be in the form of surfaces as well as planes:

```
» [A,B,C] = peaks;
» slice(X,Y,Z,V,3*A,3*B,1.2*C)
» xlabel('X-axis'), ylabel('Y-axis'), zlabel('Z-axis')
» title('Figure 27.35: Using a surface as a slice')
```

Since `slice` draws surfaces, other functions that affect surfaces can be used to customize them:

```
» shading interp; colormap(colorcube);
» title('Figure 27.36: Customized slice plot')
```

The next chapter goes into color and lighting controls in greater detail.

Figure 27.35: Using a surface as a slice

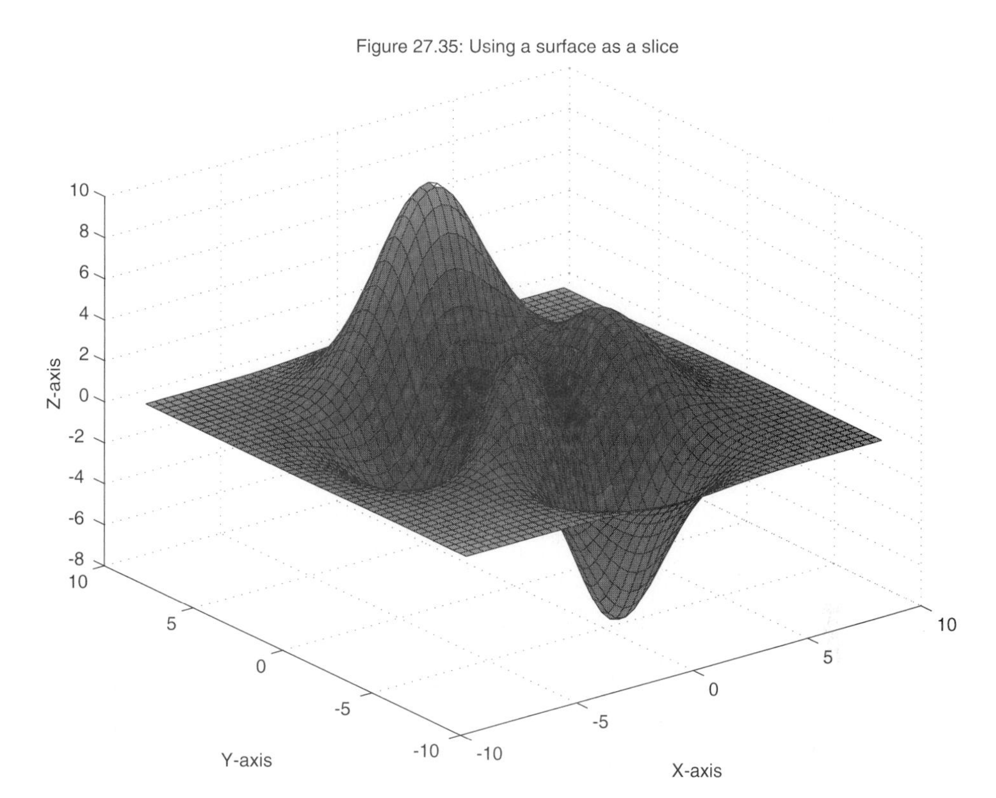


Figure 27.36: Customized slice plot

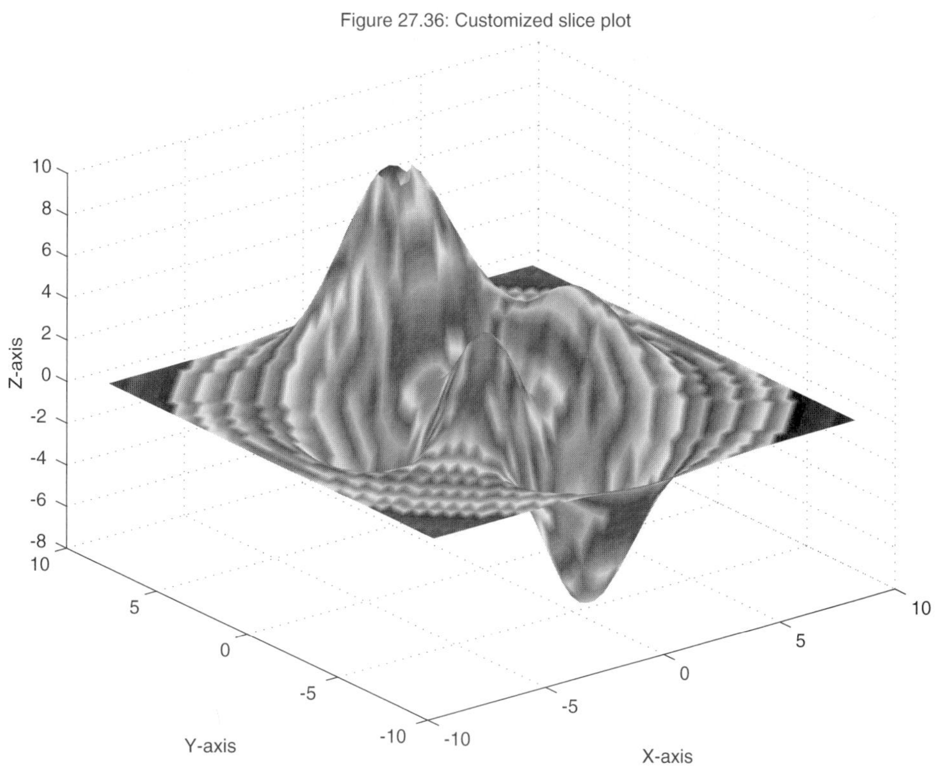

# 28

# Using Color and Light

MATLAB provides a number of tools for displaying information visually in two and three dimensions. For example, the plot of a sine curve presents more information at a glance than a set of data points could. The technique of using plots and graphs to present data sets is known as ***data visualization.*** In addition to being a powerful computational engine, MATLAB excels in presenting data visually in interesting and informative ways.

Often, however, a simple 2-D or 3-D plot cannot display all of the information you would like to present at one time. Color can provide an additional dimension. Many of the plotting functions discussed in previous chapters accept a *color* argument that can be used to add that dimension.

The discussion begins with an investigation of colormaps: how to use them, display them, alter them, and create your own. Next, techniques to simulate more than one colormap in a *Figure* window, or to use only a portion of a colormap, are illustrated. Finally, lighting models are discussed and examples are presented.

## 28.1 PLOTTING STYLES

You can select an overall color scheme or plotting style for your plots with the `colordef` command. As mentioned earlier, the default style is `colordef white`, consisting of a white *axes* background, a light gray *figure* background, black axis labels, the `jet` colormap, and plot colors starting with blue, dark green, and red. If you prefer a black background with white text, use `colordef black`. This style uses a black *axes* background, a dark gray *figure* background, white axis labels, the `jet` colormap, and plot colors starting with yellow, magenta, and cyan. `colordef none` will default to the same style that was used in previous versions of MATLAB, consisting of black *axes* and *figure* backgrounds with white axis labels, the `hsv` colormap, and plot colors starting with yellow, magenta, and cyan. You can make MATLAB default to your preferred style by adding a `colordef` command to your `startup.m` file.

## 28.2 UNDERSTANDING COLORMAPS

MATLAB uses a data structure called a colormap to represent color values. A colormap is defined as an array having three columns and some number of rows. Each row in the matrix represents an individual color using numbers in the range 0 to 1. The numbers in each row specify RGB values—that is, the intensity of red, green, and blue making up a specific color. Some representative samples are given in the following table:

| Red | Green | Blue | Color |
|---|---|---|---|
| 1 | 0 | 0 | red |
| 0 | 1 | 0 | green |
| 0 | 0 | 1 | blue |
| 1 | 1 | 0 | yellow |
| 1 | 0 | 1 | magenta |
| 0 | 1 | 1 | cyan |
| 0 | 0 | 0 | black |
| 1 | 1 | 1 | white |
| 0 | .5 | 0 | dark green |
| 0.67 | 0 | 1 | violet |
| 1 | .5 | 0 | orange |
| .5 | 0 | 0 | dark red |
| .5 | .5 | .5 | medium gray |

There are 17 MATLAB functions that generate predefined colormaps:

| Standard Colormap | Description |
|---|---|
| `hsv` | Hue saturation value (begins and ends with red). |
| `jet` | A variant of `hsv` (begins with blue and ends with red). |
| `hot` | Black to red to yellow to white. |
| `cool` | Shades of cyan and magenta. |
| `summer` | Shades of green and yellow. |
| `autumn` | Shades of red and yellow. |
| `winter` | Shades of blue and green. |
| `spring` | Shades of magenta and yellow. |
| `white` | All-white colormap. |
| `gray` | Linear gray scale. |
| `bone` | Gray scale with a tinge of blue. |
| `pink` | Pastel shades of pink. |
| `copper` | Linear copper tone. |
| `prism` | Prism, alternating red, orange, yellow, green, and blue violet. |
| `flag` | Alternating red, white, blue and black. |
| `lines` | Colormap using `plot` line colors. |
| `colorcube` | Enhanced color-cube colormap. |

By default, each of these colormaps generates a 64-by-3 array specifying the RGB descriptions of 64 colors. Each of these functions accepts an argument specifying the number of rows to generate. For example, `hot(m)` will generate an `m`-by-3 matrix containing the RGB values of colors that range from black, through shades of red, orange, and yellow, to white.

Most computers can display 256 colors at one time in an 8-bit hardware color lookup table, although many now have display cards that can handle up to 24 bits, over 16 million colors. This means that conservatively up to three or four 64-by-3 colormaps can be in use at one time in different figures using 256-color mode. If more colormap entries are used, the computer usually swaps out entries in its hardware lookup table. For example, this is what is happening if you notice the screen background colors change when plotting MATLAB figures. As a result, it is usually prudent to keep the total number of colormap entries used at any one time below 256 unless your computer has a display card that can display more colors at once.

## 28.3 USING COLORMAPS

The statement `colormap(M)` installs the matrix `M` as the colormap to be used in the current *Figure* window. For example, `colormap(cool)` installs a 64-entry version of the `cool` colormap. `colormap default` installs the default colormap, usually `hsv` or `jet` depending on the default color scheme you have chosen with the `colordef` command.

Line plotting functions such as `plot` and `plot3` do not use colormaps; they use the colors listed in the `plot` color and linestyle table. The sequence of colors used by these functions varies depending on the plotting style you have chosen. Most other plotting functions, such as `mesh`, `surf`, `contour`, `fill`, `pcolor`, and their variations, use the current colormap to determine color sequences.

Plotting functions that accept a *color* argument usually accept the argument in one of three forms: (1) a character string representing one of the colors in the `plot` color and linestyle table (e.g., `'r'` or `'red'` for red), (2) a three-entry row vector representing a single RGB value (e.g., `[.25 .50 .75]`), or (3) a matrix. If the color argument is a matrix, the elements are scaled and used as indices into the current colormap matrix. This last form will be discussed in greater detail later in this chapter.

## 28.4 DISPLAYING COLORMAPS

You can display a colormap in a number of ways. One way is to view the elements in a colormap matrix directly:

```
» hot(8)
ans =
    0.3333        0        0
    0.6667        0        0
    1.0000        0        0
    1.0000   0.3333        0
    1.0000   0.6667        0
    1.0000   1.0000        0
    1.0000   1.0000   0.5000
    1.0000   1.0000   1.0000
```

The preceding shows that the first row of an 8-by-3 `hot` colormap represents a very dark red color while the last row represents white. The `pcolor` function can be used to display a colormap as well. For example,

```
» n = 16;
» colormap(jet(n))
» pcolor([1:n+1;1:n+1]')
» title('Figure 28.1: Jet Colormap')
```

The columns of a colormap can be plotted in red, green, and blue, respectively, using the MATLAB function `rgbplot`. For example,

Figure 28.1: Jet Colormap

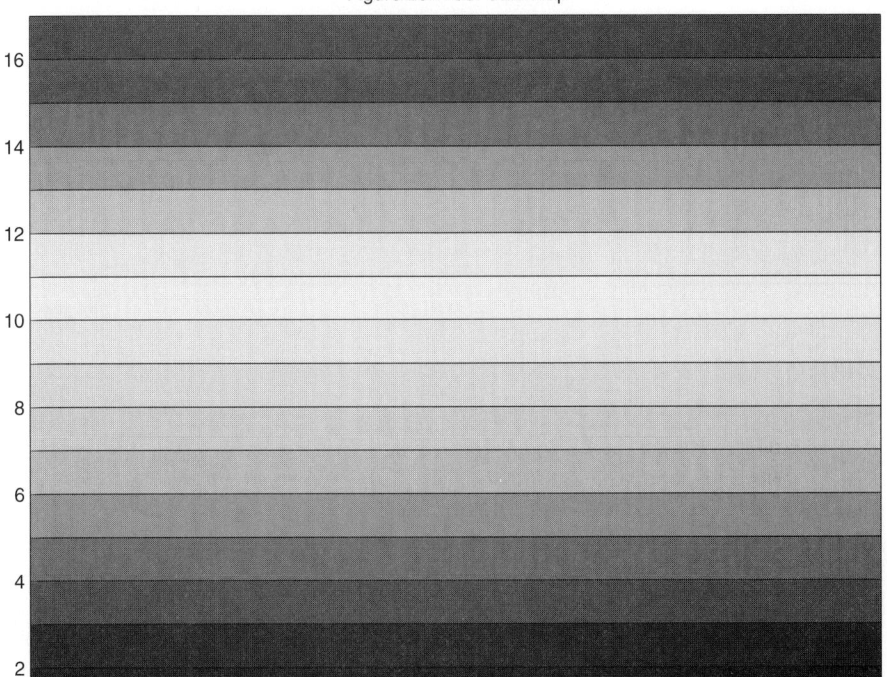


```
» rgbplot(hot)
» title('Figure 28.2: Hot Colormap')
```

This shows that the red component increases first, then the green, then the blue. `rgbplot(gray)` shows that all three columns increase linearly and equally (all three lines overlap).

Finally, the `colorbar` function adds a vertical or horizontal color scale to the current *Figure* window showing the colormap for the current axis. `colorbar('horiz')` places a color bar horizontally beneath your current plot. `colorbar('vert')` places a vertical color bar to the right of your plot. `colorbar` without arguments either adds a vertical color bar if no color bars exist, or updates an existing color bar. The following example demonstrates the use of `colorbar`:

```
» [x,y,z] = peaks;
» mesh(x,y,z);
» colormap(hsv)
» axis([-3 3 -3 3 -6 8])
» colorbar
» title('Figure 28.3: Color Bar Example')
```

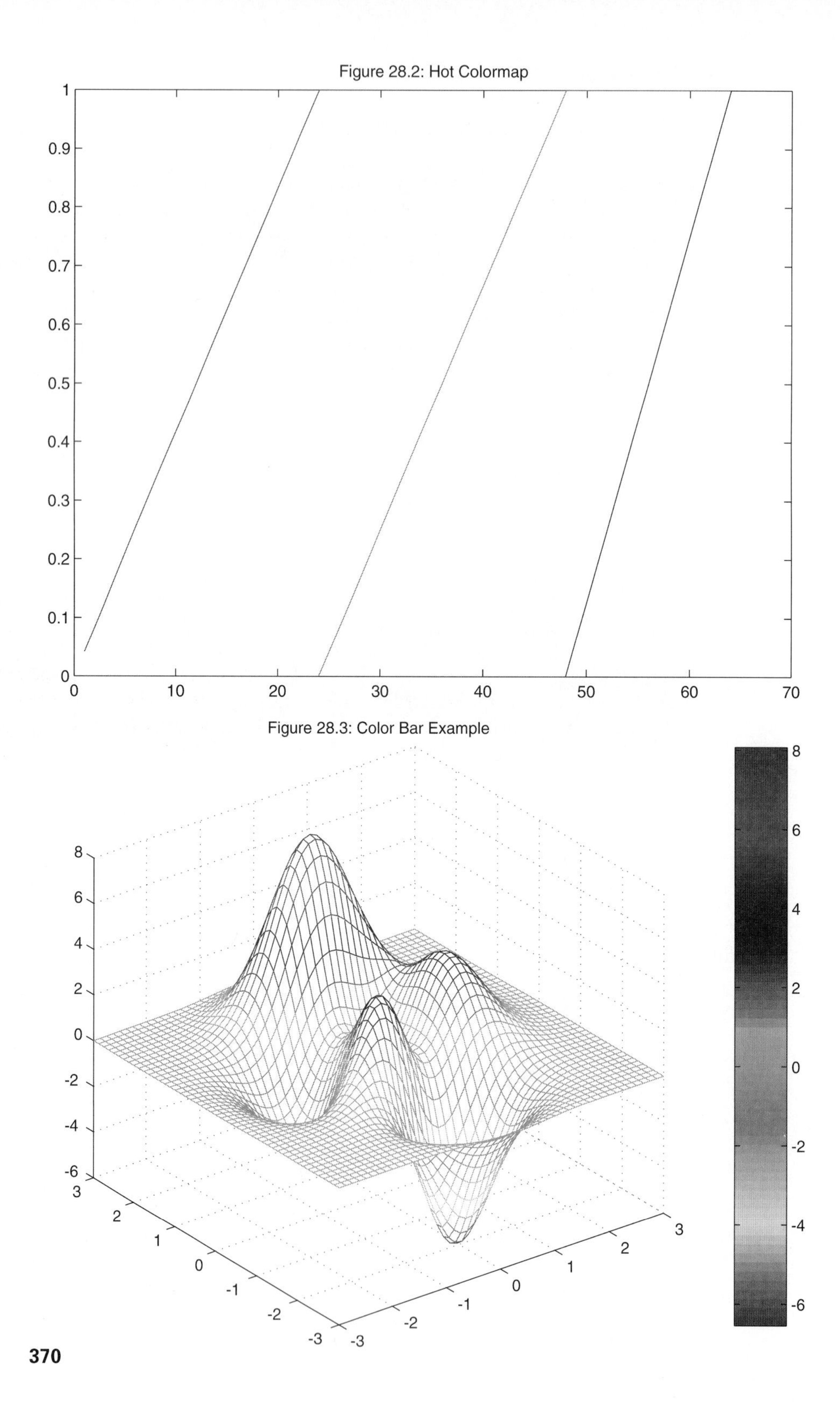


Figure 28.2: Hot Colormap

Figure 28.3: Color Bar Example

## 28.5 CREATING AND ALTERING COLORMAPS

The fact that colormaps are matrices means that you can manipulate them exactly like other arrays. The function `brighten` takes advantage of this fact to adjust a given colormap to increase or decrease the intensity of the colors. `brighten(beta)` brightens (`0 < n ≤ 1`) or darkens (`-1 ≤ n < 0`) the current colormap. `brighten(beta)` followed by `brighten(-beta)` restores the original colormap. The command `newmap= brighten(beta)` creates a brighter or darker version of the current colormap without changing the current map. The command `mymap=brighten(cmap,beta)` creates an adjusted version of the specified colormap without affecting either the current colormap or the specified colormap `cmap`. The algorithm used in `brighten` is as follows:

```
gamma = 1 - abs(beta)
if n > 0
    mymap = cmap.^(gamma)
else
    mymap = cmap.^(1/gamma)
end
```

You can create your own colormap by generating an `m`-by-3 array `mymap` and installing it with `colormap(mymap)`. Each value in a colormap matrix must be between 0 and 1. If you try to use a matrix with more or less than three columns or containing any values less than zero or greater than one, `colormap` will report an error.

Colormaps can be converted between the red-green-blue (RGB) standard and the hue-saturation-value (HSV) standard using the `rgb2hsv` and `hsv2rgb` functions. MATLAB, however, always interprets colormaps as RGB values. You can also combine colormaps arithmetically, although the results are sometimes unpredictable. For example, the map called `pink` is simply

```
pinkmap = sqrt(2/3*gray + 1/3*hot);
```

Again, the result is a valid colormap only if all elements of the `m`-by-3 matrix are between 0 and 1 inclusive.

A colormap defines the color palate that will be used to render a plot. A default colormap allows 64 distinct RGB values to be used for your data. MATLAB uses the function `caxis` to determine which data values map to individual colormap entries.

Normally, a colormap is scaled to extend from the minimum to the maximum values of your data—that is, the entire colormap is used to render your plot. You may occasionally wish to change the way these colors are used. The `caxis` function, which stands for color axis since color adds another dimension, allows you to use the entire colormap for a subset of your data range or use only a portion of the current colormap for your entire data set.

`[cmin,cmax] = caxis` returns the minimum and maximum data values that are mapped to the first and last entries of the colormap, respectively. These will normally be set to the minimum and maximum values of your data. For example, `mesh(peaks)` will create a mesh plot of the `peaks` function and set `caxis` to `[-6.5466, 8.0752]`, the minimum and maximum *z* values. Data points between these values use colors interpolated from the colormap.

`caxis([cmin,cmax])` uses the entire colormap for data in the range between `cmin` and `cmax`. Data points greater than `cmax` are rendered with the color associated with `cmax`, and data points smaller than `cmin` are rendered with the color associated with `cmin`. If `cmin` is less than `min(data)` or `cmax` is greater than `max(data)`, the colors associated with `cmin` or `cmax` will never be used. Only the portion of the colormap associated with `data` will be used. `caxis('auto')` or the command form `caxis auto` restores the default values of `cmin` and `cmax`.

While the following examples are difficult to show distinctively in gray scale in this text, they will be more apparent on your computer screen. The default color range is illustrated by

```
» pcolor([1:17;1:17]'), colormap(hsv(8))
» title('Figure 28.4: Default Color Range')
» caxis('auto')
» colorbar
» caxis
ans =
       1    17
```

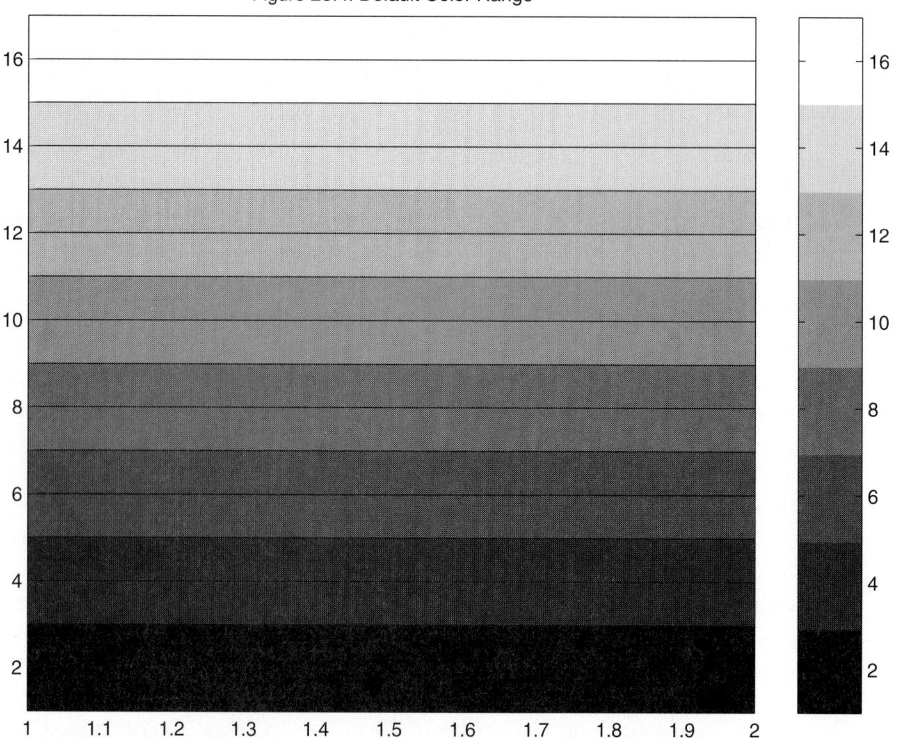

As you can see, all eight colors in the current colormap are used for the entire data set, two bars for each color. If the colors are mapped to values from $-3$ to 23, only five colors will be used in the plot. This scenario is generated by the commands

```
» title('Figure 28.5: Extended Color Range')
» caxis([-3,23]) % extend the color range
» colorbar       % redraw the color scale
```

Figure 28.5: Extended Color Range

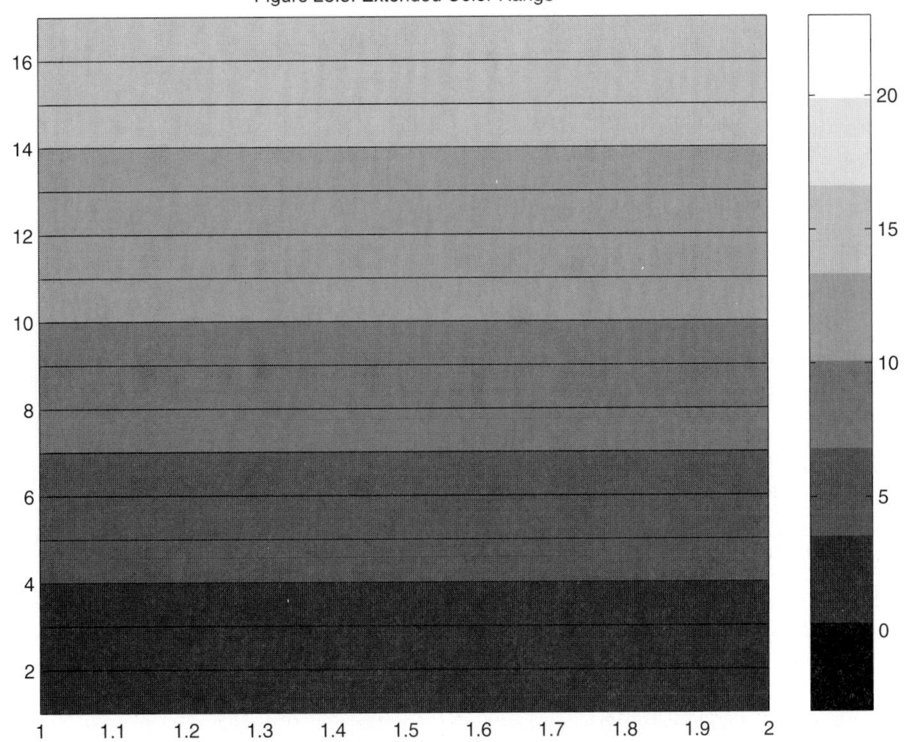


If the colors are mapped to values from 5 to 12, all colors are used. However, the data less than 5 or greater than 12 get the colors associated with 5 and 12, respectively. This is produced by the commands

```
» title('Figure 28.6: Restricted Color Range')
» caxis([5,12])  % restrict the color range
» colorbar       % redraw the color scale
```

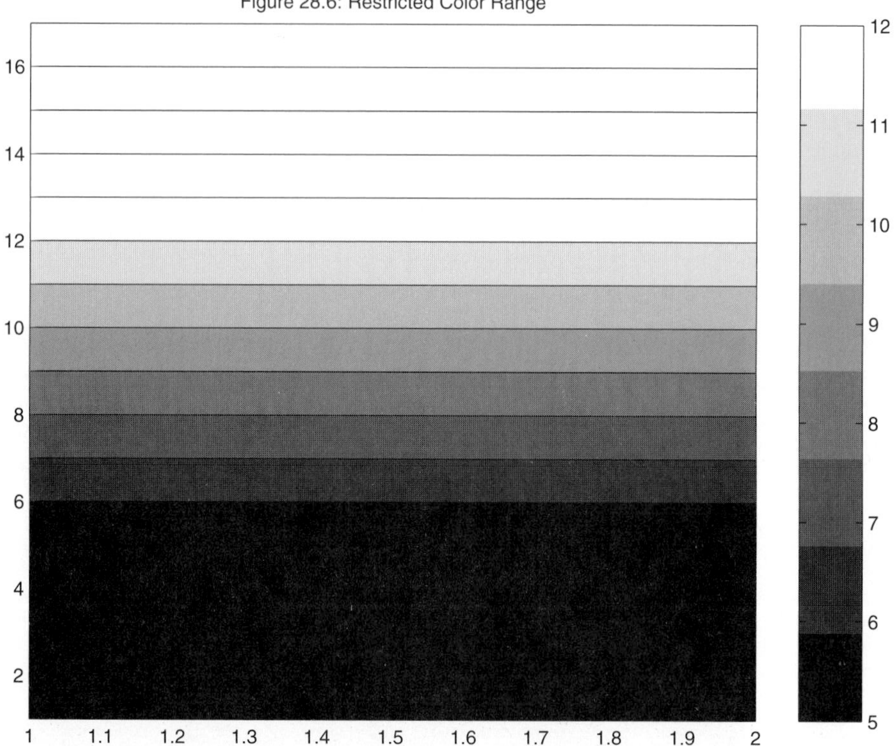


## 28.6 USING MORE THAN ONE COLORMAP

Sometimes it is helpful to use different colors for different portions of a plot. Since a colormap is a property of the *Figure* window itself, only one colormap can be used in any one *Figure* window. However, you can create your own colormap to produce the effect you want. For example,

```
» mymap = [hsv(32); copper(32)];   % stack two colormaps into one
» colormap(mymap)                  % install it
» mesh(peaks+8)                    % create two sample plots
» view(90,0)
» hold on
» mesh(peaks-8)
» colorbar                         % and add a color scale
» title('Figure 28.7: Merging Two Colormaps')
» hold off
```

Interesting effects can be generated with the `spinmap` function. `spinmap(t,inc)` rotates the colormap for about `t` seconds using the increment `inc`. If not given, `spinmap` uses `t=3` and `inc=2` as the defaults. `spinmap(inf)` is an infinite loop; use `Ctrl-C` or

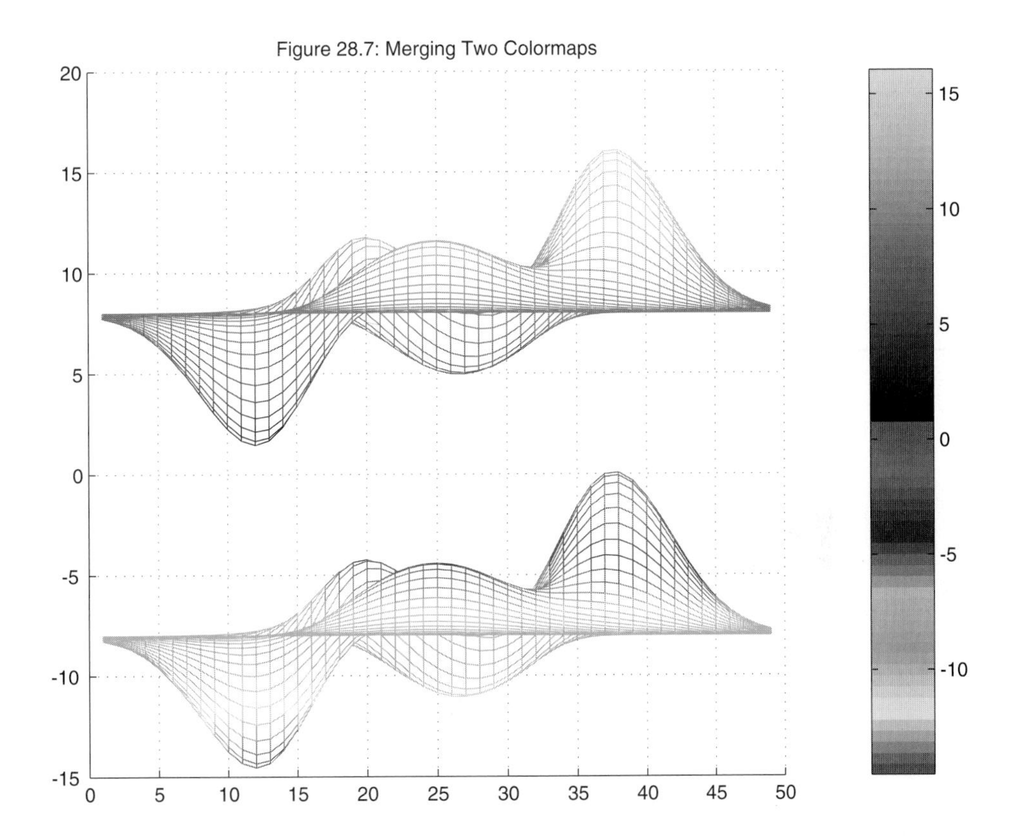


⌘-. to break. For `inc=1` the color cycle is slower; values greater than 2 cause the colors to cycle faster. Negative values rotate the colors in the opposite direction. Try this example with various increments to see the effects.

```
» sphere            % create a unit sphere with default aspect ratio
» axis off          % remove distractions
» colormap(hsv)     % choose an interesting colormap
» spinmap(10,3)     % fast color cycle for 10 seconds
```

## 28.7 USING COLOR TO DESCRIBE A FOURTH DIMENSION

Functions such as `mesh` and `surf` vary the color along the $z$-axis unless a color argument is given—for example, `surf(X,Y,Z)` is equivalent to `surf(X,Y,Z,Z)`. Applying color to the $z$-axis produces a colorful plot but does not provide additional information since the $z$-axis already exists. To make better use of color, it is suggested that color be used to describe some property of the data that is not reflected by the three axes. To do so requires specifying different data for the color argument to 3-D plotting functions.

If the color argument to a plotting function is a vector or matrix, it is scaled and used as an index into the colormap. This argument can be any real vector or matrix the same size as the other arguments. Consider the following examples:

```
» x = -7.5:.5:7.5; y = x;        % create a data set - the famous sombrero
» [X Y] = meshgrid(x,y);         % create plaid data
» R = sqrt(X.^2 + Y.^2)+eps;
» Z = sin(R)./R;

» surf(X,Y,Z,Z)             % default color order
» title('Figure 28.8: surf(X,Y,Z,Z)')
```

Figure 28.8: surf(X,Y,Z,Z)

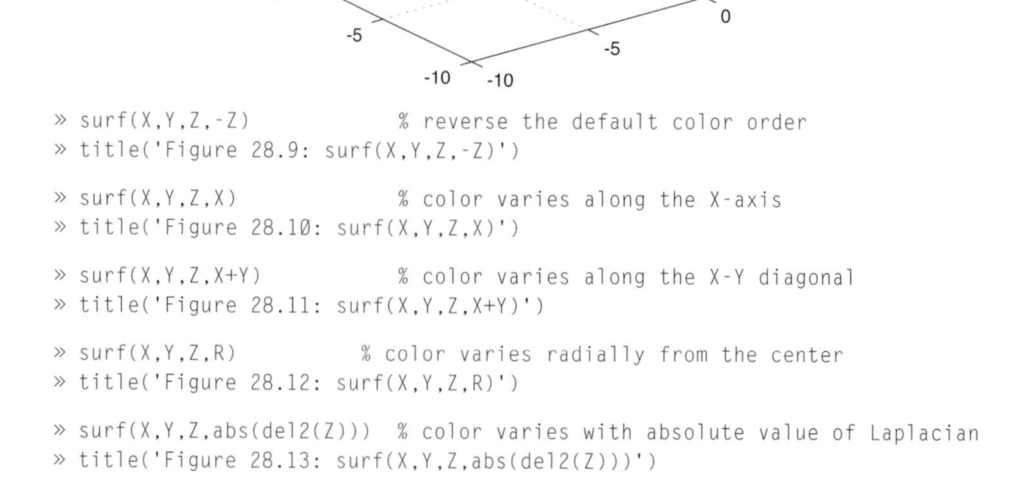


```
» surf(X,Y,Z,-Z)            % reverse the default color order
» title('Figure 28.9: surf(X,Y,Z,-Z)')

» surf(X,Y,Z,X)             % color varies along the X-axis
» title('Figure 28.10: surf(X,Y,Z,X)')

» surf(X,Y,Z,X+Y)           % color varies along the X-Y diagonal
» title('Figure 28.11: surf(X,Y,Z,X+Y)')

» surf(X,Y,Z,R)           % color varies radially from the center
» title('Figure 28.12: surf(X,Y,Z,R)')

» surf(X,Y,Z,abs(del2(Z)))  % color varies with absolute value of Laplacian
» title('Figure 28.13: surf(X,Y,Z,abs(del2(Z)))')
```

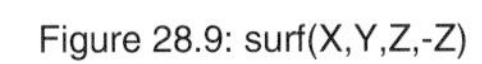

Figure 28.9: surf(X,Y,Z,-Z)

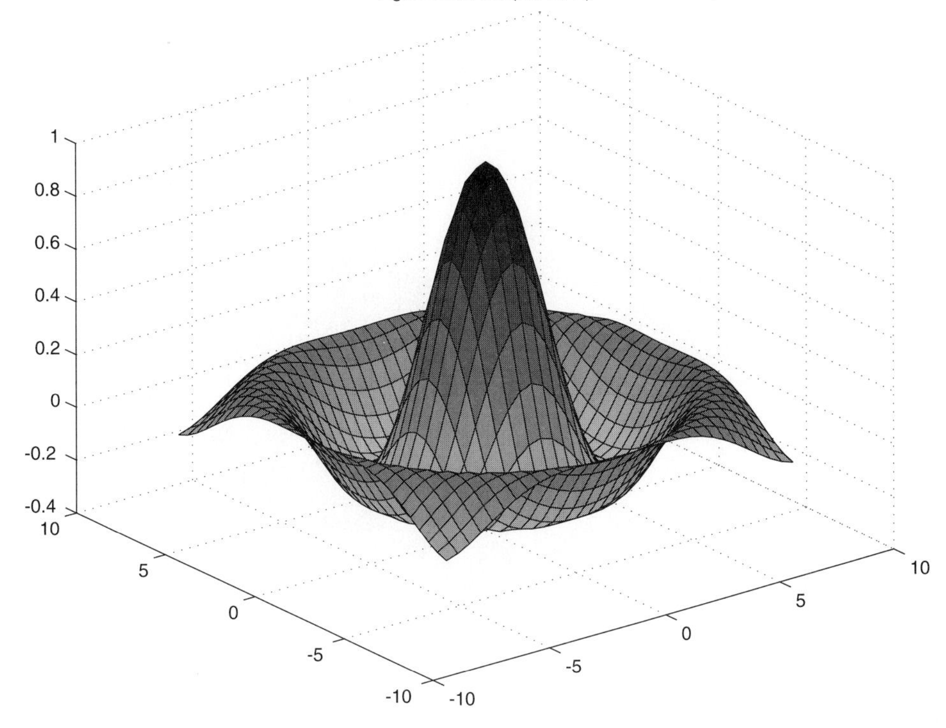


Figure 28.10: surf(X,Y,Z,X)

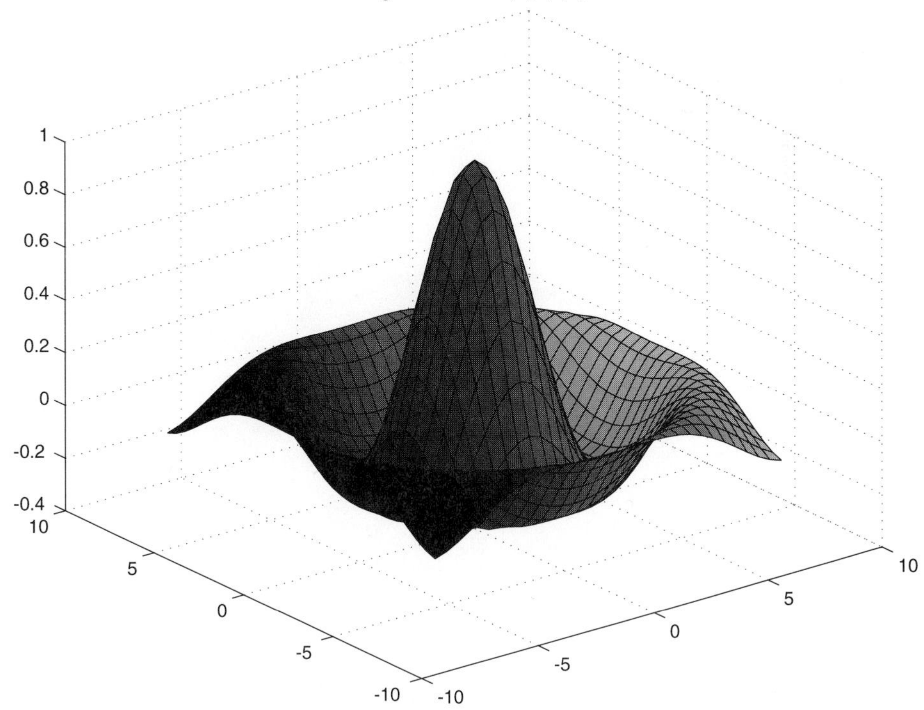

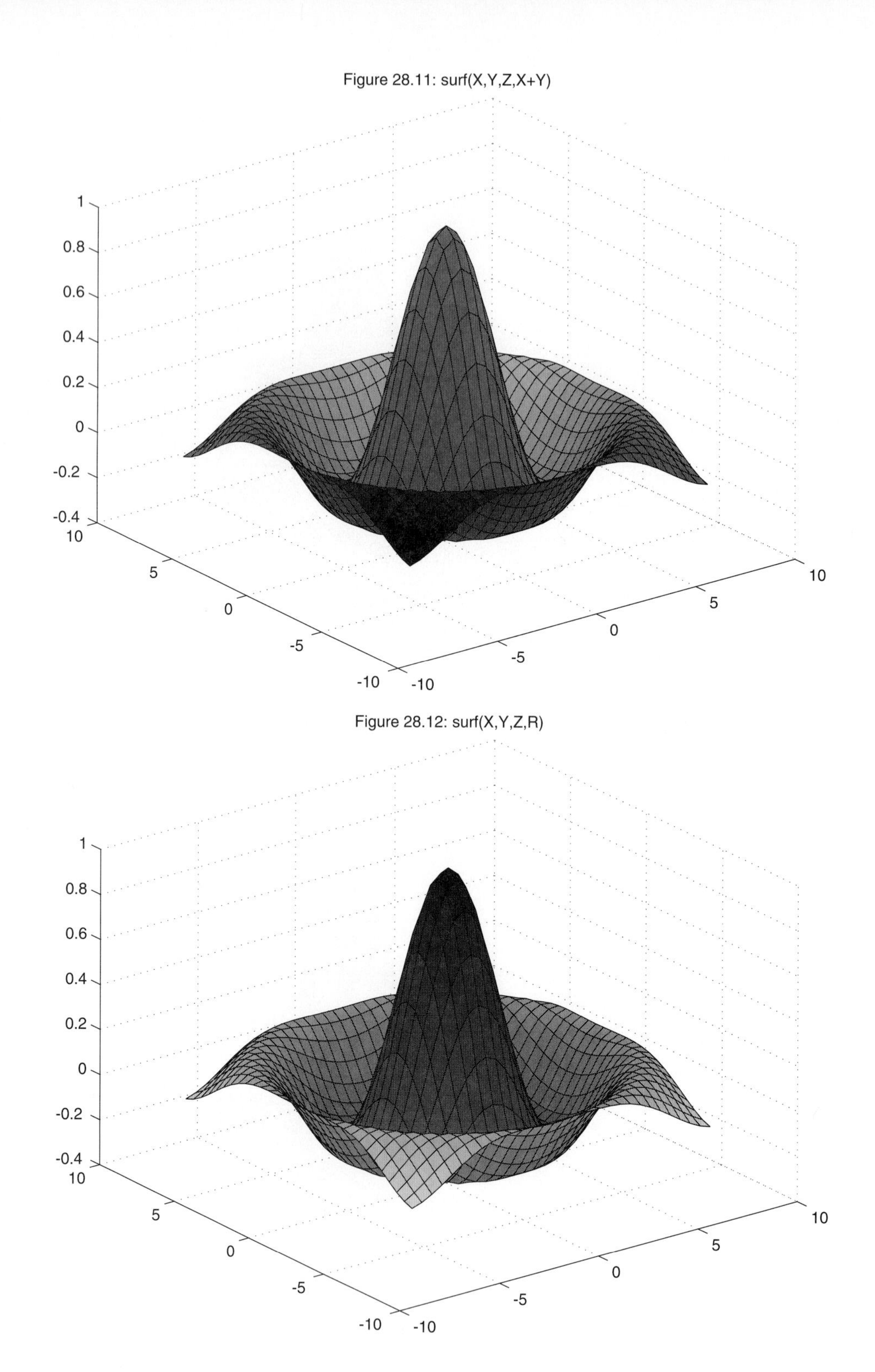


Figure 28.11: surf(X,Y,Z,X+Y)

Figure 28.12: surf(X,Y,Z,R)

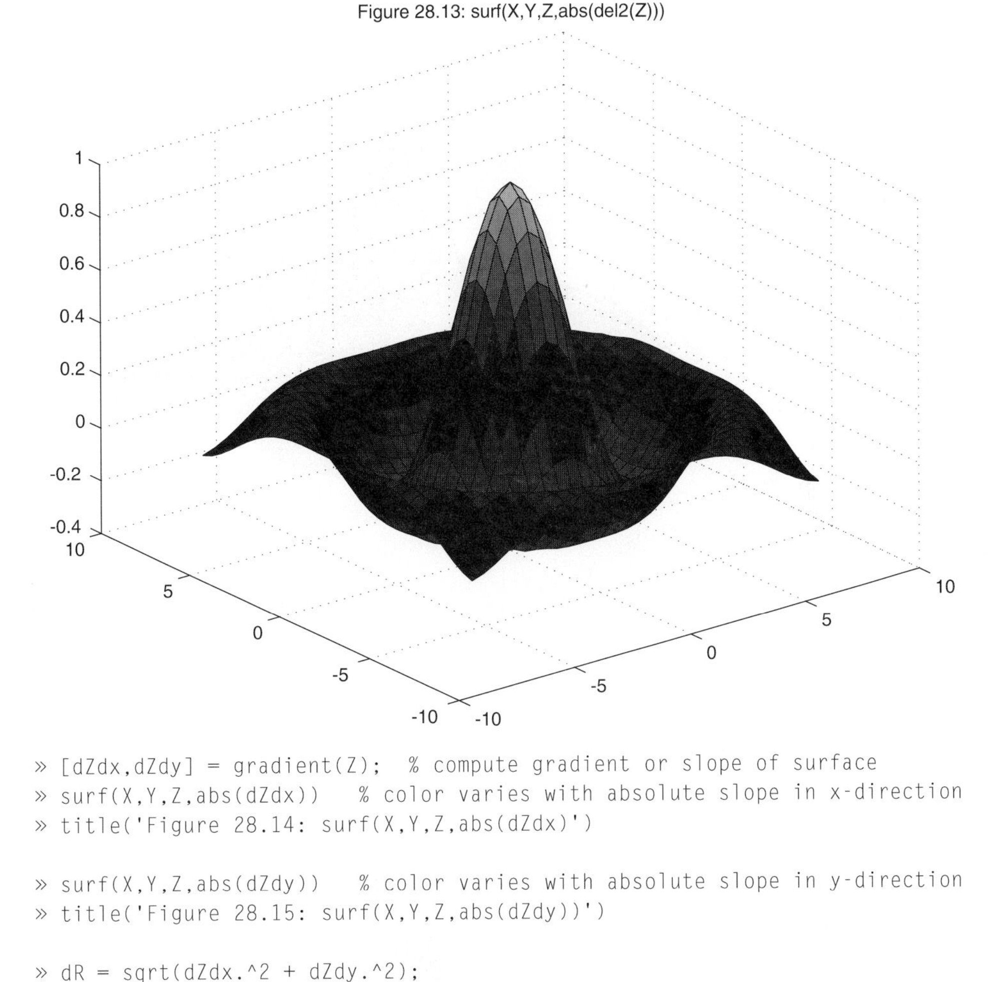


Figure 28.13: surf(X,Y,Z,abs(del2(Z)))

```
» [dZdx,dZdy] = gradient(Z);  % compute gradient or slope of surface
» surf(X,Y,Z,abs(dZdx))    % color varies with absolute slope in x-direction
» title('Figure 28.14: surf(X,Y,Z,abs(dZdx)')

» surf(X,Y,Z,abs(dZdy))    % color varies with absolute slope in y-direction
» title('Figure 28.15: surf(X,Y,Z,abs(dZdy))')

» dR = sqrt(dZdx.^2 + dZdy.^2);
» surf(X,Y,Z,dR)           % color varies with magnitude of slope
» title('Figure 28.16: surf(X,Y,Z,dR)')
```

Note how color in each of the last five examples does indeed provide an additional dimension to the plotted surface. The function `del2` is the discrete Laplacian function that applies color based on the curvature of the surface. The function `gradient` approximates the gradient or slope of the surface.

## 28.8 LIGHTING MODELS

The graphics functions `pcolor`, `fill`, `fill3`, `mesh`, and `surf`, discussed in the previous chapters, render objects that appear to be well lit from all sides by very diffuse light.

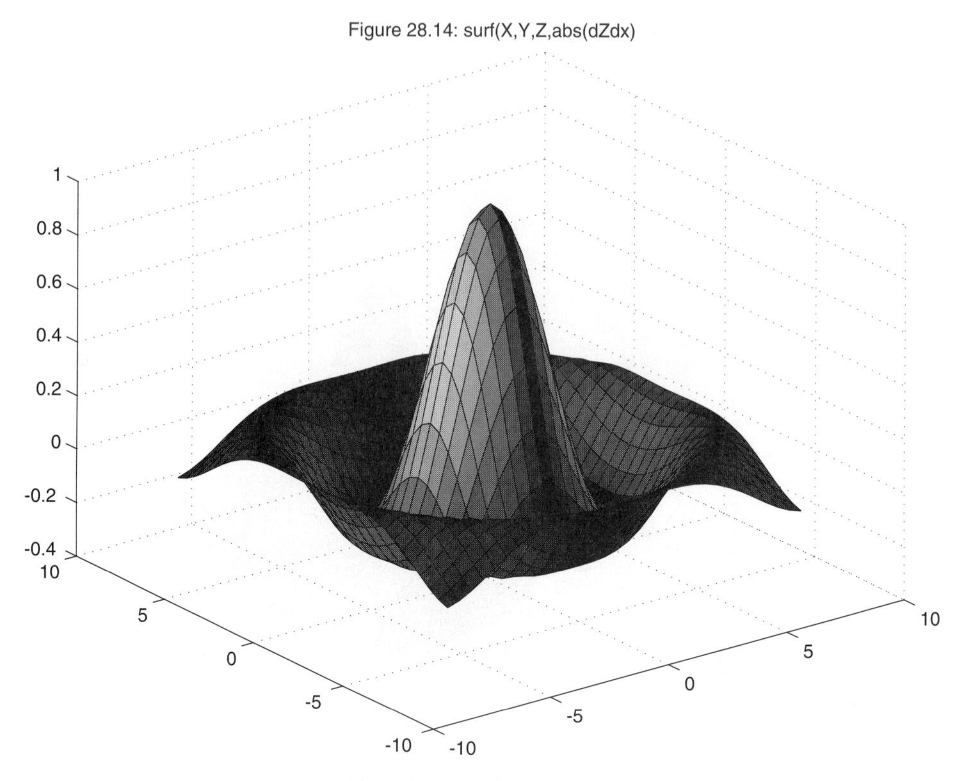


Figure 28.14: surf(X,Y,Z,abs(dZdx)

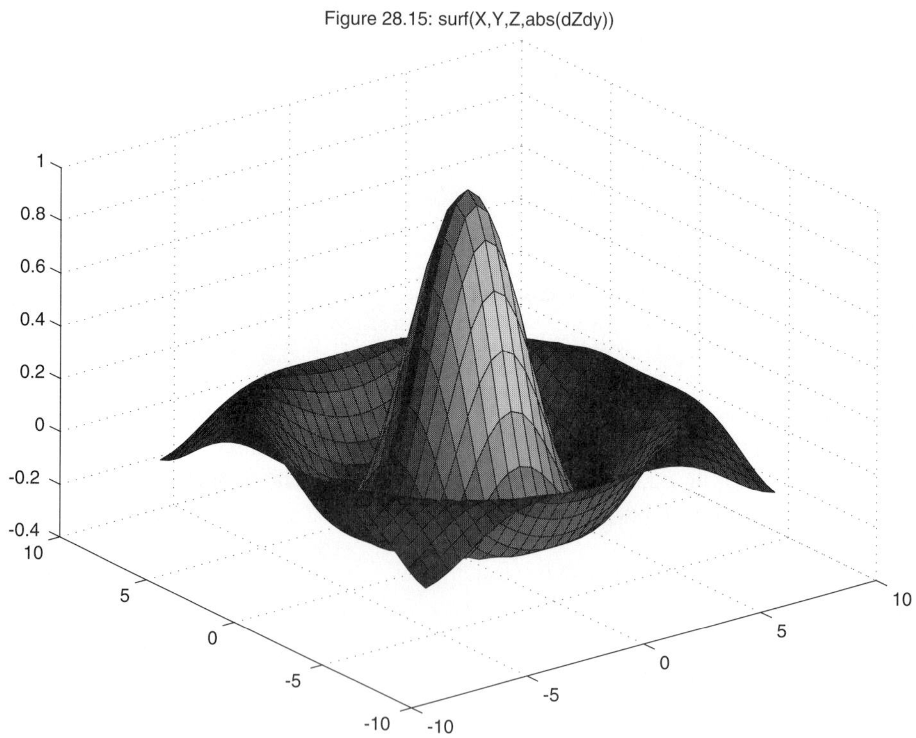


Figure 28.15: surf(X,Y,Z,abs(dZdy))

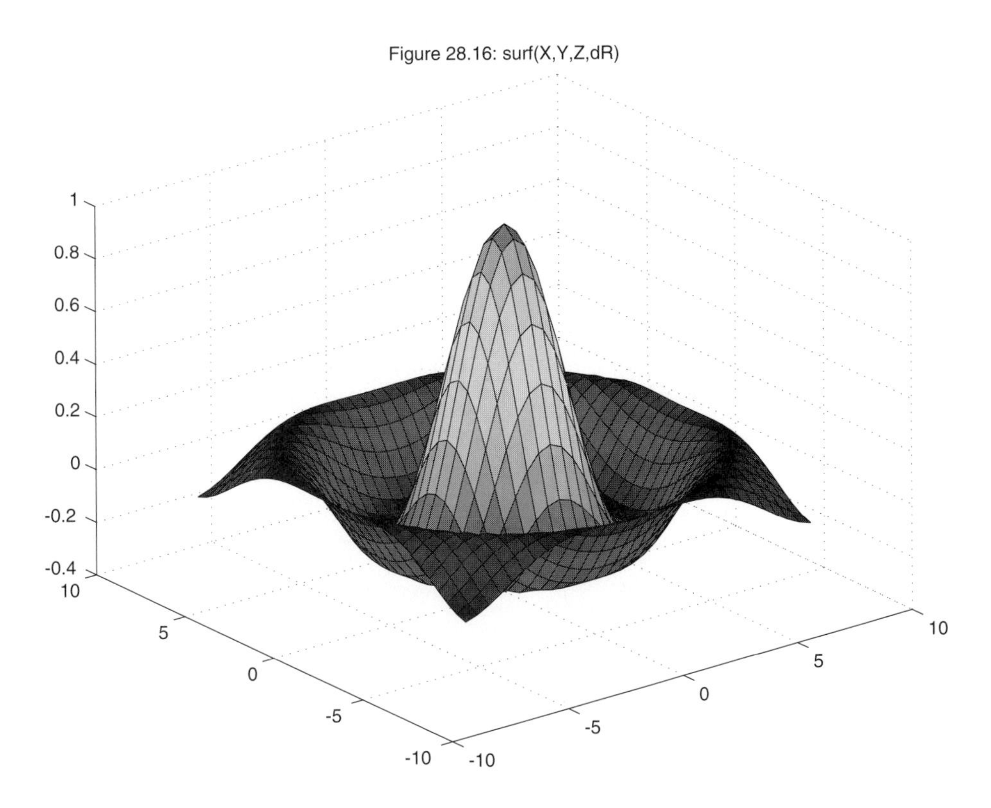


This technique emphasizes the characteristics of the objects in the *Figure* window and enhances the user's ability to visualize the data being analyzed. Although the data can be visualized quite clearly, the realism of the scene can be enhanced or diminished by creating different lighting effects.

The `shading` function selects between faceted, flat, or interpolated shading. Examples of each of these were illustrated in the 3-D graphic chapter (Chapter 27). Although requiring more computational power and subsequently more time to render, interpolated shading of the objects in a scene can enhance the realism of the scene being rendered. The `spinmap` function can also generate some interesting effects when objects, shading, and colormaps are chosen appropriately.

One or more light sources can be added to stimulate the highlights and shadows associated with directional lighting. The `light` function creates a white light source infinitely far away along the vector [1 0 1]. Once the light source has been created, the `lighting` function allows you to select from four different lighting models: `none` (which ignores any light source), `flat` (the default when a light source is created), `phong`, and `gouraud` lighting. Each of these models uses a different algorithm to change the appearance of the object. Flat lighting uses a uniform color for each face of the object. Gouraud lighting interpolates the face colors from the vertices. Phong lighting interpolates the vertex normals across each face and calculates the reflectance at each pixel. The `lighting` affects all objects in the *Figure* window. Here are some examples:

```
» sphere, colormap(gray), shading interp;   % plot a surface
» light                                     % create a light source
» title('Figure 28.17: Flat Lighting')
```

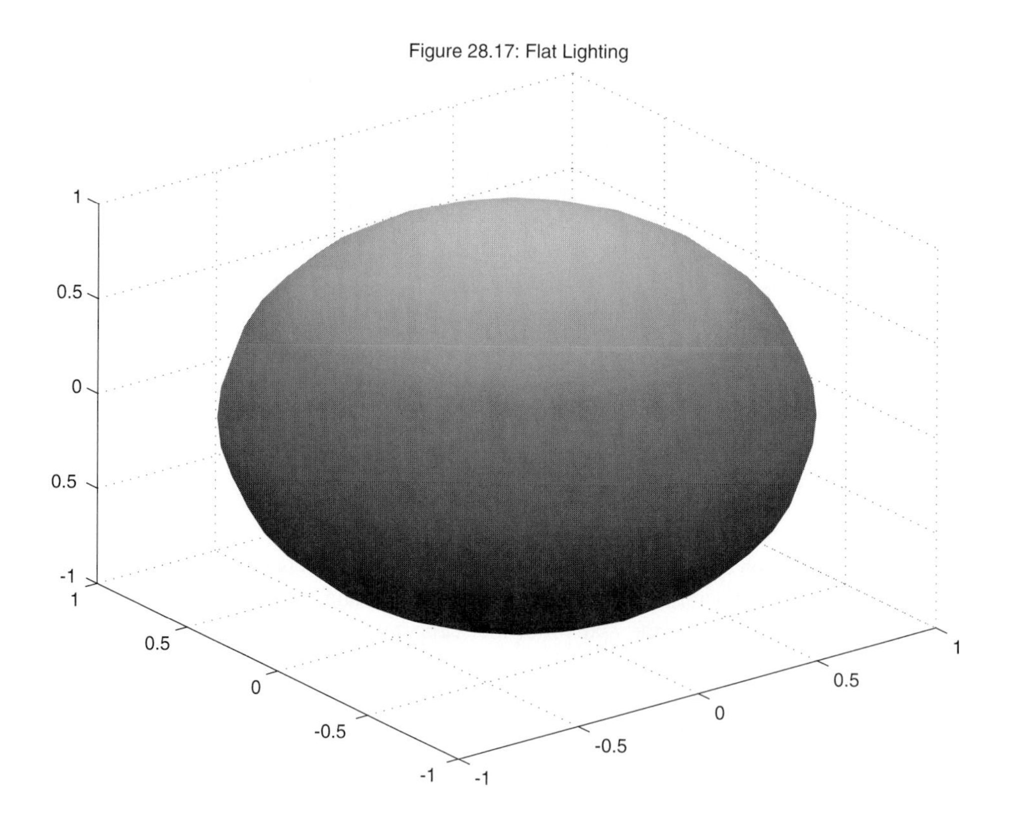


```
» lighting none
» title('Figure 28.18: Default Lighting')

» lighting gouraud
» title('Figure 28.19: Gouraud Lighting')

» lighting phong
» title('Figure 28.20: Phong Lighting')
```

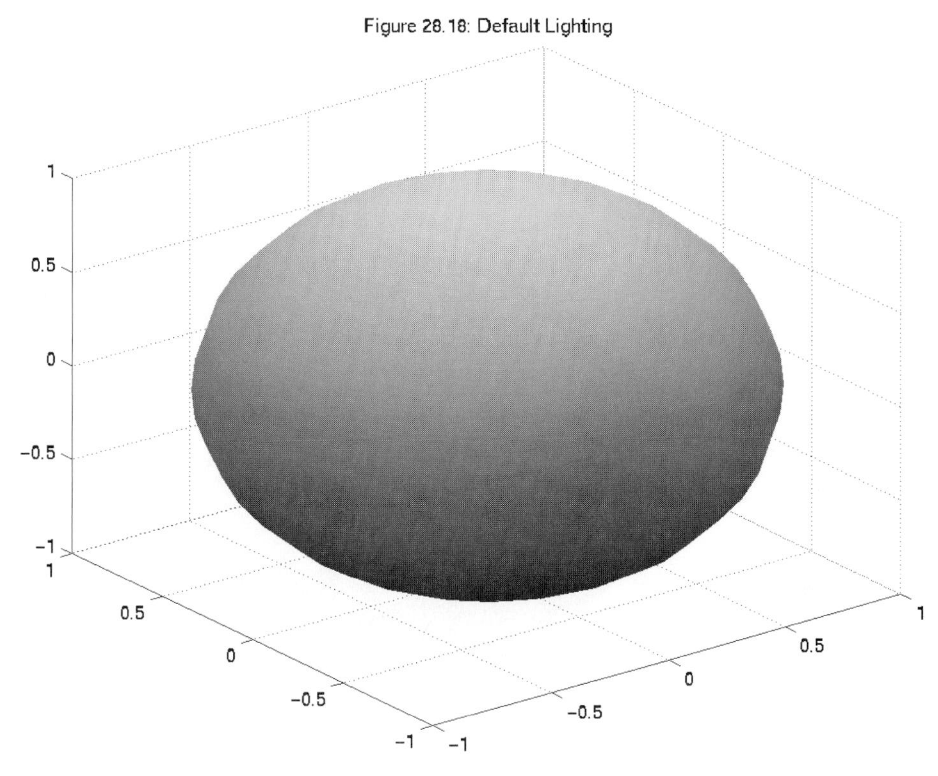

Figure 28.18: Default Lighting

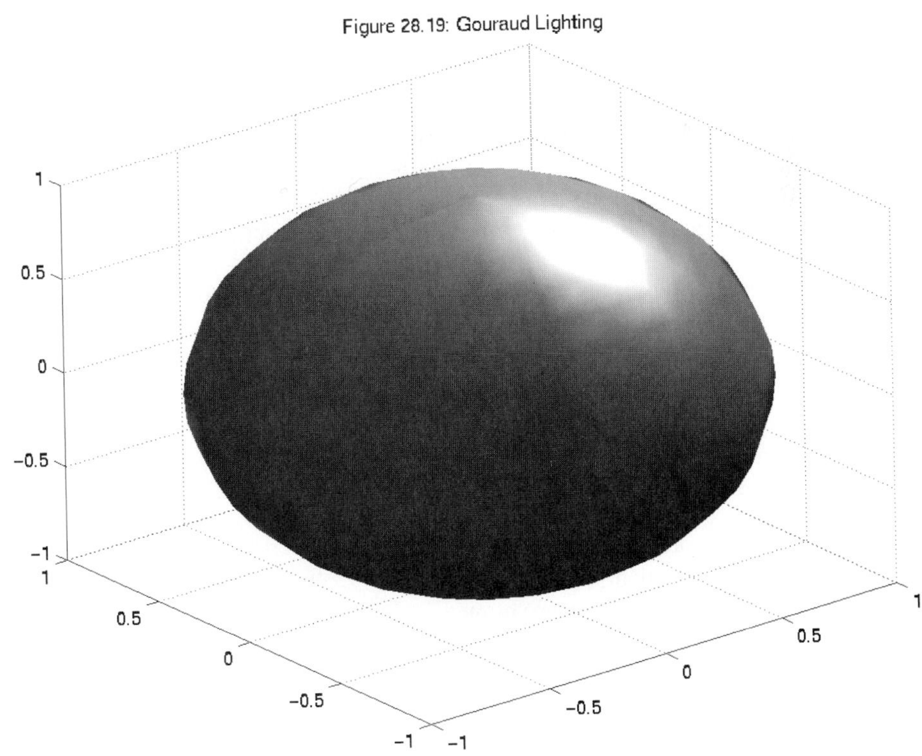

Figure 28.19: Gouraud Lighting

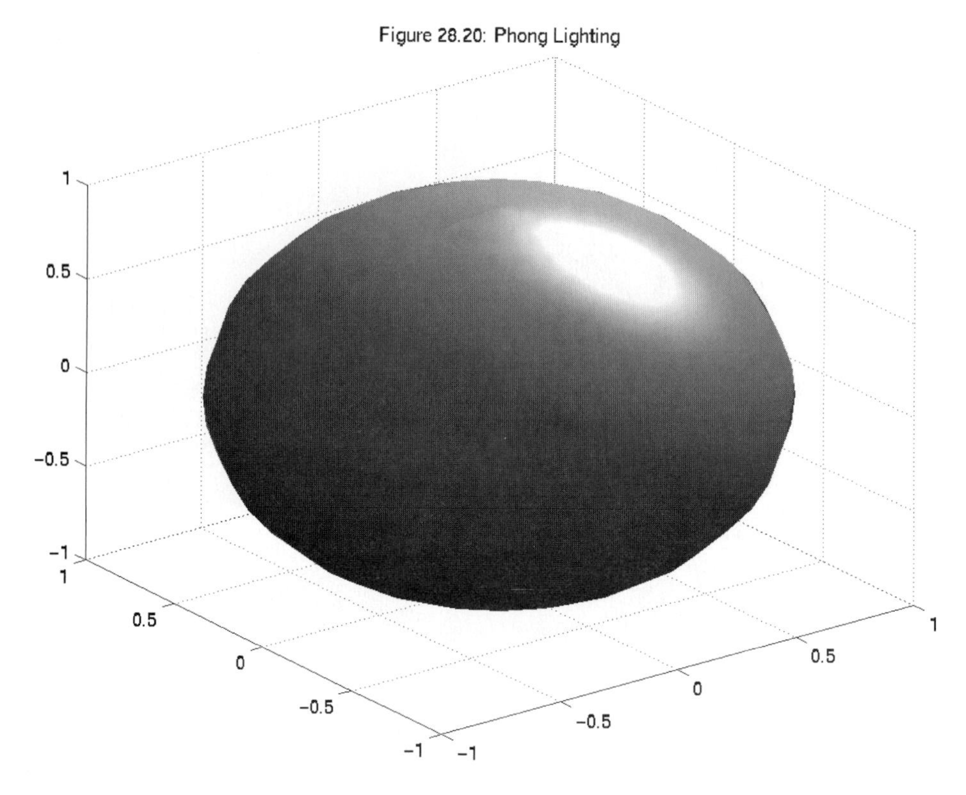

Figure 28.20: Phong Lighting

In addition to the lighting, the appearance of objects in the figure can be changed by modifying the apparent reflective characteristics or ***reflectance*** of the surfaces.

Reflectance is made up of a number of components, including

- ***ambient light***—strength of the uniform directionless light in the figure
- ***diffuse reflection***—intensity of the soft directionless reflected light
- ***specular reflection***—intensity of the hard directional reflected light
- ***specular exponent***—controls the size of the specular "hot spot" or spread
- ***specular color reflectance***—determines the contribution of the surface color to the reflectance.

Some predefined surface reflectance properties may be selected using the `material` function. Options include `shiny`, `dull`, `metal`, and `default` to restore the default surface reflectance properties. Like `lighting`, `material` affects all objects in the *Figure* window.

The form `material([ka kd ks n sc])`, where `n` and `sc` are optional, sets the ambient strength, diffuse reflectance, specular reflectance, specular exponent, and specular color reflectance of all objects in the figure.

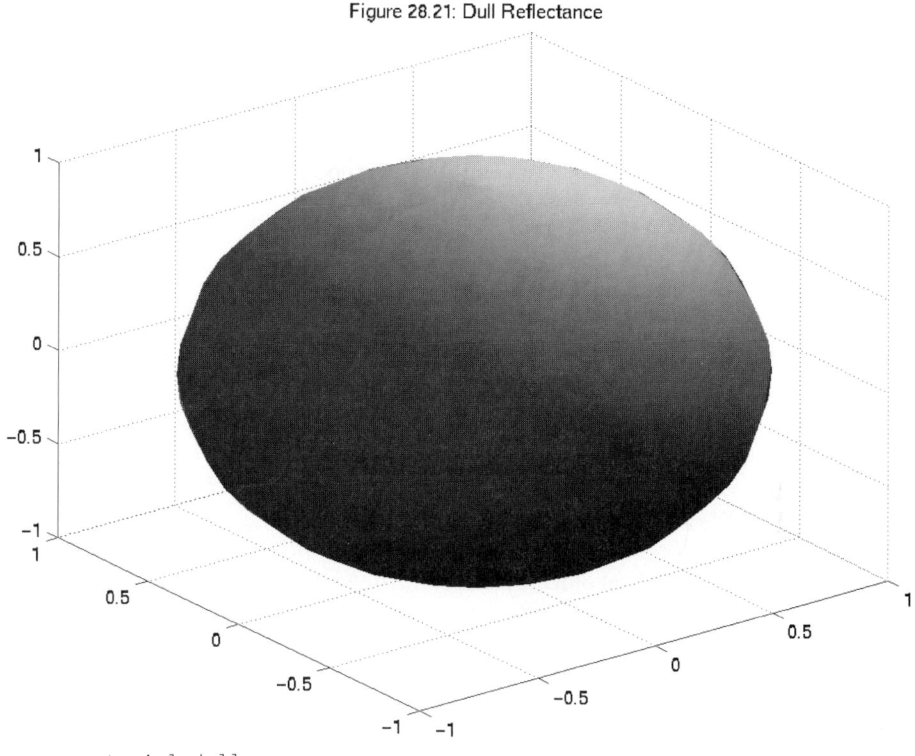


```
» material dull
» title('Figure 28.21: Dull Reflectance)
```

Two related utility functions are `diffuse` and `specular`. `Rd=diffuse(Nx,Ny,Nz,S)` returns the diffuse reflectance of a surface with normal vector components `[Nx,Ny,Nz]`. `S` is a vector that defines the direction of the light source: `[Sx,Sy,Sz]` or `[Theta,Phi]` in spherical coordinates. `Rs=specular(Nx,Ny,Nz,S,V,n)` returns the specular reflectance of a surface with normal vector components `[Nx,Ny,Nz]`. `S` defines the direction of the light source as before, and `V` defines the direction to the viewer. The optional argument `n` specifies the specular spread exponent.

The `surfl` function introduced in the previous chapter is an efficient way to add lighting effects to a surface plot. The most general form is `surfl(X,Y,Z,S,K,method)`, where `X`, `Y`, and `Z` are equivalent to `surf(X,Y,Z)`. `S` is the direction of the light source in the form `[Sx,Sy,Sz]` or `[az,el]`. If not specified, the default light source is 45 degrees counterclockwise (i.e, to the right) from the current viewpoint. `K` controls the reflectance properties of the surface, where `K=[ka kd ks n]`, as discussed previously. If omitted, the default values for `K` are `[.55 .6 .4 10]`. The final argument is a string specifying the lighting method. `surfl(...,'light')` effectively uses `light` to create a light source and `material` to set the reflectance of the surface. This produces results different from `surfl(...,'cdata')`, which is the default. The latter form simulates a light source by actually changing the color data for the surface to be the reflectance of the surface. This technique often produces a more realistic plot.

You may have noticed that something seems to be missing. The `light` function creates a light source but lacks an argument to specify the direction of the light source. Actually, `light` does allow you to specify not only the direction, but the color of the light and even the position of the light source. However, the arguments set properties of the *light* object. These will be covered in much more detail in the Handle Graphics chapter (Chapter 31). In the meantime, you can control basic properties of the light source using the following command:

```
» Hl = light('Position',[x,y,z],'Color',[r,g,b],'Style','local');
```

This command creates a light source at position `(x,y,z)` using light color `[r,g,b]`, and specifies that the position is a location `('local')` rather than a vector `('infinite')`. It also saves the handle of the *light* object (`Hl`), which can be used to change properties of the light source at a later time. For example, the following command will return the light source to the defaults:

```
» set(Hl,'Position',[1 0 1],'Color',[1 1 1],'Style','infinite');
```

The `set` function and other aspects of the *light* object are discussed in Chapter 31.

# 29

# Images, Movies, and Sound

MATLAB provides commands for displaying several types of images. Images can be created and stored as standard double-precision floating-point numbers (`double`) and optionally as 8-bit unsigned integers (`uint8`). MATLAB can read and write image files in a number of standard graphics files formats as well as using `load` and `save` to save image data in MAT-files as discussed in Chapter 5. MATLAB provides commands to create and play animations as movies (sequences of frames). Sound functions are available as well for those computers (such as Macintosh, PC with a sound card, and many Unix platforms) that support sound.

## 29.1 IMAGES

Images in MATLAB consist of a data matrix and usually an associated colormap matrix. There are three types of image data matrices that are interpreted differently: indexed images, intensity images, and truecolor or RGB images.

An ***indexed*** image requires a colormap and interprets the image data as indices into the colormap matrix. The colormap is a standard colormap: any `m`-by-3 array containing valid RGB data. Given an image data array $X_{ij}$ and a colormap array `cmap`, the color of each image pixel $P_{ij}$ is `cmap(X(i,j),:)`. This implies that the data values in `X` are integers within the range `[1 length(cmap)]`. This image can be displayed using

```
» image(X); colormap(cmap)
```

An ***intensity*** image scales the image data over a range of intensities. This form is normally used with images to be displayed in gray scale or one of the other monochromatic colormaps, but other colormaps can be used if desired. The image data is not required to be in the range `[1 length(cmap)]`, as is the case with indexed images. The data are scaled over a given range and the result is used to index into the colormap. For example,

```
» imagesc(X,[0 1]); colormap(gray)
```

associates the value `0` with the first colormap entry and `1` with the last colormap entry. Values in `X` between `0` and `1` are scaled and used as indices into the colormap. If the scale is omitted, it defaults to `[min(min(X)) max(max(X))]`.

A ***truecolor*** or ***RGB*** image is created from an `m`-by-`n`-by-3 data array containing valid RGB triples. The first two dimensions specify the pixel location and the third specifies each color component. For example, pixel $P_{ij}$ is rendered in the color specified by `X(i,j,:)`. A colormap is not required since the color data are stored within the image data array itself. If the host computer does not support truecolor images in hardware (e.g., has an 8-bit display), then MATLAB uses color approximation and dithering to display the image. For example,

```
» image(X)
```

where `X` is a `m`-by-`n`-by-3 truecolor or RGB image, will display the image.

If images are displayed on default axes, the aspect ratio will often be incorrect and the image will be distorted. If this occurs, use the command `axis image` to force the axes aspect ratio to fit the image data you are displaying.

Intensity images can sometimes benefit from increased contrast. The `contrast` command returns a grayscale colormap matched to the image. The resulting displayed image has roughly an equally distributed grayscale histogram. Try this example to view the differences:

```
» load clown               % load an image
» cmap = contrast(X);      % create a colormap
» image(X)                 % display the image
» colormap(map)            % load the image colormap to get the length
» colormap(gray)           % use a gray colormap
» colormap(cmap)           % enhance the contrast
```

## 29.2 IMAGE FORMATS

The default numeric data type in MATLAB is `double`. This refers to a double-precision, 64-bit, floating-point number. MATLAB has limited support for other formats, like the 16-bit character data type and the 8-bit unsigned integer type (`uint8`) for images.

The `image` and `imagecs` commands can display 8-bit images without first converting them to `double` format. However, the range of data values for `uint8` data is `[0 255]`, as supported in standard graphics file formats.

For indexed images, `image` maps the value 0 to the first entry in a 256-entry colormap and maps the value 255 to the last entry by automatically supplying the proper offset. Since the normal range of `double` data for indexed images is `[1 length(cmap)]`, converting between `uint8` and `double` requires shifting the values by 1. In addition, mathematical operations on `uint8` arrays are not defined. This implies the following conversions for indexed image data:

```
» Xdouble = double(Xuint8) + 1;

» Xuint8 = uint8(Xdouble - 1);
```

For 8-bit intensity and RGB images, the range of values is normally [0 255] rather than [0 1]. To display 8-bit intensity and RGB images, use the following commands:

```
» imagesc(Xuint8,[0 255]); colormap(cmap)

» image(Xuint8)
```

The following conversions are appropriate for both intensity and RGB image data:

```
» Xdouble = double(Xuint8)/255;

» Xuint8 = uint8(round(Xdouble*255));
```

The 8-bit color data contained in the RGB image data is also automatically scaled when it is displayed. For example, the color white is normally `[1 1 1]` when using doubles. If the same color is stored as 8-bit data, the color white is represented as `[255 255 255]`.

Although mathematical operations are not defined for `uint8` arrays, reading and writing files are defined, so `imread`, `imwrite`, `save` and `load` support `uint8` data. Standard MATLAB indexing and subscripting, and the `reshape`, `cat`, `permute`, `max`, `min`, and `find` functions are supported, as are the `[]` and `'` operators. To perform any other mathematical operation, first convert the data to double-precision format using the conversions suggested previously. It is also possible to create your own methods for the `uint8` data type. See the chapter on object-oriented programming (Chapter 25) for more information on classes and methods.

As an alternative, a company called MathTools, Ltd. has developed add-on data types for MATLAB that they claim will fully support single-precision `float`, `uint8`, and `uint32` data types. See the Internet Resources chapter (Chapter 35) for contact information.

## 29.3 IMAGE FILES

Image data can be saved to files and reloaded into MATLAB using many different file formats. The normal MATLAB `save` and `load` functions support image data in either `double` or `uint8` format in the same way they support any other MATLAB variable and data type. When saving indexed images or intensity images with nonstandard colormaps, be sure to save the colormap as well as the image data as shown. For example,

```
» save myimage.mat X map
```

MATLAB also supports several industry standard image file formats using the `imread` and `imwrite` functions. Information about the contents of a graphics file can be obtained using the `imfinfo` function. These three image functions support the following file formats:

| File Extension | Format | Description |
|---|---|---|
| `bmp` | BMP | MS Windows Bitmap Format |
| `hdf` | HDF | Hierarchical Data Format |
| `jpg` or `jpeg` | JPEG | Joint Photographic Experts Group Format |
| `pcx` | PCX | PC Paintbrush Format |
| `tif` or `tiff` | TIFF | Tagged Image File Format |
| `xwd` | XWD | X Window Dump Format |

The calling syntax for `imread` varies depending on the image type and file format. `[X map]=imread('`*`filename`*`','`*`fmt`*`')` reads indexed image data from the file named *`filename`* or *`filename.fmt`* into `X` in `uint8` format and reads the associated colormap into `map` as a standard `m`-by-3 colormap of class `double`. The colormap values are rescaled to the range `[0 1]` if necessary. The graphics file format is specified by the *`fmt`* string. If *`fmt`* is omitted, `imread` attempts to determine the appropriate graphics format from the contents of the file. `X=imread(`*`filename, fmt`*`)` reads a grayscale intensity image or a truecolor/RGB image into `X`. If the file contains an intensity image, `X` contains a two-dimensional array. If the file contains a truecolor or RGB image, `X` contains a three-dimensional array. The following graphics file formats can be read by `imread`:

| Format | Description |
|---|---|
| BMP | 1-, 4-, 8-, and 24-bit uncompressed images and 4- and 8-bit RLE (run-length encoded) images. |
| HDF | 8-bit raster images with or without associated colormap and 24-bit raster images. |
| JPEG | Any standard JPEG image. Some common extensions are also supported. |
| PCX | 1-, 8-, and 24-bit images. |
| TIFF | Any standard TIFF image, including 1-, 8-, and 24-bit images with or without compression |
| XWD | XYBitmaps, 1-bit XYPixmaps, and 1- and 8-bit Zpixmaps. |

Here are some examples:

```
» J = imread('cat.jpg');  % Read truecolor image data from a JPEG file

» [X,xmap] = imread('icon.bmp','bmp'); % Read bitmap image and colormap

» G = imread('grayday','tif');   % Read grayscale intensity image

» [H,hmap] = imread('hootie.ras'.'hdf');  % Read HDF image and colormap
```

A single image may be selected from multiimage TIFF or HDF files by specifying an image number. For TIFF files, the argument is the image sequence number. For HDF files, the argument is a reference number. For example,

```
» [D,dmap]=imread('dogs.tif','tif',3);   % Read third TIFF image and map

» [B,bmap]=imread('bowtie','hdf',12);  % Read HDF image labeled #12 and map
```

If the image number is not specified, `imread` will select the first image in the file.

The calling syntax for `imwrite` varies depending on the image type and file format as well. `imwrite(X,map,'`*`filename`*`','`*`fmt`*`')` writes the image in `X` and its associated colormap `map` into a file named *`filename`*. If `X` is a double-precision array, `imwrite` converts the values to `uint8` with the appropriate offset; otherwise the values are written without conversions. The colormap values must be `doubles` and are rescaled the range `[0 255]` if necessary. The graphics file format is specified by the *`fmt`* string. If *`fmt`* is omitted, `imwrite` infers the appropriate graphics format from the file name extension. `imwrite(X,'`*`filename`*`','`*`fmt`*`')` writes the grayscale intensity image or a truecolor/RGB image in `X` into *`filename`* depending on the dimensions of `X`. The following graphics file formats can be written by `imwrite`:

| Format | Description |
|---|---|
| BMP | 8-bit uncompressed images with associated colormap and 24-bit uncompressed images. |
| HDF | 8-bit raster images with or without associated colormap and 24-bit raster images. |
| JPEG | Standard JPEG images. |
| PCX | 8-bit images. |
| TIFF | Any standard TIFF image including 1-, 8-, and 24-bit images with or without compression. |
| XWD | 8-bit Zpixmaps. |

For those formats with more than one option, `imwrite` accepts parameters to control these characteristics. The available parameters are as follows:

| Format | Parameter | Value | Default |
|---|---|---|---|
| JPEG | `'Quality'` | Any number between 0 and 100. | `75` |
| TIFF | `'Compression'` | One of `'none'`, `'packbits'`, `'ccitt'`. `'ccitt'` is only valid for binary images. | `'ccitt'` for binary images, `'packbits'` for all others. |
| TIFF | `'Description'` | Any string. | empty string |
| HDF | `'Compression'` | One of `'none'`, `'rle'`, `'jpeg'`. | `'rle'` |
| HDF | `'WriteMode'` | One of `'overwrite'` or `'append'`. | `'overwrite'` |
| HDF | `'Quality'` | Only valid for `'jpeg'` compression. A number between 0 and 100. | `75` |

To use these parameters, add the appropriate parameter/value pairs to the `imwrite` command as `imwrite(...,'Parameter',Value)`. Here are some examples:

```
» imwrite(J,'cat.jpg','jpg','Quality',65); % Write RGB image to a JPEG file

» imwrite(X,xmap,'icon.bmp'); % Write image and colormap to a BMP file

» imwrite(G,'grayday.tif','tif');    % Write grayscale TIFF image

» imwrite(H,hmap,'hootie.ras','hdf','Compression','jpeg', . . .
          'Quality','65','WriteMode','append');
```

The technique used here for selecting options using parameters/value pairs is used extensively to modify and query object properties. See the Handle Graphics chapter (Chapter 31) for more details.

The `imfinfo` function obtains information about graphics files in any of these formats. The usual form is `imfinfo('filename','fmt')`, where *filename* is a string specifying the name of the file and *fmt* is a string specifying the file format as listed previously. The information obtained varies depending on the image type, but basic information such as the file name, date, format, image type, and image size are always returned. Here are some examples for files that are not included in MATLAB:

```
» imfinfo('cat.jpg')
ans =
         Filename: 'cat.jpg'
      FileModDate: '27-May-1997 20:22:23'
         FileSize: 13553
           Format: 'jpg'
    FormatVersion: ''
            Width: 270
           Height: 196
         BitDepth: 24
        ColorType: 'truecolor'

» imfinfo('grayday','tif')
ans =
                     Filename: 'grayday.tif'
                  FileModDate: '22-May-1997 16:50:33'
                     FileSize: 53118
                       Format: 'tif'
                FormatVersion: []
                        Width: 270
                       Height: 196
                     BitDepth: 8
                    ColorType: 'grayscale'
              FormatSignature: [73 73 42 0]
                    ByteOrder: 'little-endian'
               NewSubfileType: 0
                BitsPerSample: 8
                  Compression: 'Uncompressed'
    PhotometricInterpretation: 'BlackIsZero'
                 StripOffsets: 8
              SamplesPerPixel: 1
                 RowsPerStrip: 196
              StripByteCounts: 52920
                  XResolution: 1200
                  YResolution: 1200
               ResolutionUnit: 'Inch'
                     Colormap: []
          PlanarConfiguration: 'Chunky'
                    TileWidth: []
                   TileLength: []
                  TileOffsets: []
```

```
             TileByteCounts: []
                Orientation: 1
                  FillOrder: 1
           GrayResponseUnit: 0.0100
             MaxSampleValue: 255
             MinSampleValue: 0
               Thresholding: 1
```

```
» imfinfo('icon.bmp')
ans =
             Filename: 'icon.bmp'
          FileModDate: '12-May-1996 10:23:20'
             FileSize: 54338
               Format: 'bmp'
        FormatVersion: 'Version 3 (Microsoft Windows 3.x)'
                Width: 270
               Height: 196
             BitDepth: 8
            ColorType: 'indexed'
      FormatSignature: 'BM'
   NumColormapEntries: 243
             Colormap: [243x3 double]
              RedMask: []
            GreenMask: []
             BlueMask: []
      ImageDataOffset: 1026
     BitmapHeaderSize: 40
            NumPlanes: 1
      CompressionType: 'none'
           BitmapSize: 53312
       HorzResolution: 2925
       VertResolution: 2925
        NumColorsUsed: 243
   NumImportantColors: 243
```

The form `I = imfinfo(...)` returns a structure whose fields contain the information just listed. If the file contains more than one image, an array of structures is returned.

## 29.4 MOVIES

MATLAB offers the ability to save a sequence of plots of any type, 2-D or 3-D, and then play the sequence back as a movie. In a sense, the motion provided by a movie adds yet another dimension to a plot. While the sequence of plots do not have to be related in any way, they commonly are. One obvious movie type takes a 3-D plot and slowly rotates it so it can be viewed from a variety of angles. Another would be a sequence of plots showing the solution of some problem as a parameter value changes.

In MATLAB, the functions `moviein`, `getframe`, and `movie` provide the tools required to capture and play movies. `moviein` creates an array called a ***frame matrix*** to store movie frames in, `getframe` takes a snapshot of the current figure, and `movie` plays back the sequence of frames. With this understanding, the approach to capturing and playing back movies is as follows: (1) Create the frame matrix, (2) for each movie frame create the plot and capture it in the frame matrix, and (3) play the movie back from the frame matrix.

Consider the following script M-file example, where the `peaks` function is plotted and rotated about the $z$-axis:

```
% movie making example: rotate a 3-D surface plot

[X,Y,Z]=peaks(30);            % create data
surfl(X,Y,Z)                  % plot surface with lighting
axis([-3 3 -3 3 -10 10])      % fix axes so that scaling does not change
axis off                      % erase axes because they jump around
shading interp                % make it pretty with interpolated shading
colormap(hot)                 % choose a good colormap for lighting

m=moviein(15);                % choose 15 movie frames for frame matrix m

for i=1:15                    % rotate and capture each frame
  view(-37.5+24*(i-1),30)     % change the viewpoint for this frame
  m(:,i)=getframe;            % add this figure to the frame matrix
end

movie(m)  % play the movie!
```

Note that each movie frame occupies a different column in the frame matrix. The size of the frame matrix increases with the number of frames and with the size of the *Figure* window. The frame matrix size is not a function of the complexity of the plotted figure since `getframe` simply captures a bitmap image of the current axis. Optional arguments to `getframe` allow you to specify the figure or axis to capture as well as the size and position of the capture rectangle. Indexed image data and a colormap suitable for use by the `image` function may be output if two output arguments are specified—for example, `[X,map]=getframe(...)`.

By default, the function `movie` plays a movie once. By including other input arguments, it can play the movie forward, backward, a specified number of times and at a specified frame rate. Unfortunately, `moviein`, `getframe`, and `movie` only support double-precision data types; none of them support 8-bit `uint8` data sets.

## 29.5 IMAGE UTILITIES

Conversion between indexed images and movie frames is possible with the `im2frame` and `frame2im` functions. For example,

```
» [X,cmap] = frame2im(M(n,:)
```

converts the `n`th frame of the movie matrix `M` into an indexed image `X` and associated colormap `cmap`. Similarly,

```
» M(:,n) = im2frame(X,cmap)
```

converts the indexed image `X` and colormap `cmap` into the `n`th frame of the movie matrix `M`. Note that `im2frame` can be used to convert a series of images into a movie in the same way that `getframe` converts a series of figures or axes into a movie.

The `capture` function creates a bitmap copy of the contents of the current *Figure* window, including any *uicontrols* (see the Graphical User Interfaces chapter, Chapter 32, for a discussion of *uicontrols*). `capture` creates a new *Figure* window and displays the bitmap copy as an image in the new *Figure* window. The form `[X,cmap]=capture(h)` returns an image matrix `X` and colormap `cmap` from figure `h`. If `h` is omitted, the current figure is captured. The image can then be displayed using

```
» colormap(cmap)
» image(X)
```

Note that the resolution of the bitmap image is less than that which can be obtained by using the `print` command. See the chapter on printing and exporting graphics (Chapter 30) for more information.

On the Macintosh platform, MATLAB movies can be saved as QuickTime movie files using the `qtwrite` function, which requires QuickTime. The basic form `qtwrite(M,cmap,'filename')` writes the movie `M` with colormap `cmap` to the file *`filename`*. Frame rate, compressor type, and spacial quality may be specified; if omitted, the defaults are 10 frames per second, video compression, and normal spacial quality. Indexed image arrays or *decks* may be written as well.

## 29.6 SOUND

MATLAB supports sound on PC and Macintosh platforms and on any Unix platform with a `/dev/audio` device. `sound(y,f,b)` sends the signal in vector `y` to the computer's speaker at sample frequency f. Values in y outside the range [-1 1] are clipped. If f is omitted, the default sample frequency of 8192 Hz is used. MATLAB plays the sound using b bits/second if possible. Most platforms support b=8 or b=16. If b is omitted, b=16 is used.

The soundsc function is the same as sound except that the values in y are *scaled* to the range [-1 1] rather than clipped. This results in a sound that is as loud as possible without clipping. An additional argument is available that permits mapping a range of values in y to the full sound range. The format is soundsc(y,...,[smin smax]). If omitted, the default range is [min(y) max(y)]. Two industry-standard sound file formats are supported in MATLAB. NeXT/Sun audio format (file.au) files and Microsoft WAVE format (file.wav) files can be written and read.

The NeXT/Sun Audio sound storage format supports multichannel data for 8-bit mu-law, 8-bit linear, and 16-bit linear formats. The most general form of auwrite is auwrite(y,f,n,'*method*','filename'), where y is the sample data, f is the sample rate in hertz, b specifies the number of bits in the encoder, '*method*' is a string specifying the encoding method, and '*filename*' is a string specifying the name of the output file. Each column of y represents a single channel. Any value in y outside the range [-1 1] is clipped prior to writing the file. The f, n, and '*method*' arguments are optional. If omitted, f=8000, n=8, and method='mu'. The method argument must be either 'linear' or 'mu'. If the file name string contains no extension, '.au' is appended.

Conversion between mu-law and linear formats can be performed using the mu2lin and lin2mu functions. More information about the exact conversion processes involved with these two functions can be found using on-line help.

Multichannel 8-bit or 16-bit Wave sound storage format sound files can be created with the wavwrite function. The most general form is wavwrite(y,f,n, '*filename*'), where y is the sample data, f is the sample rate in hertz, b specifies the number of bits in the encoder, and '*filename*' is a string specifying the name of the output file. Each column of y represents a single channel. Any value in y outside the range [-1 1] is clipped prior to writing the file. The f and n arguments are optional. If omitted, f=8000 and n=16. If the file name string contains no extension, '.wav' is appended.

Both auread and wavread have the same syntax and options. The most general form is [y,f,b]=auread('*filename*',n), which loads a sound file specified by the string '*filename*' and returns the sampled data into y. The appropriate extension (.au or .wav) is appended to the file name if no extension is given. Values in y are in the range [-1 1]. If three outputs are requested, as illustrated previously, the same rate in hertz and the number of bits per sample are returned in f and b, respectively. If n is given, only the first n samples are returned from each channel in the file. If n=[n1 n2], only samples from n1 through n2 are returned from each channel. The form [samples, channels]=wavread('*filename*','size') returns the size of the audio data in the file rather than the data itself. This form is useful for preallocating storage or estimating resource use.

Additional sound support is provided for Macintosh versions of MATLAB only. Sound functions specific to the Macintosh include functions to display the sound capabilities of the computer, record sound data from the default sound input device, speak text if the Speech Manager is installed, and read and write SND format sound files. The functions available are described in the following table.

| Function | Description |
|---|---|
| `recordsound` | Record sound (requires a default sound input device). |
| `speak` | Speak text string (requires the Speech Manager). |
| `readsnd` | Read SND resources and files. |
| `writesnd` | Write SND resources and files. |
| `soundcap` | Display the sound capabilities of the computer. |

# 30

# Printing and Exporting Graphics

MATLAB graphics are very effective tools for data visualization and analysis. As such, it is often desirable to create hard copy output or to use these graphics in other applications. MATLAB provides a very flexible system for printing graphics and creating output in many different graphics formats, including EPS and TIFF. Most other applications can import one or more of the graphics file formats supported by MATLAB. You can even generate MATLAB M-files that can be used to re-create a *Figure* window graphic.

## 30.1 PRINTING FROM THE MENU

The simplest way to print a graphic is to use the menu. To print a graphic, make the *Figure* window the active window by clicking on it with the mouse. Then select **Print** from the **File** menu on the *Figure* window menu bar or select **Print** from the Macintosh **File** menu, or use the `printdlg` command from the command line. On PC and Macintosh platforms, a standard print dialog box will appear. Here you can select the number of copies desired, the destination (printer or file), and a number of other options. On Unix systems, a custom dialog box will appear where you can select paper type and orientation, the destination, and device driver and printer options. Select the desired options and click the **OK** or **Print** button to print the graphic, or click on the **Save** button if available to print to a file. When printing to a file, another dialog box will appear where you can enter a file name. The current plot is then sent to the printer or file in a format appropriate for the selected printer.

The **File** menu includes the standard **Page Setup** or **Print Setup** menu options on the Macintosh and PC platforms, respectively. These selections allow you to choose additional characteristics of the printed output, such as paper size and orientation. The PC version allows you to select a printer here as well. The Macintosh version uses the Chooser to select a default printer. Unix versions use the command line option `-d`*`device`* to select a device driver and `-P`*`printer`* to select a specific printer. See the section on printing from the command line (Section 30.3) for more information.

Printing from the menu on PC and Macintosh computers uses the printing software and *device drivers* you have already installed on your computer. Device drivers are used by your operating system to prepare printed output in a form that a specific printer understands. The appropriate device driver is automatically selected when you select a specific printer. If you elect to print to a file rather than to the printer, your printing software will still use the device driver for the selected printer to format the data it sends to the file. On Unix platforms the print dialog box is a front end to the `print` command, where the device driver and printer are selected separately.

## 30.2 POSITIONING AND RESIZING GRAPHICS

When a *Figure* window is active, MATLAB adds a **Page Position** menu item to the **File** menu that allows you to adjust the size and placement of the graphic on the printed page. Making this menu choice or typing `pagedlg(fig)` in the *Command* window brings up the page dialog box for the figure `fig`. If `fig` is omitted, the current figure is used. You can select the paper orientation, image size, and image location on the page for the selected *Figure* window. The mouse can be used to move and resize the graphic on the page. Additional options are available to center the image on the page, fill the page, preserve the *Figure* window size, and revert to the default settings. When the size and position have been adjusted to your satisfaction, click on the **Done** button to return to MATLAB or click on the **Print** button to bring up the print dialog box directly. Changes made in the page position dialog box remain in effect until they are changed or until the figure is closed. Each figure may be adjusted independently.

Options selected in the page position dialog box can also affect some graphics file output parameters. For example, the size of the output image can affect the size of some bitmapped or EPS output files.

## 30.3 PRINTING FROM THE COMMAND LINE

Many additional options are available using the command-line printing functions supplied by MATLAB. The `orient` command changes the print orientation mode of the current

*Figure* window. The default `portrait` mode prints vertically in the middle of the page. `orient landscape` prints horizontally and fills the page. `orient tall` prints vertically and fills the page; the figure's aspect ratio is not preserved. `orient` without arguments returns the current orientation of the current *Figure* window.

To print a *Figure* window graphic to your default printer from the command line, make the desired *Figure* window the active window by plotting a graphic, clicking on the *Figure* window with the mouse or using the `figure(n)` command, and then issuing the `print` command.

```
» plot(rand(10,3))  % plot some lines
» print       % print the current plot to your default printer
```

The `print` command has many options available. The most general form is `print -ddevice -option filename`. If a file name is specified, the graphic is printed to the file rather than sent to the printer. A device driver may be selected using the `device` option. MATLAB supplies built-in device drivers for PostScript, Encapsulated PostScript, Hewlett-Packard Graphics Language (HPGL), and Adobe Illustrator 88 format devices. The `print` command can also create MATLAB M-files. The command-line switches for these built-in devices are listed in the following table.

| Switch | Description |
|---|---|
| `-dps` | PostScript Level 1, grayscale. |
| `-dps2` | Postscript Level 2, grayscale. |
| `-deps` | Encapsulated PostScript Level 1, grayscale (EPS). |
| `-deps2` | Encapsulated PostScript Level 2, grayscale (EPS). |
| `-dpsc` | PostScript Level 1, color. |
| `-dpsc2` | PostScript Level 2, color. |
| `-depsc` | Encapsulated PostScript Level 1, color (EPS). |
| `-depsc2` | Encapsulated PostScript Level 2, color (EPS). |
| `-dhpgl` | Hewlett-Packard Graphics Language. |
| `-dill` | Adobe Illustrator 88 compatible illustration file. |
| `-dmfile` | M-file (and MAT-file if necessary), which will re-create the graphic. |

## 30.4 SELECTING A DEVICE DRIVER

There are a number of things to think about when selecting a device driver. If your printer supports PostScript, then a built-in PostScript driver should be used. Level 1 PostScript is an older specification and is required for some printers. Level 2 PostScript is smaller and faster than Level 1 and should be used if possible. If you are using a color printer, a color driver should be used. Black and white or grayscale printers can use either driver. However, when using a color driver for a black and white printer, the file will be larger and colors will be dithered, making lines and text less clear in some cases. When in doubt, plot lines with different linestyles or markers and use a black and white driver.

If you wish to import your graphic into another application, save the output to a file rather than sending it directly to a printer. Choose a device driver that will create a file format that your importing application will accept. EPS files can be imported into many applications and often may be resized, but you may not be able to edit the graphic within the application. If you wish to edit the graphic, use the Illustrator 88 format and import the graphic into Adobe Illustrator on any platform for editing.

The HP 7475A six-pen plotter and other compatibles understand the HPGL format. This format uses black and the first five plot colors as the pen colors and assumes that the graphic has a white background. HPGL files can also be imported into a number of applications, including Microsoft Word.

To save and reload a graphic in MATLAB, use the `print -dmfile` *`filename`* command. This will create the M-file *`filename.m`* that contains MATLAB commands to re-create the graphic. A MAT-file *`filename.mat`* will also be created if necessary. To display the graphic, execute the M-file. The MAT-file will be loaded if necessary, and the graphic will be displayed.

## 30.5 ADDITIONAL DEVICE DRIVERS

In addition to these built-in devices, MATLAB supports platform-specific devices as well. On Macintosh platforms, the PICT format is supported as a command-line device option.

| Switch | Description |
|---|---|
| `-dpict` | Macintosh PICT format |

The **Save As** selection on the Macintosh **File** menu also supports the EPS, Illustrator, Bitmap (BMP), PICT, and M-file formats. The EPS and Illustrator formats can use CMYK color instead of RGB and can include a PICT, EPSI, or Bitmap preview if desired. On the other platforms, the **Save As** menu selection is limited to saving the graphic in the M-file format.

On Unix and PC platforms, however, MATLAB provides support for over 40 additional output formats. MATLAB actually creates PostScript output and uses a program called Ghostscript to convert the result into one of the other available formats. Device options for these formats are listed in the following table.

| Switch | Description |
|---|---|
| -dlaserjet | HP LaserJet |
| -dljetplus | HP LaserJet+ |
| -dljet2p | HP LaserJet IIP |
| -dljet3 | HP LaserJet III |
| -dljet4 | HP LaserJet 4 (defaults to 600 dpi) |
| -ddeskjet | HP DeskJet and DeskJet Plus |
| -ddjet500 | HP DeskJet 500 |
| -dcdeskjet | HP DeskJet 500C with 1 bit/pixel color |
| -dcdjmono | HP DeskJet 500C printing black only |
| -dcdjcolor | HP DeskJet 500C with 24-bit color and Floyd-Steinberg dithering |
| -dcdj500 | HP DeskJet 500C |
| -dcdj550 | HP DeskJet 550C |
| -dpaintjet | HP PaintJet color printer |
| -dpjxl | HP PaintJet XL color printer |
| -dpjetxl | HP PaintJet XL color printer |
| -dpjxl300 | HP PaintJet XL300 color printer |
| -ddnj650c | HP DesignJet 650C |
| -dbj10e | Canon BubbleJet BJ10e |
| -dbj200 | Canon BubbleJet BJ200 |
| -dbjc600 | Canon Color BubbleJet BJC-600 and BJC-4000 |
| -dln03 | DEC LN03 laser printer |
| -depson | Epson-compatible dot matrix printers (9- or 24-pin) |
| -depsonc | Epson LQ-2550 and Fujitsu 3400/2400/1200 |
| -deps9high | Epson-compatible 9-pin, interleaved lines (triple resolution) |
| -dibmpro | IBM 9-pin Proprinter |
| -dbmp256 | 8-bit (256-color) .BMP file format |
| -dbmp16m | 24-bit .BMP file format |
| -dpcxmono | Monochrome PCX file format |
| -dpcx16 | Older color PCX file format (EGA/VGA, 16-color) |

| Switch | Description |
|---|---|
| -dpcx256 | Newer color PCX file format (256-color) |
| -dpcx24b | 24-bit color PCX file format, three 8-bit planes |
| -dpbm | Portable Bitmap (plain format) |
| -dpbmraw | Portable Bitmap (raw format) |
| -dpgm | Portable Graymap (plain format) |
| -dpgmraw | Portable Graymap (raw format) |
| -dppm | Portable Pixmap (plain format) |
| -dppmraw | Portable Pixmap (raw format) |
| -dbit | A plain "bit bucket" device |
| -dbitrgb | Plain bits, RGB |
| -dbitcmyk | Plain bits, CMYK |

The PC platform supports some additional device options as well.

| Switch | Description |
|---|---|
| -dwin | Use Windows printing services, grayscale. |
| -dwinc | Use Windows printing services, color. |
| -dmeta | Windows Enhanced Metafile format. |
| -dbitmap | Windows Bitmap format (BMP). |
| -dsetup | Display the **Print Setup** dialog box. |
| -v | Display the **Print** dialog box (normally suppressed). |

Printing on the PC uses Windows drivers by default. The menu printing method always uses Windows drivers and the default command-line driver is `-dwin`, which uses Windows drivers as well. When printing from the command line using a built-in driver or a Ghostscript driver, MATLAB writes to a file and copies the file directly to a port (such as LPT1), bypassing the Windows drivers. Therefore, if you are printing over a network, you must capture the port using the appropriate network command. See your Network Administrator for the specific procedure in your network environment.

Note that since the default PC device driver option is `-dwin` rather than `-dwinc`, all graphics printed from the command line print in black and white or gray scale even if you are printing to a color printer. To print in color from the command line, use the `-dwinc` option each time or edit the `printopt.m` file to change your default, as discussed later.

## 30.6 OTHER PRINTING OPTIONS

The `print` command accepts a number of additional option flags that allow you to select a figure or printer or to modify certain aspects of the output. These option flags are listed in the following table.

| Switch | Description |
|---|---|
| `-f`*n* | Specify the *Figure* to print. |
| `-P`*printer* | Specify the printer to use (Unix only). |
| `-append` | Append to an existing file (not valid for EPS). |
| `-noui` | Print the *Figure* without printing any *uicontrols*. |
| `-painters` | Render using painter's algorithm. |
| `-zbuffer` | Render using Z-buffer. |
| `-r`*dpi* | Specify resolution in dots per inch (Z-buffer rendering only). |
| `-epsi` | Include a black and white EPSI preview (EPS only). |
| `-loose` | Include white space around the graphics (EPS only). |
| `-cmyk` | Use CMYK colors instead of RGB (PostScript, EPS, and Illustrator). |
| `-adobecset` | Use PostScript default character set encoding (Postscript and EPS). |

MATLAB selects the current *Figure* window unless the `-f`*n* option is used. The `-P` option is used to select a specific printer on Unix systems. The `-append` flag is used to add a graphic to an existing graphic file. Note that `-append` is not valid for EPS files. The `-noui` flag is used to suppress printing any *uicontrols* that may be in the *Figure* that contains the graphic. See the graphical user interface chapter (Chapter 32) for more information on *uicontrols.*

The next two flags (`-painters` and `-zbuffer`) specify the rendering algorithm to be used. Painter's algorithm uses vector graphics while Z-buffer uses raster or bitmap graphics. MATLAB normally chooses the best rendering method based on the complexity of the graphic and the *Figure* settings. However, some graphics do not print correctly using Painter's algorithm and Z-buffer must be used. Also note that Z-buffer files can get very large if the *Figure* window is large, the output is in color, or the resolution specified by the `-r` option is high. The default Z-buffer resolution on the Macintosh is 72 dots per inch (dpi) and 150 dpi on other platforms. Using `-r0` sets the Z-buffer resolution to the screen resolution.

The last four flags specify properties of PostScript files. The `-epsi` flag adds a 1-bit EPSI preview to EPS output files. The `-loose` flag adds some white space to the outside of the EPS graphic so that it matches the EPSI preview. The `-cmyk` flag instructs MATLAB to use CMYK color values in the PostScript, EPS, or Illustrator format file rather than the default RGB color values. Use this option if the result will be printed using four-color separations. The `-adobecset` option should be used with early PostScript Level 1 printers that do not support the `ISOLatin1Encoding` operator. If you notice text problems in printed output from an older PostScript printer, try including this flag with the `print` command.

Note that any of these options can be used in the Device Option text box of the **Print** dialog box on Unix systems. The default device option (`-d`) and printer selection (`-P`) are listed in this area when the **Print** dialog box appears. You may modify these text strings and/or add additional option flags here if desired.

## 30.7 CHANGING DEFAULTS

The `printopt` function specifies the default options for the `print` command. `[pcmd,dev]=printopt` returns the print command and default device option for the operating system used on your computer. The device option is ignored if a device is specified on the command line. Default values are listed in the following table.

| Platform | pcmd | dev |
|---|---|---|
| PC MS-Windows | `Copy /B %s LPT1:` | `-dwin` |
| Macintosh | `(not applicable)` | `-dps2` |
| Silicon Graphics | `lp` | `-dps2` |
| All other Unix | `lpr -r -s` | `-dps2` |

These defaults may be changed by editing the `$MATLAB/toolbox/local/printopt.m` file. Changes should be made after the line

```
%--> Put your own changes to the defaults here (if needed)
```

For example, if you have a PC platform with a color printer and normally print graphics in color, you could add the line

```
dev = '-dwinc';
```

to use the `-dwinc` device option by default. If you are printing on a Microsoft network to the printer `laser1` served by the host `pserver`, you could add the line

```
pcmd = 'COPY /B %s \\pserver\laser1';
```

to print over the network rather than capturing the LPT1 port.

## 30.8 EXPORTING IMAGES

If your graphic is an *image,* you have some additional export options. If your graphic is not an image, you can *create* an image from your graphic. The `capture` command creates a bitmapped image of a *Figure* window. The resulting image data can then be saved to a file in one of a number of standard graphics file formats using the `imwrite` command. MATLAB supports the following formats.

| Format | Description | Options |
|---|---|---|
| BMP | MS Windows Bitmap | 8-bit uncompressed images with associated colormap and 24-bit uncompressed images. |
| HDF | Hierarchical Data Format | 8-bit raster images with or without associated colormap and 24-bit raster images. Both `rle` and `jpeg` compression are supported. |
| JPEG | Joint Photographic Experts Group | Standard JPEG images with selectable quality. |
| PCX | PC Paintbrush | 8-bit images. |
| TIFF | Tagged Image File Format | Any standard TIFF image including 1-, 8-, and 24-bit images with or without `packbits` or `ccitt` compression. |
| XWD | X Window Dump | 8-bit ZPixmaps. |

For example, to capture the graphic in the current *Figure* window as a bitmapped image and save it to a file in JPEG format with quality value 85, use the following commands:

```
» [img,cmap] = capture;
» imwrite(img,cmap,'graphic.jpg','Quality',85);
```

Since `capture` works by creating an image directly from the pixel data used to display the *Figure* window, the quality of the image created by `capture` is limited to the resolution of your display screen, typically as low as 72 pixels per inch. It is very likely that a better result can be obtained from another export method. Either the `print` command with the appropriate device option or the **Save As** menu item on the Macintosh will almost always do a better job. See the Images, Movies and Sound chapter (Chapter 29) for more information about supported image formats and the `capture` and `imwrite` commands.

## 30.9 APPLICATION NOTES

The PC and Macintosh platforms also support **Copy** and **Paste** to transfer graphics to other applications. Note that Windows Metafile format graphics can be edited when imported into some applications, such as Powerpoint. Windows Bitmap format is a raster image and can only be edited in a paint program. Applications are also available on all supported platforms to create image files from screen snapshots. Another option is to use one of the many graphics file format conversion applications available for all of these platforms to convert a graphic from a format that is supported by MATLAB (such as TIFF or JPEG) to one (such as GIF) that is not.

Some PC applications do not fully support the Windows Enhanced Metafile format and lose color information when MATLAB graphics are imported. If an imported graphic appears with a black box over the graph in applications such as Microsoft Word or PowerPoint, you might have this problem. Issue the command

```
» system_dependent(14,'on')
```

after you have displayed the plot and before exporting the graphic. This will limit Metafile rendering of the graphic to a lowest common denominator mode and allow the Metafile to be imported successfully into these applications.

If the graphic is to be imported into an application where it cannot be edited or resized, you should adjust the size of the graphic before exporting it. Bring up the page position dialog box by selecting the **Page Position** menu item from the **File** menu of the *Figure* window or use the `pagedlg` command from the *Command* window. If you select the **Match Paper Area to Figure Area** button, the exported graphic will match the size of the *Figure* window. If you then resize the *Figure* window, the graphic will be resized as well. If you resize the graphic in the dialog box using the mouse, the exported file will be resized as well. You can also type the desired graphic size into the Paper Position text box. The format is `[left,bottom,width,height]`, where `left` is the left margin, `bottom` is the bottom margin, and `width` and `height` are the dimensions of the graphic. Finally, Handle Graphics commands may be used to set the `'PaperPosition'`, `'PaperPositionMode'`, and `'PaperOrientation'` properties of the *figure,* as described in the Handle Graphics chapter (Chapter 31).

# 31

# Handle Graphics

What is Handle Graphics? Handle Graphics is the name given to the collection of low-level graphics routines that actually do the work of generating graphics in MATLAB. Handle Graphics provide a high degree of control over graphics. These functions and options are often hidden inside higher-level graphics M-files but are available if you want to use them.

The MATLAB documentation may give the impression that Handle Graphics is very complex and of use to power users only. However, this is not the case. Handle Graphics can be used by anyone to change the way MATLAB renders graphics, whether you want to make a small change in a single plot or to make global changes that affect all graphical output.

Handle Graphics lets you customize aspects of plots that cannot be addressed using the high-level commands and functions described in previous chapters. For example, suppose you wanted to plot an orange line rather than one of the colors available to the `plot` command. How would you do it? Handle Graphics provides a way.

This chapter is not an exhaustive discussion of Handle Graphics. There is just too much detail involved to do so. Here the goal is to develop a basic understanding of Handle Graphics concepts and present enough practical information so that Handle Graphics is accessible to even casual MATLAB users. With this background, the tables of Handle Graphics object properties presented in the appendices can become useful tools.

## 31.1 WHO NEEDS HANDLE GRAPHICS?

To start with, we want to emphasize that this chapter is intended primarily for those readers who want more than the basic graphics features of MATLAB. If you are satisfied with your graphics as they are, you can skip this discussion for now. Just remember that the information is here if you want to customize your graphics in the future.

For those of you who are sticking around, we want to emphasize that learning to use Handle Graphics is not difficult. If you just want to change the font in the title of a graph or change the background color of a *Figure* window, you can do so without becoming a Handle Graphics expert. On the other hand, if you like to customize your graphics and like the idea of having control over every possible aspect of your plots, Handle Graphics provides powerful tools to do just that.

The graphics features presented in previous chapters are considered high-level commands and functions. They include `plot`, `mesh`, `axis`, and others. These functions are based on a system of lower-level functions and properties called Handle Graphics.

## 31.2 OBJECTS

Handle Graphics is based on the idea that every component of a graph is an ***object,*** that each object has an identifier or ***handle*** associated with it, and that each object has ***properties*** that can be modified as desired.

One of the most popular buzzwords in computing today is the word "object." Object-oriented programming languages, database objects, and operating system and application interfaces all use the concept of objects. An object can be loosely defined as a closely related collection of data structures and functions that form a unique whole. In MATLAB, a graphics object is a distinct component of a graph that can be manipulated individually.

Everything created by a graphics command is a graphics object. These include *Figure* windows or simply *figures,* as well as *axes, lines, surfaces, text,* and others. These objects are arranged in a hierarchy with parent objects and child objects. The computer screen itself is the *root* object and the parent of everything else. *Figures* are children of the *root; axes* and *user-interface objects* (to be discussed in the next chapter) are children of *figures.* The *line, text, surface, light, patch,* and *image* objects are children of *axes.* This hierarchy is shown in the following figure.

The *root* can contain one or more *figures,* and each *figure* can contain one or more sets of *axes.* All other objects (except *uicontrol* and *uimenu* objects discussed in the next chapter) are children of *axes* and display on those *axes.* All functions that create a graphics object will create the parent object or objects if they do not exist. For example, if there are no *figures,* the `plot(rand(1,10))` function creates a new *figure* and a set of *axes* with predefined property values and then plots the *line* within those *axes.*

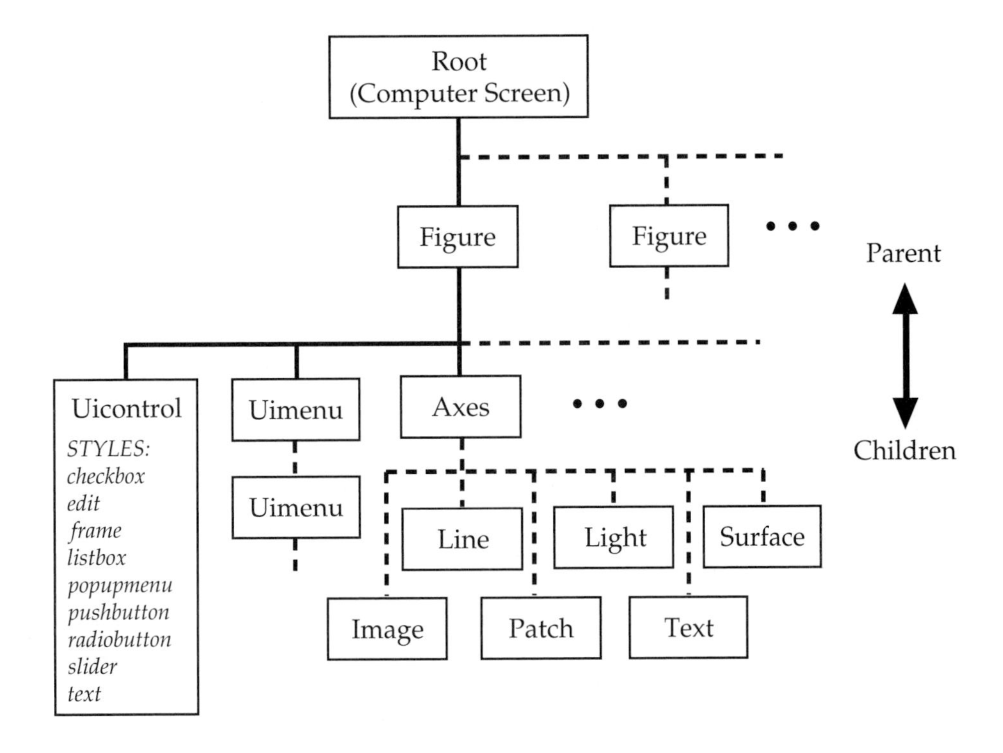


## 31.3 OBJECT HANDLES

Suppose you have three *figures* open with subplots in two of them and you want to change the color of a *line* on one of the subplot *axes*. How would you identify the *line* you wish to change? In MATLAB each object is associated with an identifier, implemented as a double-precision number, called its *handle*.

Each time an object is created, a unique handle is created for it. The handle of the *root* object, the computer screen, is always zero. The `Hf_fig=figure` command creates a new *figure* and returns its handle in the `Hf_fig` variable. *Figure* handles are normally integers and are usually displayed in the *Figure* window title bar. Other object handles are floating-point numbers in full MATLAB precision. All object creation functions return the handles of the objects they create. These functions are shown in the following table.

| Object | Description |
|---|---|
| figure | A window in which graphics are drawn. |
| axes | A rectangular coordinate system for plots in a *figure.* |
| line | Line connecting data points on an *axes.* |
| text | Character string placed relative to *axes* coordinates. |
| patch | Polygonal area defined by connected *lines.* |
| surface | Surface defined by connected faces similar to *patches.* |
| light | Directional light source in *axes* coordinates affecting *patches* and *surfaces.* |
| image | Two-dimensional picture defined by pixel colors. |
| uimenu | Programmable menu for a *Figure* window. |
| uicontrol | Programmable user-interface device such as a slider, button, or text box. |

MATLAB commands are available to obtain the handles of *figures, axes,* and other objects. For example, Hf_fig=gcf returns the handle of the current *figure,* and Ha_ax=gca returns the handle of the current *axes* in the current *figure.* These functions and other object manipulation tools are discussed later in this chapter.

To improve readability, in this text variables that contain object handles are given names beginning with a capital H, followed by a letter identifying the object type, then an underscore, and finally one or more descriptive characters. Thus, Hf_fig is a handle to a *figure,* Ha_ax1 is a handle to an *axes* object, and Ht_title is a handle to a *text* object. When an object type is unknown, the letter x is used, such as Hx_obj. While handles can be given any name, following this convention makes it easy to spot handle variables in an M-file.

All MATLAB functions that create objects return a handle or column vector of handles for each object created. Some plots are made up of more than one object. For example, a mesh plot consists of a single *surface* object with a single handle, while a waterfall plot consists of a number of *line* objects with individual handles associated with each *line.* For example, Hl_wfall=waterfall(peaks(20)) returns a column vector containing 20 handles to *lines.*

## 31.4 OBJECT PROPERTIES

All objects have a set of *properties* that define their characteristics. It is by setting these properties that one modifies how graphics are displayed. The list of properties associated with each object type (e.g., *axes, line, surface,* etc.) is unique, although a number of properties are valid for all objects. Object properties can include such things as an object's position, color, object type, parent object handle, child object handles, and many others.

Each distinct object has properties associated with it that can be changed without affecting other objects of the same type. A complete list of the properties associated with each type of object is presented in the appendices.

Object properties consist of property names and their associated values. Property names are character strings. They are typically displayed in mixed case with the initial letter of each word capitalized (e.g., 'LineStyle'). However, MATLAB recognizes a property regardless of case. In addition, you need only use enough characters to identify the property name uniquely. For example, the position property of an *axes* object can be called 'Position', 'position', or even 'pos'.

When an object is created, it is initialized with a full set of default property values that can be changed in one of two ways. The object creation function can be issued with ('Property-name',Property-value) pairs, or property values can be changed after the object is created. An example of the former method is

```
» Hf_1 = figure('Color','yellow')
```

which creates a new *figure* with default properties, except that the background color is set to yellow rather than to the default color.

## 31.5 THE UNIVERSAL FUNCTIONS get AND set

Only two functions are necessary to obtain or change Handle Graphics object properties. The get function returns the current value of one or more properties of an object. The most common syntax is get(handle,'PropertyName'). For example,

```
» p = get(Hf_1,'Position')
```

returns the position vector of the *figure* having the handle Hf_1. Similarly,

```
» c = get(Hl_a,'Color')
```

returns the color of an object identified by the handle Hl_a.

The set function changes the values of Handle Graphics object properties and uses the syntax set(handle,'PropertyName',PropertyValue). For example,

```
» set(Hf_1,'Position',p_vect)
```

sets the position of the *figure* having the handle Hf_1 to that specified by the vector p_vect. Likewise,

```
» set(Hl_a,'Color','r')
```

sets the color of the object having the handle Hl_a to red.

In general, the `set` function can have any number of `('PropertyName', PropertyValue)` pairs. For example,

```
» set(Hl_a,'Color',[1 0 0],'LineWidth',2,'LineStyle','--')
```

changes the color of the *line* having the handle `Hl_a` to red, its line width to 2 points, and its line style to dashed.

In addition to these primary purposes, the `get` and `set` functions provide help. For example, `set(handle,'PropertyName')` returns a list of values that can be assigned to the object described by `handle`. For example,

```
» set(Hf_1,'Units')
[ inches | centimeters | normalized | points | {pixels} ]
```

shows that there are five allowable character string values for the `'Units'` property of the *figure* referenced by `Hf_1` and that `'pixels'` is the default value.

If you specify a property that does not have a fixed set of values, MATLAB informs you of that fact:

```
» set(Hf_1,'Position')
A figure's 'Position' property does not have a fixed set of property values.
```

In addition to the `set` command, the Handle Graphics object creation functions accept multiple pairs of property names and values. For example,

```
» figure('Color','blue','NumberTitle','off','Name','My Figure')
```

creates a new *figure* with a blue background entitled `'My Figure'` rather than the default window title `'Figure No. 1'`.

To illustrate these concepts, consider the following example:

```
» Hf_fig = figure       % create a figure having an integer handle
Hf_fig =
     1
» Hl_light = light      % create a light object with a floating-point handle
Hl_light =
    4.0002
» set(Hl_light)         % list setable properties and potential values
   Position
   Color
   Style: [ {infinite} | local ]

   ButtonDownFcn
   Children
   Clipping: [ {on} | off ]
```

```
   CreateFcn
   DeleteFcn
   BusyAction: [ {queue} | cancel ]
   HandleVisibility: [ {on} | callback | off ]
   Interruptible: [ {on} | off ]
   Parent
   Selected: [ on | off ]
   SelectionHighlight: [ {on} | off ]
   Tag
   UserData
   Visible: [ {on} | off ]

» get(Hl_light)        % list properties and current property values
   Position = [1 0 1]
   Color = [1 1 1]
   Style = infinite

   ButtonDownFcn =
   Children = []
   Clipping = on
   CreateFcn =
   DeleteFcn =
   BusyAction = queue
   HandleVisibility = on
   Interruptible = on
   Parent = [5.00024]
   Selected = off
   SelectionHighlight = on
   Tag =
   Type = light
   UserData = []
   Visible = on
```

The *light* object was used in the preceding example because it contains the fewest properties of any object. A *figure* was created and returned a handle. A *light* object was then created that returned a handle of its own. An *axes* object was also created since *light* objects are children of *axes;* the *axes* handle is available in the `'Parent'` property of the *light.*

> Notice that the property lists as shown are divided into two groups. The first group, before the blank line, lists properties that are unique to the particular object type. The second group, after the blank line, lists the properties common to all object types. Note also that the `set` and `get` functions return slightly different property lists. `set` lists only those properties that can be changed with the `set` command, while `get` lists all visible object properties. In the previous example, `get` listed the `'Type'` property while `set` did not. This property can be read, but not changed (i.e., it is a read-only property).

The number of properties associated with each object type is fixed in each release of MATLAB, but the number varies among object types. As shown previously, a *light* object lists 3 unique and 15 common properties, or 18 properties in all. On the other hand, an *axes* object lists 89 properties. Clearly, it is beyond the scope of this text to describe thoroughly and illustrate all properties of all 11 object types. Many of them, however, are discussed in detail and all known properties are listed in the appendices.

As an example of the use of object handles, consider the problem posed earlier, which was to plot a line in a nonstandard color. In this case, the line color is specified using an RGB value of `[1 .5 0]`, a medium orange color:

```
» x = -2*pi:pi/40:2*pi;              % create data
» y = sin(x);                        % find the sine of x
» Hl_sin = plot(x,y)                 % plot sine and save line handle
Hl_sin =
   59.0002
» set(Hl_sin,'Color',[1 .5 0],'LineWidth',3) % set color and width
```

Now add a cosine curve in light blue:

```
» Z = cos(x);                       % find the cosine of x
» hold on                           % keep the sine curve
» Hl_cos = plot(x,z);               % plot the cosine and save the handle
» set(Hl_cos,'Color',[.75 .75 1]) % color it light blue
» hold off
```

It is also possible to do the same thing with fewer steps:

```
» Hl_lines = plot(x,y, x,z);       % plot and save both handles
» set(Hl_line(1),'Color',[1 .5 0],'LineWidth', 3)
» set(Hl_line(2),'Color',[.75 .75 1])
```

How about adding a title and making the font size larger than normal?

```
» title('Handle Graphics Example') % add a title
» Ht_text = get(gca,'Title')       % get the handle to the title
» set(Ht_text,'FontSize',16)      % customize the font size
```

This last example illustrates an interesting point about *axes* objects. Every object has a `'Parent'` property as well as a `'Children'` property that contains handles to descendent objects. A *line* plotted on a set of axes has the handle of the *axes* object in its `'Parent'` property, and the empty array in the `'Children'` property. At the same time, the *axes* object has the handle of its *figure* in the `'Parent'` property, and the handles of *line* objects in the `'Children'` property. Even though *text* objects created with the `text` and `gtext` commands are children of *axes* and their handles are included in the `'Children'` property, the handles associated with the title string and axis labels are not.

These handles are kept in the 'Title', 'XLabel', 'YLabel', and 'ZLabel' properties of the *axes*. These *text* objects are always created when an *axes* object is created. The title command simply sets the 'String' property of the title *text* object within the current axes. Finally, the standard MATLAB functions title, xlabel, ylabel, and zlabel return handles and accept property and value arguments. For example, the following command adds a 24-point green title to the current plot and returns the handle of the title *text* object:

```
» Ht_title = title('This is a title.','FontSize',24,'Color','green')
```

In addition to set and get, MATLAB provides several other functions to manipulate objects and their properties. Objects may be copied from one parent to another using the copyobj function. For example,

```
» Ha_new = copyobj(Ha_ax1,Hf_fig2)
```

makes a copy of the *axes* object with handle Ha_ax1 ***and all of its children,*** assigns new handles, and places the objects into the *figure* with handle Hf_fig2. A handle to the new *axes* object is returned in Ha_new. Any object can be copied into any valid parent object based on the hierarchy described earlier. Either one or both arguments to copyobj can be vectors of handles.

Note that any object can be ***moved*** from one parent to another simply by changing the 'Parent' property value to the handle of another valid parent object. For example,

```
» figure(1)
» set(gca,'Parent',2)
```

moves the current *axes* and all its children from the *figure* having handle 1 to the *figure* having handle 2. Any existing objects in *figure* 2 are not affected except that they may become obscured by the relocated objects.

Any object and all of its children can be deleted using the delete(handle) function. Similarly, reset(handle) resets all object properties associated with handle except for the 'Position' property to the defaults for that object type. If handle is a column vector of object handles, all referenced objects are affected by set, reset, copyobj, and delete.

The get and set functions return a structure when an output is assigned. For example,

```
» lprop = get(Hl_light)
lprop =
          BusyAction: 'queue'
       ButtonDownFcn: ''
            Children: []
            Clipping: 'on'
               Color: [1 1 1]
```

```
             CreateFcn: ''
             DeleteFcn: ''
      HandleVisibility: 'on'
         Interruptible: 'on'
                Parent: 5.00024
              Position: [1 0 1]
              Selected: 'off'
    SelectionHighlight: 'on'
                 Style: 'infinite'
                   Tag: ''
                  Type: 'light'
              UserData: []
               Visible: 'on'

» lopt = set(Hl_light)
lopt =
            BusyAction: {2x1 cell}
         ButtonDownFcn: {}
              Children: {}
              Clipping: {2x1 cell}
                 Color: {}
             CreateFcn: {}
             DeleteFcn: {}
      HandleVisibility: {3x1 cell}
         Interruptible: {2x1 cell}
                Parent: {}
              Position: {}
              Selected: {2x1 cell}
    SelectionHighlight: {2x1 cell}
                 Style: {2x1 cell}
                   Tag: {}
              UserData: {}
               Visible: {2x1 cell}
```

The field names of the resulting structures are the object property name strings and are assigned alphabetically. Note that even though property names are not case sensitive, these field names are. For example,

```
» lopt.BusyAction
ans =
    'queue'
    'cancel'
» lopt.busyaction
??? Reference to non-existent field 'busyaction'.
```

Combinations of property values can be set using structures as well. For example,

```
» newprop.Color = [1 0 0];
» newprop.Position = [-10 0 10];
» newprop.Style = 'local';
» set(Hl_light,newprop)
```

changes the 'Color', 'Position', and 'Style' properties but has no effect on any other properties of the *light* object. Note that you cannot simply obtain a structure of property values and use the same structure to reset the values. For example,

```
» light_prop = get(Hl_light);
» light_prop.Color = [1 0 0];      % change the light color to red
» set(Hl_light,light_prop);        % reapply the property values
??? Error using ==> set
Attempt to modify read-only light property: 'Type'.
```

Since 'Type' is the only read-only property of a *light* object, you can work around the problem by removing the 'Type' field from the structure.

```
» light_prop = rmfield(light_prop,'Type');
» set(Hl_light,light_prop)
```

A cell array can also be used to query a selection of property values. To do so, create a cell array containing the desired property names in the desired order, and pass the cell array to get. The result will be returned as a cell array as well. For example,

```
» plist = {'Color','Position','Style'}
plist =
    'Color'     'Position'    Style'
» get(Hl_light,plist)
ans =
    [ double]    [1x3 double]    'local'
» class(ans)
ans =
cell
```

One more point about the get function. If H is a vector of handles, get(H, 'Property') returns a cell array rather than a vector. For example, given a *Figure* window with four subplots,

```
» Ha = get(gcf,'Children')         % get axes handles
Ha =
      15.0002
      13.0002
      11.0002
       9.0002
```

```
» Ha_kids = get(Ha,'Children')      % get handles of axes children
Ha_kids =
    [   16.0002]
    [4x1 double]
    [   12.0002]
    [2x1 double]]
» class(Ha_kids)
ans =
    cell
» Hx = cat(1,Ha_kids{:})      % convert to column vector
Hx =
      16.0002
      26.0002
      24.0002
      18.0002
      22.0002
      12.0002
      14.0002
      10.0002
» class(Hx)
ans =
    double
```

Now Hx can be used as an argument to Handle Graphics functions expecting a vector of object handles.

## 31.6 FINDING OBJECTS

As you have seen, Handle Graphics provides access to the objects in a *figure* and allows you to customize graphics using the get and set commands. However, what if you forgot to save a handle or handles of objects in a *figure?* Or maybe your handle variables were overwritten. How can you change object properties if you cannot identify their handles? To solve this problem, MATLAB provides tools for finding object handles.

Two of these tools, gcf and gca, were introduced earlier.

```
» Hf_fig = gcf
```

returns the handle of the current *figure,* and

```
» Ha_ax = gca
```

returns the handle of the current *axes* in the current *figure.*

In addition, MATLAB includes gco, a function to get the handle of the current object.

```
» Hx_obj = gco
```

returns the handle of the current object in the current *figure,* or alternatively

```
» Hx_obj = gco(Hf_fig)
```

returns the handle of the current object in the *figure* associated with handle `Hf_fig`.

The current object is defined as the last object clicked on with the mouse within a *figure.* This object can be any graphics object except that *root* (computer screen). When a *figure* is initially created, however, the current object has not yet been selected and `gco` returns an empty matrix. The mouse button must be clicked while the pointer is within a *figure* before `gco` will return an object handle.

Once an object handle has been obtained, the object type can be found by querying an object's `'Type'` property, which is a character string object name such as `'figure'`, `'axes'`, or `'text'`. The `'Type'` property is common to all objects. For example,

```
» x_type = get(Hx_obj,'Type')
```

is guaranteed to return a valid object string for all objects.

The MATLAB functions `gcf`, `gca`, and `gco` are good examples illustrating how MATLAB uses Handle Graphics to get information about objects. The function `gcf` gets the `'CurrentFigure'` property value of the *root* object, which is the handle of the current *figure.* The `gcf` M-file contains the following:

```
function h = gcf
%GCF Get handle to current figure.
%   H = GCF returns the handle of the current figure. The current
%   figure is the window into which graphics commands like PLOT,
%   TITLE, SURF, etc. will draw.
%
%   The handle of the current figure is stored in the root
%   property CurrentFigure, and can be queried directly using GET,
%   and modified using FIGURE or SET.
%
%   Clicking on uimenus and uicontrols contained within a figure,
%   or clicking on the drawing area of a figure cause that
%   figure to become current.
%
%   The current figure is not necessarily the frontmost figure on
%   the screen.
%
%   GCF should not be relied upon during callbacks to obtain the
%   handle of the figure whose callback is executing - use GCBO
%   for that purpose.
%
%   See also FIGURE, CLOSE, CLF, GCA, GCBO, GCO, GCBF.

%   Copyright (c) 1984-96 by The MathWorks, Inc.
%   $Revision: 5.11 $  $Date: 1996/10/25 19:34:46 $

h = get(0,'CurrentFigure');

% 'CurrentFigure' is no longer guaranteed to return a figure,
% so we might need to create one (because gcf IS guaranteed to
% return a figure)
if isempty(h)
  h = figure;
end
```



Similarly, the function `gca` returns the `'CurrentAxes'` property of the current *figure* as described by its M-file:

```
function h = gca
%GCA Get handle to current axis.
%   H = GCA returns the handle to the current axis. The current
%   axis is the axis that graphics commands like PLOT, TITLE,
%   SURF, etc. draw to if issued.
%
%   Use the commands AXES or SUBPLOT to change the current axis
%   to a different axis, or to create new ones.
%
%   See also AXES, SUBPLOT, DELETE, CLA, HOLD, GCF, GCBO, GCO, GCBF.

%   Copyright (c) 1984-96 by The MathWorks, Inc.
%   $Revision: 5.11 $ $Date: 1996/10/25 19:34:37 $

h = get(gcf,'CurrentAxes');

% 'CurrentAxes' is no longer guaranteed to return an axes,
% so we might need to create one (because gca IS guaranteed to
% return an axes)
if isempty(h)
  h = axes;
end
```



The `gco` function is similar, except that it tests for the existence of a *figure* before attempting to get the current object. Note that `gcf` and `gca` cause the associated object to be created if none existed. In `gco` as shown next, the existence of `'Children'` is checked first so that no *figure* is created if it does not already exist.

```
function object = gco(fig)
%GCO Get handle to current object.
%   OBJECT = GCO returns the current object in the current figure.
%
%   OBJECT = GCO(FIG) returns the current object in the figure FIG.
%
%   The current object is the last object clicked on, excluding
%   uimenus. If the click was not over a figure child, the
%   figure itself will be the current object.
%
%   The handle of the current object is stored in the figure
%   property CurrentObject, and can be accessed directly using GET
%   and SET.
%
%   Use GCO in a callback to obtain the handle of the object that
%   was clicked on. MATLAB updates the current object before
%   executing each callback, so the current object may change if
%   one callback is interrupted by another. To obtain the right
%   handle during a callback, get the current object early, before
%   any calls to DRAWNOW, WAITFOR, PAUSE, FIGURE, or GETFRAME which
%   provide opportunities for other callbacks to interrupt.
%
%   If no figures exist, GCO returns [].
%
%   See also GCBO, GCF, GCA, GCBF.

%   Copyright (c) 1984-96 by The MathWorks, Inc.
%   $Revision: 5.15 $  $Date: 1996/10/25 19:34:49 $

if isempty (get (0, 'Children')),
    object = [];
    return;
end;
if(nargin == 0)
    fig = get(0,'CurrentFigure');
end
object = get( fig, 'CurrentObject')
```



When something other than the 'CurrentFigure', 'CurrentAxes', or 'CurrentObject' is desired, the function get can be used to obtain a vector of handles to the children of an object. For example,

```
» Hx_kids = get(gcf,'Children');
```

returns a vector containing handles of the children of the current *figure*. This technique of getting 'Children' handles can be used to search through the Handle Graphics hierarchy to find specific objects. For example, consider the problem of finding the handle of a green *line* object after plotting some data:

```
» x = -pi:pi/20::pi;                    % create some data
» y = sin(x);
» z = cos(x);
» plot(x,y,'r',x,z,'g');                % plot two lines in red and green

» Hl_lines = get(gca,'Children');  % get the line handles
» for k=1:size(Hl_lines)           % find the green line
»     if get(Hl_lines(k),'Color') == [0 1 0]
»         Hl_green = Hl_lines(k)
»     end
» end
Hl_green =
    58.0001
```

Although this technique is effective, it gets complicated if many objects exist. This technique also misses *text* objects in titles and axis labels, unless these objects are tested individually. Conversion from cell arrays is also necessary when get(H) is issued with vector H.

Consider the problem of finding the handles of all blue *line* objects when there are multiple *figures* with multiple *axes* on each. You would have to traverse the object tree testing the 'Type' and 'Color' properties until all of the objects had been tested.

To simplify the process of finding object handles, MATLAB contains the built-in function findobj, which returns the handles of objects with specified property values. Hx=findobj(Handles,'flat','PropertyName',PropertyValue) returns the handles of all objects in Handles whose 'PropertyName' property contains the value PropertyValue. Multiple ('PropertyName',PropertyValue) pairs are allowed and all must match. If Handles is omitted, the root object is assumed. If no (PropertyName',PropertyValue) pairs are given, all objects match and all Handles are returned. If 'flat' is omitted, all objects in Handles ***and all decendents of these objects,*** including axes titles and labels, are searched. If no objects are found to match the specified criteria, findobj returns an empty matrix.

The solution to the preceding example problem becomes one line using findobj:

```
» Hl_blue = findobj(0,'Type','line','Color',[0 0 1]);
```

## 31.7 SELECTING OBJECTS WITH THE MOUSE

The gco command returns the handle of the current object, which is the last object clicked on with the mouse. However, how does MATLAB know which object was selected? For example, what handle is returned if you click on the intersection of two lines on a plot? Or how far can the pointer be from a line when the button is clicked and still select the line?

The answers depend on the rules MATLAB uses to select objects and on something called *stacking order*.

Stacking order determines which overlapping object is on *top* of the others. Initially, the stacking order is determined when the object is created, with the newest object at the top of the stack. For example, if you issue two `figure` commands, two *figures* are created. The second *figure* is drawn on top of the first. The resulting stacking order has *figure* 2 on top of *figure* 1 and the handle returned by `gcf` is 2. If the `figure(1)` command is issued or if *figure* 1 is clicked on, the stacking order changes. *Figure* 1 moves to the top of the stack and becomes the current *figure*.

In the preceding example, the stacking order was apparent from the window overlap on the computer screen. However, this is not always the case. If two lines are plotted, the second line drawn is on top of the first at points where they intersect. If the first line is clicked on with the mouse at some other point, the first line is now the current object, but the stacking order has not changed. A click on the intersecting point will continue to select the second line until the stacking order is explicitly changed.

> The current stacking order is given by the order in which `'Children'` handles appear for a given object. That is, `Hx_kids=get(handle,'Children')` returns handles of child objects in stacking order. The first element in the vector `Hx_kids` is at the top of the stack and the last element is at the bottom of the stack. The stacking order can be changed by changing the order of the `'Children'` property value of the parent object.

For example,

```
» Hf = get(0,'Children');
» if length(Hf) > 1
     set(0,'Children',Hf([end 1:end-1]);
  end
```

moves the bottom *figure* to the top of the stack, where it becomes the new current *figure*.

In addition to stacking order, when you click *near* an object with the mouse, the object is selected. Each type of object has its own selection region associated with it. For example, a *line* is selected if the mouse pointer is within 5 pixels of the *line*. The selection region of a *surface, patch,* or *text* object is the smallest rectangle that contains the object. The selection region of an *axes* object is the *axes* box itself plus the areas where labels and titles appear. Objects within *axes,* such as *lines* and *surfaces,* are higher in the stacking order, and clicking on them selects the associated object rather than the *axes*. Selecting an area outside the *axes* selection region selects the *figure* itself.

Note that the `'Layer'` property of *axes* objects determines if axis lines and tick marks are drawn above (`'top'`) or below (`'bottom'`) *axes* children objects when the view is two dimensional. This permits drawing tick marks on top of images. The default value is `'bottom'`.

## 31.8 POSITION AND UNITS

The 'Position' property of *figure* objects and most other Handle Graphics objects is a four-element row vector. As shown in the following figure, the values in this vector are [left, bottom, width, height], where [left, bottom] is the position of the lower left corner of the object relative to its parent and [width, height] is the width and height of the object.

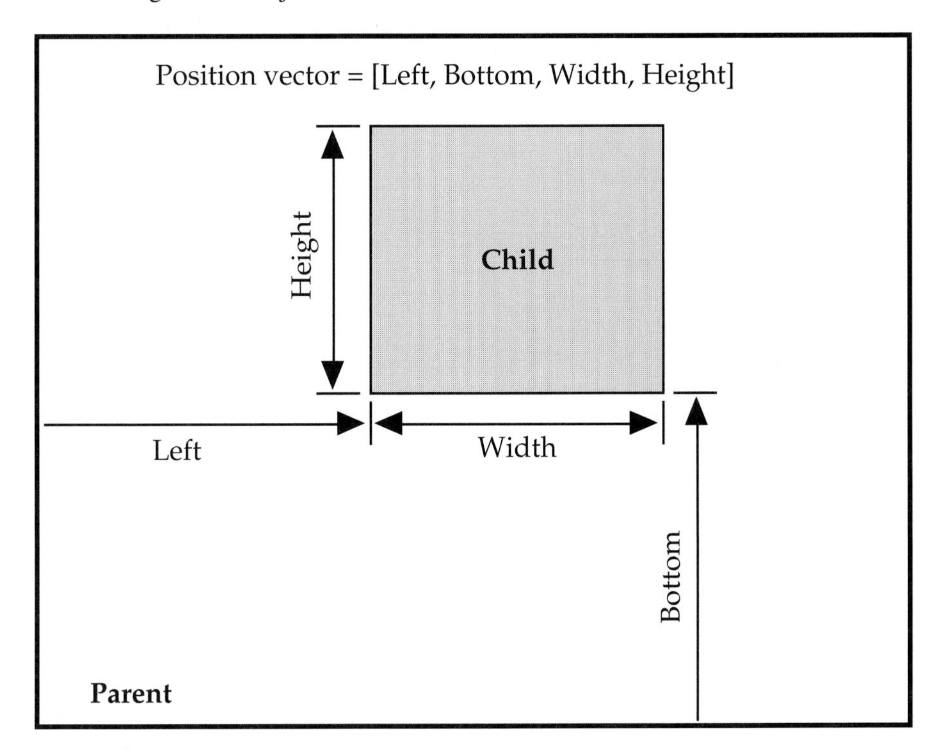


These position vector values are in units specified by the 'Units' property of the object. For example,

```
» get(gcf,'Position')
ans =
      360    544    560    420
» get(gcf,'Units')
ans =
      pixels
```

shows that the lower left corner of the current *figure* object is 360 pixels to the right and 544 pixels above the lower left corner of the screen, and the *figure* object is 560 pixels wide and 420 pixels high. ***Note that the position vector for a figure gives the drawable area***

***within the figure object itself and does not include window borders, scroll bars, menus, or title bar of the Figure window.***

The 'Units' property defaults to pixels but can be inches, centimeters, points, or normalized coordinates. Pixels represent screen pixels, the smallest rectangular object that can be represented on a computer screen. For example, a computer display set to a resolution of 800 by 600 is 800 pixels wide by 600 pixels high. Points are a typesetting standard, where one point is equal to 1/72 of an inch. Normalized coordinates are in the range zero to one. In normalized coordinates, the lower left corner of the parent is at [0, 0] and the upper right corner is at [1, 1]. Inches and centimeters are self-explanatory.

To illustrate various 'Units' property values, reconsider the preceding example:

```
» set(gcf,'Units','inches');
» get(gcf,'Position')
ans =
       3.7764    5.7203    5.8907    4.4245
» set(gcf,'Units','centimeters')
» get(gcf,'Position')
ans =
       9.5847    14.5185    14.9511    11.2297
» set(gcf,'Units','points')
» get(gcf,'Position')
ans =
       271.9001    411.8596    424.1339    318.5655
» set(gcf,'Units','normalized')
» get(gcf,'Position')
ans =
       0.2805    0.5303    0.4375    0.4102
```

All of these values represent the same *figure* position relative to the computer screen for a particular monitor and screen resolution.

The position of *axes* objects are also four-element vectors having the same form [left, bottom, width, height] but specify the object position relative to the lower left corner of the parent *figure*. ***In general, the* 'Position' *property of a child is relative to the position of its parent.***

To be more descriptive, the computer screen or *root* object position property is not called 'Position', but rather 'ScreenSize'. In this case, [left, bottom] is always [0, 0], and [width, height] are the dimensions of the computer screen in units specified by the 'Units' property of the *root* object.

## 31.9 PRINTING FIGURES

In addition to *figure* placement on the computer screen, MATLAB provides properties to control *figure* placement on the printed page as well as properties to specify attributes of the paper itself. For example, MATLAB orient command uses get and set to change the values of paper properties of the current *figure*.

| Figure Property | Description |
|---|---|
| PaperUnits | [ *inches* \| centimeters \| normalized \| points ] |
| PaperOrientation | [ *portrait* \| landscape ] |
| PaperPosition | A position vector in the form [left,bottom,width,height], where left and bottom are offsets from the lower-left corner of the page and width and height are figure dimensions. |
| PaperPositionMode | [ auto \| *manual* ] |
| PaperSize | A two-element vector containing the paper size in units specified by PaperUnits. |
| PaperType | [*usletter* \| uslegal \| a3 \| a4letter \| a5 \| b4 \|tabloid] |

The 'PaperPosition' and 'PaperSize' properties are returned in units specified by the 'PaperUnits' property. Changing the 'PaperPosition' property changes the size and location of the plot on a printed page in the same way that changing the 'Position' property of a *figure* object changes the size and location of the *figure* on the screen. The 'PaperPositionMode' property determines if the 'PaperPosition' property is used to size the graphic. If it is 'manual', the 'PaperPosition' property is used to size the graphic; if it is 'auto', the *figure* 'Position' property is used. In 'manual' mode, the width and height of the graphic is the width and height of the *figure* and is centered on the page. Resizing the *Figure* window resizes the graphic.

Selecting **Page Position** from the *Figure* window **File** menu or using the pagedlg command in the *Command* window lets you set most of these properties interactively. See the Printing and Exporting Graphics chapter (Chapter 30) for more information. As with other *figure* properties, paper properties apply only to individual *figures.* Modifying properties for all *figures* is discussed in the next section on default properties.

## 31.10 DEFAULT PROPERTIES

MATLAB assigns default properties to each object as it is created. The built-in defaults are referred to as *Factory* defaults. Whenever you want to override these defaults, you must set the values using Handle Graphics tools. In those cases where you want to change the same properties every time, MATLAB allows you to set your own default properties. MATLAB lets you change the default properties for individual objects and for object types at any point in the object hierarchy.

When an object is created, MATLAB looks for a default value at the ***parent*** level. If no default is found, the search continues up the object hierarchy until a default value is found or it reaches the built-in factory defaults.

You can set your own default values at any level of the object hierarchy by using a special property name string consisting of `'Default'` followed by the object type and the property name. The handle you use in the `set` command determines the point in the object parent–child hierarchy at which the default is applied. For example,

```
» set(0,'DefaultFigureColor',[.5 .5 .5])
```

sets the default background color for all new *figure* objects to medium gray rather than the MATLAB default. This property applies to the *root* object (whose handle is always zero), so all new *figures* will have a gray background.

Here are some other examples of defaults you can change:

```
» set(0,'DefaultAxesFontSize',14)      % larger axes fonts - all figures
» set(gcf,'DefaultAxesLineWidth',2)    % thick axis lines - this figure only
» set(gcf,'DefaultAxesXColor','y')     % yellow X axis lines and labels
» set(gcf,'DefaultAxesYGrid','on')     % Y axis grid lines - this figure
» set(0,'DefaultAxesBox','on')         % enclose axes - all figures
» set(gca,'DefaultLineLineStyle',':') % dotted linestyle - these axes only
```

When a default property is changed, it only affects objects created after the change is made. Existing objects already have property values assigned and do not change.

When working with existing objects, it is always a good idea to restore them to their original state after they are used. If you change default properties of objects in a routine, save the previous settings and restore them when exiting the routine. For example, consider this function fragment:

```
oldunits = get(0,'DefaultFigureUnits');
set(0,'DefaultFigureUnits','normalized');
    <MATLAB statements>
set(0,'DefaultFigureUnits',oldunits);
```

If you want to customize MATLAB to use your default values at all times, simply include the desired `set` commands in your `startup.m` file. For example, if you want grids and axis boxes on all *axes* by default and you usually print on A4-size paper, add the lines

```
set(0,'DefaultAxesXGrid','on')
set(0,'DefaultAxesYGrid','on')
set(0,'DefaultAxesZGrid','on')
set(0,'DefaultAxesBox','on')
set(0,'DefaultFigurePaperType','a4paper')
```

to your `startup.m` file. Any defaults set at the *root* level affect every object in every *Figure* window.

There are three special property value strings that reverse, override, or query user-defined default properties. They are `'remove'`, `'factory'`, and `'default'`. If you've changed a default property, you can reverse the change, thereby resetting it to the original defaults using `'remove'`, as illustrated here:

```
» set(0,'DefaultFigureColor',[.5 .5 .5]  % set a new default
» set(0,'DefaultFigureColor','remove')   % return to MATLAB defaults
```

To override a default temporarily and use the original MATLAB default value for a particular object, use the special property value `'factory'`. For example,

```
» set(0,'DefaultFigureColor',[.5 .5 .5])  % set a new user default
» figure('Color','factory')  % create a new figure using the MATLAB default
```

The third special property value string is `'default'`. This value forces MATLAB to search up the object hierarchy until it encounters a default value for the desired property. If found, it uses that default value. If the *root* object is reached and no user-defined default is found, the MATLAB factory default value is used. This concept is useful when you want to set an object property to a default property value after it was created with different property value. To illustrate the use of `'default'`, consider the following example:

```
» set(0,'DefaultLineColor','r')          % set the default at the root level
» set(gcf,'DefaultLineColor','g')        % current figure level default

» Hl_rand = plot(rand(1,10)); % plot a line using 'ColorOrder' color

» set(Hl_rand,'Color','default')         % the line becomes green
» close(gcf)                             % close the window
» Hl_rand = plot(rand(1,10)); % plot a line using 'ColorOrder' color again
» set(Hl_rand,'Color','default')         % the line becomes red
```

Note that the `plot` command does not use *line* object defaults for the line color. If a color argument is not specified, the `plot` command uses the *axes* `'ColorOrder'` property to specify the color of each *line* it generates.

A list of all of the factory defaults can be obtained with the command

```
» get(0,'factory')
```

Any default properties that have been set at any level in the object hierarchy can be listed using the command

```
» get(handle,'default')
```

The *root* object contains default values for a number of color properties and the *figure* position at startup.

```
» get(0,'default')
    figureColormap : [ (64 by 3) double array]
    axesColorOrder : [ (7 by 3) double array]
    axesColor : [1 1 1]
    figureInvertHardcopy : on
    figureColor : [0.8 0.8 0.8]
    lineColor : [0 0 0]
    surfaceEdgeColor : [0 0 0]
    patchEdgeColor : [0 0 0]
    patchFaceColor : [0 0 0]
    axesZColor : [0 0 0]
    axesYColor : [0 0 0]
    axesXColor : [0 0 0]
    textColor : [0 0 0]
    figurePosition : [40 390 560 420]
```

Other defaults are not listed until they have been created by the user. For example,

```
» get(gcf,'default')
» set(gcf,'DefaultLineMarkerSize',10)
» get(gcf,'default')
    lineMarkerSize : [10]
```

## 31.11 COMMON PROPERTIES

All Handle Graphics objects share a common set of object properties.

| | |
|---|---|
| ButtonDownFcn | Callback routine that executes when button press occurs. |
| Children | Handles of all this object's children. |
| Clipping | Enable or disable clipping for children of *axes*. |
| CreateFcn | Callback routine that executes just after this object is created. |
| ChangeFcn* | Callback routine that executes when any property of this object is changed (*not included in MATLAB 5.0 but expected to be included in later versions). |
| DeleteFcn | Callback routine that executes just before this object is deleted. |
| BusyAction | Controls callback routine interruption mode. |
| HandleVisibility | Determines whether the object handle is visible to the command line and/or to callback routines. |
| Interruptible | Determines if a callback routine is interruptible by a later callback routine. |

| | |
|---|---|
| Parent | Contains the handle of the parent object. |
| Selected | Indicates whether the object has been selected. |
| SelectionHighlight | Determines if the object changes appearance when selected. |
| Tag | User-specified object label. |
| Type | Object type string ('figure','axes','line', etc.). |
| UserData | Available storage area for a variable containing any data type. |
| Visible | Determines whether object is visible. |

Four of these properties contain callback routines: 'ButtonDownFcn', 'CreateFcn', 'ChangeFcn', and 'DeleteFcn'. Callback routines or *callbacks* are MATLAB strings passed to the eval function to be executed in the *Command* window workspace. A button click over the object or some other event, such as object creation, change, or deletion, is used to trigger a callback. Callbacks are discussed in detail later. Note that the 'ChangeFcn' property is included in the MATLAB documentation but is not included in MATLAB 5.0. It may or may not be included in subsequent releases.

The 'Parent' and 'Children' properties contain handles of other objects in the hierarchy. Objects drawn on the *axes* are clipped at the *axes* limits if 'Clipping' is 'on' (the default for all except *text* objects). 'Interruptible' and 'BusyAction' control the behavior of callbacks if a later callback is triggered. 'Type' is a string specifying the object type. 'Selected' is 'on' if this object is the 'CurrentObject' of the *figure,* and 'SelectionHighlight' determines if the object changes appearance when selected.

'HandleVisibility' specifies whether the object handle is visible, invisible, or visible only to callbacks. The *root* 'ShowHiddenHandles' property can override the 'HandleVisibility' property of all objects if needed. If 'Visible' is set to 'off', the object disappears from view. It is still there and the object handle is still valid, but it is not rendered. Setting 'Visible' to 'on' restores the object to view.

The 'Tag' and 'UserData' properties are reserved for the user. The 'Tag' property is typically used to tag an object with a user-defined text string. For example,

```
» set(gca,'Tag','My Axes')
```

attaches the string 'My Axes' to the current *axes* in the current *figure.* This string does not display in the *axes* or in the *figure,* but you can query the 'Tag' property to identify the object. For example, if there are numerous *axes* you can find the handle to the preceding *axes* object by using

```
» Ha_myaxes = findobj(0,'Tag','My Axes');
```

> The 'UserData' property can contain any variable you wish to put into it. A character string, a number, a structure, or even a multidimensional cell array can be stored in any object's 'UserData' property. Both of these properties are available for your exclusive use. No MATLAB function or M-file changes or makes assumptions about the values contained in these properties. However, as will be discussed in the next chapter, some user-contributed M-files and several *Mastering MATLAB Toolbox* functions make use of the 'UserData' property of some objects to store temporary data.

The properties listed for each object using the get and set commands are the documented properties. There are also undocumented or hidden properties used by MATLAB developers. Some of them can be modified, but others are read-only. Undocumented properties are simply hidden from view. The properties still exist and can be modified. The undocumented *root* property 'HideUndocumented' controls whether get returns all properties or only documented properties. Use

```
» set(0,'HideUndocumented','off')
```

to see all properties.

Since undocumented properties have been purposely left undocumented, one must be very cautious when using them. They are sometimes less robust than documented properties and are always subject to change. Undocumented properties may appear, disappear, change functionality, or even become documented in future versions of MATLAB.

## 31.12 NEW PLOTS

When a new graphics object is created using a low-level command such as line or text, the object will appear on the current *axes* in the current *figure* by default. High-level graphics functions like mesh and plot, however, clear the current *axes* and reset most *axes* properties to their defaults before displaying the plot. As discussed earlier, the hold command can be used to change this default behavior.

Both *figures* and *axes* have a 'NextPlot' property that is used to control how MATLAB reuses existing *figures* and *axes*. The hold, newplot, clf, and cla functions all affect the 'NextPlot' property of *figures* and *axes*.

```
» set(gcf,'NextPlot')
    [ {add} | replace | replacechildren ]
» set(gca,'NextPlot')
    [ add | {replace} | replacechildren ]
```

The default setting for *figures* is 'add'; the default for *axes* is 'replace'. When 'NextPlot' is set to 'add', a new plot is added without clearing or resetting the current *figure* or *axes*. When the value is 'replace', a new object causes the *figure* or *axes* to remove all child objects and reset their properties to the defaults (except 'Position' and 'Units') before drawing the new plot. The default settings clear and reset the current *axes* and reuse the current *figure*.

The third possible setting is 'replacechildren'. This setting removes all child objects but does not change current *figure* or *axes* properties. The command hold on sets both *figure* and *axes* 'NextPlot' properties to 'add'. The command hold off sets the *axes* 'NextPlot' property to 'replace'.

The newplot function is available to provide the default MATLAB new plot behavior to graphics M-files. You should use the command

```
Ha_ax = newplot;
```

early in your M-file to obtain an appropriate *axes* handle for your plot. newplot is a built-in command but could be implemented as the following M-file function:

```
function Ha = newplot

Hf = gcf;                          % get current figure or create one
next = lower(get(Hf,'NextPlot'));
switch next
  case 'replacechildren', clf;     % delete figure children
  case 'replace', clf('reset');    % delete children and reset properties
end

Ha = gca;                          % get current axes or create one
next = lower(get(Ha,'NextPlot'));
switch next
  case 'replacechildren', cla;     % deletes axes children
  case 'replace', cla('reset');    % delete children and reset properties
end
```

You should also check the hold state before changing any properties of the *axes* or *figure* to remain consistent with other high-level graphics commands and honor the existing hold state. If ishold returns TRUE, then you should not change axis limits, aspect ratio, or view properties.

## 31.13 M-File EXAMPLES

The *Mastering MATLAB Toolbox* is a collection of function M-files written in conjunction with this book to illustrate some of the concepts discussed here and to provide utilities and other useful functions that are not included in MATLAB. Some of these M-files are used here as examples of Handle Graphics concepts.

One of the simplest *Mastering MATLAB Toolbox* functions addresses a common problem. The MATLAB function `gcf` returns the handle of the current *figure*. However, if no *figure* exists, `gcf` creates one and then returns its handle. What if you want to find out if a *figure* exists in the first place, but not create an empty one if there happens to be none? The function `mmgcf` does just that as described by its contents:

```
function Hf=mmgcf(flag)
%MMGCF Get Current Figure if it Exists. (MM)
% MMGCF returns the handle of the current figure if it exists
% If no current figure exists, MMGCF returns an empty handle.
%
% MMGCF(Flag) when Flag=TRUE, generates an error message if
% no current figure exists.
%
% Note that the function GCF is different. It creates a figure
% and returns its handle if it does not exist.
%
% See also MMGCA, GCF.

Hf = get(0,'CurrentFigure');
if isempty(Hf) & nargin == 1 & flag ~=0
   error('No Figure Window Exists.')
end
```

The function `mmgcf` first obtains the *root* `'CurrentFigure'` property value, which is either a *figure* handle or the empty matrix if no *figure* exists. `mmgcf` also accepts an optional logical flag that enables generation of an error message if no current figure exists.

The function `mmgca` does the same thing for *axes* as described in its M-file:

```
function Ha=mmgca(Hf,flag)
%MMGCA Get Current Axes if it Exists. (MM)
% MMGCA returns the handle of the current axes
% in the current figure if it exists.
% MMGCA(Hf) returns the handle of the current axes in the
% figure having handle Hf.
%
% If no current axes exists, MMGCA returns an empty handle.
%
% MMGCA(Hf,Flag) when Flag=TRUE, generates an error message if
% no current axes exists
%
% Note that the function GCA is different. It creates a figure
% and an axes and returns the axes handle if it does not exist.
%
% See also GCA, MMGCF, GCF

if nargin == 0
   Hf = mmgcf;
end
Ha=get(Hf,'CurrentAxes');
if isempty(Ha) & nargin == 2 & flag ~=0
   error('No Axes Exists in the Current Figure.')
end
```

There is no need for an `mmgco` function since `gco` already exhibits the behavior of returning an empty matrix when no object exists.

Another function in the *Mastering MATLAB Toolbox* is `mmzap`. As shown in its M-file, it uses `mmgcf` for error checking, along with `findobj` and `get` to delete a specified graphics object.

```
function mmzap(arg)
%MMZAP Delete Graphics Object Using Mouse. (MM)
% MMZAP waits for a mouse click on an object in
% a figure window and deletes the object.
% MMZAP or MMZAP text erases text objects.
% MMZAP axes erases axes objects.
% MMZAP line erases line objects.
% MMZAP surf erases surface objects.
% MMZAP patch erases patch objects.
%
% Clicking on an object other than the selected type, or striking
% a key on the keyboard aborts the command.

if nargin < 1, arg = 'text'; end

Hf = mmgcf(1);
if length(findobj(0,'Type','figure')) == 1
   figure(Hf) % bring only figure forward
end
key = waitforbuttonpress;
if key  % key on keyboard pressed
   return
else    % object selected
   object = gco;
   type = get(object,'Type');
   if strncmp(type,arg,4)
      delete(object)
   end
end
```

The `mmzap` function illustrates a technique that is very useful when writing Handle Graphics function M-files. It uses the combination of `waitforbuttonpress` and `gco` to get the handle to an object selected using the mouse. `waitforbuttonpress` is a built-in MATLAB function that waits for a mouse click or key press. Its help text is as follows:

```
» help waitforbuttonpress
WAITFORBUTTONPRESS Wait for key/buttonpress over figure.
    T = WAITFORBUTTONPRESS stop program execution until a key or
    mouse button is pressed over a figure window. Returns 0
    when terminated by a mouse buttonpress, or 1 when terminated
    by a keypress. Additional information about the terminating
    event is available from the current figure.

    See also GINPUT, GCF.
```

After a mouse button is pressed with the mouse pointer over a *figure,* `gco` returns the handle of the selected object. This handle is then used to manipulate the selected object.

## 31.14 CALLBACKS

Callbacks are the character strings passed to `eval` in the *Command* window work-space when a specific event occurs. Callbacks are stored in object properties whose property names usually end with the string `'Fcn'`, such as `'ButtonDownFcn'` or `'CreateFcn'`. The object property name usually describes the event that triggers the callback. The process that takes place when a *figure* is closed is a good example of a callback.

```
» get(gcf, 'CloseRequestFcn')
ans =
    closereq

» class(ans)
ans =
    char
```

The `'CloseRequestFcn'` property contains a callback that is executed when the *figure* is closed, either by issuing `close` from the command line, selecting **Close** from the *Figure* window **File** menu, closing the *Figure* window using a close box or button, or quitting MATLAB while a *figure* is visible. The default callback is an M-file called `closereq` containing the following commands.

```
function closereq
%CLOSEREQ  Figure close request function.
%   CLOSEREQ deletes the current figure window. By default, CLOSEREQ is
%   the CloseRequestFcn for new figures.

%   Copyright (c) 1984-96 by The MathWorks, Inc.
%   $Revision: 1.2 $  $Date: 1996/10/29 14:05:02 $

shh=get(0,'ShowHiddenHandles');
set(0,'ShowHiddenHandles','on');
delete(get(0,'CurrentFigure'));
set(0, 'ShowHiddenHandles',shh);
```



This function overrides the `'HandleVisibility'` property of the current *figure,* deletes the *figure,* and restores the *root* `'ShowHiddenHandles'` property. By replacing the callback in the *figure* `'CloseRequestFcn'` property, you can customize the actions taken when a *figure* is closed. You can save data before closing, request confirmation using a dialog box or in the *Command* window, or protect a *figure* from accidental deletion. Note that this process is circumvented when the `delete(fig)` command is used. The `delete` command deletes the *figure* without executing any callbacks.

Callbacks are also used in the *Mastering MATLAB Toolbox* function mmtext. This function lets you place text on a graph and drag it to the desired location using the mouse. The M-file will be described in sections to illustrate the concepts involved. First, define the function, include some help text, and set defaults:

```
function mmtext(arg)
%MMTEXT Place and Drag Text with Mouse. (MM)
% MMTEXT waits for a mouse click on a text object,
% in the current figure then allows it to be dragged
% while the mouse button remains down.
% MMTEXT('whatever') places the string 'whatever' on
% the current axes and allows it to be dragged.
%
% MMTEXT becomes inactive after the move is complete or
% no text object is selected.

% Calls: mmgcf

if ~nargin, arg = 0; end
```

If the input argument is a string, create a *text* object from the string and make a recursive call to mmtext to let the user drag it around:

```
if ischar(arg)  % user entered text to be placed
   Ht = text('Units','normalized',...
         'Position',[.05 .05],...
         'String',arg,...
         'HorizontalAlignment','left',...
         'VerticalAlignment','baseline');
   mmtext(0)  % call mmtext again to drag it
```

Set the *figure* 'WindowButtonDownFcn' callback to call mmtext again when the mouse button is pressed over the *figure:*

```
elseif arg ==0  % initial call, select text for dragging
   Hf = mmgcf(1);
   set(Hf,'BackingStore','off',...
         'WindowButtonDownFcn','mmtext(1)')
   figure(Hf)  % bring figure forward
```

The next section implements the callback for the `'WindowButtonDownFcn'`. The mouse button has been pressed. If a *text* object has been selected, specify some properties of the *text* object for smoother movement, change the pointer shape to show that dragging is in progress, and set up callbacks for the `'WindowButtonMotionFcn'` and `'WindowButtonUpFcn'` properties.

```
elseif arg == 1 & strcmp(get(gco,'type'),'text') % text object selected
   set(gco,'Units','data',...
         'HorizontalAlignment','left',...
         'VerticalAlignment','baseline',...
         'EraseMode','xor');
   set(gcf,'Pointer','topr',...
         'WindowButtonMotionFcn','mmtext(2)',...
         'WindowButtonUpFcn','mmtext(99)')
```

The next section implements the callback for the `'WindowButtonMotionFcn'` property. It simply updates the `'Position'` property of the *text* object so that it follows the mouse around the *figure* as long as the mouse button remains down:

```
elseif arg == 2  % dragging text object
   cp = get(gca,'CurrentPoint');
   set(gco,'Position',cp(1,1:3))
```

The `'WindowButtonUpFcn'` callback resets the *figure* callbacks and restores the `'EraseMode'` property of the *text* object when the mouse is released.

```
elseif arg == 99  % mouse button up, reset everything
   set(gco,'EraseMode','normal')
   set(gcf,'WindowButtonDownFcn','',...
         'WindowButtonMotionFcn','',...
         'WindowButtonUpFcn','',...
         'Pointer','arrow',...
         'Units','pixels',...
         'BackingStore','on')
```

Finally, if the selected object is not a *text* object, restore the *figure* properties and exit:

```
else      % incorrect object selected, reset and abort
   set(gcf,'WindowButtonDownFcn','',...
         'WindowButtonMotionFcn','',...
         'WindowButtonUpFcn','',...
         'Pointer','arrow',...
         'Units','pixels',...
         'BackingStore','on')
end
```

More callbacks and other examples of Handle Graphics programming techniques are presented in the next chapter, where we discuss the use of *uimenu* and *uicontrol* user-interface objects to add controls and menus to *figures* and to create dedicated graphical user interfaces or GUIs.

## 31.15 SUMMARY

Handle Graphics functions provide the ability to fine-tune the appearance of your graphs and plots. Each graphics object has a handle associated with it that can be used to manipulate the object. Object properties can be viewed and changed using `get` and `set` to customize your graphics.

The MATLAB Handle Graphics functions discussed in this chapter are summarized in the following table.

| Handle Graphics Function | Description |
|---|---|
| `set(handle,'PropertyName',Value)` | Set object properties. |
| `get(handle,'PropertyName')` | Get object properties. |
| `reset(handle)` | Reset object properties to defaults. |
| `delete(handle)` | Delete object and all its children. |
| `copyobj(handle,parent)` | Copy object and all its children to another parent object. |
| `gcf` | Get handle to the current *figure*. |
| `gca` | Get handle to the current *axes*. |
| `gco` | Get handle to the current object. |
| `findobj('PropertyName',Value)` | Get handles to objects with specified property values. |
| `waitforbuttonpress` | Wait for key or button press over a figure. |

| Handle Graphics Function | Description |
|---|---|
| `figure('PropertyName',Value)` | Create figure object. |
| `axes('PropertyName',Value)` | Create axes object. |
| `line(X,Y,'PropertyName',Value)` | Create line object. |
| `text(X,Y,S,'PropertyName',Value)` | Create text object. |
| `patch(X,Y,C,'PropertyName',Value)` | Create patch object. |
| `light('PropertyName',Value)` | Create light object. |
| `surface(X,Y,Z,'PropertyName',Value)` | Create surface object. |
| `image(C,'PropertyName',Value)` | Create image object. |

# 32

# Creating Graphical User Interfaces

## 32.1 GUI?: WHAT IS A GUI?

A user interface is the point of contact or method of interaction between a person and a computer or computer program. It is the method used by the computer and the user to communicate information. The computer displays text and graphics on the computer screen and may generate sounds with a speaker. The user communicates with the computer using input devices such as a keyboard, mouse, trackball, drawing pad, and microphone. The user interface defines the look and feel of the computer, operating system, or application. Often a computer or program is chosen on the basis of pleasing design and functional efficiency of its user interface.

A graphical user interface, or GUI is a user interface incorporating graphical objects such as windows, icons, buttons, menus, and text. Selecting or activating these objects in some way usually causes an action or change to occur. The most common activation method is to use a mouse or other pointing device to control the movement of a pointer on the screen, and to press a mouse button to signal object selection or some other action.

In the same way that the Handle Graphics capabilities of MATLAB discussed in the previous chapter let you customize the way MATLAB displays information, the Handle Graphics user interface functions described in this chapter let you customize the way you communicate with MATLAB. The *Command* window is not the only way to interact with MATLAB.

This chapter illustrates the use of Handle Graphics *uicontrol* and *uimenu* objects to add graphical user interfaces to MATLAB functions and M-files. *uimenu* objects create drop-down menus and submenus in *Figure* windows. *uicontrol* objects create objects such as buttons, sliders, popup menus, and text boxes.

MATLAB includes an excellent example of its GUI capabilities in the `demo` command:

```
» demo
```

Explore this command to see how *uimenus* and *uicontrols* provide interactive input to MATLAB functions.

## 32.2 WHO SHOULD CREATE GUIs—AND WHY?

After running `demo` you are likely to be asking yourself, "Why would I want to create a GUI in MATLAB?" Good question! The short answer is that you may not. Many people who use MATLAB primarily to analyze data, solve problems, and plot results may not consider GUI tools to be worth the effort.

On the other hand, GUIs can be used to create very effective tools and utilities in MATLAB or to build interactive demonstrations of your work. The most common reasons to create a graphical user interface are as follows:

1. You are writing a utility function that you will use over and over again and menus, buttons, or text boxes make sense as input methods.
2. You are writing a function or developing an application for others to use.
3. You want to create an interactive demonstration of a process, technique, or analysis method.
4. You think GUIs are neat and you want to experiment with them.

A number of GUI-based tools and utility functions are included in the *Mastering MATLAB Toolbox*. Other tools and utilities have been written by MATLAB users incorporating MATLAB's GUI functions. Most of these tools are available from The MathWorks anonymous FTP site and from other sources. The Internet Resources chapter (Chapter 35) explains how to access the FTP site as well as a number of other Internet resources.

Before we begin, remember that a basic understanding of Handle Graphics is a prerequisite to designing and implementing a GUI in MATLAB. If you skipped the previous chapter, you should go back and read it now.

## 32.3 GUI OBJECT HIERARCHY

As we demonstrated in the previous chapter, everything created by a graphics command is a graphics object. These include *uimenu* and *uicontrol* objects as well as *figures, axes,* and their children. Let's take another look at the object hierarchy. The computer screen itself is the *root* object and *figures* are children of the *root; axes, uimenus,* and *uicontrols* are children of *figures*. This hierarchy is as follows:

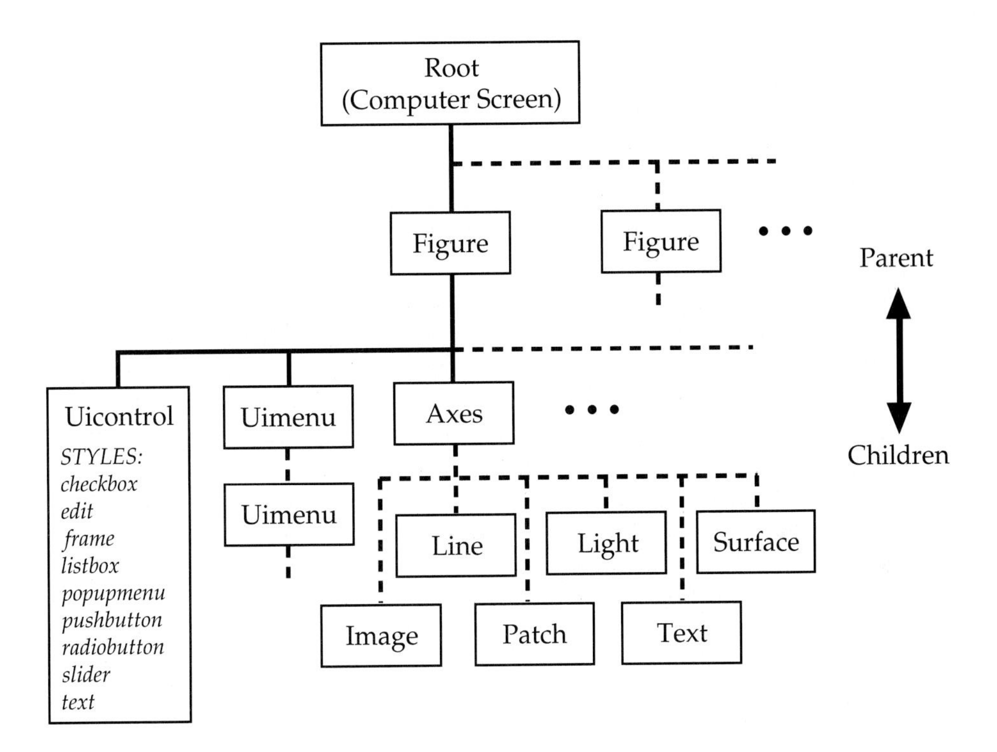


The *root* can contain one or more *figures,* and each *figure* can contain one or more sets of *axes* and their children. Each *figure* can also contain one or more *uimenus* and *uicontrols* that are independent of *axes*. While *Uicontrol* objects have no children, they do have a variety of styles. However, *uimenu* objects often have other *uimenu* objects as children.

Graphic displays are generated differently on each computer platform that runs MATLAB. Unix workstations use the X-Window System, which has a number of window managers, such as `mwm` or `twm`, to control the layout of the display. PCs rely on Microsoft Windows or Windows NT for window management. Macintosh computers use code in the Macintosh Toolbox for window management. Although displays may be visually different on each platform, in most cases Handle Graphics code is identical. MATLAB takes care of platform and window system differences internally. Functions incorporating Handle Graphics routines, including those using *uimenu* and *uicontrol* objects, usually run on all platforms. Where known differences exist, they are pointed out later in this chapter.

## 32.4 MENUS

Menus are used in each windowing system to let users select commands and options. Commonly, there is a menu bar across the top of the window or display. When one of these top-level menus is selected by moving the pointer over the menu label and pressing mouse button, a list of menu items drop down from the menu label. This type of menu is often called a pull-down menu. Menu items are selected by holding down the mouse button, moving the pointer over a menu item, and releasing the mouse button. MS-Windows and some X-Window System platforms provide an additional method of selecting a menu item. Pressing and releasing the mouse button, or clicking, on a top-level menu opens the pull-down menu. Menu items can then be selected by moving the pointer over the menu item and clicking again. Selecting one of the menu items in the pull-down menu causes some action to occur.

A menu item can also act as a submenu with its own list of menu items. Submenu items display a small triangle or arrowhead to the right of the submenu label to indicate that there are more menu items available for this choice. When a submenu item is selected, another menu with more menu items is displayed in a pull-down menu to the right of the submenu item. This is sometimes called a walking menu. Selecting one of these items also causes some action to occur.

Submenus may be nested, with the number of levels limited only by the windowing system used and available system resources.

### Menu Placement

The Macintosh uses a menu bar containing the titles of each pull-down menu located at the very top of the display. When a *Figure* window is selected, the menu bar changes to reflect the options available for the active *Figure* window. The standard Macintosh menu titles include **Apple, File, Edit,** and **Window.** Menu titles added by *uimenu* objects are positioned after the **Window** title. If you want to remove the **File, Edit,** and **Window** menu titles from the menu bar, you can issue the following `set` command:

```
» set(gcf,'MenuBar','none')
```

The **Apple** menu title cannot be removed from the menu bar. Likewise the standard menus are restored by the command

```
» set(gcf,'MenuBar','figure')
```

On PC platforms, the menu bar is located at the top of the *Figure* window. Each *Figure* window has its own menu bar containing **File, Edit, Window,** and **Help** titles. Menu titles added by *uimenu* objects are positioned after the **Help** title. You can remove or restore all standard menus from the *Figure* window by issuing the same `set` commands shown previously.

When you wish to build a cross-platform GUI that hides the menu bar on PC and X-Window *figure* windows, but not hide the top-of-screen Macintosh menu bar, use the utility function `menubar` as the value for the `'MenuBar'` property:

```
» set(gcf,'MenuBar',menubar)
```

This utility function simply checks the computer platform running MATLAB and returns `'none'` for PCs and workstations, and `'figure'` for the Macintosh.

MATLAB uses its own menu bar and menu titles for *Figure* windows on X-Window System workstations. The window manager may place its own menu bar above each window on the screen, but this menu bar is independent of the MATLAB menus. MATLAB creates its own menu bar along the top edge of the *Figure* window as long as the *figure's* `'MenuBar'` property is `'figure'`. You can remove and restore all three of MATLAB's standard **File, Windows,** and **Help** menus by issuing the same `set` commands shown previously.

## Creating Menus and Submenus

Menu items are created with the `uimenu` function. The general syntax is similar to other object-creation functions. For example,

```
» Hm_1 = uimenu(Hx_parent,'PropertyName',PropertyValue,...)
```

where `Hm_1` is the handle of the menu item created by `uimenu`, and (`'PropertyName'`, `PropertyValue`) pairs define the characteristics of the menu by setting *uimenu* object property values. `Hx_parent` is the handle of the parent object, which must be either a *figure* or another *uimenu.*

The most important properties of a *uimenu* object are the `'Label'` and `'CallBack'` properties. The `'Label'` property value is the text string that is placed on the menu bar or in the pull-down menu to identify the menu item. The `'CallBack'` property contains a MATLAB string that is passed to `eval` for execution when the menu item is selected. These and other properties are discussed in detail later in this chapter.

## Menu Example

The following example uses the `uimenu` function to add simple menus to the current *Figure* window. It is presented here to illustrate how working menus can be created with just a few MATLAB commands. Later examples will discuss *uimenu* commands and properties in more detail.

This example adds a menu bar with two pull-down menus to the current *Figure* window. First a top-level menu having the name `Example` is created by entering

```
» HM_ex = uimenu(gcf,'Label','Example');
```

Two menu items are then added under this menu. The first item is labeled **Grid** and toggles the state of the axis grid:

```
» Hm_exgrid = uimenu(Hm_ex,'Label','Grid','CallBack','grid');
```

Note that the handle Hm_ex is associated with the top-level menu. The **Grid** *uimenu* will appear under the top-level menu. Note also that the value for the 'CallBack' property is a quoted string, as required by eval.

The second item under **Example** is a menu item labeled **View** with a submenu:

```
» Hm_exview = uimenu(Hm_ex,'Label','View');
```

The **View** submenu contains items to select from 2-D or 3-D views:

```
» Hm_exv2d = uimenu(Hm_exview,'Label','2-D','CallBack','view(2)');
» HM_exv3d = uimenu(Hm_exview,'Label','3-D','CallBack','view(3)');
```

Note here that these are submenus to the **View** menu because they specify Hm_exview as their parent.

Now add a second top-level menu in the menu bar entitled **Close**:

```
» Hm_close = uimenu(gcf,'Label','Close');
```

From this top-level menu add two menu items. The first item closes the *Figure* window, and the second leaves the *Figure* window open but removes the user menus.

```
» Hm_clfig  = uimenu(Hm_close,'Label','Close Figure','CallBack','close');
» Hm_clmenu = uimenu(Hm_close,'Label','Remove Menus',...
        'CallBack','delete(findobj(gcf,''Type'',''uimenu'')); drawnow');
```

### Menu properties

As with all Handle Graphics object creation functions, *uimenu* properties can be defined at the time the object is created, as shown previously, or changed with the set command. All settable properties can be changed by set, including the label text, text colors, and even the callback string. This ability is very useful for customizing menus and attributes on-the-fly.

The following table lists some of the object properties of *uimenu* objects. A number of common properties, such as 'Interruptible', 'Tag', and 'UserData', have been omitted. See the appendices for a complete list of *uimenu* properties.

| Uimenu Properties | Description |
|---|---|
| `Accelerator` | A character specifying the keyboard equivalent or shortcut key for the menu item. For PC and Unix platforms, the key sequence is **Control-*Char***; for Macintosh systems the sequence is **⌘-*Char***. |
| `Callback` | MATLAB callback string, passed to the `eval` function in the base MATLAB workspace whenever the menu item is selected. |
| `Checked` | Enables a check mark next to the menu item label. |
| `Enable` | Normally `'on'`. When set to `'off'`, the label string is dimmed, and the callback is disabled. |
| `ForegroundColor` | Label text color for platforms that support colored text. |
| `Label` | A text string containing the label of the menu item. |
| `Position` | Relative position of the *uimenu* object. Top-level menus are numbered from left to right, and submenus are numbered top to bottom. |
| `Separator` | If set to `'on'`, a dividing line is drawn above the menu item. |
| `Children` | Object handles of other *uimenu* objects (submenus). |
| `Type` | Read-only object identification string; always `'uimenu'`. |
| `Visible` | When set to `'off'`, the menu item disappears from view. It still exists, but is not rendered. |

Property values simply define the attributes of *uimenu* objects and control how the menus are displayed. They also determine actions that result from selecting menu items. Some of these properties are discussed in more detail next.

### Menu Shortcut Keys

The `'Label'` property contains the text string that appears on the menu or menu item. It can also be used to define a shortcut key for all platforms, except the Macintosh, by preceding the desired character in the label string with the ampersand (&) character. For example,

```
» Hm_top = uimenu('Label','Example');
» uimenu(Hm_top,'Label','&Grid','CallBack','grid');
```

defines the shortcut key as the letter **G** on the keyboard. The menu item label will appear as **Grid** on the menu. To invoke the shortcut key, press the **G** key while the **Grid** menu item is visible. For a top-level *uimenu,* hold down the **Alt** key along with the accelerator key while *Figure* window is active. The shortcut key need not be the first character in the string. The following example defines the **R** key as the shortcut key:

```
» uimenu(Hm_top,'Label','G&rid','CallBack','grid');
```

and the label will appear as **Grid** on the menu. On Macintosh platforms, the **&** character in label strings is ignored. To use an ampersand in a text string, use two of them in succession, as in `'Black && White'` to produce a **Black & White** menu label.

The `'Accelerator'` property can be used rather than the `'Label'` property to define shortcut key combinations on all platforms. Only *leaf* menu items (those without children) can use an accelerator. These key combinations can be used whenever the *Figure* window is the current window; it is not necessary to make the menu item visible.

```
» uimenu(Hm_top,'Label','Grid','Accelerator','G','CallBack','grid');
```

defines the accelerator key as the letter **G**. On most platforms, the menu item label will appear as **Grid <Ctrl>-G.** The Macintosh uses the standard ⌘ symbol rather than **<Ctrl>-**. To invoke the accelerator, hold down the **Control** key (the ⌘ key on a Macintosh) and press the **G** key while the *Figure* window is selected.

An accelerator key cannot be defined for top-level menus on the Macintosh. In addition, any accelerator keys already defined on standard menus cannot be reassigned without removing the standard *figure* menu first. If the standard *figure* menus are subsequently restored, the redefined shortcut keys override the standard shortcuts.

## Menu Appearance

Three properties that affect the placement and appearance of your menus are `'Position'`, `'Checked'`, and `'Separator'`. The `'Position'` property value for *uimenu* objects is an integer that specifies the menu location relative to other menus or menu items. The `'Position'` property value is set when the object is created.
Position 1 refers to the leftmost menu in a menu bar as well as the top menu item in a pull-down menu or submenu at each level.

Menus can be reordered by setting the `'Position'` property. Consider the following example:

```
» Hm_1 = uimenu('Label','First');        % Create two menus
» Hm_2 = uimenu('Label','Second');
» get(Hm_1,'Position')                   % Check the locations
ans =
    1
» get(Hm_2,'Position')
ans =
    2
» set(Hm_2,'Position',1)                 % Change menu order
» get(Hm_1,'Position')                   % Check the locations
ans =
    2
» get(Hm_2,'Position')
ans =
    1
```

Note that when the 'Position' property of one menu is changed, the other menus are shifted to accommodate the change, and their 'Position' property values are updated. Menu items in submenus can be reordered in the same way.

The 'Checked' property value controls the appearance of a check mark to the left of the menu item label. The default value is 'off'. The command

```
» set(Hm_item,'Checked','on')
```

causes a check mark to appear next to the label of the Hm_item *uimenu*. This property is useful for creating menu items representing attributes. For example,

```
» Hm_top = uimenu('Label','Example');
» Hm_box = uimenu(Hm_top,'Label','Axis Box',...
           'CallBack', [...
              'if strcmp(get(gca,''Box''),''on''),',...
                 'set(gca,''Box'',''off''),',...
                 'set(Hm_box,''Checked'',''off''),',...
              'else,',...
                 'set(gca,''Box'',''on''),',...
                 'set(Hm_box,''Checked'',''on''),',...
              'end']);
```

creates a pull-down menu item labeled **Axis Box.** When selected, the callback string is evaluated. The callback string determines the present value of the current *axes'* 'Box' property and sets the *axes* 'Box' property and the *uimenu* 'Checked' property appropriately.

The *uimenu* 'Label' property can also be changed to reflect the current status of a menu item. The following example changes the menu item label itself rather than adding a check mark:

```
» Hm_top = uimenu('Label,'Example');
» Hm_box = uimenu(Hm_top,'Label','Axis Box',...
           'CallBack', [...
              'if strcmp(get(gca,''Box''),''on''),',...
                 'set(gca,''Box'',''off''),',...
                 'set(Hm_box,''Label'',''Set Box On''),',...
              'else,', . . .
                 'set(gca,''Box'',''on''),', . . .
                 'set(Hm_box,''Label'',''Set Box Off''),',...
              'end']);
```

Pull-down menu items can be separated into logical groups by using the 'Separator' property. If 'Separator' is set to 'on' for a *uimenu* item, that item will appear in the pull-down menu with a horizontal line drawn above it to separate it visually from previous items. The default value is 'off' but can be changed when the object is created:

```
» Hm_box = uimenu(Hm_top,'Label','Box','Separator','on');
```

or later using the `set` command:

```
» set(Hm_box,'Separator','on');
```

Top-level menus ignore the value of the `'Separator'` property.

## Color Control

Only one color property can be set for *uimenu* objects. The `'ForeGroundColor'` property determines the color of the text of the menu item label on those platforms that support colored text. The default text color is black. The background color is now determined by the platform and operating system and can no longer be set by the user.

Colored text can be used equally well with top-level menus and pull-down menu items. Colors can be used to indicate status information, or simply to add interest to your menus. For example, each menu item in a submenu for selecting line color properties can be labeled with the appropriate color text:

```
» Hm_green = uimenu(Hm_1color,'Label','Green','ForeGroundColor','g', . . .
             'CallBack','set(Hl_line,''Color'',''g'')');
```

## Disabling Menu Items

A menu item may be temporarily disabled by changing the value of the `'Enable'` or the `'Visible'` property of a *uimenu* object. The `'Enable'` property is normally set to `'on'`. When `'Enable'` is set to `'off'`, the label string is dimmed and the associated *uicontrol* callback is disabled. In this state, the menu item remains visible but cannot be selected. This property can be used to disable an inappropriate menu choice.

The following example illustrates another way to set the axes `'Box'` property using two menu items and the `'Enable'` property:

```
» Hm_top = uimenu('Label','Example');
» Hm_boxon = uimenu(Hm_top,'Label','Set Box on',...
             'CallBack', [...
                'set(gca,''Box'',''on''),',...
                'set(Hm_boxon,''Enable'',''off''),',...
                'set(Hm_boxoff,''Enable'',''on'')']);
» Hm_boxoff = uimenu(Hm_top,'Label','Set Box Off',...
             'Enable','off',...
             'CallBack', [...
                'set(gca,''Box'',''off''),',...
                'set(Hm_boxon,''Enable'',''on''),',...
                'set(Hm_boxoff,''Enable'',''off'')']);
```

A menu item can be completely hidden by setting the `'Visible'` property to `'off'`. The menu item seems to disappear from the menu, and the other menu items change position on the display to accommodate the loss of the invisible menu item. However, the invisible menu item still exists, and the `'Position'` property values of the *uimenu* objects do not change. The menu item reappears in its normal position when the `'Visible'` property is set to `'on'` again.

This property can be used to temporarily remove a menu or menu item. The following example creates two top-level menus and two menu items:

```
» Hm_control = unimenu('Label','Control');
» Hm_extra = uimenu('Label','Extra');
» Hm_limit = uimenu(Hm_control,'Label','Limited Menus',...
                'CallBack','set(Hm_extra,''Visible'',''off'')');
» Hm_full = uimenu(Hm_control,'Label','Full Menus',...
                'CallBack','set(Hm_extra,''Visible'',''on'')');
```

When the **Limited Menus** item is selected, the **Extra** menu disappears from the menu bar. When the **Full Menus** menu item is selected, the **Extra** menu reappears on the menu bar in its original location.

### The CallBack Property

The `'CallBack'` property value is a MATLAB string that is passed to the function `eval` to be executed in the *Command* window workspace. This has important implications for function M-files and is addressed later in this chapter.

Since the `'CallBack'` property value must be a string, multiple MATLAB commands, continuation lines, and strings within strings can make the necessary syntax quite complicated. If there is more than a single command to be executed, the commands must be separated appropriately. For example,

```
» uimenu('Label','Test','CallBack','grid on; set(gca,''Box'',''on'')');
```

passes a single string to `eval` and causes the command

```
grid on; set(gca,'Box','on')
```

to be executed in the *Command* window workspace. This is legal syntax, since multiple commands can be entered on the same command line if they are separated by a comma or semicolon. The MATLAB convention of using two single quotes to represent a single quote within a quoted string is also followed when defining the callback string.

Strings can be concatenated to create a legal MATLAB string by enclosing them in square brackets.

```
» uimenu('Label','Test','CallBack',['grid on,','set(gca,''Box'',''on'')']);
```

Note that the string `'grid on,'` contains the required comma to separate the two commands.

If continuation lines are used, the preceding command can become

```
» uimenu('Label','Test', . . .
        'CallBack', [ . . .
                'grid on,', . . .
                'set(gca,''Box'',''on'')' . . .
                    ]);
```

Here the lines are broken appropriately and three periods are added to the end of each of the lines to show that the statement is continued. Note that all elements of the preceding single-line example are preserved, including the comma within the strings to separate the commands. The comma after the final quote in the line `'grid on,', . . .` is optional; the spaces at the beginning of the next line serve the same purpose. Refer to the discussion in Chapter 6 about creating row vectors for more details.

MATLAB will complain if the quotes, commas, and parentheses are not entered correctly, but an error in a complicated callback string can be very difficult to find. To minimize errors, remember these guidelines for callback strings containing multiple MATLAB statements:

1. Enclose the entire callback in square brackets, and do not forget the final close parentheses ')'.
2. Enclose each statement in `'single quotes'`.
3. Quotes within a quoted string double up: `'quoted'`;`'a ''quoted'' string'`; `'Quote ''a '''quoted''' string'' now'`.
4. Each statement except the last ends with a comma or semicolon within the quotes and a comma or space after the quotes.
5. Each line that is to be continued ends with three periods ( `. . .` ).

One of the earlier examples is a good illustration of the syntax of more involved callback strings:

```
» Hm_top = uimenu('Label','Example');
» Hm_boxon = uimenu(Hm_top, . . .
                'Label','Set Box On',...
                'CallBack',[...
                   'set(gca,''Box'',''on''),',...
                   'set(Hm_boxon,''Enable'',''off''),',...
                   'set(Hm_boxoff,''Enable'',''on'')']);
» Hm_boxoff = uimenu(Hm_top,...
                'Label','Set Box Off',...
                'Enable','off',...
                'CallBack',[...
                   'set(gca,''Box'',''off''),',...
                   'set(Hm_boxon,''Enable'',''on''),',...
                   'set(Hm_boxoff,''Enable'',''off'')']);
```

The preceding example brings up another important point about callback strings. Hm_boxoff is used in the callback string for Hm_boxon before the variable Hm_boxoff is defined. MATLAB does not complain because the callback string is simply a string and is not executed by MATLAB until the *uimenu* object is activated and the string is passed to eval. This has implications for function M-file design and testing that will be discussed later in this chapter.

### Example M-file

The following example demonstrates the creation of a simple set of menus using M-file fragments. First create a menu bar with a top-level **Line** menu in the current *figure,* containing three submenus for **Line Style, Line Width,** and **Line Color:**

```
Hm_line   = uimenu(gcf, 'label', 'Line');
Hm_lstyle = uimenu(Hm_line, 'label', 'Line Style');
Hm_lwidth = uimenu(Hm_line, 'label', 'Line Width');
Hm_lcolor = uimenu(Hm_line, 'label', 'Line Color');
```

Next, use waitforbuttonpress and gco to get a handle to the current object, make sure it is a *line* object, and apply the appropriate 'LineStyle' value. Note that since the handles to these menu items are never used, they are not saved.

```
uimenu(Hm_lstyle, 'Label', 'Solid',...
    'CallBack', ['waitforbuttonpress;',...
        'if get(gco,''Type'') == ''line'',',...
            'set(gco,''LineStyle'',''-''),',...
        'end']);
uimenu(Hm_lstyle, 'Label', 'Dotted',...
    'CallBack', ['waitforbuttonpress;',...
        'if get(gco,''Type'') == ''line'',',...
            'set(gco,''LineStyle'','':''),',...
        'end']);
uimenu(Hm_lstyle, 'Label', 'Dashed',...
    'CallBack', ['waitforbuttonpress;',...
        'if get(gco,''Type'') == ''line'',',...
            'set(gco,''LineStyle'',''--''),',...
        'end']);
uimenu(Hm_lstyle, 'Label', 'DashDot',...
    'CallBack', ['waitforbuttonpress;',...
        'if get(gco,''Type'') == ''line'',',...
        'set(gco,''LineStyle'',''-.''),',...
    'end']);
```

Now do the same for the **Line Width** submenu items:

```
uimenu(Hm_lwidth, 'Label', 'Default',...
    'CallBack', ['waitforbuttonpress;',...
        'if get(gco,''Type'') == ''line'',',...
            'set(gco,''LineWidth'',0.5),',...
        'end']);
uimenu(Hm_lwidth, 'Label', 'Thick',...
    'CallBack', ['waitforbuttonpress;',...
        'if get(gco,''Type'') == ''line'',',...
            'set(gco,''LineWidth'',2.0),',...
        'end']);
uimenu(Hm_lwidth, 'Label', 'Thicker',...
    'CallBack', ['waitforbuttonpress;',...
        'if get(gco,''Type'') == ''line'',',...
            'set(gco,''LineWidth'',3.0),',...
        'end']);
```

and some **Line Color** submenu items. Color the text of the menu items as appropriate.

```
uimenu(Hm_lcolor, 'Label', 'Red',...
    'ForegroundColor','r',...
    'CallBack', ['waitforbuttonpress;',...
        'if get(gco,''Type'') == ''line'',',...
            'set(gco,''Color'',''r''),',...
        'end']);
uimenu(Hm_lcolor,' Label', 'Green',...
    'ForegroundColor', 'g',...
    'CallBack', ['waitforbuttonpress;',...
        'if get(gco,''Type'') == ''line'',',...
            'set(gco,''Color'',''g''),',...
        'end']);
uimenu(Hm_lcolor, 'Label', 'Blue',...
    'ForegroundColor','b',...
    'CallBack', ['waitforbuttonpress;',...
        'if get(gco,''Type'') == ''line'',',...
            'set(gco,''Color'',''b''),',...
        'end']);
```

Additional menus can be added in the same way to change *axes, surface, patch,* and *figure* properties. For example, the following adds a **Colormap** menu for changing the *figure* colormap:

```
Hm_cmap = uimenu(gcf, 'Label', 'Colormap');
uimenu(Hm_cmap, 'Label', 'Lighter', 'CallBack', 'brighten(.3)');
uimenu(Hm_cmap, 'Label', 'Darker',  'CallBack', 'brighten(-.3)');
uimenu(Hm_cmap, 'Label', 'Default', 'CallBack', 'colormap(''default'')');
uimenu(Hm_cmap, 'Label', 'Prism',   'CallBack', 'colormap(prism)');
uimenu(Hm_cmap, 'Label', 'Jet',     'CallBack', 'colormap(jet)');
uimenu(Hm_cmap, 'Label', 'Flag',    'CallBack', 'colormap(flag)');
uimenu(Hm_cmap, 'Label', 'HSV',     'CallBack', 'colormap(hsv)');
```

Finally, a **Quit** menu is added with two items. **Close Figure** closes the *figure*. **Remove Menus** leaves the *figure* open but removes the top-level user menus and all of their children. The `drawnow` command removes the menus immediately.

```
Hm_quit = uimenu(gcf, 'Label', 'Quit');
uimenu(Hm_quit,'Label', 'Close Figure', 'CallBack', 'close; return');
uimenu(Hm_quit,'Label', 'Remove Menus',...
    'CallBack', [...
        'delete(findobj(gcf,''Type'',''uimenu'')),',...
        'drawnow']);
```

The *Mastering MATLAB Toolbox* contains a number of additional functions that use these techniques and others to create useful object-based tools.

## 32.5 CONTROLS

Controls and menus are used by windowing systems on each computer platform to let users perform some action or to set an option or attribute. Controls are graphic objects such as icons, text boxes, and scroll bars that are used along with menus to create the graphical user interface implemented by the windowing system and window manager on your computer.

MATLAB controls are very similar to those used by your window manager. They are graphical objects that can be placed anywhere in a MATLAB *Figure* window and activated with the mouse. MATLAB controls include buttons, sliders, text boxes, and popup menus.

Controls created by MATLAB have a slightly different appearance on Macintosh, MS Windows, and X-Window System computer platforms due to differences in the way the windowing systems render graphical objects. However, the functionality is essentially the same, so the same MATLAB code will create the same objects that perform the same functions across platforms.

Controls are created with the function `uicontrol`. The general syntax is similar to `uimenu` discussed earlier:

```
» Hc_1 = uicontrol(Hf_fig,'PropertyName',PropertyValue,...)
```

Here `Hc_1` is the handle of the *uicontrol* object created by the `uicontrol` function, and `('PropertyName',PropertyValue)` pairs define the characteristics of the control by setting *uicontrol* object property values. `Hf_fig` is the handle of the parent object, which must be a *figure*. If the *figure* handle is omitted, the current *figure* is used.

### Creating Different Control Types

There are nine different types or styles of MATLAB controls in MATLAB versions 5.0 and 5.1. Contextual menus and toggle buttons are scheduled to appear in version 5.2. They are all created using the `uicontrol` function. The `'Style'` property value determines the type of control that is created. The `'CallBack'` property value is the MATLAB string passed to `eval` to execute in the *Command* window workspace when the control is activated.

In the following sub-sections, each *uicontrol* object style is discussed individually and is illustrated with an example. A more thorough discussion of the properties of *uicontrol* objects and more complete examples of their use will follow later.

### Push Buttons

Push buttons, sometimes called *command buttons* or just *buttons,* are small rectangular screen objects that are usually labeled with text on the object itself. Selecting the push button with the mouse by moving the pointer over the object and clicking the mouse button causes MATLAB to perform the action defined by the object's callback string. The `'Style'` property value of a push button is `'pushbutton'`.

Push buttons are typically used to perform an action rather than change a state or set an attribute. The following example creates a push button labeled **Close** that closes the current *figure* when activated. The `'Position'` property defines the size and location of the push button in pixels, which is the default `'Units'` property value. The `'String'` property specifies the button's label.

```
» Hc_close = uicontrol(gcf, . . .
           'Style',     'pushbutton', . . .
           'Position', [10 10 100 25], . . .
           'String',    'Close', . . .
           'CallBack', 'close');
```

### Radio Buttons

Radio buttons, sometimes called *option buttons,* consist of a button containing a label and a small circle or diamond to the left of the label text. When selected, the circle or diamond is filled and the `'Value'` property is set to the value specified by the `'Max'` property, which is `1` by default; when unselected, the indicator is cleared and the `'Value'` property is set to the value specified by the `'Min'` property, which is `0` by default. The `'Style'` property value of a radio button is `'radiobutton'`.

Radio buttons are typically used to select one of a group of mutually exclusive options. To enforce this exclusivity, however, the callback string for each radio button must unselect all other buttons in the group by setting the 'Value' of each of them to 0 or the value assigned to the 'Min' property. This is only a convention, however. Radio buttons can be used interchangeably with check boxes, if desired.

The following example creates two mutually exclusive radio buttons that change the *axes* 'Box' property.

```
» Hc_boxon = uicontrol(gcf,...
          'Style',     'radiobutton',...
          'Position', [20 45 100 20],...
          'String',    'Set Box On',...
          'Value',     0,...
          'CallBack', [...
              'set(Hc_boxon, ''Value'',1),'...
              'set(Hc_boxoff,''Value'',0),'...
              'set(gca,''Box'',''on'')']);
» Hc_boxoff = uicontrol(gcf,...
          'Style',     'radiobutton',...
          'Position', [20 20 100 20],...
          'String',    'Set Box Off',...
          'Value',     1,...
          'CallBack', [...
              'set(Hc_boxon, ''Value'',0),'...
              'set(Hc_boxoff,''Value'',1),'...
              'set(gca,''Box'',''off'')']);
```

### Check Boxes

Check boxes consist of a button with a label and a small square box to the left of the label text. When activated, the control is toggled between checked and cleared. When checked, the box is filled or contains an X depending on the platform and the 'Value' property is set to the value specified by the 'Max' property, which is 1 by default; when cleared, the box becomes empty and the 'Value' property is set to the value specified by the 'Min' property, which is 0 by default. The 'Style' property value of a check box is 'checkbox'.

Check boxes are typically used to indicate the state of an option or attribute. Check boxes are usually independent objects but can be used interchangeably with radio buttons, if desired.

The following example creates a check box to set the *axes* 'Box' property. When activated, the 'Value' property of the check box is tested to determine if the check box had just been checked or cleared, and the *axes* 'Box' property is set appropriately.

```
» Hc_box = uicontrol(gcf,...
          'Style',     'checkbox',...
          'Position', [100 50 100 20],...
```

```
'String',    'Axis Box',...
'CallBack', [...
    'if get(Hc_box,''Value'') == 1,'...
        'set(gca,''Box'',''on''),'...
    'else,'...
        'set(gca,''Box'',''off''),'...
    'end']);
```

### Static Text Boxes

Static text boxes are controls that simply display a text string as determined by the `'String'` property. The `'Style'` property value of a static text box is `'text'`. Static text boxes are typically used to display labels, user information, or current values. Static text boxes are static in the sense that the user cannot dynamically change the text displayed. The text can only be changed by changing the `'String'` property.

Text strings are centered at the top of the text box. Text strings that are longer than the width of the text box will *word wrap;* that is, multiple lines will be displayed with the lines broken between words where possible. If the height of the text box is too small for the text string, some of text will not be visible.

This example creates a text box containing the MATLAB version number:

```
» Hc_ver = uicontrol(gcf,...
        'Style',      'text',...
        'Position', [10 10 150 20],...
        'String',     ['MATLAB Version ', version]);
```

The `'FontName'`, `'FontSize'`, `'FontWeight'`, `'FontAngle'`, and `'FontUnits'` properties of *uicontrols* provide a wide range of options for all *uicontrol* text strings, including button label strings and text boxes. The `'HorizontalAlignment'` property is also available to customize your text. However, the LaTeX interpreter is **not** available for *uicontrol* text strings. A string such as `'x^2'` will be displayed exactly as specified.

### Editable Text Boxes

Editable text boxes, like static text boxes, display text on the screen. Unlike static text, however, editable text boxes allow the user to modify or replace the text string dynamically just as you would using a text editor or word processor. The new text string then becomes available in the `'String'` property of the *uicontrol.* The `'Style'` property value of an editable text box is `'edit'`. Editable text boxes are typically used to let the user enter a text string or a specific value.

Editable text boxes may contain one or more lines of text. A single-line editable text box will accept one line of text from the user, while a multiline text box will accept more than one line of text. Single-line text entry is terminated by pressing the **Return** key. Multi-line text entry is terminated with **Control-Return** or **⌘-Return** on the Macintosh.

This example creates a static text label and a single line editable text box. The user enters a colormap name string in the text box and the callback string applies it to the *figure:*

```
» Hc_label = uicontrol(gcf, . . .
           'Style',    'text', . . .
           'Position', [10 10 70 20], . . .
           'String',   'Colormap:');
» Hc_map = uicontrol(gcf, . . .
           'Style',    'edit', . . .
           'Position', [80 10 60 20], . . .
           'String',   'default', . . .
           'Callback', 'colormap(eval(get(Hc_map,''String'')))');
```

Multi-line text boxes are created by setting the `'Max'` and `'Min'` property values to numbers such that `Max-Min>1`. The `'Max'` property does **not** specify the maximum number of lines. Multiline text boxes can have an unlimited number of lines. A multiline editable text box is as follows:

```
» Hc_multi = uicontrol(gcf, . . .
           'Style',    'edit', . . .
           'Position', [20 50 75 75], . . .
           'String',   'Line1|Line 2|Line 3', . . .
           'Max',      2);
```

Multiline strings can be specified as a cell array of strings, a padded string matrix, or a single quoted string using a vertical bar character `'|'` or linefeed `'\n'` to designate where the lines are broken.

### Sliders

Sliders, or *scroll bars,* consist of three distinct parts. These parts are the *trough,* or rectangular area representing the range of valid object values, the *indicator* within the trough representing the current value of the slider, and *arrows* at each end of the trough. The `'Style'` property value of a slider is `'slider'`.

Sliders are typically used to select a value from a range of values. Slider values can be set in three ways. First, the indicator can be moved by positioning the mouse pointer over the indicator, holding down the mouse button while moving the mouse, and releasing the mouse button when the indicator is in the desired location. The second method is to click the mouse button while the pointer is in the trough, but to one side of the indicator. The indicator moves in that direction over a default distance equal to about 10% of the total range of the slider. Finally, clicking on one of the arrows at the ends of the slider by default moves the indicator about 1% of the slider range in the direction of the arrow. Sliders are often accompanied by separate text objects used to display labels, the current slider value, and range limits.

The following example implements a slider that can be used to set the azimuth of a viewpoint. Three text boxes are used to indicate the maximum, minimum, and current value of the slider.

```
» vw = get(gca,'View');
» Hc_az = uicontrol(gcf,...
        'Style',     'slider',...
        'Position', [10 5 140 20],...
        'Min',-90,   'Max',90, 'Value',vw(1),...
        'CallBack', [...
            'set(Hc_cur,''String'',num2str(get(Hc_az,''Value''))),'...
            'set(gca,''View'',[get(Hc_az,''Value'') vw(2)])']);
» Hc_min = uicontrol(gcf,...
        'Style',    'text',...
        'Position', [10 25 40 20],...
        'String',    num2str(get(Hc_az,'Min')));
» Hc_max = uicontrol(gcf,...
        'Style',     'text',...
        'Position',  [110 25 40 20],...
        'String',    num2str(get(Hc_az,'Max')));
» Hc_cur = uicontrol(gcf,...
        'Style',     'text',...
        'Position',  [60 25 40 20],...
        'String',    num2str(get(Hc_az,'Value')));
```

The `'Position'` property of *uicontrols* contains the familiar `[left bottom width height]` vector in units designated by the `'Units'` property. The orientation of a slider depends on the aspect ratio of `width` to `height`. A horizontal slider will be drawn if `width > height`, and a vertical slider will be drawn if `width < height`. On X-Window System platforms only, the arrows will only appear if one dimension is greater than four times the other. All sliders have arrows on other platforms.

The `'SliderStep'` property is a two-element vector `[arrow_step trough_step]` that controls the change in slider value when a slider arrow is clicked or the mouse is clicked in the slider trough. Default values are `[0.01 0.10]` for a change of 1% and 10% of the maximum slider value, respectively.

Note that MATLAB now enforces the requirement that `Max > Min` for sliders and that `Max` is at the top of a vertical slider and at the right end of a horizontal slider. If you require something different (e.g., a vertical slider with values of `0` at the top and `10` at the bottom), use negative numbers for `'Min'`, `'Max'`, and `'Value'` and convert as necessary using the `abs` function.

### Popup Menus

Popup menus are typically used to present a list of mutually exclusive choices to the user. Popup menus, unlike the pull-down menus discussed previously, are not located on a menu bar. A popup menu can be positioned anywhere in the *Figure* window. The `'Style'` property value of a popup menu is `'popupmenu'`.

When closed, a popup menu appears as a rectangle or button containing the label of the current selection with a small raised rectangle or downward-pointing arrow to the right of the label to indicate that the object is a popup menu. When the pointer is positioned over a popup control and the mouse button is pressed, other choices appear. Moving the pointer to a different choice and releasing the mouse button closes the popup menu and displays the new selection. MS-Windows and some X-Window System platforms allow the user to click on a popup to open it, and then click on another choice to select it.

When a popup item is selected, the `'Value'` property is set to the index of the selected element of the vector of choices (the index of the top element is 1). Choice labels can be specified as a cell array of strings, a padded string matrix, or a single string separated by vertical bar `'|'` characters, similar to the way multiline text strings are specified. However, the `'\n'` option is **not** allowed. The following example creates a popup menu of *figure* colors. The callback sets the figure's `'Color'` property to the chosen value. The RGB values associated with each color are stored in the `'UserData'` property of the popup control. The `'UserData'` property of all Handle Graphics objects simply provides isolated and persistent storage for a single variable of any class.

```
» Hc_fcolor = uicontrol(gcf,...
    'Style',     'popupmenu',...
    'Position', [20 20 80 20],...
    'String',    'Black|Red|Yellow|Green|Cyan|Blue|Magenta|White',...
    'Value',     1,...
    'UserData', [[0 0 0];...
                 [1 0 0];...
                 [1 1 0];...
                 [0 1 0];...
                 [0 1 1];...
                 [0 0 1];...
                 [1 0 1];...
                 [1 1 1]],...
    'CallBack', [...
        'UD = get(Hc_fcolor,''UserData'');',...
        'set(gcf,''Color'', UD(get(Hc_fcolor,''Value''),:))']);
```

The `'Position'` property of a popup menu contains the familiar `[left bottom width height]` vector, where the `width` and `height` values determine the dimensions of the popup object. On X-Window Systems and Macintosh systems, these are the dimensions of the closed popup menu. When opened, the menu expands to display all of the choices that can fit on the screen. On MS-Windows systems, the `height` value is essentially ignored. These platforms create a popup that is tall enough to display one line of text regardless of the `height` value.

## Frames

Frame *uicontrol* objects are simply shaded rectangular regions. Frames are analogous to the `'Separator'` property of *uimenu* objects in the sense that they provide visual separation. Frames are typically used to form groups of radio buttons or other *uicontrol* objects. Frames should be defined before other objects that are placed within the frames. Otherwise, the frame may cover the other *uicontrols.*

The following example creates a frame and places two push buttons and a text label within it.

```
» Hc_frame = uicontrol(gcf,...
      'Style',    'frame',...
      'Position', [250 200 95 65]);
» Hc_pb1 = uicontrol(gcf,...
      'Style',    'pushbutton',...
      'Position', [255 205 40 40],...
      'String',   'OK');
» Hc_pb2 = uicontrol(gcf,
      'Style',    'pushbutton',...
      'Position', [300 205 40 40],...
      'String',   'NOT');
» Hc_lbl = uicontrol(gcf,...
      'Style',    'text',...
      'Position', [255 250 85 10],...
      'String',   'Push Me');
```

## List Boxes

List box *uicontrol* objects look like multiline text boxes that allow users to select individual or multiple list entries with mouse clicks. Individual list entries are specified by a cell arrays of strings, a padded string matrix, or a single string with a vertical bar `'|'` character used to separate list entries. When a list entry is selected with a mouse click, the `'Value'` property is updated with the index of the selected item.

If `Max-Min>1`, multiple list entries may be selected. Contiguous list entries are selected by dragging the mouse pointer over the desired entries or by selecting one item with a mouse click and selecting a second item while pressing a **Shift** key on the keyboard. Noncontiguous entries may be selected by pressing the **Control** or ⌘ key and selecting individual list entries. If multiple entries *are* selected, the `'Value'` property is updated with a vector of indices of the selections.

The size and placement of the list box is specified using the 'Position' property. If the listbox 'String' values exceed the width or height of the listbox *uicontrol,* scroll bars are added to the listbox as necessary.

The listbox callback string is evaluated whenever a mouse-button-up event changes the contents of the 'Value' property. If you wish to allow multiple selections and only take action after all selections have been made, do not use the listbox 'Callback' string. In this case, add a **Done** or **Apply** push button and use its 'Callback' string instead.

Alternatively, a double click can be used. Listbox *uicontrols* set the parent *figure* 'SelectionType' property to 'normal' for a single mouse click and 'open' for a double click. The *uicontrol* callback can be designed to perform its function only after a double click to indicate that the last selection has been made.

The following example creates a frame and places a text label and listbox within it.

```
» Hc_frame = uicontrol(gcf,...
      'Style',    'frame',...
      'Position', [15 15 800 120]);
» Hc_text = uicontrol(gcf,...
      'Style',      'text',...
      'Position',   [20 110 70 20],...
      'String',     'Select',...
      'FontWeight', 'bold');
» Hc_list = uicontrol(gcf,
      'Style',    'listbox',...
      'Position', [20 20 70 90],...
      'String',   'One|Two|Three|Four|Five|Six',...
      'Max'       2);
```

Try selecting single, multiple, and multiple noncontiguous entries.

## Control Properties

As with all Handle Graphics object-creation functions, *uicontrol* properties can be defined at the time the object is created, as shown previously, or changed with the set command. All settable properties can be changed by set, including the string text, callback string, and even the style of control. Some examples appear later in this chapter.

The following table lists some of the object properties of *uicontrol* objects. A number of common properties, such as 'Interruptible', 'Tag', and 'UserData', have been omitted. See the appendices for a complete list of *uicontrol* properties.

| Uicontrol Properties | Description |
|---|---|
| Style | String that defines the type of *uicontrol* object. |
| BackgroundColor | Background color of the *uicontrol.* |
| ForegroundColor | Foreground (text) color on platforms that support colored text. The default text color is black. |

| Uicontrol Properties | Description |
|---|---|
| `String` | Text string specifying the text in text boxes or the label on push buttons, radio buttons, and check boxes. For multiple items in a popup menu, a listbox, or an editable text box, this is a cell array of strings, a string matrix, or a single string with a vertical bar characters `'|'` separating individual items. Not used for frames or sliders. |
| `HorizontalAlignment` | Horizontal alignment of label string. |
| `FontName` | String specifying the name of the font used to display the contents of the `String` property. |
| `FontUnits` | Units used by the `FontSize` property. |
| `FontSize` | Specifies the size of the font used. |
| `FontAngle` | Specifies the font slant (e.g., `italic`) |
| `FontWeight` | Specifies the weight of the font (e.g., `bold`). |
| `Max` | The value depends on the *uicontrol* Style. Radio buttons and check boxes set `Value` to `Max` while the *uicontrol* is in the `on` state. This value defines the maximum value of a slider. On X-Windows platforms, it contains the maximum index value of a popup menu. Editable text boxes are multiline text when `Max-Min>1`. Listboxes allow multiple selections when `Max-Min>1`. The default value is `1`. |
| `Min` | The value depends on the *uicontrol* Style. Radio buttons and check boxes set `Value` to `Min` while the *uicontrol* is in the `off` state. This value defines the minimum value of a slider. On X-Windows platforms, it contains the value 1 for popup menus. Editable text boxes are multiline text when `Max-Min>1`. Listboxes allow multiple selections when `Max-Min>1`. The default value is `0`. |
| `Value` | Current value of the *uicontrol.* Radio buttons and check boxes set `Value` to `Max` when `on` and `Min` when `off`. Sliders set `Value` to the number set by the slider between `Min` and `Max`. Popup menus and list boxes set `Value` to the index of the item selected. Text objects, frames, and push buttons do not set this property. |
| `Position` | Position vector `[left bottom width height]` where `[left bottom]` represents the location of the lower left corner of the *uicontrol* with respect to the lower left corner of the *figure* object and `[width height]` are the *uicontrol* dimensions. Units are specified by the `Units` property. |
| `Units` | Unit of measurement for position property values. |
| `SliderStep` | Controls the amount of movement caused by a mouse click on an arrow or in the trough of a slider. |
| `ListboxTop` | Index of the topmost string displayed in a listbox. |

| Uicontrol Properties | Description |
|---|---|
| CallBack | MATLAB callback string passed to eval whenever the *uicontrol* object is activated; initially an empty matrix. |
| ButtonDownFcn | Callback string evaluated when the mouse is clicked within a 5-pixel border of the *uicontrol* or on the *uicontrol* when Enable is either off or inactive. |
| Enable | Normally on. When set to off, the text string is grayed out and the *uicontrol* is disabled. When set to Inactive, the *uicontrol* is disabled, but the label is not dimmed. |
| Visible | Visibility of *uicontrol* object. |
| Type | Read-only object identification string; always 'uicontrol'. |

## Control Placement and Font Considerations

The 'Position' and 'Units' properties of *uicontrols* are used to locate control objects in the *Figure* window. The default 'Position' vector for *uicontrols* is [20 20 60 20] represented in pixels, which is the default 'Units' value. This is a 60 by 20 pixel rectangle with the lower left corner of the *uicontrol* positioned 20 pixels to the right and 20 pixels above the lower left corner of the parent *figure*. The default *figure* size is around 560 by 420 pixels, placed in the upper middle of the display. The specific default size and placement of a *figure* on your platform can be found using

```
» deffigpos = get(0,'DefaultFigurePosition')
```

The actual size and position of any *figure* can be obtained from the *figure's* 'Position' property. Using this information, placing *uicontrols* becomes a layout problem in two-dimensional geometry.

Some restrictions apply. For example, on MS-Windows platforms, popup menus ignore the height value of the position vector and use just enough vertical space to display one line of text. In all other situations, one must make sure that the *uicontrol* is large enough to contain the label string or some of the text will be clipped. The default font used to display the 'String' property value of *uicontrols* is usually different from the font used in the *Command* window. Different fonts and font sizes are used as defaults on different platforms. The values used on your platform can be found using

```
» cfname = get(0,'DefaultUicontrolFontName')
» cfsize = get(0,'DefaultUicontrolFontSize')
```

MATLAB's default *uicontrol* fonts were chosen to give the best appearance on each platform, so use of the default *uicontrol* font is recommended unless you have special requirements. Portable or platform-independent code also works best using the default fonts.

Often sizing and placing controls is a process of trial and error. Even when you are satisfied with the result, the appearance of the *figure* on another platform may be sufficiently different to require more adjustments. Often it is desirable to make the control a bit larger than appears necessary, simply to ensure that the label is readable on all platforms.

Just because a *figure* has a default size, there is no guarantee that all *figures* are that size. If you add controls or menus to an existing *figure,* it may be smaller or larger than the default. In addition, the user can resize any *figure* at any time unless prevented from doing so by setting the *figure's* `'Resize'` property to `'off'`.

Two things to consider when adding controls to a *figure* that may be resized are the `'Units'` property and the restrictions imposed by fixed-font label strings. If the position of each control is specified in absolute units such as pixels, inches, centimeters, or points, resizing the window will not change the size or placement of the controls. The controls will maintain the same position relative to the bottom and left sides of the *figure.* If the *figure* is made smaller, some of the controls may become located outside of the *figure* and will no longer be visible.

If the position of controls is specified in normalized units, the *uicontrols* will maintain their relative relationship to each other and to the *figure* itself when the *figure* is resized. There is one drawback to using normalized units, however. If the *figure* is made smaller and the controls are resized as a result, control labels may become unreadable since the font size is fixed. Any portion of the label that is outside the dimensions of the resized control is clipped.

## 32.6 PROGRAMMING AND CALLBACK CONSIDERATIONS

Now that the basics of Handle Graphics user-interface functions have been covered, it is time to apply them. As you have seen, creating menus and controls by typing them in on the command line is not very efficient. Script or function M-files are a lot easier to use. Suppose you have an idea for an M-file that you want to implement. The first decision is whether to write a script or a function.

### Scripts versus Functions

Scripts seem to be an obvious choice. In scripts, everything executes in the *Command* window workspace, so all MATLAB functions and all object handles are available at all times. There is no difficulty passing information to callback strings, but there are a number of tradeoffs. The first is that while all variables are available, the workspace can become cluttered with variable names and values even after they are no longer useful. On the other hand, if the user issues the `clear` command, important object handles may be lost. Another disadvantage is that defining callback strings can become very complicated using scripts. For example, here is a slider definition fragment from a script M-file.

```
Hc_rsli = uicontrol(Hf_fig,'Style','slider', . . .
  'Position',[.10 .55 .35 .05], . . .
  'Min',0,'Max',1,'Value',initrgb(1), . . .
  'CallBack',[ . . .
    'set(Hc_nfr,''BackgroundColor'',', . . .
      '[get(Hc_rsli,''Val''),get(Hc_gsli,''Val''),get(Hc_bsli,''Val'')]),', . . .
    'set(Hc_ncur,''String'',', . . .
      'sprintf(''[%.2f %.2f %.2f]'',get(Hc_nfr,''BackgroundColor''))),',...
    'hv=rgb2hsv(get(Hc_nfr,''BackGroundColor''));', . . .
    'set(Hc_hsli,''Val'',hv(1)),', . . .
    'set(Hc_hcur,''String'',sprintf(''%.2f'',hv(1))),', . . .
    'set(Hc_ssli,''Val'',hv(2)),', . . .
    'set(Hc_scur,''String'',sprintf(''%.2f'',hv(2))),', . . .
    'set(Hc_vsli,''Val'',hv(3)),', . . .
    'set(Hc_vcur,''String'',sprintf(''%.2f'',hv(3))),', . . .
    'set(Hc_rcur,''String'',sprintf(''%.2f'',get(Hc_rsli,''Val'')))']);
```

Another problem is that scripts execute more slowly than functions that are compiled the first time they are run. Finally, scripts are less flexible than functions. Functions can accept input arguments and can return values. Functions, therefore, can be used as arguments to other functions.

Functions keep the *Command* window workspace uncluttered, execute rapidly when called repeatedly, and accept input arguments and return values; furthermore, their callback strings are less complicated to write. Therefore, in many situations function M-files are a better choice.

Consider the previous example of a slider definition from a script file. Here is the equivalent fragment taken from a function M-file called `mmsetc`.

```
Hc_rsli = uicontrol(Hf_fig,'Style','slider', . . .
     'Position',[.10 .55 .35 .05], . . .
     'Min',0,'Max',1,'Value',initrgb(1), . . .
     'CallBack','mmsetc(0,''rgb2new'')');
```

Note that the callback string calls `mmsetc` again with different arguments. This is an example of using recursive function calls in callbacks.

Functions also have their own problems. The major difficulty with functions results from the fact that callback strings are passed to `eval` and executed in the *Command* window workspace, while the rest of the code executes within the function workspace. The scope rules for variables and functions discussed in Chapter 14 apply here. Variables defined only within the function are not available in the *Command* window workspace, and therefore cannot be used in callback strings. At the same time, variables used in the *Command* window workspace are not available inside the function itself.

There are a number of ways to work around this problem while gaining the advantages of using functions. Global variables, 'UserData' and 'Tag' properties, special function M-files used only for callbacks, private directories, subfunctions, and recursive function calls can all be useful techniques for creating GUI function M-files.

### Callback Functions

One effective technique for creating GUI functions is to write separate M-file functions designed specifically to execute one or more callbacks. Object handles and other variables used by the callback function can be passed as arguments, and the callback function can return values if necessary.

Consider an earlier example that creates an azimuth slider implemented as a script:

```
% setview.m script file
vw = get(gca,'View');
Hc_az = uicontrol(gcf,'Style','slider',...
        'Position',[10 5 140 20],...
        'Min',-90,'Max',90,'Value',vw(1),...
        'CallBack',[...
            'set(Hc_cur,''String'',num2str(get(Hc_az,''Value''))),'...
            'set(gca,''View'',[get(Hc_az,''Value'') vw(2)])']);
Hc_min = uicontrol(gcf,'Style','text',...
        'Position',[10 25 40 20],...
        'String',num2str(get(Hc_az,'Min')));
Hc_max = uicontrol(gcf,'Style','text',...
        'Position',[110 25 40 20],...
        'String',num2str(get(Hc_az,'Max')));
Hc_cur = uicontrol(gcf,'Style','text',...
        'Position',[60 25 40 20],...
        'String',num2str(get(Hc_az,'Value')));
```

Here is the same example as a function using the 'Tag' property to identify controls and a separate function M-file to execute the callback:

```
function setview()
vw=get(gca,'View');
Hc_az = uicontrol(gcf,'Style','slider',...
        'Position',[10 5 140 20],...
        'Min',-90,'Max',90,'Value',vw(1),...
        'Tag','AZslider',...
        'CallBack','svcback');
Hc_min = uicontrol(gcf,'Style','text,...
        'Position',[10 25 40 20],...
        'String',num2str(get(Hc_az,'Min')));
Hc_max = uicontrol(gcf,'Style','text',...
        'Position',[110 25 40 20],...
        'String',num2str(get(Hc_az,'Max')));
Hc_cur = uicontrol(gcf,'Style','text',...
        'Position',[60 25 40 20],...
        'Tag','AZcur',...
        'String',num2str(get(Hc_az,'Value')));
```

The callback function itself is given by

```
function svcback()
vw = get(gca,'View');
Hc_az = findobj(gcf,'Tag','AZslider');
Hc_cur = findobj(gcf,'Tag','AZcur');
str = num2str(get(Hc_az,'Value'));
newview = [get(Hc_az,'Value') vw(2)];
set(Hc_cur,'String',str)
set(gca,'View',newview)
```

The previous example does not save much coding but gains all the advantages of using functions rather than scripts. The callback function can use temporary variables without cluttering up the *Command* window workspace, and the syntax of the commands in callback functions becomes much simpler without all those quotes and strings required for `eval`. More complicated callbacks are much simpler using this technique.

The disadvantage of separate callback function M-files is the potentially large number of M-files required to implement a single GUI function that uses a number of controls or menu items. All of the M-files must be available in the MATLAB path, and each must have a distinctive name. On platforms that have filename size limits or are case insensitive, the chances of a filename conflict are increased. Also, the callback functions should only be called by the GUI function itself and not by the user.

One way to resolve the filename conflict problem is to place all of the callback functions into a private directory. Once there, these functions are not visible to the command line or any function except M-file functions in the same or parent directory. For example, the function M-file /mydir/fmain.m can call the function M-file /mydir/private/fback.m as a callback, but fback.m is not visible to any function outside of the mydir and mydir/private directories. The disadvantage is that the directory structure must be maintained for this technique to work.

### Recursive Function Calls

You can avoid the complexity of multiple M-files while maintaining the advantages of functions by using a single M-file and calling it recursively to execute callbacks. The callback functions can be incorporated into the calling function by using switch/case or if/elseif statements. The generic structure of such a function call is

```
guifunc(flag)
```

where flag is the argument that determines which function switch is executed. This could be a string such as 'startup', 'close', 'setcolor', etc. It could also be a code or number. If flag is a string, the function can be coded as shown in the following M-file fragment:

```
if nargin < 1, flag = 'startup'; end;
if ~ischar(flag), error('Invalid argument'), end;
switch flag
  case 'startup'
    <statements to create controls or menus>
    <statements to implement the GUI function>
  case 'setcolor'
    <statements to perform the callback associated with setcolor>
  case 'close'
    <statements to perform the callback associated with close>
end
```

If codes or numbers are used, the switch can be coded in a similar manner:

```
if nargin < 1, flag = 0; end;
if ischar(flag), error('Invalid argument'), end;
if flag == 0
    <statements to create controls or menus>
    <statements to implement the GUI function>
elseif flag == 1
    <statements to perform the callback associated with setcolor>
elseif flag == 2
    <statements to perform the callback associated with close>
end
```

The following example shows how the azimuth slider example could be implemented as a single function M-file:

```
function setview(flag)

if nargin < 1, flag = 'startup'; end;
if ~ischar(flag), error('Invalid argument.'); end;
vw = get(gca,'View');  % This information is needed in both sections

if strcmp(flag,'startup')  % Define the controls and tag them
    Hc_az = uicontrol(gcf,'Style','slider',...
        'Position',[10 5 140 20],...
        'Min',-90,'Max',90,'Value',vw(1),...
        'Tag','AZslider',...
        'CallBack','setview(''set'')');
Hc_min = uicontrol(gcf,'Style','text',...
        'Position',[10 25 40 20],...
        'String',num2str(get(Hc_az,'Min')));
Hc_max = uicontrol(gcf,'Style','text',...
        'Position',[110 25 40 20],...
        'String',num2str(get(Hc_az,'Max')));
Hc_cur = uicontrol(gcf,'Style','text',...
        'Position',[60 25 40 20],...
        'Tag','AZcur',...
        'String',num2str(get(Hc_az,'Value')));

elseif strcmp(flag,'set')    % Execute the callback
    Hc_az = findobj(gcf,'Tag','AZslider');
    Hc_cur = findobj(gcf,'Tag','AZcur');
    str = num2str(get(Hc_az,'Value'));
    newview = [get(Hc_az,'Value') vw(2)];
    set(Hc_cur,'String',str)
    set(gca,'View',newview)
end
```

Both of the preceding examples set the `'Tag'` property and use this property with the `findobj` function to find handles to desired objects for the callbacks. Two alternative methods are described in the next subsections.

### Global Variables

Global variables can be used in a function to make specific variables available to all parts of the GUI function. The global variables are declared in the common area of the function and are therefore available to all recursive calls to the function. This example shows how the azimuth slider function can be coded using global variables:

```
function setview(flag)

global HC_AZ HC_CUR    % Create global variables
if nargin < 1, flag = 'startup'; end;
if ~ischar(flag), error('Invalid argument.'); end;
vw = get(gca,'View');  % This information is needed in both sections

if strcmp(flag,'startup')  % Define the controls
    HC_AZ = uicontrol(gcf,'Style','slider',...
        'Position',[10 5 140 20],...
        'Min',-90,'Max',90,'Value',vw(1),...
        'CallBack','setview(''set'')');
    Hc_min = uicontrol(gcf,'Style','text',...
        'Position',[10 25 40 20],...
        'String',num2str(get(HC_AZ,'Min')));
    Hc_max = uicontrol(gcf,'Style','text',...
        'Position',[110 25 40 20],...
        'String',num2str(get(HC_AZ,'Max')));
    HC_CUR = uicontrol(gcf,'Style','text',...
        'Position',[60 25 40 20],...
        'String',num2str(get(HC_AZ,'Value')));

elseif strcmp(flag,'set')    % Execute the callback
    str = num2str(get(HC_AZ,'Value'));
    newview = [get(HC_AZ,'Value') vw(2)];
    set(HC_CUR,'String',str)
    set(gca,'View',newview)
end
```

The global variables follow the MATLAB convention and use uppercase variable names. The `'Tag'` property is not needed and is not used. In addition, the callback code is simpler because the object handles are available without using `findobj` to obtain them.

One word of caution, however: Just because a variable is declared global within the function, the variable is not automatically available in the *Command* window workspace and cannot be used within a callback string unless the `assignin` function is used to create an appropriate *Command* window workspace variable. In addition, if the user issues the command

```
» clear global
```

on the command line, **all** global variables are cleared, including those within the function.

Global variables and recursive function calls are effective techniques when there is a single *figure* or a limited number of variables that need to be available to all callbacks. For more complex functions involving multiple *figures* or for implementations using separate callback functions, the 'UserData' property may be more appropriate. In addition, the 'UserData' property value of an object is available in the *Command* window workspace as long as the object handle can be obtained.

### UserData **Storage**

As is the case with the 'Tag' property, the 'UserData' property is available to pass information between functions or between different parts of a recursive function. If many variables are needed, they can be passed in the 'UserData' property of an easily identified object. As described earlier, 'UserData' provides storage for a single variable of any class that stays with a Handle Graphics object. The data can be a single value, a matrix, a string, a structure, or even a multidimensional array of structures or cells. The following code implements the azimuth slider using the 'UserData' property of the current *figure:*

```
function setview(flag)

if nargin < 1, flag = 'startup'; end;
if ~ischar(flag), error('Invalid argument.'); end;
vw = get(gca,'View');  % This information is needed in both sections

if strcmp(flag,'startup')  % Define the controls
    Hc_az = uicontrol(gcf,'Style','slider', . . .
        'Position',[10 5 140 20], . . .
        'Min',-90,'Max',90,'Value',vw(1), . . .
        'CallBack','setview(''set'')');
    Hc_min = uicontrol(gcf,'Style','text', . . .
        'Position',[10 25 40 20], . . .
        'String',num2str(get(Hc_az,'Min')));
    Hc_max = uicontrol(gcf,'Style','text', . . .
        'Position',[110 25 40 20], . . .
        'String',num2str(get(Hc_az,'Max')));
    Hc_cur = uicontrol(gcf,'Style','text', . . .
        'Position',[60 25 40 20], . . .
        'String',num2str(get(Hc_az,'Value')));
    set(gcf,'UserData',[Hc_az Hc_cur]);  % Store the object handles

elseif strcmp(flag,'set')    % Execute the callback
    Hc_all = get(gcf,'UserData'); % retrieve the object handles
    Hc_az  = Hc_all(1);
    Hc_cur = Hc_all(2);
    str = num2str(get(Hc_az,'Value'));
    newview = [get(Hc_az,'Value') vw(2)];
    set(Hc_cur,'String',str)
    set(gca,'View',newview)
end
```

The handles are stored in the *figure's* 'UserData' property at the end of the 'startup' section and are retrieved before the callback is executed. If there are many callbacks, the 'UserData' value needs to be retrieved only once, as shown in the following code fragments:

```
if strcmp(flag,'startup')  % Define the controls and tag them
    % <The 'startup' code is here>
    set(gcf,'UserData',[Hc_az Hc_cur]);  % Store the object handles

else    % This must be a callback
    Hc_all = get(gcf,'UserData'); % Retrieve the object handles
    Hc_az  = Hc_all(1);
    Hc_cur = Hc_all(2);

    switch flag
      case 'set',
        % <The 'set' callback code is here>
      case 'close',
        % <The 'close' callback code is here>
   % <Any other callback code uses additional case or otherwise statements>
    end
end
```

## Debugging GUI M-files

The fact that the callback strings are evaluated and executed in the *Command* window workspace has certain implications for writing and debugging GUI functions and scripts. Callback strings can be very complex, especially in scripts, introducing many opportunities for syntax errors. Keeping track of all the single quotes, commas, and parentheses can be a daunting task. MATLAB complains if the syntax is not correct, but as long as the value of the 'CallBack' property of an object is an actual text string, MATLAB is satisfied. The string is not checked for internal syntax errors until the object is activated and the callback string is passed to eval.

This property lets you create *forward references* or define callback strings that reference variables and object handles that have not yet been defined, making it much easier to write MATLAB code that cross-references other objects. However, each callback must be tested individually to make sure that the callback string is a legal MATLAB command and that all variables referenced in the callback string are available in the *Command* window workspace.

Coding callbacks as function M-files or as switches within the GUI function itself lets you change and test callbacks individually without running the entire GUI function.

Because callback strings are evaluated in the *Command* window workspace rather than within the function itself, passing data between the function and each callback can become complex. For example, if a function test contains the following code:

```
function test()
tpos1 = [20 20 50 20];
tpos2 = [20 80 50 20];
Hc_text = uicontrol('Style','text','String','Hello','Position',tpos1);
Hc_push = uicontrol('Style','push','String','Move Text', . . .
    'Position',[15 50 100 25], . . .
    'CallBack','set(Hc_text,''Position'',tpos2)');
```

all of the statements are valid MATLAB commands, and the callback string evaluates to a valid MATLAB statement as well. The *text* object and the push button *uicontrol* appear on the *figure*, but when the push button is activated, MATLAB complains:

```
» test
»
??? Undefined function or variable Hc_text.
??? Error while evaluating callback string.
```

If `test` were a script, there would be no problem, since all variables would be available in the *Command* window workspace. Since `test` is a function, neither `Hc_text` nor `tpos2` is defined in the *Command* window workspace and the callback string execution fails.

One apparent solution is to create the callback string using individual string elements created from values rather than from variables. For example, changing the callback string as follows:

```
'CallBack',['set(', . . .
            sprintf('%.15g',Hc_text), . . .
            ',''Position'',', . . .
            sprintf('[%.15g %.15g %.15g %.15g]',tpos2), . . .
            ')']);
```

creates a string that includes the value of the `Hc_text` object handle converted to a string with up to 15 digits of precision, and the value of the `tpos2` variable converted to the string representation of a matrix. The `sprintf` statements are evaluated within the function and the resulting strings are then used in the callback definition. The actual command that is executed in the *Command* window workspace looks something like this:

```
eval('set(87.000244140625,''Position'',[20 80 50 100])')
```

Full precision must be maintained when converting an object handle into a string. The preceding conversion uses up to 15 digits after the decimal point. This is the precision that should be used for all object handle conversions in MATLAB.

Remember that subsequent changes to the variables will not change the callback string. In the previous example, changing the value of `tpos2` after the control is defined has no effect. For example, adding the command

```
tpos2 = [20 200 50 20];
```

to the end of the function has no effect, since the callback string is created by evaluating `tpos2` before `tpos2` is redefined. In any case, it is always better to retain the numeric object handle and avoid string conversions if at all possible.

> To conclude this section, while numerous acceptable techniques exist, the most efficient, general, easily debugged, and easily maintained GUI functions are those that
>
> **1.** use the `'Tag'` property and `findobj` to retrieve handles in callbacks,
> **2.** use `'UserData'` properties for storage of data to be used by callbacks, and
> **3.** define callbacks as recursive calls to the GUI M-file itself.

## 32.7 POINTER AND MOUSE BUTTON EVENTS

GUI functions use the location of the mouse pointer and the status of mouse buttons to control MATLAB actions. This section discusses the interaction between pointer and object locations and mouse button actions, and how MATLAB responds to changes or events, such as a button press, button release, or pointer movement.

### Callback Properties, Selection Regions, and Stacking Order

All Handle Graphics objects have a `'ButtonDownFcn'` property that has been mentioned briefly but will be more fully discussed here. Both *uimenus* and *uicontrols* have a `'CallBack'` property that is central to the use of menus and controls. In addition, *figures* have `'WindowButtonDownFcn'`, `'WindowButtonUpFcn'`, and `'WindowButtonMotionFcn'` properties as well as `'KeyPressFcn'`, `'CloseRequestFcn'`, and `'ResizeFcn'` properties. All graphics objects also have `'CreateFcn'` and `'DeleteFcn'` properties. The value associated with each of these properties is a callback string that is passed to `eval` when the property is invoked. The pointer location determines which callbacks are involved and the order in which they are invoked when an event occurs.

The previous chapter contained a discussion of stacking order and object selection regions that is relevant to this discussion. MATLAB determines which callback will be invoked based on three regions within a *figure*. If the pointer is within a Handle Graphics object as determined by its `'Position'` property, the pointer is considered to be `on` the object. If the pointer is not `on` an object but is within an object's selection region, the pointer is `near` the object. Finally, if the pointer is within the *figure* but not `on` or `near` another object, the pointer is `off` the other objects. When objects or their selection regions overlap, the stacking order determines which object is selected.

The selection region of Handle Graphics *line, surface, patch, test,* and *axes* objects was discussed in the previous chapter. *Uimenu* objects have no external selection region. The pointer is either `on` a *uimenu* object or is not. *Uicontrols* have a selection region that extends about 5 pixels beyond the control's position in all directions. The pointer can be either `on` or `near` a control.

## Button Click

A button click can be defined as the press and subsequent release of a mouse button while the mouse pointer is over the same object. If the mouse pointer is over a *uimenu* or *uicontrol* object, a button click triggers the execution of the object's `'Callback'` property string as long as the `'Enable'` property is set to `'on'`. The button press prepares the control and often changes the control or menu visually, and the button release triggers the callback. If the mouse pointer is not `on` a control or menu, both Button Press and Button Release events are triggered as explained next.

## Button Press

When the mouse button is pressed with the pointer located within a *Figure* window, a number of different actions can occur, based on the location of the pointer and the proximity of Handle Graphics objects. If an object is selected, it becomes the current object. If no object is selected, the *figure* itself becomes the current object. The *figure's* `'CurrentPoint'` and `'SelectionType'` properties are also updated. The appropriate callbacks are then invoked.

The following table lists the pointer location options and the callbacks that are invoked for a button press event.

| Pointer Location | Property Invoked |
|---|---|
| `on` a *uimenu* item if the `'Enable'` property is `'on'` | Change menu appearance and prepare for a release event. |
| `on` a *uicontrol* if the `'Enable'` property is `'on'` | Change control appearance and prepare for a release event. |
| `on` a *uimenu* if the `'Enable'` property is `'off'` | Button press is ignored. |
| `on` a *uicontrol* if the `'Enable'` property is `'off'` or `'inactive'` | *Figure's* `WindowButtonDownFcn` and then the *uicontrol's* `ButtonDownFcn`. |
| `on` or `near` a Handle Graphics object, or `near` a *uicontrol* | *Figure's* `WindowButtonDownFcn` and then the object's `ButtonDownFcn`. |
| within the *figure,* but not `on` or `near` any other object | *Figure's* `WindowButtonDownFcn` and then the *figure's* `ButtonDownFcn`. |

Note that a button press event always invokes the *figure's* `'WindowButtonDownFcn'` callback before the selected object's `'ButtonDownFcn'` callback except when the pointer is `on` a *uicontrol* or *uimenu* object. When the pointer is `near` a *uicontrol* or `on` a *uicontrol* with the `'Enable'` property `'off'` or `'inactive'`, the *uicontrol's* `'ButtonDownFcn'` callback is invoked rather than the `'CallBack'` property callback after the *figure's* `'WindowButtonDownFcn'` callback has finished. The `'ButtonDownFcn'` callback of a *uimenu* is never invoked.

### Button Release

When the mouse button is released, the *figure's* `'CurrentPoint'` property is updated, and the *figure's* `'WindowButtonUpFcn'` callback is invoked. If the `'WindowButtonUpFcn'` callback is not defined, the `'CurrentPoint'` property is not updated when the button is released.

### Pointer Movement

When the pointer is moved within a *figure,* the *figure's* `'CurrentPoint'` property is updated, and the *figure's* `'WindowButtonMotionFcn'` callback is invoked. If the `'WindowButtonMotionFcn'` callback is not defined, the `'CurrentPoint'` property is not updated when the pointer moves.

Combinations of callbacks can produce many interesting effects. If you have the *Signal Processing Toolbox,* try the `sigdemo1` and `sigdemo2` function M-files to see examples of some of these effects.

## 32.8 RULES FOR INTERRUPTING CALLBACKS

Once a callback begins executing, it normally runs to completion before the next callback event is processed. If another callback is triggered before the first callback finishes, however, the executing callback will be interrupted the next time the event queue is processed. You can change this default behavior by setting the object's `'Interruptible'` property to `'off'`, preventing pending callback events from being processed until the executing callback finishes.

### Event Queue

MATLAB processes commands that perform computations or set object properties at the time they are issued. Commands that involve *Figure* window input or output generate events. Events include pointer movements and mouse button actions that generate callbacks, the `waitfor` and `waitforbuttonpress` functions, and commands that redraw graphics, such as `drawnow`, `figure`, `getframe`, or `pause` commands. Any pending events are placed in an event queue.

### Callback Processing

A callback will execute until it reaches a waitfor, waitforbuttonpress, drawnow, getframe, pause, or figure command. (Note that gcf and gca can spawn figure commands.) ***Callbacks that do not contain any of these commands will not be interrupted.***

When one of these special commands is reached, MATLAB suspends execution of the callback and examines each of the pending events in the event queue. If the 'Interruptible' property of the object with the suspended callback is set to 'on', its default value, all pending events are processed before the suspended callback is resumed. If the 'Interruptible' property is set to 'off', then only pending redraw events are processed. If the 'BusyAction' property of the object with the suspended callback is set to 'cancel', any pending callback events are discarded. If the 'BusyAction' property is set to 'queue', then pending callback events are held in the event queue until the original callback finishes. Resizefcn and Deletefcn callbacks are exceptions to these rules. These callbacks are inherently not interruptible.

### Preventing Interruptions

Even if the executing callback is not interruptible, pending redraw events are still processed when the callback reaches a waitfor, waitforbuttonpress, drawnow, figure, getframe, or pause command. These events can be suppressed by avoiding the use of all of these special commands in your callback. Another option is to set the *figure's* 'WindowStyle' property to 'modal', which traps (discards) all mouse and keyboard events. The 'WindowStyle' property will be discussed in detail in the next chapter.

## 32.9 M-FILE EXAMPLES

One good example of callback processing is mmtext, a short function M-file that uses 'WindowButtonDownFcn', 'WindowButtonMotionFcn', and 'WindowButtonUpFcn' callbacks to place and drag text. This function was presented as an example in the Handle Graphics chapter and is worth reviewing now.

The next example, which we will call mmcxy, creates a small text box at the lower left corner of a *figure* to display the [X,Y] coordinates of the mouse pointer while the pointer is in the *figure*. Clicking the mouse button anywhere in the *figure* erases the coordinate display.

Although mmcxy is a short function, it still uses many of the elements of effective GUI functions, including recursive function calls, global variables, and 'UserData' properties. It also illustrates the use of the *figure* 'WindowButtonDownFcn' and 'WindowButtonMotionFcn' properties to initiate callbacks.

```
function out=mmcxy(arg)
%MMCXY Show x-y Coordinates Using Mouse.
% MMCXY places the x-y coordinates of the mouse in the
% lower left hand corner of the current 2-D figure window.
% When the mouse is clicked, the coordinates are erased.
% XY=MMCXY returns XY=[x,y] coordinates where mouse was clicked.
% XY=MMCXY returns XY=[] if a key press was used.

global MMCXY_OUT

if ~nargin
    Hf=mmgcf(1); % get current figure if available
    Ha=findobj(Hf,'Type','axes');
    if isempty(Ha), error('No Axes in Current Figure.'),end

    Hu=uicontrol(Hf,'Style','text',...
                    'units','pixels',...
                    'Position',[1 1 140 15],...
                    'HorizontalAlignment','left');
    set(Hf,'Pointer','crossh',...
           'WindowButtonMotionFcn','mmcxy(''move'')',...
           'WindowButtonDownFcn','mmcxy(''end'')',...
           'Userdata',Hu)
    figure(Hf)  % bring figure forward
    if nargout  % must return x-y data
        key=waitforbuttonpress; % pause until mouse is pressed
        if key,
            out=[];          % return empty if aborted
            mmcxy('end')     % clean things up
        else
            out=MMCXY_OUT;   % now that move is complete return point
        end
        return
    end

elseif strcmp(arg,'move')  % mouse is moving in figure window
    cp=get(gca,'CurrentPoint');  % get current mouse position
    MMCXY_OUT=cp(1,1:2);
    xystr=sprintf('[%.3g, %.3g]',MMCXY_OUT);
    Hu=get(gcf,'Userdata');
    set(Hu,'String',xystr)  % put x-y coordinates in text box

elseif strcmp(arg,'end')   % mouse click occurred, clean things up
    Hu=get(gcf,'Userdata');
    delete(Hu)
    set(gcf,'Pointer','arrow',...
            'WindowButtonMotionFcn','',...
            'WindowButtonDownFcn','',...
            'Userdata',[])
end
```

When called the first time, mmcxy creates the text *uicontrol,* changes the pointer shape, sets up the 'WindowButtonMotionFcn' and 'WindowButtonDownFcn' callbacks, and then waits for a key or button press. If a key is pressed, the cleanup routine is called to delete the text box, restore the mouse pointer, and clean up the *figure* callbacks and 'UserData' property. If a mouse button is clicked, the 'WindowButtonDownFcn' callback takes care of the cleanup. While waiting, pointer movement in the *figure* triggers the 'WindowButtonMotionFcn' callback, which updates the text string in the *uicontrol.*

mmdraw is another useful GUI function from the *Mastering MATLAB Toolbox* that is similar to mmtext but is a bit more complex. This function lets you draw a *line* on the current *axes* with the mouse and set *line* properties as well. Consider the following description of mmdraw by breaking it up into fragments. First, define the function, add some help text, declare the global variables, and initialize variables.

```
function mmdraw(arg1,arg2,arg3,arg4,arg5,arg6,arg7)
%MMDRAW Draw a Line and Set Properties Using Mouse. (MM)
% MMDRAW draws a line in the current axes using the mouse,
% Click at the starting point and drag to the end point.
% In addition, properties can be given to the line.
% Properties are given in pairs, e.g., MMDRAW name value ...
% Properties:
% NAME       VALUE       {default}
% color      [y m c r g b {w} k] or an rgb in quotes:'[r g b]'
% style      [-  --  {:}  -. none]
% mark       [o  +  .  * x square diamond ^ v > < pentagram hexagram none]
% width      points for linewidth {0.5}
% size       points for marker size {6}
% Examples:
% MMDRAW color r width 2   sets color to red and width to 2 points
% MMDRAW mark + size 8     sets marker type to + and size to 8 points
% MMDRAW color '[1 .5 0]'  sets color to orange

% Calls: mmgcf mmgca

global MMDRAW_HL MMDRAW_EVAL

ni=nargin;
if ni==0
   arg1='color';arg2='w';arg3='style';arg4=':';ni=4;
end
```

Next, handle the initial call. Change the pointer shape, define a callback for the 'ButtonDownFcn' property, and stuff a command string into a global variable.

```
if ischar(arg1)  % initial call, set things up
   Hf=mmgcf(1);
   Ha=mmgca(Hf,1);
   set(Hf,'Pointer','crossh',...   % set up callback for line start
         'BackingStore','off',...
         'WindowButtonDownFcn','mmdraw(1)')
figure(Hf)
MMDRAW_EVAL='mmdraw(99';  % set up string to set attributes
for i=1:ni
   argi=eval(sprintf('arg%.0f',i));
   MMDRAW_EVAL=[MMDRAW_EVAL ',''' argi ''''];
end
MMDRAW_EVAL=[MMDRAW_EVAL ')'];
```

The next section implements the button down event callback. Store the current point in the *axes* `'UserData'` property and define two more callbacks.

```
elseif arg1==1  % callback is line start point
   fp=get(gca,'CurrentPoint');    % start of line point
   set(gca,'Userdata',fp(1,1:2))  % store in axes userdata
   set(gcf,'WindowButtonMotionFcn','mmdraw(2)',...
          'WindowButtonUpFcn','mmdraw(3)')
```

This section implements the button motion event callback. Create or extend the *line.*

```
elseif arg1==2  % callback is mouse motion
   cp=get(gca,'CurrentPoint');cp=cp(1,1:2);
   fp=get(gca,'Userdata');
   if isempty(MMDRAW_HL)
      MMDRAW_HL=line('Xdata',[fp(1);cp(1)],'Ydata',[fp(2);cp(2)],...
            'EraseMode','xor',...
            'Color','w','LineStyle',':',...
            'Clipping','off');
   else
      set(MMDRAW_HL,'Xdata',[fp(1);cp(1)],'Ydata',[fp(2);cp(2)])
   end
```

This section handles the button up event callback. Clear out the callbacks, restore the pointer, and release the *axes* `'UserData'` property.

```
elseif arg1==3  % callback is line end point, finish up
   set(gcf,'Pointer','arrow',...
         'BackingStore','on',...
         'WindowButtonDownFcn','',...
         'WindowButtonMotionFcn','',...
         'WindowButtonUpFcn','')
   set(gca,'Userdata',[])
   set(MMDRAW_HL,'EraseMode','normal') % render line better
   eval(MMDRAW_EVAL)
   MMDRAW_EVAL=[];
```

The last section applies the original `'PropertyName'`, `PropertyValue` arguments to the *line* that was just created.

```
elseif arg1==99  % process line properties
   for i=2:2:ni-1
      name=eval(sprintf('arg%.0f',i),[]); % get name argument
      value=eval(sprintf('arg%.0f',i+1),[]); % get value argument
      if strcmp(name,'color')
         if value(1)=='[', value=eval(value);end
         set(MMDRAW_HL,'Color',value)
      elseif strcmp(name,'style')
         set(MMDRAW_HL,'Linestyle',value)
      elseif strcmp(name,'mark')
         set(MMDRAW_HL,'Marker',value)
      elseif strcmp(name,'width')
         value=abs(eval(value));
         set(MMDRAW_HL,'LineWidth',value)
      elseif strcmp(name,'size')
         value=abs(eval(value));
         set(MMDRAW_HL,'MarkerSize',value)
      else
         disp(['Unknown Property Name: ' name])
      end
   end
   MMDRAW_HL=[];
end
```

A final example illustrates many of the topics discussed in this chapter. The *Mastering MATLAB Toolbox* function `mmcd` implements a GUI version of the `cd` command. Frame, text, edit, listbox, and push button *uicontrols* are used as are subfunctions, cell arrays, structures, the `'UserData'` property, and the `waitfor` command. As usual with longer functions, the function M-file will be discussed in sections. First we define the function, add help text, and initialize some variables.

```
function newdir=mmcd
%MMCD Change Working Directory Using a Graphical Interface. (MM)
% NEWDIR=MMCD returns the new directory path to the string NEWDIR.
%
% GUI Pushbuttons:
% Matlab - Go to the MATLAB root directory.
% Toolbox - Go to the MATLAB Toolbox directory.
% Restore - Return to the original directory.
% Close - Quit and return to the command window.
%
% See also CD, PWD, DIR.

% Calls: mmfitpos.

done = 0;
initdir = cd;
lastdir = initdir;
```

Now the main loop begins. Place the current directory contents into a structure and do some error checking.

```
while ~done

  d = dir;
  if isempty(d)
    disp(['Cannot get a directory list in ',cd,'.']);
    cd(lastdir);
    d = dir;
  end
  lastdir = cd;
```

The next section works around a PC bug in versions 5.0 and 5.1 of MATLAB, removes the '.' directory from the list, and makes sure the '..' directory exists.

```
  switch computer
    case 'PCWIN'
      d(strmatch('.',char(d.name),'exact')) = [];
      d(strmatch('..',char(d.name),'exact')).isdir = 1;
    case 'MAC2'
      didx = length(d)+1;
      d(didx).name = '..'; d(didx).isdir = 1;
    otherwise
      d(strmatch('.',char(d.name),'exact')) = [];
  end
```

Discard file names and keep directory entries. First create an index of directory entries and use the index to create a cell array of directory names. Then sort them alphabetically. Since `sort` will not work with cells, we convert to a string array, sort the array, and convert back to cells.

```
f = {d.isdir};
ff = logical(cat(1,f{:}));
dlist = {d(ff).name};

dlist = shiftdim(cellstr(sortrows(char(dlist{:}))),1);
```

Now we have a sorted list of directory names. Call the subfunction `local_mmcd` to get a selection. `done` is True when the **Close** button has been selected. The variable `s` contains the selected directory name string or a flag representing a button press.

```
[s,done] = local_mmcd(dlist);
```

If a button was pressed, change to the appropriate directory. If a directory was selected, change to the new directory. If we are done, return the new directory string if requested and quit.

```
    if ~isempty(s)
      switch char(s)
        case 'CDmatlab',  cd(matlabroot);
        case 'CDtoolbox', cd(fullfile(matlabroot,'toolbox'));
        case 'CDrestore', cd(initdir);
        otherwise, eval('cd(char(s))',...
           'disp(''Invalid directory or permission denied.'')');

      end %switch
    end   %if
  end     %while

  if nargout == 1, newdir = cd; end
  return
```

The subfunction `local_mmcd` actually puts up the window containing the *uicontrols* and waits for input from the user. A subfunction is contained within the same M-file as the calling function and is only visible to the functions in the same M-file.

First define the function arguments, initialize some variables, and calculate some values.

```
function [selection,dflag] = local_mmcd(liststring)

dflag = 0;
selection = [];
promptstring = cd;
switch computer
  case 'PCWIN', ex_const = 2.8;
  case 'MAC2',  ex_const = 2.0;
  otherwise,    ex_const = 1.8;
end
ex = get(0,'defaultuicontrolfontsize')*ex_const;
listsize = [max(length(promptstring)*ex/2.8,186) ...
            max(100,min(300,length(liststring)*ex))];
runit = get(0,'units');
set(0,'units','pixels');
dfp = get(0,'defaultfigureposition');
set(0,'units',runit);
w = listsize(1)+32;
h = ex+listsize(2)+104;
fp = [dfp(1)-w/2 dfp(2)+0.6*dfp(4)-h w h];
fp = mmfitpos(fp,0,'pixels');
w = fp(3); h = fp(4);
btn_wid = (w-40)/2;
```

The `mmfitpos` function used in the preceding fragment is part of the *Mastering MATLAB Toolbox* and simply ensures that the position vector `fp`, which is the GUI window position vector, fits within the position vector of the root object (handle `0`).

Search for an existing dialog box. If one does not yet exist, create one. Name it, control the size, use a noninteger handle, add an identification string and an initial `'UserData'` string, and specify a callback for a close request.

```
fig = findobj(0,'type','figure','tag','CDDLG');
if isempty(fig)
fig = figure('name','Change Directory','resize','off',...
             'numbertitle','off','windowstyle','modal',...
             'createfcn','','integerhandle','off',...
             'position',fp,'tag','CDDLG','userdata','CDnotyet',...
             'closerequestfcn','set(gcf,''userdata'',''CDdone'')');
```

Add some frames to group related objects visually.

```
figframe = uicontrol('style','frame','tag','figframe',...
           'position',[0 0 fp([3 4])]);

butframe = uicontrol('style','frame','tag','butframe',...
           'position',[8 8 fp(3)-16 60]);

listframe = uicontrol('style','frame','tag','listframe',...
           'position',[8 72 fp(3)-16 fp(4)-80]);
```

List the current directory in an edit box so the user can change directories by typing in a directory path.

```
prompt_text = uicontrol('style','edit','string',promptstring,...
           'tag','prompt_text',...
           'horizontalalignment','left','units','pixels',...
           'callback','set(gcf,''userdata'',''CDedit'')',...
           'position',[16 fp(4)-ex-16 listsize(1) ex]);
```

Add the listbox containing the directory names including the `'..'` directory.

```
listbox = uicontrol('style','listbox','string',liststring,...
           'backgroundcolor','w','position',[16 80 listsize],...
           'callback','set(gcf,''userdata'',''CDselected'')',...
           'tag','listbox','value',1);
```

Now add some buttons to **Restore** the original directory, go to the MATLAB home directory, go the MATLAB **Toolbox** directory, and **Close** the window.

```
rs_btn = uicontrol('style','pushbutton','string','Restore',...
           'tag','rs_btn', 'position',[16 14 btn_wid 22],...
           'callback','set(gcf,''userdata'',''CDrestore'')');

ok_btn = uicontrol('style','pushbutton','string','Close',...
           'tag','ok_btn', 'position',[24+btn_wid 14 btn_wid 22],...
           'callback','set(gcf,''userdata'',''CDdone'')');

ml_btn = uicontrol('style','pushbutton','string','Matlab',...
           'tag','ml_btn','position',[16 40 btn_wid 22],...
           'callback','set(gcf,''userdata'',''CDmatlab'')');

tb_btn = uicontrol('style','pushbutton','string','Toolbox',...
           'tag','tb_btn','position',[24+btn_wid 40 btn_wid 22],...
           'callback','set(gcf,''userdata'',''CDtoolbox'')');
```

If an existing dialog box was found, reuse it but resize the components as necessary depending on the string width and number of directory strings in the listbox.

```
else
  fp = get(fig,'position');
  fp = [fp(1) fp(2)+fp(4)-h w h];
  fp = mmfitpos(fp,0,'pixels');
  set(fig,'position',fp);

  figframe = findobj(fig,'tag','figframe');
  set(figframe,'position',[0 0 fp([3 4])]);

  butframe = findobj(fig,'tag','butframe');
  set(butframe,'position',[8 8 fp(3)-16 60]);

  rs_btn = findobj(fig,'tag','rs_btn');
  set(rs_btn,'position',[16 14 btn_wid 22]);

  ok_btn = findobj(fig,'tag','ok_btn');
  set(ok_btn,'position',[24+btn_wid 14 btn_wid 22]);

  ml_btn = findobj(fig,'tag','ml_btn');
  set(ml_btn,'position',[16 40 btn_wid 22]);

  tb_btn = findobj(fig,'tag','tb_btn');
  set(tb_btn,'position',[24+btn_wid 40 btn_wid 22]);

  listframe = findobj(fig,'tag','listframe');
  set(listframe,'position',[8 72 fp(3)-16 fp(4)-80]);

  prompt_text = findobj(fig,'tag','prompt_text');
  set(prompt_text,'string',promptstring,...
    'position',[16 fp(4)-ex-16 listsize(1) ex]);

  listbox = findobj(fig,'tag','listbox');
  set(listbox,'string',liststring,'value',1,...
    'position',[16 80 listsize]);

end
```

Now wait around for something to happen. The `waitfor(handle,'Property')` form of the `waitfor` function suspends M-file execution until the specified property value changes, but continues to allow callbacks to execute. As can be seen in the preceding code, each of the *uicontrol* objects defines a callback that changes the *figure's* `'UserData'` property value. When the user selects a *uicontrol* push button, edit box, or listbox entry, the `'UserData'` property changes and M-file execution continues.

```
waitfor(fig,'UserData');
```

Find the dialog box, get the `'UserData'` string, and determine the action to take based on the contents of this string. Return appropriate values to the calling function.

```
fig = findobj(0,'type','figure','tag','CDDLG');
if isempty(fig)
  error('Dialog box misplaced.')
else
  ud = get(fig,'userdata');

  switch ud

    case 'CDdone'
      delete(fig); selection = []; dflag = 1;

    case {'CDrestore','CDmatlab','CDtoolbox'}
      set(fig,'userdata','CDnotyet'); selection = ud;

    case 'CDselected'
      set(fig,'userdata','CDnotyet');
      listbox = findobj(fig,'tag','listbox');
      selection = liststring(get(listbox,'value'));

    case 'CDedit'
      set(fig,'userdata','CDnotyet');
      prompt_text = findobj(fig,'tag','prompt_text');
      selection = get(prompt_text,'string');

    otherwise, disp('Unknown userdata value');

  end % switch
end   % else
```

This example illustrates how a complete GUI function can be created using *uicontrols*. Because the function keeps looping until the **Close** button is pressed or the *figure* itself is closed, the *Command* window is not available while the *figure* is open.

Previous examples used the `waitforbuttonpress` and `gco` functions to obtain user input. This example uses a different technique. All of the *uicontrol* callbacks as well as the *figure's* `'CloseRequestFcn'` callback simply change the value of the *figure* `'UserData'` property. The `waitfor` function is then used to detect such a change. Each technique is appropriate in different situations, but both approaches are equally valid.

## 32.10 UTILITY FUNCTIONS

MATLAB supplies a large number of graphics utility functions that can help you manage Handle Graphics objects and GUIs. Some of the most useful are briefly discussed here. For detailed information about calling syntax and occasional examples, see MATLAB documentation or use on-line help.

`selectmoveresize`

This built-in function allows you to resize and move *axes* and *uicontrol* objects (in a manner similar to the way you move objects in a drawing or drafting program) by interactively changing the object's `'Position'` vector. For example.

```
» set(gca,'ButtonDownFcn','selectmoveresize')
```

adds corner and side hot spots and a bounding box to the current *axes* when the mouse button is pressed over the *axes*. You can then resize or move the *axes* using the mouse. Holding the **Shift** key and pressing the mouse button over the object (or `near` a *uicontrol*) removes the resize hot spots and bounding box. You should then issue the command

```
» set(gca,'ButtonDownFcn','')
```

from the command line or use it in another callback to disable the `'selectmoveresize'` callback.

`setptr`

This function provides some predefined custom pointer shapes, including various hand shapes, a looking glass, horizontal and vertical drag cursors, and a pointer with a plus sign or question mark. You can also use `setptr` to switch between MATLAB's standard pointers or to help you create and use a custom pointer. A companion function, `getptr`, returns the `'PropertyName'`, `PropertyValue` pairs you can use to recreate the current pointer.

`rotate`

This function rotates any child object of an *axes* through a specified angle about a vector. The vector can be specified in rectangular or spherical coordinates. The vector and angle are referenced to the *axes* coordinate system.

`refresh`

This function refreshes a *figure* by setting the `'Color'` property and resetting it in the same call. This is useful when a *Figure* window gets trashed, which seems to occur most often on PCs with inexpensive graphics cards.

### clf, cla, shg

`clf` and `cla` clear the current *figure* and *axes,* respectively, by deleting any child objects. `clf reset` and `cla reset` delete the children and reset all object properties, except position and units, to their default values. `shg` brings the current *figure* forward using the `figure(gcf)` command.

### gcbo, gcbf

These functions return handles to the current callback object and callback *figure.* They query the *root* `'CallBackObject'` property and can be used within callback strings to get the handle of the object with the currently executing callback.

### dragrect, rbbox

These functions simplify the process of dragging a rectangle with the mouse within a *figure.* Both functions require that the mouse button be down when the function is called. They can be used in `'ButtonDownFcn'` properties or in M-files in conjunction with `waitforbuttonpress`. For example, `rbbox` is used by the function `zoom` to drag a rubberband box for zooming in on an area in the current axes.

### makemenu

You can create an entire menu structure consisting of menus and submenus and specify `'Label'`, `'Callback'`, and `'Tag'` property values with a single call to `makemenu` with appropriate arguments.

### umtoggle

This function toggles the `'Checked'` property of a *uimenu* object and returns `1` if the new status is `'on'` and `0` if the new status is `'off'`.

### hidegui

This function can be used as an interface to the `'HandleVisibility'` property of a *figure* or other object. It can be used to set and/or query the hidden state of the object.

### edtext

This function creates an editable text box over a *text* object, allows the user to edit the text string, and updates the string of the *text* object. You can use `edtext` as the callback string of a `'ButtonDownFcn'` property to make a *text* object editable. For example,

```
» set(findobj(gcf,'type','text'),'buttondownfcn','edtext')
» set(gcf,'windowbuttondownfcn','edtext(''hide'')')
```

permits editing of all text objects in the current *figure.*

This function has been superseded by the `'Editing'` property of all *text* objects, including *axes* labels and titles. Setting this property `'on'` for a *text* object allows it to be edited directly without the creation of a *uicontrol* editable text box. See the *Mastering MATLAB Toolbox* function `mmedit` for an example showing usage of this property.

### `ishandle`

This is a logical function that returns TRUE if the argument is a valid Handle Graphics object handle.

### `uiwait, uiresume, waitfor`

These functions are used to suspend and resume M-file execution and are normally used with dialog boxes. `uiwait` and `uiresume` are often used in callbacks. They will be covered in more detail in the next chapter.

### `btngroup, btnstate, btnpress, btndown, btnup, btnicon, iconsdisp`

These functions are used to create and manage icon buttons in a *Figure* window similar to those used by MATLAB in the *Command* window and the *Editor* window on the PC and Macintosh platforms. These functions are replaced by `togglebutton` *uicontrols* in MATLAB version 5.2.

### `setuprop, getuprop, clruprop`

These functions allow you to define, set, and query user-defined properties for *figure* and *axes* objects. What they really do is create and manage invisible *uicontrol* or *text* objects, respectively, and store the property name in the `'Tag'` property of the created objects and the property value in the objects' `'UserData'` property.

### `popupstr`

Given the handle of a popup *uicontrol,* `popupstr` returns the currently selected string rather than just the index provided in the popup's `'Value'` property.

### `textwrap`

This function wraps text strings so that they will fit in a given *uicontrol.* The input text strings are contained in a cell array where each cell contains a paragraph of text.

## 32.11 GUIDE

GUIDE is a set of interactive GUI tools supplied with MATLAB that have been designed to make building your own GUIs easier and faster. The tool set includes a Control Panel for creating, placing, and resizing *uicontrol* objects and launching the other tools. A Property Editor is included that lists object properties and lets you modify these properties interactively. The CallBack Editor permits editing callback strings without the hassle of quoting everything. The Alignment Tool lets you align selected objects horizontally and vertically. The Menu Editor permits interactive editing and rearranging user-defined pull-downs menus.

When called from the command line, `guide` seizes control of the current *figure* or opens a new *figure* if none exists, creates **File, Edit, Options,** and **Tools** menus on the controlled *figure,* and opens the Control Panel. A typical session consists of placing, moving, and resizing *axes* and *uicontrol* objects on the controlled *figure* and using the Alignment Tool to align them to the *figure* and each other. If user-defined menus are desired, the Menu Editor can be used to create and arrange them. The Property Editor can then be used to add text strings and labels and to modify other properties of the *axes, uimenu,* and *uicontrol* objects. The Callback Editor can then be used to add callbacks to the *uimenu* and *uicontrol* objects.

When done, the *figure* can be saved using the **Save As . . .** menu item in the **File** menu of the *Figure* window, or by using the `print -mfile` *`filename`* command in the *Command* window. This creates *`filename`*`.m` and *`filename`*`.mat` files that are used to recreate the *figure.* To finish or polish the GUI you can edit the *`filename`*`.m` file.

Each of the tools can be launched individually from the command line, as shown in the following table.

| Command | Tool | Function |
|---|---|---|
| `guide` | Control Panel | Object layout editor. |
| `align` | Alignment Tool | Align objects in the figure. |
| `menuedit` | Menu Editor | User-defined menu layout editor. |
| `propedit` | Property Editor | Interactive object property editor. |
| `cbedit` | Callback Editor | Interactive callback editor. |

Each of these tools can be useful for specific tasks. You can quickly prototype a GUI and create a shell function with `guide` and `align`. This bypasses the time-consuming process of calculating position vectors for your GUI objects. The `propedit` and `cbedit` tools are useful for testing different property values and callback strings interactively. The authors of this text find GUIDE to be rather limited and do not recommend that you try to use it to create that killer GUI from beginning to end, but the tools themselves can be useful aids for creating your own GUIs.

## 32.12 USER-CONTRIBUTED GUI M-FILES

A number of MATLAB users have taken advantage of the GUI tools available in MATLAB and have written some interesting GUI functions. Most of them are available to the user community on the MATLAB anonymous FTP site. M-file functions written for MATLAB version 4 are located in the `/pub/contrib/v4/graphics` directory. Note that these functions may not work with MATLAB version 5. Function M-files that have been tested with MATLAB version 5 are located in the `/pub/contrib/v5/graphics` directory. The Internet Resources chapter (Chapter 35) contains detailed instructions for obtaining files from this repository.

The files in these directories are just a few of the many examples of GUI functions available. Check them out. Maybe you have an idea for a GUI application that others would like to use. Chapter 35 contains detailed instructions to help you access existing MATLAB resources and even become an M-file contributor.

## 32.13 SUMMARY

Graphical user interface design is not for everyone. However, if you have a need for a graphical interface, MATLAB makes it possible to create impressive GUI functions from M-files alone. There is no need to use MEX files, high-level languages, or library calls to create a pleasing interface for a useful MATLAB function.

# 33

# Dialog Boxes and Requesters

MATLAB contains a number of useful GUI tools for creating *dialog boxes* and *requesters.* Dialog boxes are individual windows that pop up on the screen and display a message string. Dialog boxes contain one or more push buttons for specific user responses to the message. Requesters are individual windows that pop up on the display, obtain user input using the mouse and/or keyboard, and return information to the calling function.

## 33.1 DIALOG BOXES

This discussion begins with a description of the low-level tools for creating dialog boxes and proceeds to higher-level functions used to easily create some commonly used dialog boxes. If you are eager to use dialog boxes, you can skip ahead to the built-in dialog box functions.

## Low-level Functions

Dialog boxes in MATLAB are not unique Handle Graphics objects. They are simply *figures* with certain properties set appropriately for dialog boxes. The `dialog` function is provided as a convenient way to create basic dialog boxes. Use `dialog` in the same way you would use the `figure` command to create a new *figure*. The calling syntax is the same:

```
» Hf_dialog = dialog('PropertyName',PropertyValue,...)
```

The default properties and values set by the `dialog` function are listed in the following table.

| | |
|---|---|
| 'BackingStore' | 'off' |
| 'ButtonDownFcn' | 'if isempty(allchild(gcbf)), close(gcbf), end' |
| 'Colormap' | [] |
| 'Color' | The root 'DefaultUicontrolBackgroundColor' value. |
| 'HandleVisibility' | 'callback' |
| 'IntegerHandle' | 'off' |
| 'InvertHardcopy' | 'off' |
| 'MenuBar' | 'figure' for the Macintosh, 'none' for the others. |
| 'NumberTitle' | 'off' |
| 'PaperPositionMode' | 'auto' |
| 'Resize' | 'off' |
| 'Visible' | 'on' |
| 'WindowStyle' | 'modal' |

The `'ButtondownFcn'` callback is a failsafe device. If for some reason a blank dialog box is created, a mouse click in the dialog box will close it. The `'HandleVisibility'` and `'IntegerHandle'` properties are set to protect the dialog box and any children from the command line and from other functions. The `'Color'`, `'MenuBar'`, and `'NumberTitle'` properties affect the appearance of the dialog box, while the `'BackingStore'`, `'Colormap'`, `'InvertHardcopy'`, and `'PaperPositionMode'` properties are set to minimize system resources. The `'Resize'` property is set to `'off'` to maintain control of the size and shape of the dialog box.

The `'WindowStyle'` property deserves a closer look. Possible values are `'normal'` or `'modal'`, where the latter is the default. When `'WindowStyle'` is set to `'modal'`, the modal *figure* is stacked above the *Command* window and all normal *figures* and traps all mouse and keyboard events over **all** MATLAB windows until 'WindowStyle' is set to 'normal', the 'Visible' property is set to 'off', or the *figure* is deleted. Windows of other applications are unaffected.

Modal *figures* are normally used to create dialog boxes and requesters that force the user to respond to the dialog box before interacting with other windows. Typing **Control-C** on the keyboard causes all modal figures to change their 'WindowStyle' property back to 'normal' to allow input from the *Command* window. This is a **very** useful feature for debugging purposes.

Modal dialog boxes that are also invisible do not behave modally. If you are likely to reuse a dialog box, you may want to set 'Visible' to 'off' rather than closing the *figure* when the dialog box is dismissed so that it can be reused without being re-created. The MATLAB *Help* window created by the helpwin command uses this technique.

Computer platform built-in menus, and *unimenu* objects in a *figure* are only visible if the 'WindowStyle' is 'normal'. Modal *figures* do not display menus although they can still exist. When the 'WindowStyle' is reset to 'normal', any existing *uimenus* reappear.

The msgbox function is the low-level generic dialog box creation function. It uses dialog to create the *figure* and uses a standard layout for the dialog box. Icon data is included in the msgbox M-file for creating simple icons, although you can create a custom icon if desired. msgbox is used by the higher-level dialog box functions discussed later to create standard dialog boxes.

The simplest calling syntax is msgbox('MessageString'). This creates a small nonmodal dialog box with no title or icon containing the text string MessageString and a pushbutton labeled 'OK'. Clicking on the button dismisses the dialog box. Long message strings are text wrapped as required.

The form msgbox('String','Title') adds a window title to the message box, and msgbox('String,''Title','Icon') adds an icon as well. The icon string can be one of 'help', 'warn', 'error' or 'none', the default. The form msgbox('String','Title','custom',Idata,Imap) uses a custom icon. Idata is a square numeric data matrix and Imap is a color data matrix. For example,

```
» Idata = 1:64;
» Idata = (Idata'*Idata)/64;
» Hf_1 = msgbox('MyString','BoxTitle','custom',Idata,hsv(16))
» Hf_2 = msgbox('MyString','BoxTitle','custom',Idata/2,gray(2))
```

A final argument msgbox(...,'modal') can be used to specify a modal dialog box for each of the forms shown.

### High-level Functions

MATLAB supplies some predefined dialog boxes that are simple to use. The simplest are the helpdlg and warndlg functions. The command

```
» Hd_help = helpdlg('String','DBName')
```

creates a nonmodal dialog box with handle `Hd_help` containing a black and white help icon, the text string `'String'`, window title `'DBName'` and a push button labeled `'OK'`. Both arguments are optional, and default strings will be used if they are omitted, as shown in the following figure. The `'String'` argument may be a text string, string matrix, or cell array. The `'DBName'` argument specifies the title of the dialog box. If a dialog box already exists with the same title, it will be replaced.

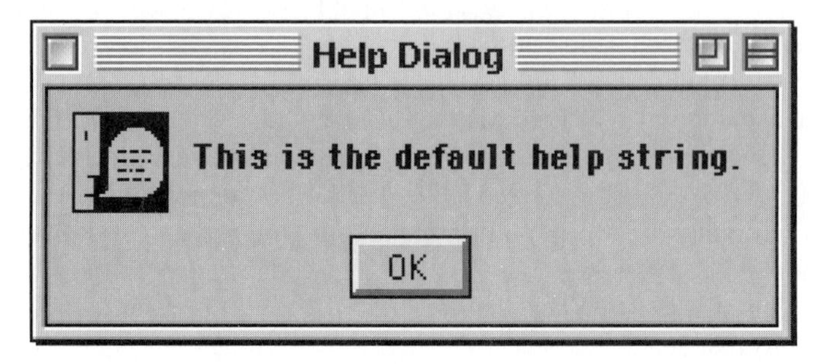


The `warndlg` function is exactly the same as the `helpdlg` function except for the icon design and the default text strings.

The `errordlg` function by default does **not** replace any existing dialog box of the same name. A third optional argument `'replace'` overrides the default.

```
» Hd_error = errordlg('String','DBName','replace')
```

The default `errordlg` dialog box is shown in the following figure.

The `questdlg('Question')` function creates a **modal** dialog box with up to three pushbuttons. The most general form is

```
» replystring = questdlg('Question','DBName','But1','But2','But3','But2')
```

where `replystring` is the label string from the selected push button, `'Question'` is the question text as a string, string array, or cell array, and `'DBName'` is the dialog box title string. The next one to three arguments are optional label strings for one to three push buttons. The final argument is the *default* string. This is the string that is returned by `questdlg` if the **Return** key is pressed while the mouse pointer is in the dialog box. Only the `'Question'` argument is required, in which case the arguments default to

```
» replystring = questdlg('Question', '', 'Yes', 'No', 'Cancel', 'Yes')
```

The following function fragment illustrates the use of a question dialog box from within a function:

```
qstring = 'Change colormap to copper?';
reply = questdlg(qstring,'Select Colormap','Sure','Nope','Maybe','Nope');
if strcmp(reply,'Sure')
    colormap(copper);
elseif strcmp(reply,'Maybe')
    reply = questdlg({'Please make up your mind!', qstring},...
                      'Select Colormap','Yes','No','No');
    if strcmp(reply,'Yes')
        colormap(copper);
    end
end
```

## 33.2 REQUESTERS

Dialog boxes normally return a single response from a limited selection of predefined responses. Requesters are essentially enhanced dialog boxes that are used to obtain variable input or multiple inputs from the user. The simplest MATLAB requester is the GUI version of the `input` function. `inputdlg` creates a modal dialog box that requests a text string from the user in an edit *uicontrol* and returns the string it receives. Accepted formats are as follows:

```
» replystring = inputdlg('Prompt')
» replystring = inputdlg('Prompt','Title')
» replystring = inputdlg('Prompt','Title',NoLines)
» replystring = inputdlg('Prompt','Title',Nolines,'IniStr')
```

where `'Prompt'` is the prompt string, `'Title'` is the title of the requester *figure*, `NoLines` specifies the expected number of response lines, and `'IniStr'` specifies the initial contents of the edit box and must be a cell string or cell array of strings. Each argument except `'Title'` may be an array if multiple inputs are desired. For example,

```
» prompstr ={'Enter the colormap name:',...
             'Enter the colormap size:'};
» inistr = {'hsv', '64'};
» titlestr = 'Colormap Info';
» nlines = 1;
» result = inputdlg(promptstr,titlestr,nlines,inistr);
» if ~isempty(result)
    colormap(eval([result{1},'(',result{2},')']))
  end
```

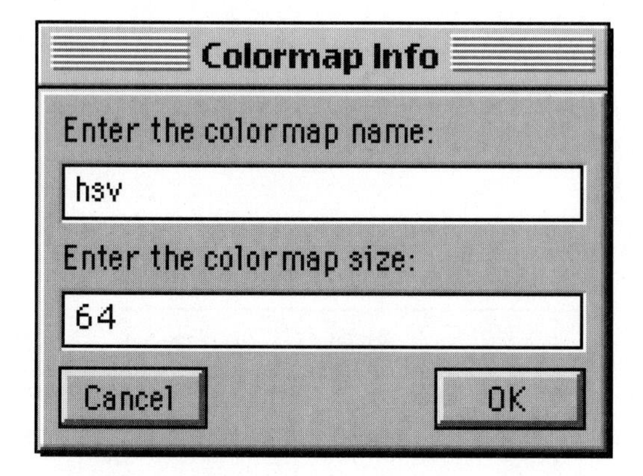


The input requester contains two push buttons labeled `'OK'` and `'Cancel'`. If the `'OK'` button is selected, `result` contains a cell array of response strings. If `'Cancel'` is selected, `result` contains an empty cell. Note that the user input is not limited to the number of lines specified in `nlines` if `nlines > 1`. This parameter simply determines the `'Max'` property of the edit *uicontrol* and the initial height of the input edit box.

Another simple requester is the `menu` function that puts up a dialog box with a variable number of buttons and returns the index of the user's selection. The format of the `menu` function is

```
» selection = menu('Prompt','But1','But2',...)
```

where `'Prompt'` is a text string displayed at the top of the requester, and `'But1'`, `'But2'`, etc. are push button *uicontrol* label strings. `menu` returns the index of the selected button in `selection`. The form `menu('Prompt',ButstrCell)` can also be used, where `ButstrCell` is a cell array of button label strings.

`listdlg` is a utility function included in MATLAB that puts up a modal dialog box consisting of a listbox, an optional prompt string and *figure* title, and two push buttons labeled `'OK'` and `'Cancel'` by default. If multiple selections are allowed, a `'Select all'` pushbutton is available as well. The syntax is

```
» [selection,ok] = listdlg('PropertyName',PropertyValue,...)
```

where the only required property is `'ListString'`, a cell array of strings for the list box. `listdlg` returns the indices of selected strings in `selection` and `1` in `ok` when the `'OK'` button is selected. If the `'Cancel'` button is selected, `selection` contains the empty matrix and `ok` contains `0`. Valid properties for `listdlg` are listed in the following table.

| Parameter | Description |
|---|---|
| `'ListString'` | Cell array of strings for the listbox. |
| `'SelectionMode'` | `'single'` or `'multiple'`; defaults to `'multiple'`. |
| `'ListSize'` | `[width height]` of listbox in pixels; defaults to `[160 300]`. |
| `'InitialValue'` | Vector of indices of the items in the listbox that are initially selected; defaults to `1`. |
| `'Name'` | String for the *figure's* title. Defaults to an empty string. |
| `'PromptString'` | Prompt string displayed above the listbox; defaults to `{}`. |
| `'OKString'` | String for the `OK` button; defaults to `'OK'`. |
| `CancelString'` | String for the `Cancel` button; defaults to `'Cancel'`. |
| `'uh'` | *uicontrol* button height in pixels; default = `18`. |
| `'fus'` | *frame/uicontrol* spacing in pixels; default = `8`. |
| `'ffs'` | *frame/figure* spacing, in pixels; default = `8`. |

The following example uses a `listdlg` dialog box to select a file name.

```
» d = dir;
» dstr = {d.name};
» [s,ok] = listdlg('PromptString', 'Select a file:',...
                   'SelectionMode','single',...
                   'ListString',dstr)
» if ok, disp(['You selected: ' dstr{s}]), end
```

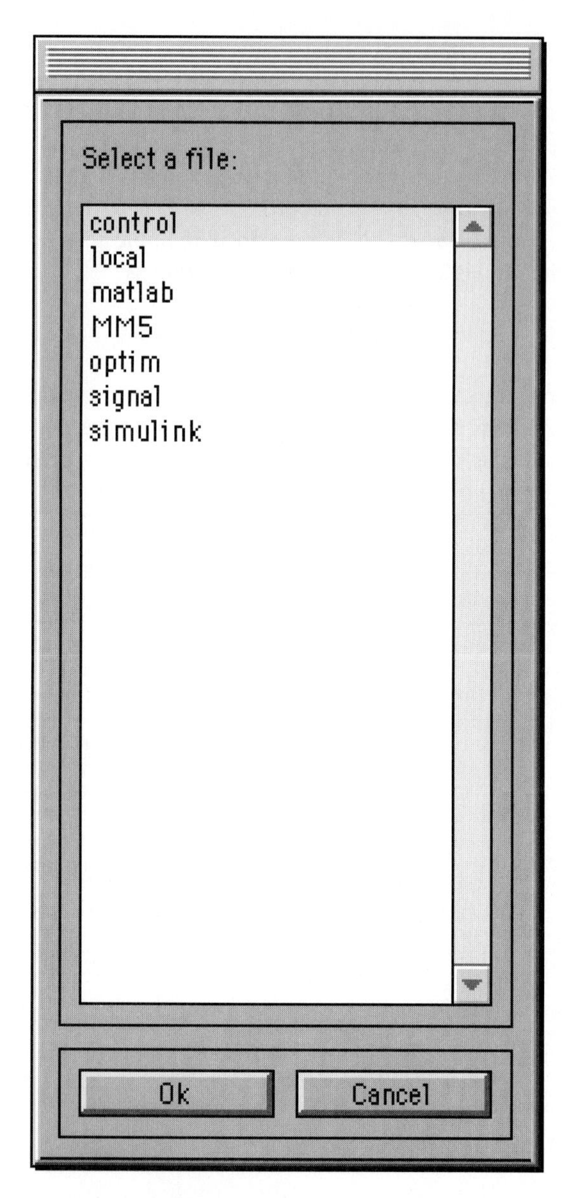


The `listdlg` function can be used to put up a requester quickly and easily to allow the user to select one or more items from a selection of strings. If the listbox is to be incorporated into a more complex requester, a listbox *uicontrol* should be used instead. See the `mmcd` function example in the previous chapter for a requester example using a listbox *uicontrol.*

MATLAB supplies two very useful built-in requesters for file management. `uigetfile` and `uiputfile` can be used to obtain file names and file path names from the user. The format of `uigetfile` is

```
» [filename, pathname] = uigetfile('Filter','Title',x,y)
```

where *filename* is a string containing the name of the selected file, *pathname* is a string containing the path to the selected file (the directory containing the file), 'Filter' is an optional string containing wildcards that is used to select matching file names (the default 'Filter' is '*.m'), and 'Title' is an optional string used as the title of the requester. The initial position of the requester is specified in the optional x and y arguments as the number of pixels to the right and down from the upper left corner of the screen. The position arguments are ignored on PC platforms and default to the upper center of the screen on Macintosh platforms if the requester would otherwise be partially off screen.

The following example shows how uigetfile can be used within a function to retrieve an ASCII data file interactively and plot the sine of the data:

```
% Ask the user for a file name.
[datafile,datapath] = uigetfile('*.dat','Choose a data file');
% If the user selected an existing file, read the data.
% (The extra quotes avoid problems with spaces in file or path names.)
% Then determine the variable name from the file name,
% copy the data to a variable, and plot the data.
if datafile
    eval(['load(''' datapath datafile ''')']);
    x = eval(strtok(datafile,'.'));
    plot(x,sin(x));
end
```

The requester will not accept the name of a file that does not exist. The only way this situation can occur is if the user types a file name into the text box in the requester. The Macintosh version, however, has no text box. The user must choose an existing file or press the **Cancel** button to exit the requester. If the **Cancel** button is pressed on any platform, the output arguments contain zero.

The uiputfile function is very similar to the uigetfile function. The arguments are similar, and they both return file names and path strings as shown follows:

```
» [filename, pathname] = uiputfile('Filter','Title',x,y)
```

However, the 'Filter' argument for uigetfile can be a wildcard specification, as shown previously, or it can be a default file name. For example, if '*.m' is used, there is no default filename, and all files with a '.m' extension are listed. If 'myfile.m' is used, the default file name is 'myfile.m' and all files with an '.m' extension are listed.

If the user selects a file that already exists, a dialog box appears asking if the user wants to delete the existing file. A **No** response returns to the original requester for another try. A **Yes** response closes the requester and dialog box and returns the file name to the calling function. The file is not deleted by the requester. The calling function must delete or overwrite the file if necessary.

> It is important to remember that neither of these functions actually reads or writes any files. The calling function must do that. The `uigetfile` and `uiputfile` functions simply return a file name and path to the calling function.

Two other excellent examples of MATLAB requesters are the familiar `printdlg` and `pagedlg` functions discussed in the Printing and Exporting Graphics chapter (Chapter 30).

MATLAB supplies two additional functions that can be used interactively to select a color and a font. `uisetcolor` lets the user choose a color value interactively and optionally apply that color to a graphics object. MATLAB version 5.0, X-Window platforms do not support `uisetcolor`. However, MATLAB version 5.1 does support `uisetcolor`.

Two forms are acceptable:

```
» RGB_out = uisetcolor([r g b],'Title')
```

uses the color specification `[r g b]` as the initial color in the requester and the optional `'Title'` as the title of the requester. `RGB_out` contains the RGB triple of the selected color. The form

```
» RGB_out = uisetcolor(Hx_in, 'Title')
```

uses the `'Color'` or `'ForegroundColor'` property of the graphics object with handle `Hx_in` as the initial color in the requester and applies the selected color to the object if the user selects the **OK** button. If the **Cancel** button is selected, the color is not applied and `RGB_out` contains the initial color in the requester. If no input arguments are supplied, the initial color is black and the selected color is returned.

The *Mastering MATLAB Toolbox* includes a function called `mmscolor` that works on all platforms and provides an interface into `uisetcolor`. `mmscolor select` lets the user select an object with a mouse click and applies the chosen color to the selected object.

The font-selection requester `uisetfont` lets the user interactively choose font attributes and apply them to an object. The syntax is

```
» Hx_out = uisetfont(Hx_in,'Title')
```

where `Hx_in` is the handle of an *axes, text,* or *uicontrol* graphics object, `'Title'` is the title of the requester, and `Hx_out` is equal to `Hx_in`. The selected font properties are applied to the object with handle `Hx_in`. If `Hx_in` is omitted, a new text object is created with the selected font attributes and the handle is returned in `Hx_out`.

The sample text string displayed in the `uisetfont` requester shows the effect of each attribute change. If the text string does not appear as you had expected after making a selection, one or more of the font attributes you have chosen (font name, size, etc.) is not available on your computer. What you see is what you get.

The *Mastering MATLAB Toolbox* includes a function called `mmsfont` that provides a front end to `uisetfont`. `mmsfont select` lets the user select an appropriate graphics object with a mouse click and applies the chosen font attributes to the selected object. *Axes, uicontrols,* and all *text* objects may be selected including *axes* labels and titles.

## 33.3 UTILITY FUNCTIONS

MATLAB supplies several utility functions that can be useful tools in requesters. See on-line help for more information about calling syntax and options for any of these functions.

### waitbar

`waitbar` is a handy little GUI function you can use to supply feedback to the user when some lengthy operation is in process.

```
» Hf_wait = waitbar(start, 'String')
```

creates a small *figure* containing the text string `'String'` over a horizontal progress meter bar. `start` is a number between 0 and 1 specifying the starting position of the progress meter. Subsequent calls to `waitbar(val)` update the progress meter to the position specified by `val`. The `waitbar` function is normally used inside a calculation loop that may take a relatively long time to complete. For example,

```
Hf_wait = waitbar(0,'Please wait. Calculating inverses...')
for k=1:bignum
  B(k)=inv(A(k));
  waitbar(k/bignum);
end
close(Hf_wait);
```

### axlimdlg

The `axlimdlg` function uses `inputdlg` to build an *axis* limit requester. Properties obtained from the user are applied to one or more *axes*. `axlimdlg` uses property values as well to control many of the options available. Simply invoking `axlimdlg` without arguments from the command line illustrates the basic functionality of this requester. Many different options are available.

### tabdlg

`tabdlg` is a special utility function for building a *tabbed* dialog box similar to the tabs on file folders or a box of index cards. The MATLAB preferences dialog box on PC and

Macintosh platforms is an example of a tabbed dialog box. This function can be useful for building specialty dialog boxes but is limited to managing the tabs themselves. `tabdlg` is a rather complex function with many arguments and options.

## 33.4 MAINTAINING FOCUS

Visible modal dialog boxes (i.e., *figures* with the `'WindowStyle'` property set to `'modal'` and the `'Visible'` property set to `'on'`) intercept **all** keyboard and mouse input over **any** MATLAB window. Modal dialog boxes and requestors are appropriate whenever you want to prevent the user from issuing any command or performing any operation in MATLAB before responding to your dialog box or requester.

The built-in `waitfor` function can be used in a function M-file that creates a modal dialog box to block M-file execution and wait for a specific event to occur. The `mmcd` function example in the previous chapter illustrates one use of the `waitfor` function. There are three different forms, as shown next.

```
» waitfor(Hx)
```

returns when the graphics object with handle `Hx` is closed or deleted.

```
» waitfor(Hx,'PropertyName')
```

returns if the object `Hx` is closed or deleted or when the value of the property `'PropertyName'` changes.

```
» waitfor(Hx,'PropertyName',PropertyValue)
```

returns if the object `Hx` is closed or deleted or when `'PropertyName'` is set to `PropertyValue`.

Although `waitfor` blocks execution, it does not block callbacks. Any pending callback events are processed as they occur. *Uicontrol* `'Callback'` or object `'ButtonDownFcn'` callbacks can be used to satisfy the conditions `waitfor` is blocking on and resume execution.

The `uiwait` and `uiresume` functions use `waitfor` and the undocumented *figure* property `'WaitStatus'` as a convenient way to use `waitfor` to block all execution until the user responds to a dialog box. `uiwait(Hf)` blocks execution until the `uiresume(Hf)` function is called or the *figure* with handle `Hf` is closed or deleted. In normal usage, a modal dialog box is created with a *uicontrol* that has a callback that calls `uiresume` or closes the *figure*. The `uiwait` function is then called to block execution. When the *uicontrol* callback is executed, execution can resume.

# 34

# Help

You probably have the sense that MATLAB has many more commands than you could possibly remember. To help you find commands, MATLAB provides assistance through its very extensive *on-line help* capabilities. These capabilities include MATLAB commands to get quick help in the *Command* window, a separate mouse-driven *Help* window, and a browser-based help system. More information, answers to frequently asked questions, *The MathWorks* technical support, and user-contributed utilities are accessible over the Internet if a network connection is available. See the next chapter for details on these Internet resources.

## 34.1 COMMAND WINDOW HELP

MATLAB offers a number of commands to help you get quick information about a MATLAB command or function within the *Command* window, including `help`, `lookfor`, `whatsnew`, and `info`.

## The `help` Command

The MATLAB `help` command is the simplest way to get help if you know the topic you want help on. Typing `help` *`topic`* displays help about that topic if it exists. For example,

```
» help sqrt
 SQRT    Square root.
     SQRT(X) is the square root of the elements of X. Complex
     results are produced if X is not positive.

     See also SQRTM.
```

Here we received help on MATLAB's square root function. On the other hand,

```
» help cows
cows.m not found.
```

simply says that MATLAB knows nothing about `cows`.

Note in the preceding `sqrt` example that SQRT is capitalized in the help text. However, when used, `sqrt` is never capitalized. In fact, because MATLAB is case sensitive, SQRT is unknown and produces an error:

```
» SQRT(2)
??? Undefined variable or capitalized internal function SQRT;
    Caps Lock may be on.
```

**To summarize, function names are capitalized in help text solely to aid readability, but in use, functions are called using lowercase characters.**

The `help` command works well if you know the exact topic you want help on. Since many times this is not true, `help` provides guidance to direct you to the exact topic you want by simply typing `help` with no topic:

```
» help
HELP topics:
matlab:general       - General purpose commands.
matlab:ops           - Operators and special characters.
matlab:lang          - Programming language constructs.
matlab:elmat         - Elementary matrices and matrix manipulation.
matlab:elfun         - Elementary math functions.
matlab:specfun       - Specialized math functions.
matlab:matfun        - Matrix functions - numerical linear algebra.
matlab:datafun       - Data analysis and Fourier transforms.
matlab:polyfun       - Interpolation and polynomials.
matlab:funfun        - Function functions and ODE solvers.
matlab:sparfun       - Sparse matrices.
```

```
matlab:graph2d        - Two dimensional graphs.
matlab:graph3d        - Three dimensional graphs.
matlab:specgraph      - Specialized graphs.
matlab:graphics       - Handle Graphics.
matlab:uitools        - Graphical user interface tools.
matlab:strfun         - Character strings.
matlab:iofun          - File input/output.
matlab:timefun        - Time and dates.
matlab:datatypes      - Data types and structures
matlab:MacOS          - Macintosh specific functions.
matlab:demos          - Examples and demonstrations.
Toolbox:control       - Control System Toolbox.
Toolbox:local         - Preferences.
Toolbox:mm5           - Mastering MATLAB Toolbox.
Toolbox:signal        - Signal Processing Toolbox.
For more help on directory/topic, type "help topic".
```

Your display may differ slightly from the preceding. In any case, this display describes primary topics or categories from which you can ask for additional help. Each of these primary topics corresponds to a directory name on the MATLAB Search Path. For example, `help general` returns a list (too long to show here) of general MATLAB topics that you can use the `help` *`topic`* command to get help on.

The form `s=help('topic')` returns the help text for `topic` in `s`. The help text is in the form of a `1`-by-`n` character string containing line feed characters (`'\n'`) as appropriate to wrap the text when displayed. This property is used in the `helpwin` command discussed later.

While the `help` command allows you to access help quickly, it is not the most convenient way to do so unless you know the exact topic you are seeking help on. When you are uncertain about the spelling or existence of a topic, other approaches to obtaining help are often more productive.

### The `lookfor` Command

The `lookfor` command provides help by searching through the first lines (the H1 lines) of all MATLAB help topics and M-files on the MATLAB Search path and returning a list of those that contain the key word you specify. *Most important is that the key word need not be a MATLAB command and case does not matter.* For example,

```
» lookfor shift
BITSHIFT Bit-wise shift.
FFTSHIFT Shift DC component to center of spectrum.
SHIFTDIM Shift dimensions.
MMPSHIFT Shift Polynomial, (A(x) -> A(x+b).
MMSHIFTD Shift or Circularly Shift Matrix Rows.
MMSHIFTR Shift or Circularly Shift Matrix Columns.
```

The key word `shift` is not a MATLAB command, but was found in the help descriptions of three MATLAB commands and three *Mastering MATLAB Toolbox* commands. You may find more or less. Given this information, the `help` command can be used to display help about a specific command. For example,

```
» help mmpshift
 MMPSHIFT Shift Polynomial, (A(x) -> A(x+b).
  MMPSHIFT(A,b) shifts the Polynomial A(x) by b giving A(x+b).

  See also MMPADD, MMPSIM, MMPSCALE, MMPOLY, MMP2STR, POLY, CONV.
```

Adding the qualifier `-all` to the `lookfor` command (e.g., `lookfor shift -all`) searches the entire help text of all M-files, not just the first line, and lists the functions containing keyword matches. The search takes a little longer, but returns many more matches. In summary, the `lookfor` command provides a way to find MATLAB commands and help topics given a general key word.

## The `whatsnew` and `info` Commands

As the names suggest, the `whatsnew` and `info` commands display information about changes and improvements to MATLAB and the toolboxes. If used without an argument, `info` displays general information about MATLAB, how to contact *The MathWorks,* how to become a subscribing user, and how to join the MATLAB User's Group. On X-Window platforms, the `whatsnew` command without arguments describes the location on your computer of Postcript versions of some of the MATLAB documentation. The help text for `whatsnew` is then displayed on all platforms. If either function is used with an argument (e.g., `whatsnew matlab` or `info signal`), the `Readme.m` file for the specified toolbox is displayed if it exists. This file lists the very latest information about changes and additions to the toolboxes.

## Utility Functions

MATLAB includes a number of functions for obtaining file information. In addition to the functions `dir` and `cd`, which return file listings and directory information, respectively, the `what` function lists MATLAB-related files and directories. `what` with no arguments lists M-files, MAT-files, MEX-files, P-files, MDL-files, and object classes (i.e., directories beginning with the `'@'` symbol) in the current directory. `what('`*`path`*`')`, where `'`*`path`*`'` is a MATLAB path or partial path argument, lists the MATLAB-related files in the specified directory. For example, `what uitools` lists the M-files, MEX-files, and P-files in the `toolbox/matlab/uitools` directory.

When a variable or function name (e.g., `cow`) is used on the command line or in an M-file, MATLAB uses predefined rules to search for the variable or function. MATLAB first looks for a variable by that name, then checks for a built-in MATLAB function, looks in the current directory for an M-file, and finally searches the MATLAB path for an M-file named `cow.m`. Actually, P-files, MEX-files, subfunctions, private functions, and class directories are searched as appropriate, but the simplified procedure above is followed in most cases. MATLAB evaluates the first instance of `cow` that it finds.

The `which arg` command can be used to determine the specific instance of the argument `arg` that MATLAB will evaluate. The form `which arg -all` lists all instances of the argument `arg`. For example,

```
» which char
/usr/local/matlab/toolbox/matlab/strfun/@char/char.m      % char method

» which char -all
/usr/local/matlab/toolbox/matlab/strfun/@char/char.m      % char method
char is a built-in function.
/usr/local/matlab/toolbox/matlab/strfun/char.m            % Shadowed
/usr/local/matlab/toolbox/matlab/funfun/@inline/char.m    % inline method
/usr/local/matlab/toolbox/matlab/strfun/@cell/char.m      % cell method
```

The `which` function can also be used with an M-file argument to determine which instance is called from the M-file. `which fun1 in fun2` returns a string specifying the instance of `fun1` that is called from within the `fun2.m` M-file. The form `which fun(a,b,c)` determines which instance of `fun` is used when the arguments `a`, `b`, and `c` are passed to `fun`.

The `exist` function can be used to check for the existence of a variable or function. `exist` returns `0` if the argument is not found. If an instance is found, an integer is returned specifying that the argument is a variable, a built-in MATLAB function, a P-file, an M-file, a directory, etc. An optional argument can be used to restrict the search to variables, built-in functions, files, or directories.

## 34.2 THE HELP WINDOW

A major enhancement to the MATLAB help system is the new *Help* window. The `helpwin` command opens a new window on your computer screen that can be used to view the help text for MATLAB commands in an interactive *Help* window that uses a mouse for navigation. This is in addition to the *Command* window and any *Figure* windows you have open. If `helpwin` is used without arguments, the *Help* window initially contains the list of primary help topics as shown:

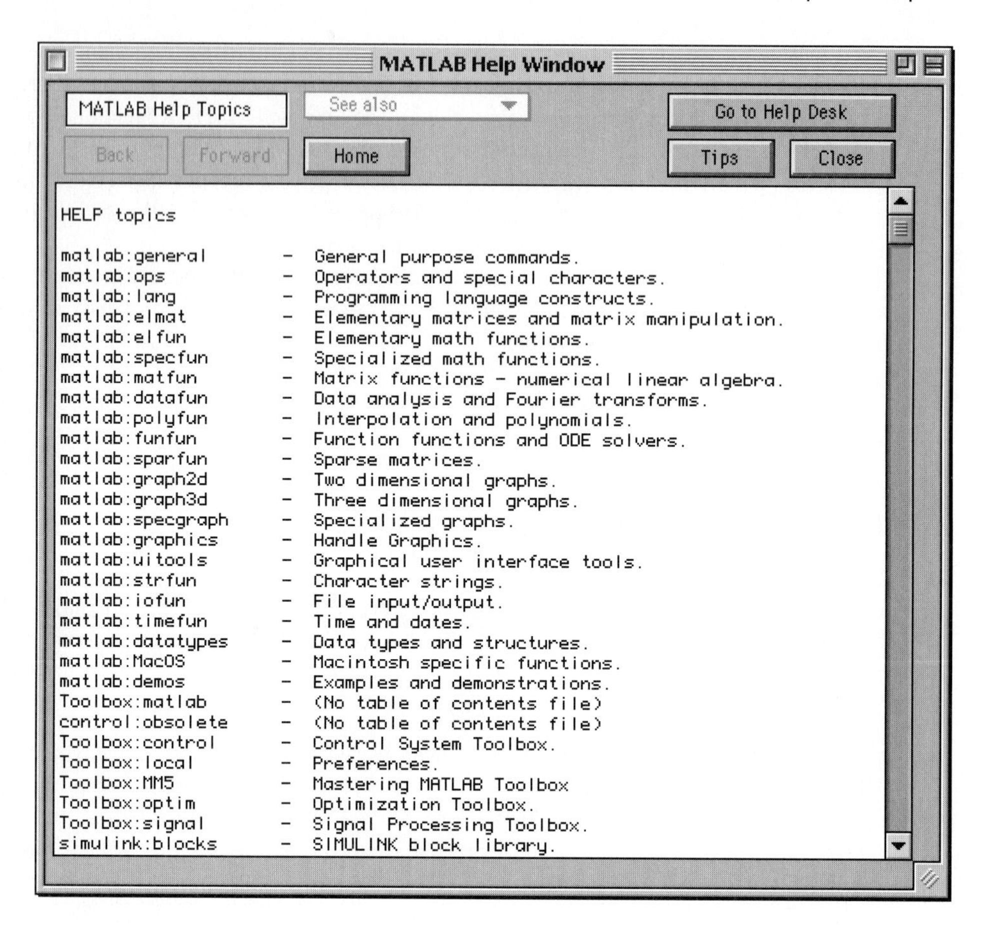
MATLAB Help Window
MATLAB Help Topics
See also
Go to Help Desk
Back
Forward
Home
Tips
Close
HELP topics
matlab:general - General purpose commands.
matlab:ops - Operators and special characters.
matlab:lang - Programming language constructs.
matlab:elmat - Elementary matrices and matrix manipulation.
matlab:elfun - Elementary math functions.
matlab:specfun - Specialized math functions.
matlab:matfun - Matrix functions - numerical linear algebra.
matlab:datafun - Data analysis and Fourier transforms.
matlab:polyfun - Interpolation and polynomials.
matlab:funfun - Function functions and ODE solvers.
matlab:sparfun - Sparse matrices.
matlab:graph2d - Two dimensional graphs.
matlab:graph3d - Three dimensional graphs.
matlab:specgraph - Specialized graphs.
matlab:graphics - Handle Graphics.
matlab:uitools - Graphical user interface tools.
matlab:strfun - Character strings.
matlab:iofun - File input/output.
matlab:timefun - Time and dates.
matlab:datatypes - Data types and structures.
matlab:MacOS - Macintosh specific functions.
matlab:demos - Examples and demonstrations.
Toolbox:matlab - (No table of contents file)
control:obsolete - (No table of contents file)
Toolbox:control - Control System Toolbox.
Toolbox:local - Preferences.
Toolbox:MM5 - Mastering MATLAB Toolbox
Toolbox:optim - Optimization Toolbox.
Toolbox:signal - Signal Processing Toolbox.
simulink:blocks - SIMULINK block library.

Double click on any of the help topics displayed in the central text window, as shown, to list the associated functions or subtopics.

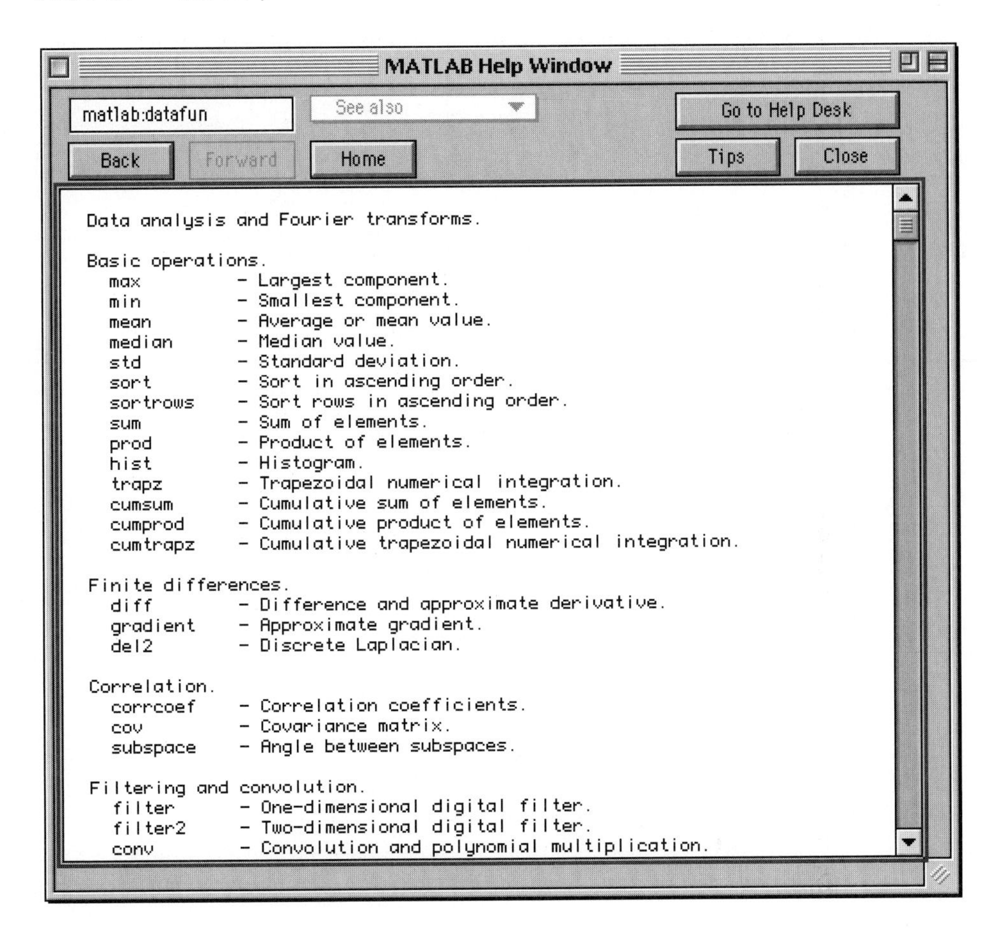


Double click on any function to visit the help text for that function.

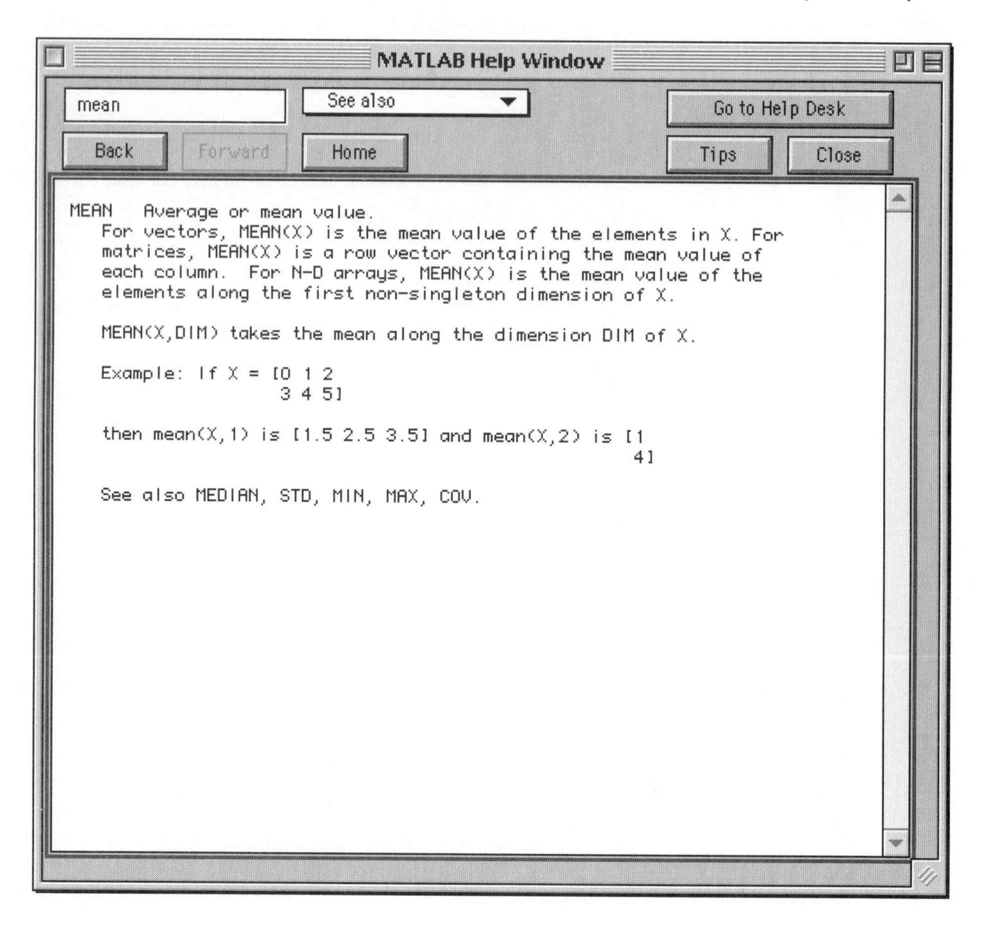


If invoked with an argument (e.g., `helpwin plot`), the *Help* window will appear with the help text for the argument (the `plot` command in this case) in the central text window.

Scroll bars at the right and bottom of the central text window are available to move through the text. The **Back** and **Forward** buttons help you navigate between the help screens you have visited, and the **Home** button returns to the primary help topics, as shown previously. The **Tips** button on the right of the *Help* window displays general information about `help` and `helpwin`, as well as the `helpdesk` command to be discussed later. The **Close** button makes the *Help* window disappear.

The **See also** popup lists related functions; they can be visited by selecting a function name from the **See also** popup menu. You can also enter a command or topic in the text box at the top left of the *Help* window to jump to that command or topic.

The *Help* window can also be accessed from the *Help* menu on the *Command* and *Figure* windows on the PC version. You can also click in the ? button icon on the *Command* window toolbar on the PC or Macintosh platforms to open the *Help* window.

## 34.3 THE HELP DESK

Your existing Web browser can also be used to display a wide range of help and reference information. All of this information is stored on disk or CD-ROM on your local system, so a network connection is optional. Many of the documents use HyperText Markup Language (HTML) to take advantage of graphics and the hot links used in HTML documents for navigation between pages. All of MATLAB's operators and functions have on-line reference pages in HTML format and many provide more details and examples than the basic help entries. The reference index can be searched for a key word as well.

Browser-based help can be accessed by using the `helpdesk` or `doc` commands from the *Command* window or by selecting the **Help Desk (HTML)** option from the PC *Command* or *Figure* window **Help** menu, as described previously. The **Go To Help Desk** button on the *Help* window and the **See also** popup item beginning with the word **More** [e.g., **More helpwin help (HTML)**] also launch your favorite Web browser. On X-Window systems the menu item **MATLAB Help . . .** under the **Help** menu on the *Figure* window works as well. You can specify the Web browser to use in MATLAB preferences.

Your browser will open on a page similar to the following:

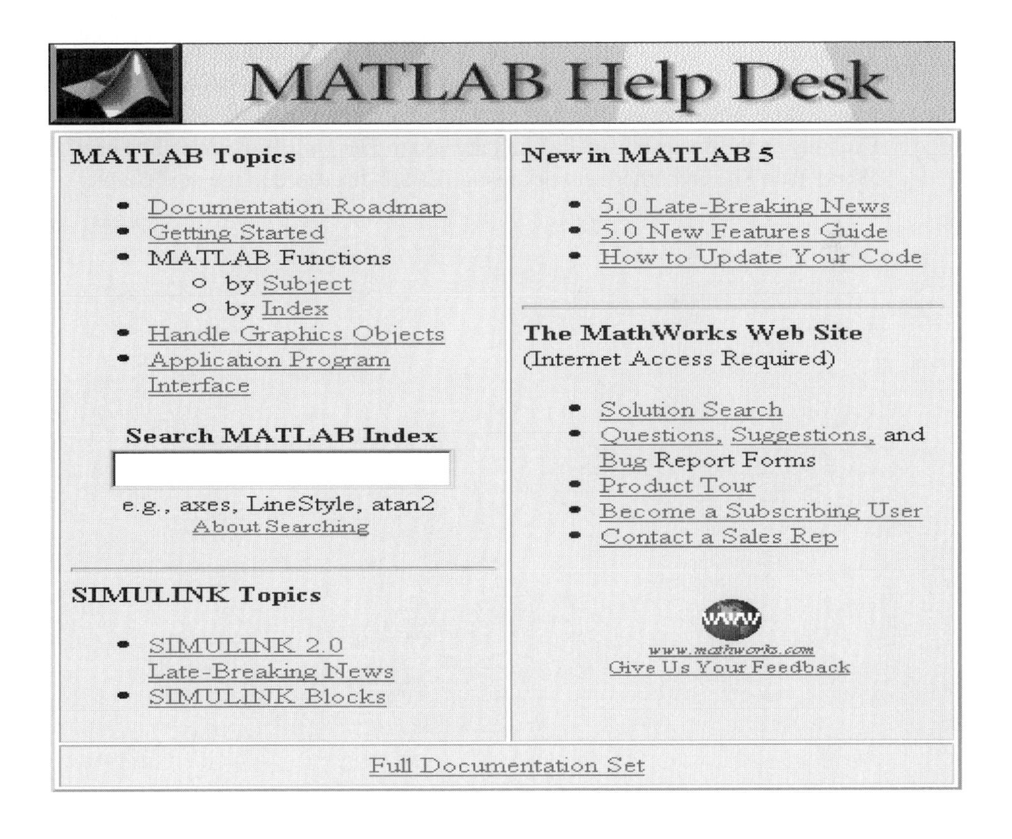

All of the highlighted links (underlined text) can be used to access more information on various topics. One section of the page contains links to tutorials, a function index, help for new users, and information about Handle Graphics objects. A search engine running on your own computer can query an index of the on-line reference material. Simply enter the desired key word in the text box and press the **Return** or **Enter** key. Other sections contains links to information on new features, release notes, and help with updating your MATLAB code or M-files to work with version 5.

A link to the **Full Documentation Set** of on-line reference pages in the Portable Document Format (PDF) used by Adobe's Acrobat reader is also available. These pages reproduce the look and feel of the printed page, complete with fonts, graphics, formatting, and images. This is the best way to get printed copies of selected portions of the reference material.

Please note that although Web browsers and PDF viewers can present useful information in familiar formats, they are large programs and use significant amounts of computer system resources. If you are running MATLAB on a computer with limited memory or a slow processor, starting a Web browser or PDF viewer while MATLAB is running can result in poor response time. The `help` and `lookfor` commands and the *Help* window are recommend in these situations.

If your computer is connected to the Internet, the Help Desk provides a connection to *The MathWorks* Web Site, the Internet home of MATLAB. You can use electronic mail to ask questions, make suggestions, and report possible bugs. You can get product information or register to become a subscribing user. You can also use the Solution Search Engine at *The MathWorks* Web Site to query an up-to-date data base of technical support information. Clicking on the image of a globe will take you to *The MathWorks* Web Site itself. *The MathWorks* Web Site and other network resources are discussed in the next chapter.

# 35

# Internet Resources

There are a wealth of MATLAB resources available for those who have access to the Internet. These resources are accessible through direct Internet connections from educational institutions, government offices, and commercial firms, as well as through commercial Internet access providers and on-line services such as America Online, Delphi, CompuServe, and others.

## 35.1 MATHWORKS WEB SITE

The easiest and most flexible way to get information, images, and files over the internet is called the World Wide Web, WWW, or simply the Web. The Web is a collection of server sites that deliver hypertext documents to your local computer upon request. Hypertext documents can incorporate imbedded graphics and links to other documents. These links can be activated simply by clicking on the highlighted text or graphic with the mouse. The linked document is then retrieved for viewing and can be saved to your local computer, if desired. Some of the links can connect to Gopher sites (text-based information servers) and to FTP sites around the world.

If you have access to a full-service Internet connection, you have the option of this kind of graphical interface to MATLAB information. *The MathWorks* operates a World Wide Web server that can be accessed with a Web browser such as Mosaic, Internet Explorer, Netscape, or the text-based Lynx. Click on the globe icon or any link in the **Help Desk** window in the section labeled **The MathWorks Web Site** or use your browser's **File** menu to select **Open Location...** or **Open URL...** and enter the string `http://www.mathworks.com` to connect to *The MathWorks* home page. The home page is the top-level hypertext document and contains links to many other documents and files on the site. Once you have connected, you can save the access information as a bookmark to save time and effort the next time you connect.

From *The MathWorks* home page, hypertext links can be used to retrieve information on *The MathWorks, Inc.* and its products, on-line copies of the quarterly newsletter and the e-mail news digest, a listing of books based on MATLAB (over 150 at this writing) along with their associated M-files, information about trade shows that the staff will be attending, and on-line copies of the latest MATLAB conference proceedings. There is an archive of Frequently Asked Questions (FAQs) and responses about MATLAB and SIMULINK. You can also read the Technical Notes section contributed by *The MathWorks* Technical Support Staff containing tips and techniques relating to a number of topics, including memory management, graphics and printing, MEX files, toolboxes, and integrating MATLAB with other software.

Anonymous FTP sites are repositories of text files and software on the Internet that allow users to connect to them and transfer files without requiring a user account and password on the host computer. *The MathWorks* has a very extensive anonymous FTP site containing thousands of files you can access from their Web site. You can search the repository, browse the directories, read the index files, and retrieve files using your Web browser. If you prefer to access *The MathWorks* FTP site directly using an FTP program such as WinFTP, Fetch, or command-line `ftp`, simply connect to `ftp.mathworks.com`, log in as `anonymous`, and use your electronic mail address (e.g., `user@my.domain.net`) as a password.

## 35.2 OTHER MATHWORKS RESOURCES

*The MathWorks* maintains a number of other services that provide information for MATLAB users.

### MATLIB Automated Electronic Mail Response System

Those without FTP capability but who have an electronic mail connection to the Internet are not left out. *The MathWorks* maintains an automated file-by-mail service that sends files from the FTP site by e-mail. Send e-mail to `matlib@mathworks.com` with the single word `help` in the body of the message to get more detailed instructions about accessing the server.

### The MathWorks MATLAB Digest

The MATLAB Digest is a monthly electronic newsletter sent to subscribers via e-mail. This newsletter contains MATLAB news, articles contributed by MATLAB developers and users, hints and tips, and responses to user questions. To subscribe to the Digest, send e-mail to `subscribe@mathworks.com` with a request to be added to the mailing list or type `subscribe` at the MATLAB prompt. The `subscribe` command asks questions and uses your answers to generate an e-mail message or a printed request form that can be mailed or faxed to *The MathWorks*. Back issues of the newsletter are available on *The MathWorks* Web Site.

### The MATLAB Newsletter

The quarterly written publication *MATLAB News & Notes* is available to anyone who uses MATLAB. If you become a subscribing user, you will receive the Digest and the quarterly newsletter, as well as free technical support. You do not have to be a registered user to become a subscribing user, and it is a free service. Use the form on *The MathWorks* Web Site, type `subscribe` at the MATLAB prompt, or send e-mail to `subscribe@mathworks.com` with your request. The newsletter usually has news, tips, articles by Cleve Moler and other *MathWorks* staff members, and a calendar of events.

## 35.3 OTHER NETWORK RESOURCES

Other Web resources include a category on the **Yahoo!** Web site devoted to MATLAB. Enter the URL (Universal Resource Locator)

```
http://www.yahoo.com/Science/Mathematics/Software/MATLAB
```

into your Web browser and check out the current listings. This page has links to *The MathWorks* Web Site as well as a number of other sites. On-line tutorials and course announcements are accessible from some of the sites listed here.

There are two sources for compilers to turn MATLAB M-files into stand-alone applications. One source is *The MathWorks* itself, which offers a compiler and a math library. Each is sold as a separate toolbox and both are required for stand-alone applications. The MATCOM compiler from *MathTools, Ltd.* is an alternative. This product includes a matrix library along with an M-file-to-C++ compiler. Evaluation copies may be downloaded from `http://www.mathtools.com`.

One of the most popular information sources and forums on the Internet is a bulletin-board system called a NetNews Feed, USENET, or simply News. This is a huge collection of messages or postings that is circulated from computer to computer around the globe. A News feed consists of thousands of individual postings collected in thousands of newsgroups or topics. Many sites make all newsgroups available, while others are more selective due to disk space limitations or company policy concerns.

If you have access to Internet News, you can read postings from other MATLAB users and post comments and questions yourself. The MATLAB newsgroup is called `comp.soft-sys.matlab`. By subscribing to this newsgroup you can browse through questions, comments, and answers from other users and from the staff at The MathWorks. In fact, MATLAB developers and authors (including the authors of this text) frequently post to this forum. If you get a limited selection of newsgroups at your site and cannot find this one, contact your local News administrator and ask that `comp.soft-sys.matlab` be included. Be aware that this newsgroup is unmoderated. While this means that you or anyone else can post your comments and questions, there is no method to filter out inappropriate postings. Thousands of people around the world read this newsgroup, so please post carefully and appropriately.

## 35.4 INTERNET e-MAIL AND NETWORK ADDRESSES

The following is a list of useful e-mail addresses and network resources.

### e-mail Addresses at The MathWorks, Inc.

```
         support@mathworks.com - Technical support.
            bugs@mathworks.com - Bug reports.
             doc@mathworks.com - Documentation error reports.
      conference@mathworks.com - MATLAB conference.
          access@mathworks.com - Subscribe to the News and Notes newsletter.
      news-notes@mathworks.com - News and Notes newsletter feedback.
         suggest@mathworks.com - Product enhancement suggestions.
         service@mathworks.com - Order status, license renewals, passcodes.
       subscribe@mathworks.com - Subscribing user information.
            info@mathworks.com - Sales, pricing, and general information.
   micro-updates@mathworks.com - PC and Macintosh updates.
          matlib@mathworks.com - File-by-mail server.
          digest@mathworks.com - MATLAB Digest submissions.
        ftpadmin@mathworks.com - The Mathworks FTP site maintainer.
      webmasters@mathworks.com - The Mathworks World Wide Web maintainer.
```

### MathWorks Network Resources

```
          www.mathworks.com - World Wide Web site.
    education.mathworks.com - MATLAB For Education Web site.
      finprod.mathworks.com - MATLAB For Finance Web site.
          ftp.mathworks.com - Anonymous FTP site.
             144.212.100.10 - WWW and FTP internet address.
             144.212.100.20 - Finance Web site address.
             144.212.100.21 - Education Web site address.
```

### Mastering MATLAB

E-mail address for questions and comments to the authors about *Mastering MATLAB 5* and the *Mastering MATLAB Toolbox.*

```
mm@eece.maine.edu
```

The *Mastering MATLAB* Web Site is:

```
http://www.eece.maine.edu/mm
```

### Other MATLAB Resources

```
http://www.yahoo.com/Science/Mathematics/Software/MATLAB
news://news.my.domain/comp.soft-sys.matlab
http://www.mathtools.com
```

### Browsers and viewers

Mosaic for PC, Macintosh, and Unix:

```
http://www.ncsa.uiuc.edu/SDG/Software/Mosaic
```

Netscape Navigator or Communicator Web browsers for PC, Mac, and Unix:

```
ftp://ftp.netscape.com/pub
```

Microsoft Internet Explorer for PC and Macintosh:

```
http://www.microsoft.com/ie/download
```

Adobe Acrobat PDF viewer for PC, Mac, and Unix:

```
http://www.adobe.com
```

Other Browsers:

```
http://www.yahoo.com/Computers_and_Internet/Software/Internet/...
      World_Wide_Web/Browsers
```

## Software Download Sites

```
http://www.yahoo.com/Computers_And_Internet/Software
http://www.download.com
http://www.browsers.com
http://www.hotfiles.com
ftp://sumex-aim.stanford.edu/pub
http://gatekeeper.dec.com
http://sunsite.unc.edu
http://ftp.wustl.edu
ftp://ftp.funet.fi
ftp://ftp.luth.se
http://www.cdrom.com/archive
http://www.simtel.net/simtel.net
```

36

# The Mastering MATLAB Toolbox

The *Mastering MATLAB Toolbox* is a collection of approximately 140 function M-files written by the authors of this text. The functions in the *Toolbox* range from commonly used utilities, to numerical analysis and optimization functions, to powerful high-level plotting routines, to polished graphical user interface functions for setting axes, line, surface, paper, and other properties. If you downloaded or otherwise received a copy of the *Mastering MATLAB Toolbox* that was associated with the first edition of this text, you have a general idea of what is available in the *Toolbox*. However, since then a lot has changed. The changes include the following:

1. The size of the *Toolbox* has nearly doubled, with many of the new functions providing significant new capabilities.
2. Many of the original functions were modified to improve functionality.
3. All original functions have been reviewed and revised to make them MATLAB version 5 compatible and to make use of new features in version 5.
4. Some function names have been changed to improve naming consistency.

The *Mastering MATLAB Toolbox* is available at:

```
http://www.eece.maine.edu/mm
```

Questions and comments regarding this text and the *Mastering MATLAB Toolbox* can be sent by e-mail to

```
mm@eece.maine.edu
```

The following tables list the functions available in the *Mastering MATLAB Toolbox* in an order that coincides with the chapter ordering of this text.

CHAPTER 2 BASIC FEATURES

| Function | Description |
|---|---|
| `mmdigit` | Round array to given significant digits. |
| `mmlimit` | Limit array values between given extremes. |
| `mmlog10` | Dissect floating-point numbers, i.e., find mantissa and exponent of array contents. |

CHAPTER 5 FILES AND DIRECTORY MANAGEMENT

| Function | Description |
|---|---|
| `mmcd` | Change working directory using a GUI. |
| `mmload` | Load matrix and header from an ASCII file using a GUI. |
| `mmsave` | Save matrix and header to an ASCII file using a GUI. |

CHAPTER 6 ARRAYS AND ARRAY OPERATIONS

| Function | Description |
|---|---|
| `mmbrowse` | Graphical matrix browser. |
| `mmdeal` | Deal array or string matrix into individual arguments. |
| `mmrand` | Uniformly distributed random arrays, with specified mean and variation. |
| `mmrandn` | Normally distributed random arrays, with specified mean and variance. |
| `mmshiftr` | Shift or circularly shift matrix columns. |
| `mmshiftd` | Shift or circularly shift matrix rows. |

### CHAPTER 8 RELATIONAL AND LOGICAL OPERATIONS

| Function | Description |
|---|---|
| mmempty | Substitute value if empty. |
| mmisv5 | True for MATLAB Version 5. |
| mmono | Test for monotonic vector. |

### CHAPTER 9 SET, BIT, AND BASE FUNCTIONS

| Function | Description |
|---|---|
| mmismem | True for set members. Much faster than ismem. |

### CHAPTER 10 CHARACTER STRINGS

| Function | Description |
|---|---|
| mmdeal | Deal array or string matrix into individual arguments. |
| mmonoff | String ON/OFF to/from logical TRUE/FALSE conversion. |
| mmprintf | Data array to string matrix conversion. Much faster than num2str, but more restrictive. |

### CHAPTER 16 NUMERICAL LINEAR ALGEBRA

| Function | Description |
|---|---|
| mmrwls | Recursive weighted least squares. |

### CHAPTER 17 DATA ANALYSIS

| Function | Description |
|---|---|
| mmax | Matrix maximum values. |
| mmin | Matrix minimum values. |
| mmpeaks | Find indices of relative maxima. |

## CHAPTER 18 POLYNOMIALS

| Function | Description |
|---|---|
| mmp2str | Polynomial vector to string conversion. |
| mmpadd | Polynomial addition. |
| mmpintrp | Inverse interpolate polynomial. |
| mmpoly | Real-valued polynomial construction. |
| mmpscale | Scale polynomial. |
| mmpshift | Shift polynomial. |
| mmpsim | Polynomial simplification, strip leading zero terms. |
| mmrwpfit | Recursive weighted polynomial curve fitting. |
| mm2dpfit | Two-dimensional polynomial curve fitting. |
| mm2dpval | Two-dimensional polynomial evaluation. |
| mm2dpstr | Two-dimensional polynomial vector to string conversion. |
| mmp2pm | Polynomial to polynomial matrix conversion. |
| mmpm2p | Polynomial matrix to polynomial conversion. |
| mmpmder | Polynomial matrix derivative. |
| mmpmfit | Ploynomial matrix curve fitting. |
| mmpmint | Polynomial matrix integral. |
| mmpmsel | Select subset of a polynomial matrix. |
| mmpmeval | Polynomial matrix evaluation. |

## CHAPTER 19 INTERPOLATION

| Function | Description |
|---|---|
| mmgetpt | Get graphical point with interpolation. |
| mminterp | 1-D monotonic linear interpolation. Faster than interp1 for nonequally spaced arrays. |
| mmsearch | 1-D NON-monotonic linear interpolation. |

## CHAPTER 20 CUBIC SPLINES

| Function | Description |
|---|---|
| mmsparea | Cubic spline area, i.e., compute area under spline given bounds. |
| mmspcut | Extract or cut out part of a spline to form another spline. |

## CHAPTER 20 CUBIC SPLINES

| Function | Description |
|---|---|
| `mmspdata` | Cubic spline data extraction, i.e., extract data used to form spline. |
| `mmspder` | Cubic spline derivative interpolation, i.e., find spline representation of slope of data. |
| `mmspii` | Inverse interpolate cubic spline, e.g., find $x$ given $y = s(x)$. |
| `mmspint` | Cubic spline integral interpolation, i.e., find spline representaion of integral of data. |
| `mmspline` | Cubic spline construction with method choice. Similar to the function `spline`, but offers a choice of seven endpoint conditions as well as the ability to specify selected breakpoints as joints where the slope is discontinuous. |
| `mmspxtrm` | Cubic spline extremes, i.e., find relative minima and maxima of a spline. |

## CHAPTER 21 FOURIER ANALYSIS

| Function | Description |
|---|---|
| `fshelp` | Help for Fourier series. |
| `fsangle` | Angle between two Fourier series. |
| `fsderiv` | Fourier series derivative. |
| `fsdelay` | Add time delay to a Fourier series. |
| `fseval` | Fourier series function evaluation. |
| `fsfind` | Find Fourier series approximation. |
| `fsform` | Fourier series format conversion. |
| `fsharm` | Fourier series harmonic component selection. |
| `fsintgrl` | Fourier series integral. |
| `fsmsv` | Fourier series mean square value. |
| `fspeak` | Fourier series peak value. |
| `fspf` | Fourier series power factor computation. |
| `fsprod` | Fourier series of a product of time functions. |
| `fsresize` | Resize a Fourier series. |
| `fsresp` | Round Fourier series coefficients. |
| `fssize` | Highest harmonic in a Fourier series. |
| `fssum` | Fourier series of a sum of time functions. |
| `fssym` | Enforce symmetry constraints. |

## CHAPTER 21 FOURIER ANALYSIS

| Function | Description |
|---|---|
| fstable | Generate common Fourier series. |
| fsthd | Total harmonic distortion of a Fourier series. |
| mmfftbin | FFT bin frequencies. |
| mmfftpfc | FFT positive frequency components. |
| mmftfind | Find Fourier transform approximation. |
| mmwindow | Generate window functions. |

## CHAPTER 22 OPTIMIZATION

| Function | Description |
|---|---|
| mmfminc | Minimization with inequality constraints. |
| mmfminc_ | Helper function for mmfminc. |
| mmfminu | Minimize a function of several variables. Same algorithm as that used by fminu in the *Optimization Toolbox.* |
| mmfsolve | Solve a set of nonlinear equations, i.e., multidimensional equivalent to fzero. |
| mmlceval | Evaluate a linear combination of functions. |
| mmlcfit | Curve fit to a linear combination of functions. |
| mmnlfit | Nonlinear curve fitting. |
| mmnlfit2 | 2-D Nonlinear curve fitting. |
| mmnlfit_ | Helper function for mmnlfit and mmnlfit2. |
| mmsneval | Simple nonlinear curve fit evaluation. |
| mmsnfit | Simple nonlinear curve fit by transformation. |

## CHAPTER 23 INTEGRATION AND DIFFERENTATION

| Function | Description |
|---|---|
| mmderiv | Compute derivative using weighted central differences. Much better than using diff to approximate derivatives. |
| mmintgrl | Cumulative integral using Simpson's rule. Improvement over cumtrapz. |
| mmvolume | Cumulative volume integral using trapezoidal rule. 2-D equivalent to cumtrapz. |

### CHAPTER 24 ORDINARY DIFFERENTIAL EQUATIONS

| Function | Description |
|---|---|
| `mmlsim` | Linear system simulation using `mmodess`. Computes the response of a linear state space system. |
| `mmlsim_` | Helper function for `mmlsim`. |
| `mmode45` | ODE solution using 4–5th order `mmodess`. |
| `mmode45p` | ODE plotted solution using 4–5th order `mmodess`. |
| `mmodechi` | ODE cubic hermite interpolation. |
| `mmodeini` | Initialize ODE parameters for `mmodess`. |
| `mmodess` | Single step ODE solution, 4th–5th order. Alternative to `ode45` when you want to get closer to the integrator. This function propagates the solution a single step at a time. |

### CHAPTER 26 2-D GRAPHICS

| Function | Description |
|---|---|
| `mmarrow` | Plot moveable arrows on current axes. Great for annotating plots. |
| `mmedit` | Edit axes text using mouse. |
| `mmfill` | Fill plot of area between two curves. |
| `mmplot2` | Plot two *Y* arrays versus one *X* with right side axis scale. Much improved over `plotyy`. |
| `mmplotc` | 2-D Plot with an ASCII character marker at data points. |
| `mmploti` | Incremental 2-D line plotting. Plots data as they are generated. |
| `mmplotz` | Plot with axes drawn through ZERO. |
| `mmpolar` | Linear or logarithmic polar coordinate plot. Complete with settable parameters. Dramatic improvement over `polar`, `polarhg`, etc. |
| `mmginput` | Graphical input using mouse. |
| `mmprobe` | Probe data on 2-D axis using mouse. Like using a digital oscilloscope. |
| `mmgui` | Double-click activation of plotting GUIs. |
| `mmsaxes` | Set axes specifications using a GUI. |
| `mmscolor` | Set RGB specification using a GUI. |
| `mmsfont` | Set font characteristics using a GUI. |
| `mmsline` | Set line specifications using a GUI. |

### CHAPTER 26 2-D GRAPHICS

| Function | Description |
|---|---|
| mmtext | Place and drag text with mouse. Great for annotating plots. |
| mmtile | Tile figure windows on screen. |
| mmzoom | Simple 2-D zoom-in function. |

### CHAPTER 27 3-D GRAPHICS

| Function | Description |
|---|---|
| mmhole | Create hole in 3-D graphics data. |
| mminxy | Minima of 3-D data along *X* and *Y* axes. |
| mmxtract | Extract subset of 3-D graphics data. |
| mmssurf | Set surface specifications using a GUI. |
| mmsview | Set azimuth and elevation using a GUI. |
| mmzoom3 | Simple 3-D X-Y plane zoom-in function. |

### CHAPTER 28 USING COLOR AND LIGHT

| Function | Description |
|---|---|
| mmap | Single color colormap. |
| mmsmap | Set and manipulate figure colormap using a GUI. |
| rainbow | Colormap variant to `hsv`. |

### CHAPTER 30 PRINTING AND EXPORTING GRAPHICS

| Function | Description |
|---|---|
| mmspage | Set figure paper position when printed using a GUI. Alternative to `pagedlg`. |
| mmpaper | Set default paper properties. |

## CHAPTER 31 HANDLE GRAPHICS

| Function | Description |
|---|---|
| `mmbox` | Get position vector of a rubberband box. |
| `mmgca` | Get current axes if it exists. |
| `mmgcf` | Get current figure if it exists. |
| `mmget` | Get multiple object properties. |
| `mmgetpos` | Get object position vector in specified units. |
| `mmgetsiz` | Get font size in specified units. |
| `mmfitpos` | Fit position within another object. |
| `mmsetpos` | Set object position relative to another object. |
| `mmsetptr` | Set mouse pointer location over object or at axis location. |
| `mminrect` | True when point is inside position rectangle. |
| `mmrgb` | Color specification conversion and substitution. |
| `mmzap` | Delete graphics object using mouse. |

## CHAPTER 32 GRAPHICAL USER INTERFACES

| Function | Description |
|---|---|
| `mmcd` | Change working directory using a GUI. |
| `mmload` | Load matrix and header from an ASCII file using a GUI. |
| `mmsave` | Save matrix and header to an ASCII file using a GUI. |
| `mmgui` | Double-click activation of plotting GUIs. |
| `mmsaxes` | Set axes specifications using a GUI. |
| `mmscolor` | Set RGB specification using a GUI. |
| `mmsfont` | Set font characteristics using a GUI. |
| `mmsline` | Set line specifications using a GUI. |
| `mmsmap` | Set and manipulate figure colormap using a GUI. |
| `mmspage` | Set figure paper position when printed using a GUI. |
| `mmssurf` | Set surface specifications using a GUI. |
| `mmsview` | Set aximuth and elevation using a GUI. |

# A

# MATLAB Function Listing

This appendix lists all MATLAB functions (except demo functions) based on their placement on the MATLAB search path.

**Directory:** `Toolbox/MATLAB/datafun` (Data analysis and Fourier transforms)

*Basic operations*

| | |
|---|---|
| `max` | Largest component. |
| `min` | Smallest component. |
| `mean` | Average or mean value. |
| `median` | Median value. |
| `std` | Standard deviation. |
| `sort` | Sort in ascending order. |
| `sortrows` | Sort rows in ascending order. |

| | |
|---|---|
| `sum` | Sum of elements. |
| `prod` | Product of elements. |
| `hist` | Histogram. |
| `trapz` | Trapezoidal numerical integration. |
| `cumsum` | Cumulative sum of elements. |
| `cumprod` | Cumulative product of elements. |
| `cumtrapz` | Cumulative trapezoidal numerical integration. |

### *Finite differences*

| | |
|---|---|
| `diff` | Difference and approximate derivative. |
| `gradient` | Approximate gradient. |
| `del2` | Discrete Laplacian. |

### *Correlation*

| | |
|---|---|
| `corrcoef` | Correlation coefficients. |
| `cov` | Covariance matrix. |
| `subspace` | Angle between subspaces. |

### *Filtering and convolution*

| | |
|---|---|
| `filter` | One-dimensional digital filter. |
| `filter2` | Two-dimensional digital filter. |
| `conv` | Convolution and polynomial multiplication. |
| `conv2` | Two-dimensional convolution. |
| `convn` | N-dimensional convolution. |
| `deconv` | Deconvolution and polynomial division. |

### *Fourier transforms*

| | |
|---|---|
| `fft` | Discrete Fourier transform. |
| `fft2` | Two-dimensional discrete Fourier transform. |
| `fftn` | N-dimensional discrete Fourier Transform. |
| `ifft` | Inverse discrete Fourier transform. |
| `ifft2` | Two-dimensional inverse discrete Fourier transform. |
| `ifftn` | N-dimensional inverse discrete Fourier Transform. |
| `fftshift` | Move zeroth lag to center of spectrum. |

### *Sound and audio*

| | |
|---|---|
| `sound` | Play vector as sound. |
| `soundsc` | Autoscale and play vector as sound. |
| `speak` | Convert input string to speech (Macintosh only). |

| | |
|---|---|
| `recordsound` | Record sound (Macintosh only). |
| `soundcap` | Sound capabilities (Macintosh only). |
| `mu2lin` | Convert mu-law encoding to linear signal. |
| `lin2mu` | Convert linear signal to mu-law encoding. |

*Audio file import/export*

| | |
|---|---|
| `auwrite` | Write NeXT/SUN (".au") sound file. |
| `auread` | Read NeXT/SUN (".au") sound file. |
| `wavwrite` | Write Microsoft WAVE (".wav") sound file. |
| `wavread` | Read Microsoft WAVE (".wav") sound file. |
| `readsnd` | Read SND resources and files (Macintosh only). |
| `writesnd` | Write SND resources and files (Macintosh only). |

*Utilities*

| | |
|---|---|
| `playsnd` | Implementation for SOUND. |

**Directory:** `Toolbox/MATLAB/datatypes` (Data types and structures)

*Data types (classes)*

| | |
|---|---|
| `double` | Convert to double precision. |
| `sparse` | Create sparse matrix. |
| `char` | Create character array (string). |
| `cell` | Create cell array. |
| `struct` | Create or convert to structure array. |
| `uint8` | Convert to unsigned 8-bit integer. |
| `inline` | Construct INLINE object. |

*Multidimensional array functions*

| | |
|---|---|
| `cat` | Concatenate arrays. |
| `ndims` | Number of dimensions. |
| `ndgrid` | Generate arrays for N-D functions and interpolation. |
| `permute` | Permute array dimensions. |
| `ipermute` | Inverse permute array dimensions. |
| `shiftdim` | Shift dimensions. |
| `squeeze` | Remove singleton dimensions. |

*Cell array functions*

| | |
|---|---|
| `cell` | Create cell array. |
| `celldisp` | Display cell array contents. |

| | |
|---|---|
| `cellplot` | Display graphical depiction of cell array. |
| `num2cell` | Convert numeric array into cell array. |
| `deal` | Deal inputs to outputs. |
| `cell2struct` | Convert cell array into structure array. |
| `struct2cell` | Convert structure array into cell array. |
| `iscell` | True for cell array. |

### *Structure functions*

| | |
|---|---|
| `struct` | Create or convert to structure array. |
| `fieldnames` | Get structure field names. |
| `getfield` | Get structure field contents. |
| `setfield` | Set structure field contents. |
| `rmfield` | Remove structure field. |
| `isfield` | True if field is in structure array. |
| `isstruct` | True for structures. |

### *Object-oriented programming functions*

| | |
|---|---|
| `class` | Create object or return object class. |
| `struct` | Convert object to structure array. |
| `methods` | Display class method names. |
| `isa` | True if object is a given class. |
| `isobject` | True for objects. |
| `inferiorto` | Inferior class relationship. |
| `superiorto` | Superior class relationship. |

### *Overloadable operators*

| | |
|---|---|
| `minus` | Overloadable method for `a-b`. |
| `plus` | Overloadable method for `a+b`. |
| `times` | Overloadable method for `a.*b`. |
| `mtimes` | Overloadable method for `a*b`. |
| `mldivide` | Overloadable method for `a\b`. |
| `mrdivide` | Overloadable method for `a/b`. |
| `rdivide` | Overloadable method for `a./b`. |
| `ldivide` | Overloadable method for `a.\b`. |
| `power` | Overloadable method for `a.^b`. |
| `mpower` | Overloadable method for `a^b`. |
| `uminus` | Overloadable method for `-a`. |
| `uplus` | Overloadable method for `+a`. |
| `horzcat` | Overloadable method for `[a b]`. |
| `vertcat` | Overloadable method for `[a;b]`. |
| `le` | Overloadable method for `a<=b`. |

| | |
|---|---|
| lt | Overloadable method for a<b. |
| gt | Overloadable method for a>b. |
| ge | Overloadable method for a>=b. |
| eq | Overloadable method for a==b. |
| ne | Overloadable method for a~=b. |
| not | Overloadable method for ~a. |
| and | Overloadable method for a&b. |
| or | Overloadable method for a \| b. |
| subsasgn | Overloadable method for a(i)=b, a{i}=b, and a.field=b. |
| subsref | Overloadable method for a(i), a{i}, and a.field. |
| colon | Overloadable method for a:b. |
| transpose | Overloadable method for a.' |
| ctranspose | Overloadable method for a'. |
| subsindex | Overloadable method for x(a). |

## **Directory:** Toolbox/MATLAB/dde (Dynamic data exchange: PC specific functions)

### *DDE Client Functions*

| | |
|---|---|
| ddeadv | Set up advisory link. |
| ddeexec | Send string for execution. |
| ddeinit | Initiate DDE conversation. |
| ddepoke | Send data to application. |
| ddereq | Request data from application. |
| ddeterm | Terminate DDE conversation. |
| ddeunadv | Release advisory link. |

## **Directory:** Toolbox/MATLAB/elfun (Elementary math functions)

### *Trigonometric*

| | |
|---|---|
| sin | Sine. |
| sinh | Hyperbolic sine. |
| asin | Inverse sine. |
| asinh | Inverse hyperbolic sine. |
| cos | Cosine. |
| cosh | Hyperbolic cosine. |
| acos | Inverse cosine. |
| acosh | Inverse hyperbolic cosine. |
| tan | Tangent. |
| tanh | Hyperbolic tangent. |
| atan | Inverse tangent. |

| | |
|---|---|
| `atan2` | Four quadrant inverse tangent. |
| `atanh` | Inverse hyperbolic tangent. |
| `sec` | Secant. |
| `sech` | Hyperbolic secant. |
| `asec` | Inverse secant. |
| `asech` | Inverse hyperbolic secant. |
| `csc` | Cosecant. |
| `csch` | Hyperbolic cosecant. |
| `acsc` | Inverse cosecant. |
| `acsch` | Inverse hyperbolic cosecant. |
| `cot` | Cotangent. |
| `coth` | Hyperbolic cotangent. |
| `acot` | Inverse cotangent. |
| `acoth` | Inverse hyperbolic cotangent. |

*Exponential*

| | |
|---|---|
| `exp` | Exponential. |
| `log` | Natural logarithm. |
| `log10` | Common (base 10) logarithm. |
| `log2` | Base 2 logarithm and dissect floating point number. |
| `pow2` | Base 2 power and scale floating point number. |
| `sqrt` | Square root. |
| `nextpow2` | Next higher power of 2. |

*Complex*

| | |
|---|---|
| `abs` | Absolute value. |
| `angle` | Phase angle. |
| `conj` | Complex conjugate. |
| `imag` | Complex imaginary part. |
| `real` | Complex real part. |
| `unwrap` | Unwrap phase angle. |
| `isreal` | True for real array. |
| `cplxpair` | Sort numbers into complex conjugate pairs. |

*Rounding and remainder*

| | |
|---|---|
| `fix` | Round towards zero. |
| `floor` | Round towards minus infinity. |
| `ceil` | Round towards plus infinity. |
| `round` | Round towards nearest integer. |
| `mod` | Modulus (signed remainder after division). |
| `rem` | Remainder after division. |
| `sign` | Signum. |

## Directory: `Toolbox/MATLAB/elmat` (Elementary matrices and manipulation)

### *Elementary matrices*

| | |
|---|---|
| `zeros` | Zeros array. |
| `ones` | Ones array. |
| `eye` | Identity matrix. |
| `repmat` | Replicate and tile array. |
| `rand` | Uniformly distributed random numbers. |
| `randn` | Normally distributed random numbers. |
| `linspace` | Linearly spaced vector. |
| `logspace` | Logarithmically spaced vector. |
| `meshgrid` | X and Y arrays for 3-D plots. |
| : | Regularly spaced vector and index into matrix. |

### *Basic array information*

| | |
|---|---|
| `size` | Size of matrix. |
| `length` | Length of vector. |
| `ndims` | Number of dimensions. |
| `disp` | Display matrix or text. |
| `isempty` | True for empty matrix. |
| `isequal` | True if arrays are identical. |
| `isnumeric` | True for numeric arrays. |
| `islogical` | True for logical arrays. |
| `logical` | Convert numeric values to logical. |

### *Matrix manipulation*

| | |
|---|---|
| `reshape` | Change size. |
| `diag` | Diagonal matrices and diagonals of matrix. |
| `tril` | Extract lower triangular part. |
| `triu` | Extract upper triangular part. |
| `fliplr` | Flip matrix in left/right direction. |
| `flipud` | Flip matrix in up/down direction. |
| `flipdim` | Flip matrix along specified dimension. |
| `rot90` | Rotate matrix 90 degrees. |
| : | Regularly spaced vector and index into matrix. |
| `find` | Find indices of nonzero elements. |
| `end` | Last index. |
| `sub2ind` | Linear index from multiple subscripts. |
| `ind2sub` | Multiple subscripts from linear index. |

*Special variables and constants*

| | |
|---|---|
| `ans` | Most recent answer. |
| `eps` | Floating point relative accuracy. |
| `realmax` | Largest positive floating point number. |
| `realmin` | Smallest positive floating point number. |
| `pi` | 3.1415926535897... |
| `i, j` | Imaginary unit. |
| `inf` | Infinity. |
| `NaN` | Not-a-Number. |
| `isnan` | True for Not-a-Number. |
| `isinf` | True for infinite elements. |
| `isfinite` | True for finite elements. |
| `flops` | Floating point operation count. |
| `why` | Succinct answer. |

*Specialized matrices*

| | |
|---|---|
| `compan` | Companion matrix. |
| `gallery` | Higham test matrices. |
| `hadamard` | Hadamard matrix. |
| `hankel` | Hankel matrix. |
| `hilb` | Hilbert matrix. |
| `invhilb` | Inverse Hilbert matrix. |
| `magic` | Magic square. |
| `pascal` | Pascal matrix. |
| `rosser` | Classic symmetric eigenvalue test problem. |
| `toeplitz` | Toeplitz matrix. |
| `vander` | Vandermonde matrix. |
| `wilkinson` | Wilkinson's eigenvalue test matrix. |

*Other shared functions*

| | |
|---|---|
| `freqspace` | Frequency spacing for frequency response. |

**Directory:** `Toolbox/MATLAB/funfun` (Function functions and ODE solvers)

*Optimization and root finding*

| | |
|---|---|
| `fmin` | Minimize function of one variable. |
| `fmins` | Minimize function of several variables. |
| `fzero` | Find zero of function of one variable. |

### Numerical integration (quadrature)

| | |
|---|---|
| `quad` | Numerically evaluate integral, low order method. |
| `quad8` | Numerically evaluate integral, higher order method. |
| `dblquad` | Numerically evaluate double integral. |

### Plotting

| | |
|---|---|
| `ezplot` | Easy to use function plotter. |
| `fplot` | Plot function. |

### Inline function object

| | |
|---|---|
| `inline` | Construct INLINE object. |
| `argnames` | Argument names. |
| `formula` | Function formula. |
| `char` | Convert INLINE object to character array. |

### Ordinary differential equation solvers

| | |
|---|---|
| `ode45` | Solve non-stiff differential equations, medium order method. |
| `ode23` | Solve non-stiff differential equations, low order method. |
| `ode113` | Solve non-stiff differential equations, variable order method. |
| `ode15s` | Solve stiff differential equations, variable order method. |
| `ode23s` | Solve stiff differential equations, low order method. |
| `odefile` | ODE file syntax. |

### ODE Option handling

| | |
|---|---|
| `odeset` | Create/alter ODE OPTIONS structure. |
| `odeget` | Get ODE OPTIONS parameters. |

### ODE output functions

| | |
|---|---|
| `odeplot` | Time series ODE output function. |
| `odephas2` | 2-D phase plane ODE output function. |
| `odephas3` | 3-D phase plane ODE output function. |
| `odeprint` | Command window printing ODE output function. |

### Utility functions

| | |
|---|---|
| `foptions` | Set default parameters used by the optimization routines. |
| `fcnchk` | Check FUNFUN function argument. |
| `innerlp` | Used with `dblquad` to evaluate inner loop of integral. |

### *Numerical Jacobian matrix helper functions for the ODE solvers*

| | |
|---|---|
| `numjac` | Numerically compute the Jacobian dF/dY of function F(t,y). |
| `colgroup` | Helper function used by `numjac` for sparse Jacobians. |

### *Event location and output helper functions for the ODE solvers*

| | |
|---|---|
| `odezero` | Locate any zero-crossings of event functions in a time step. |
| `ntrp45` | Interpolation helper function for ODE45. |
| `ntrp15s` | Interpolation helper function for ODE15S. |
| `ntrp23` | Interpolation helper function for ODE23. |
| `ntrp23s` | Interpolation helper function for ODE23S. |
| `ntrp113` | Interpolation helper function for ODE113. |

## **Directory:** `Toolbox/MATLAB/general` (General-purpose commands)

### *General information*

| | |
|---|---|
| `help` | On-line help, display text at command line. |
| `helpwin` | On-line help, separate window for navigation. |
| `helpdesk` | Comprehensive hypertext documentation and troubleshooting. |
| `demo` | Run demonstrations. |
| `ver` | MATLAB, SIMULINK, and toolbox version information. |
| `whatsnew` | Display Readme files. |
| `Readme` | What's new in MATLAB 5. |

### *Managing the workspace*

| | |
|---|---|
| `who` | List current variables. |
| `whos` | List current variables, long form. |
| `clear` | Clear variables and functions from memory. |
| `pack` | Consolidate workspace memory. |
| `load` | Load workspace variables from disk. |
| `save` | Save workspace variables to disk. |
| `quit` | Quit MATLAB session. |

### *Managing commands and functions*

| | |
|---|---|
| `what` | List MATLAB-specific files in directory. |
| `type` | List M-file. |
| `edit` | Edit M-file. |
| `lookfor` | Search all M-files for keyword. |
| `which` | Locate functions and files. |
| `pcode` | Create pre-parsed pseudo-code file (P-file). |

| | |
|---|---|
| `inmem` | List functions in memory. |
| `mex` | Compile MEX-function. |

*Managing the search path*

| | |
|---|---|
| `path` | Get/set search path. |
| `addpath` | Add directory to search path. |
| `rmpath` | Remove directory from search path. |
| `editpath` | Modify search path. |

*Controlling the command window*

| | |
|---|---|
| `echo` | Echo commands in M-files. |
| `more` | Control paged output in command window. |
| `diary` | Save text of MATLAB session. |
| `format` | Set output format. |

*Operating system commands*

| | |
|---|---|
| `cd` | Change current working directory. |
| `pwd` | Show (print) current working directory. |
| `dir` | List directory. |
| `delete` | Delete file. |
| `getenv` | Get environment variable. |
| `!` | Execute operating system command. |
| `dos` | Execute DOS command and return result. |
| `unix` | Execute UNIX command and return result. |
| `vms` | Execute VMS DCL command and return result. |
| `web` | Open Web browser on site or files. |
| `computer` | Computer type. |

*Debugging M-files*

| | |
|---|---|
| `debug` | List debugging commands. |
| `dbstop` | Set breakpoint. |
| `dbclear` | Remove breakpoint. |
| `dbcont` | Continue execution. |
| `dbdown` | Change local workspace context. |
| `dbstack` | Display function call stack. |
| `dbstatus` | List all breakpoints. |
| `dbstep` | Execute one or more lines. |
| `dbtype` | List M-file with line numbers. |
| `dbup` | Change local workspace context. |
| `dbquit` | Quit debug mode. |
| `dbmex` | Debug MEX-files (UNIX only). |

### *Profiling M-files*

| | |
|---|---|
| `profile` | Profile M-file execution time. |

### *Others*

| | |
|---|---|
| `binpatch` | Patch binary file. |
| `doc` | A utility to load HTML documentation into a Web browser. |
| `exit` | Exit from MATLAB. |
| `helpinfo` | Information about help options. |
| `info` | Information about MATLAB and The MathWorks. |
| `isstudent` | True for the student edition of MATLAB. |
| `isunix` | True for the UNIX version of MATLAB. |
| `isvms` | True for the VMS version of MATLAB. |
| `isppc` | True for Macintosh PowerPC. |
| `isieee` | True for computers with IEEE arithmetic. |
| `ls` | List directory. |
| `memory` | Help for memory limitations. |
| `notebook` | Open an m-book in Microsoft Word (Windows only). |
| `nnload` | Netscape Navigator load. |
| `profsumm` | Summarize profile information. |
| `subscribe` | Subscribe to the MathWorks Newsletter. |
| `whichcls` | Return classes of inputs (used by `which`). |

### *GUI utilities*

| | |
|---|---|
| `openvar` | Edit an array graphically. |
| `maeasgn` | Assign the result of an expression into an array subrange. |
| `maeresize` | Change the size of a matrix to be [m n]. |
| `mauifindexe` | Return the absolute pathname to a MAUI executable. |
| `mdbstatus` | DBSTATUS for the Debugger/Editor. |
| `miport` | Get the port which MATLAB is listening on. |
| `genpath` | Generate reasonable path based on toolbox. |
| `pathedit` | Modify current MATLAB path (GUI implementation) |
| `path2rc` | Save the current MATLAB path in the `pathdef.m` file. |
| `pathtool` | Path Browser for Macintosh and PC. |
| `workspace` | Workspace Browser for Macintosh and PC. |

## **Directory:** `Toolbox/MATLAB/graph2d` (Two-dimensional graphs)

### *Elementary X-Y graphs*

| | |
|---|---|
| `plot` | Linear plot. |
| `loglog` | Log-log scale plot. |

| | |
|---|---|
| `semilogx` | Semi-log scale plot. |
| `semilogy` | Semi-log scale plot. |
| `polar` | Polar coordinate plot. |
| `plotyy` | Graphs with y tick labels on the left and right. |

### *Axis control*

| | |
|---|---|
| `axis` | Control axis scaling and appearance. |
| `zoom` | Zoom in and out on a 2-D plot. |
| `grid` | Grid lines. |
| `box` | Axis box. |
| `hold` | Hold current graph. |
| `axes` | Create axes in arbitrary positions. |
| `subplot` | Create axes in tiled positions. |

### *Graph annotation*

| | |
|---|---|
| `legend` | Graph legend. |
| `title` | Graph title. |
| `xlabel` | X-axis label. |
| `ylabel` | Y-axis label. |
| `text` | Text annotation. |
| `gtext` | Place text with mouse. |

### *Hardcopy and printing*

| | |
|---|---|
| `print` | Print graph or SIMULINK system; or save graph to file. |
| `printopt` | Printer defaults. |
| `orient` | Set paper orientation. |

### *Utilities*

| | |
|---|---|
| `lscan` | Scan for good legend location. |
| `moveaxis` | Used to grab and move legend axis. |

## **Directory:** `Toolbox/MATLAB/graph3d` (Three-dimensional graphs)

### *Elementary 3-D plots*

| | |
|---|---|
| `plot3` | Plot lines and points in 3-D space. |
| `mesh` | 3-D mesh surface. |
| `surf` | 3-D colored surface. |
| `fill3` | Filled 3-D polygons. |

### *Color control*

| | |
|---|---|
| `colormap` | Color look-up table. |
| `caxis` | Pseudocolor axis scaling. |
| `shading` | Color shading mode. |
| `hidden` | Mesh hidden line removal mode. |
| `brighten` | Brighten or darken color map. |

### *Lighting*

| | |
|---|---|
| `surfl` | 3-D shaded surface with lighting. |
| `lighting` | Lighting mode. |
| `material` | Material reflectance mode. |
| `specular` | Specular reflectance. |
| `diffuse` | Diffuse reflectance. |
| `surfnorm` | Surface normals. |

### *Colormaps*

| | |
|---|---|
| `hsv` | Hue-saturation-value color map. |
| `hot` | Black-red-yellow-white color map. |
| `gray` | Linear gray-scale color map. |
| `bone` | Gray-scale with tinge of blue color map. |
| `copper` | Linear copper-tone color map. |
| `pink` | Pastel shades of pink color map. |
| `white` | All white color map. |
| `flag` | Alternating red, white, blue, and black color map. |
| `lines` | Color map with the line colors. |
| `colorcube` | Enhanced color-cube color map. |
| `jet` | Variant of `hsv`. |
| `prism` | Prism color map. |
| `cool` | Shades of cyan and magenta color map. |
| `autumn` | Shades of red and yellow color map. |
| `spring` | Shades of magenta and yellow color map. |
| `winter` | Shades of blue and green color map. |
| `summer` | Shades of green and yellow color map. |

### *Axis control*

| | |
|---|---|
| `axis` | Control axis scaling and appearance. |
| `zoom` | Zoom in and out on a 2-D plot. |
| `grid` | Grid lines. |
| `box` | Axis box. |
| `hold` | Hold current graph. |

| | |
|---|---|
| `axes` | Create axes in arbitrary positions. |
| `subplot` | Create axes in tiled positions. |

*Viewpoint control*

| | |
|---|---|
| `view` | 3-D graph viewpoint specification. |
| `viewmtx` | View transformation matrix. |
| `rotate3d` | Interactively rotate view of 3-D plot. |

*Graph annotation*

| | |
|---|---|
| `title` | Graph title. |
| `xlabel` | X-axis label. |
| `ylabel` | Y-axis label. |
| `zlabel` | Z-axis label. |
| `colorbar` | Display color bar (color scale). |
| `text` | Text annotation. |
| `gtext` | Mouse placement of text. |

*Hard copy and printing*

| | |
|---|---|
| `print` | Print graph or SIMULINK system; or save graph to file. |
| `printopt` | Printer defaults. |
| `orient` | Set paper orientation. |

**Directory:** `Toolbox/MATLAB/graphics` (Handle Graphics)

*Figure window creation and control*

| | |
|---|---|
| `figure` | Create figure window. |
| `gcf` | Get handle to current figure. |
| `clf` | Clear current figure. |
| `shg` | Show graph window. |
| `close` | Close figure. |
| `refresh` | Refresh figure. |

*Axis creation and control*

| | |
|---|---|
| `subplot` | Create axes in tiled positions. |
| `axes` | Create axes in arbitrary positions. |
| `gca` | Get handle to current axes. |
| `cla` | Clear current axes. |
| `axis` | Control axis scaling and appearance. |

| | |
|---|---|
| `box` | Axis box. |
| `caxis` | Control pseudocolor axis scaling. |
| `hold` | Hold current graph. |
| `ishold` | Return hold state. |

*Handle Graphics objects*

| | |
|---|---|
| `figure` | Create figure window. |
| `axes` | Create axes. |
| `line` | Create line. |
| `text` | Create text. |
| `patch` | Create patch. |
| `surface` | Create surface. |
| `image` | Create image. |
| `light` | Create light. |
| `uicontrol` | Create user interface control. |
| `uimenu` | Create user interface menu. |

*Handle Graphics operations*

| | |
|---|---|
| `set` | Set object properties. |
| `get` | Get object properties. |
| `reset` | Reset object properties. |
| `delete` | Delete object. |
| `gco` | Get handle to current object. |
| `gcbo` | Get handle to current callback object. |
| `gcbf` | Get handle to current callback figure. |
| `drawnow` | Flush pending graphics events. |
| `findobj` | Find objects with specified property values. |
| `copyobj` | Make copy of graphics object and its children. |

*Hard copy and printing*

| | |
|---|---|
| `print` | Print graph or SIMULINK system; or save graph to file. |
| `printopt` | Printer defaults. |
| `orient` | Set paper orientation. |

*Utilities*

| | |
|---|---|
| `closereq` | Figure close request function. |
| `newplot` | M-file preamble for NextPlot property. |
| `ishandle` | True for graphics handles. |

### *Printing utilities*

| | |
|---|---|
| `bwcontr` | Contrasting black or white color. |
| `hardcopy` | Save figure window to file. |
| `nodither` | Modify figure to avoid dithered lines. |
| `savtoner` | Modify figure to save printer toner. |

## **Directory:** `Toolbox/MATLAB/iofun` (File input/output)

### *File opening and closing*

| | |
|---|---|
| `fopen` | Open file. |
| `fclose` | Close file. |

### *Binary file I/O*

| | |
|---|---|
| `fread` | Read binary data from file. |
| `fwrite` | Write binary data to file. |

### *Formatted file I/O*

| | |
|---|---|
| `fscanf` | Read formatted data from file. |
| `fprintf` | Write formatted data to file. |
| `fgetl` | Read line from file, discard newline character. |
| `fgets` | Read line from file, keep newline character. |
| `input` | Prompt for user input. |

### *String conversion*

| | |
|---|---|
| `sprintf` | Write formatted data to string. |
| `sscanf` | Read string under format control. |

### *File positioning*

| | |
|---|---|
| `ferror` | Inquire file error status. |
| `feof` | Test for end-of-file. |
| `fseek` | Set file position indicator. |
| `ftell` | Get file position indicator. |
| `frewind` | Rewind file. |

### *Filename handling*

| | |
|---|---|
| `matlabroot` | Root directory of MATLAB installation. |
| `filesep` | Directory separator for this platform. |
| `pathsep` | Path separator for this platform. |
| `mexext` | MEX filename extension for this platform. |
| `fullfile` | Build full filename from parts. |
| `partialpath` | Partial pathnames. |
| `tempdir` | Get temporary directory. |
| `tempname` | Get temporary file. |

### *File import/export functions*

| | |
|---|---|
| `load` | Load workspace from MAT-file. |
| `save` | Save workspace to MAT-file. |
| `dlmread` | Read ASCII delimited file. |
| `dlmwrite` | Write ASCII delimited file. |
| `wk1read` | Read spreadsheet (WK1) file. |
| `wk1write` | Write spreadsheet (WK1) file. |

### *Image file import/export*

| | |
|---|---|
| `imread` | Read image from graphics file. |
| `imwrite` | Write image to graphics file. |
| `imfinfo` | Return information about graphics file. |

### *Audio file import/export*

| | |
|---|---|
| `auwrite` | Write NeXT/SUN (".au") sound file. |
| `auread` | Read NeXT/SUN (".au") sound file. |
| `wavwrite` | Write Microsoft WAVE (".wav") sound file. |
| `wavread` | Read Microsoft WAVE (".wav") sound file. |

### *Command window I/O*

| | |
|---|---|
| `clc` | Clear command window. |
| `home` | Send cursor home. |
| `disp` | Display array. |
| `input` | Prompt for user input. |
| `pause` | Wait for user response. |

### *Utilities*

| | |
|---|---|
| `str2rng` | Convert spreadsheet range string to numeric array. |
| `wk1const` | WK1 record type definitions. |

| | |
|---|---|
| `wk1wrec` | Write a WK1 record header. |
| `hdf` | MEX-file interface to the HDF library. |

## Directory: `Toolbox/MATLAB/lang` (Programming language constructs)

### *Control flow*

| | |
|---|---|
| `if` | Conditionally execute statements. |
| `else` | `if` statement condition. |
| `elseif` | `if` statement condition. |
| `end` | Terminate scope of `for`, `while`, `switch` and `if` statements. |
| `for` | Repeat statements a specific number of times. |
| `while` | Repeat statements an indefinite number of times. |
| `break` | Terminate execution of `while` or `for` loop. |
| `switch` | Switch among several cases based on expression. |
| `case` | `switch` statement case. |
| `otherwise` | Default `switch` statement case. |
| `return` | Return to invoking function. |

### *Evaluation and execution*

| | |
|---|---|
| `eval` | Execute string with MATLAB expression. |
| `feval` | Execute function specified by string. |
| `evalin` | Evaluate expression in workspace. |
| `builtin` | Execute built-in function from overloaded method. |
| `assignin` | Assign variable in workspace. |
| `run` | Run script. |

### *Scripts, functions, and variables*

| | |
|---|---|
| `script` | About MATLAB scripts and M-files. |
| `function` | Add new function. |
| `global` | Define global variable. |
| `mfilename` | Name of currently executing M-file. |
| `lists` | Comma separated lists. |
| `exist` | Check if variables or functions are defined. |
| `isglobal` | True for global variables. |

### *Argument handling*

| | |
|---|---|
| `nargchk` | Validate number of input arguments. |
| `nargin` | Number of function input arguments. |
| `nargout` | Number of function output arguments. |

| | |
|---|---|
| `varargin` | Variable length input argument list. |
| `varargout` | Variable length output argument list. |
| `inputname` | Input argument name. |

*Message display*

| | |
|---|---|
| `error` | Display error message and abort function. |
| `warning` | Display warning message. |
| `lasterr` | Last error message. |
| `errortrap` | Skip error during testing. |
| `disp` | Display an array |
| `fprintf` | Display formatted message. |
| `sprintf` | Write formatted data to a string. |

*Interactive input*

| | |
|---|---|
| `input` | Prompt for user input. |
| `keyboard` | Invoke keyboard from M-file. |
| `pause` | Wait for user response. |
| `uimenu` | Create user interface menu. |
| `uicontrol` | Create user interface control. |

**Directory:** `Toolbox/MATLAB/MacOS` (Macintosh specific functions)

*Macintosh operating system commands*

| | |
|---|---|
| `gestalt` | Macintosh GESTALT function (Macintosh only). |
| `applescript` | Load and execute AppleScript file. |
| `!` | Execute Application or Toolserver command. |

*Multimedia functions*

| | |
|---|---|
| `recordsound` | Record sound. |
| `speak` | Speak text string. |
| `readsnd` | Read SND resources and files. |
| `writesnd` | Write SND resources and files. |
| `soundcap` | Sound capabilities. |
| `qtwrite` | Write QuickTime movie file to disk. |

*AppleScript-based functions*

| | |
|---|---|
| `xlsetrange` | Set range of cells in Microsoft Excel worksheet. |
| `xlgetrange` | Get range of cells from Microsoft Excel worksheet. |

| | |
|---|---|
| `areveal` | Reveal file on the Macintosh Desktop. |
| `arename` | Rename file. |
| `amove` | Move file from one folder to another. |
| `acopy` | Copy file from one folder to another. |

### *Utilities*

| | |
|---|---|
| `macurl` | A utility for opening a URL into a Web browser. |

## **Directory:** `Toolbox/MATLAB/matfun` (Matrix functions)

### *Matrix analysis*

| | |
|---|---|
| `norm` | Matrix or vector norm. |
| `normest` | Estimate the matrix 2-norm. |
| `rank` | Matrix rank. |
| `det` | Determinant. |
| `trace` | Sum of diagonal elements. |
| `null` | Null space. |
| `orth` | Orthogonalization. |
| `rref` | Reduced row echelon form. |
| `subspace` | Angle between two subspaces. |

### *Linear equations*

| | |
|---|---|
| `\ and /` | Linear equation solution; use `help slash`. |
| `inv` | Matrix inverse. |
| `cond` | Condition number with respect to inversion. |
| `condest` | 1-norm condition number estimate. |
| `chol` | Cholesky factorization. |
| `cholinc` | Incomplete Cholesky factorization. |
| `lu` | LU factorization. |
| `luinc` | Incomplete LU factorization. |
| `qr` | Orthogonal-triangular decomposition. |
| `nnls` | Non-negative least-squares. |
| `pinv` | Pseudoinverse. |
| `lscov` | Least squares with known covariance. |

### *Eigenvalues and singular values*

| | |
|---|---|
| `eig` | Eigenvalues and eigenvectors. |
| `svd` | Singular value decomposition. |
| `eigs` | A few eigenvalues. |

| | |
|---|---|
| `svds` | A few singular values. |
| `poly` | Characteristic polynomial. |
| `polyeig` | Polynomial eigenvalue problem |
| `condeig` | Condition number with respect to eigenvalues. |
| `hess` | Hessenberg form. |
| `qz` | QZ factorization for generalized eigenvalues. |
| `schur` | Schur decomposition. |

*Matrix functions*

| | |
|---|---|
| `expm` | Matrix exponential. |
| `logm` | Matrix logarithm. |
| `sqrtm` | Matrix square root. |
| `funm` | Evaluate general matrix function. |

*Factorization utilities*

| | |
|---|---|
| `qrdelete` | Delete column from QR factorization. |
| `qrinsert` | Insert column in QR factorization. |
| `rsf2csf` | Real block diagonal form to complex diagonal form. |
| `cdf2rdf` | Complex diagonal form to real block diagonal form. |
| `balance` | Diagonal scaling to improve eigenvalue accuracy. |
| `planerot` | Given's plane rotation. |

*Other shared functions*

| | |
|---|---|
| `expm1` | Matrix exponential via Pade approximation. |
| `expm2` | Matrix exponential via Taylor series. |
| `expm3` | Matrix exponential via eigenvalues and eigenvectors. |

**Directory:** `Toolbox/MATLAB/ops` (Operators and special characters)

*Arithmetic operators*

| | | |
|---|---|---|
| `plus` | Plus | + |
| `uplus` | Unary plus | + |
| `minus` | Minus | - |
| `uminus` | Unary minus | - |
| `mtimes` | Matrix multiply | * |
| `times` | Array multiply | .* |
| `mpower` | Matrix power | ^ |
| `power` | Array power | .^ |
| `mldivide` | Backslash or left matrix divide | \ |
| `mrdivide` | Slash or right matrix divide | / |

| | | |
|---|---|---|
| ldivide | Left array divide | .\ |
| rdivide | Right array divide | ./ |
| kron | Kronecker tensor product | kron |

*Relational operators*

| | | |
|---|---|---|
| eq | Equal | == |
| ne | Not equal | ~= |
| lt | Less than | < |
| gt | Greater than | > |
| le | Less than or equal | <= |
| ge | Greater than or equal | >= |

*Logical operators*

| | | |
|---|---|---|
| and | Logical AND | & |
| or | Logical OR | \| |
| not | Logical NOT | ~ |
| xor | Logical Exclusive OR | |
| any | True if any element of vector is nonzero | |
| all | True if all elements of vector are nonzero | |

*Special characters*

| | | |
|---|---|---|
| colon | Colon | : |
| paren | Parentheses and subscripting | ( ) |
| paren | Brackets | [ ] |
| paren | Braces and subscripting | { } |
| punct | Decimal point | . |
| punct | Structure field access | . |
| punct | Parent directory | .. |
| punct | Continuation | ... |
| punct | Separator | , |
| punct | Semicolon | ; |
| punct | Comment | % |
| punct | Invoke operating system command | ! |
| punct | Assignment | = |
| punct | Quote | ' |
| transpose | Transpose | .' |
| ctranspose | Complex conjugate transpose | ' |
| horzcat | Horizontal concatenation | [,] |
| vertcat | Vertical concatenation | [;] |
| subsasgn | Subscripted assignment | (),{},. |
| subsref | Subscripted reference | (),{},. |
| subsindex | Subscript index | |

### *Bitwise operators*

| | |
|---|---|
| `bitand` | Bit-wise AND. |
| `bitcmp` | Complement bits. |
| `bitor` | Bit-wise OR. |
| `bitmax` | Maximum floating point integer. |
| `bitxor` | Bit-wise XOR. |
| `bitset` | Set bit. |
| `bitget` | Get bit. |
| `bitshift` | Bit-wise shift. |

### *Set operators*

| | |
|---|---|
| `union` | Set union. |
| `unique` | Set unique. |
| `intersect` | Set intersection. |
| `setdiff` | Set difference. |
| `setxor` | Set exclusive-or. |
| `ismember` | True for set member. |

## **Directory:** `Toolbox/MATLAB/polyfun` (Interpolation and polynomials)

### *Data interpolation*

| | |
|---|---|
| `interp1` | 1-D interpolation (table lookup). |
| `interp1q` | Quick 1-D linear interpolation. |
| `interpft` | 1-D interpolation using FFT method. |
| `interp2` | 2-D interpolation (table lookup). |
| `interp3` | 3-D interpolation (table lookup). |
| `interpn` | N-D interpolation (table lookup). |
| `griddata` | Data gridding and surface fitting. |

### *Spline interpolation*

| | |
|---|---|
| `spline` | Cubic spline interpolation. |
| `ppval` | Evaluate piecewise polynomial. |

### *Geometric analysis*

| | |
|---|---|
| `delaunay` | Delaunay triangulation. |
| `dsearch` | Search Delaunay triangulation for nearest point. |
| `tsearch` | Closest triangle search. |
| `convhull` | Convex hull. |

| | |
|---|---|
| `voronoi` | Voronoi diagram. |
| `inpolygon` | True for points inside polygonal region. |
| `rectint` | Rectangle intersection area. |
| `polyarea` | Area of polygon. |

### *Polynomials*

| | |
|---|---|
| `roots` | Find polynomial roots. |
| `poly` | Convert roots to polynomial. |
| `polyval` | Evaluate polynomial. |
| `polyvalm` | Evaluate polynomial with matrix argument. |
| `residue` | Partial-fraction expansion (residues). |
| `polyfit` | Fit polynomial to data. |
| `polyder` | Differentiate polynomial. |
| `conv` | Multiply polynomials. |
| `deconv` | Divide polynomials. |

### *Utilities*

| | |
|---|---|
| `xychk` | Check arguments to 1-D and 2-D data routines. |
| `xyzchk` | Check arguments to 3-D data routines. |
| `xyzvchk` | Check arguments to 3-D volume data routines. |
| `automesh` | True if inputs should be automatically meshgridded. |
| `mkpp` | Make piecewise polynomial. |
| `unmkpp` | Supply details about piecewise polynomial. |
| `resi2` | Residue of a repeated pole. |
| `tzero` | Transmission zeros. |
| `abcdchk` | Check consistency of A,B,C,D matrices. |
| `ss2tf` | Convert state-space system to transfer function. |
| `ss2zp` | Convert state-space system to zero-pole. |
| `tf2ss` | Convert transfer function to state-space. |
| `tf2zp` | Convert transfer function to zero-pole. |
| `tfchk` | Check for proper transfer function. |
| `zp2ss` | Convert zero-pole system to state-space. |
| `zp2tf` | Convert zero-pole system to transfer function. |

## **Directory:** `Toolbox/MATLAB/sparfun` (Sparse matrices)

### *Elementary sparse matrices*

| | |
|---|---|
| `speye` | Sparse identity matrix. |
| `sprand` | Sparse uniformly distributed random matrix. |
| `sprandn` | Sparse normally distributed random matrix. |

| | |
|---|---|
| `sprandsym` | Sparse random symmetric matrix. |
| `spdiags` | Sparse matrix formed from diagonals. |

*Full to sparse conversion*

| | |
|---|---|
| `sparse` | Create sparse matrix. |
| `full` | Convert sparse matrix to full matrix. |
| `find` | Find indices of nonzero elements. |
| `spconvert` | Import from sparse matrix external format. |

*Working with sparse matrices*

| | |
|---|---|
| `nnz` | Number of nonzero matrix elements. |
| `nonzeros` | Nonzero matrix elements. |
| `nzmax` | Amount of storage allocated for nonzero matrix elements. |
| `spones` | Replace nonzero sparse matrix elements with ones. |
| `spalloc` | Allocate space for sparse matrix. |
| `issparse` | True for sparse matrix. |
| `spfun` | Apply function to nonzero matrix elements. |
| `spy` | Visualize sparsity pattern. |

*Reordering algorithms*

| | |
|---|---|
| `colmmd` | Column minimum degree permutation. |
| `symmmd` | Symmetric minimum degree permutation. |
| `symrcm` | Symmetric reverse Cuthill-McKee permutation. |
| `colperm` | Column permutation. |
| `randperm` | Random permutation. |
| `dmperm` | Dulmage-Mendelsohn permutation. |

*Linear algebra*

| | |
|---|---|
| `eigs` | A few eigenvalues. |
| `svds` | A few singular values. |
| `luinc` | Incomplete LU factorization. |
| `cholinc` | Incomplete Cholesky factorization. |
| `normest` | Estimate the matrix 2-norm. |
| `condest` | 1-norm condition number estimate. |
| `sprank` | Structural rank. |

*Linear equations, iterative methods*

| | |
|---|---|
| `pcg` | Preconditioned Conjugate Gradients Method. |
| `bicg` | BiConjugate Gradients Method. |

| | |
|---|---|
| `bicgstab` | BiConjugate Gradients Stabilized Method. |
| `cgs` | Conjugate Gradients Squared Method. |
| `gmres` | Generalized Minimum Residual Method. |
| `qmr` | Quasi-Minimal Residual Method. |

*Operations on trees*

| | |
|---|---|
| `treelayout` | Lay out tree or forest. |
| `treeplot` | Plot picture of tree. |
| `etree` | Elimination tree. |
| `etreeplot` | Plot elimination tree. |
| `gplot` | Plot graph, as in "graph theory". |

*Miscellaneous*

| | |
|---|---|
| `symbfact` | Symbolic factorization analysis. |
| `spparms` | Set parameters for sparse matrix routines. |
| `spaugment` | Form least squares augmented system. |

*Utility functions*

| | |
|---|---|
| `rjr` | Random Jacobi rotation. |
| `sparsfun` | Sparse auxiliary functions and parameters. |

**Directory:** `Toolbox/MATLAB/specfun` (Specialized math functions)

*Specialized math functions*

| | |
|---|---|
| `airy` | Airy functions. |
| `besselj` | Bessel function of the first kind. |
| `bessely` | Bessel function of the second kind. |
| `besselh` | Bessel functions of the third kind (Hankel function). |
| `besseli` | Modified Bessel function of the first kind. |
| `besselk` | Modified Bessel function of the second kind. |
| `beta` | Beta function. |
| `betainc` | Incomplete beta function. |
| `betaln` | Logarithm of beta function. |
| `ellipj` | Jacobi elliptic functions. |
| `ellipke` | Complete elliptic integral. |
| `erf` | Error function. |
| `erfc` | Complementary error function. |
| `erfcx` | Scaled complementary error function. |
| `erfinv` | Inverse error function. |

| | |
|---|---|
| `expint` | Exponential integral function. |
| `gamma` | Gamma function. |
| `gammainc` | Incomplete gamma function. |
| `gammaln` | Logarithm of gamma function. |
| `legendre` | Associated Legendre function. |
| `cross` | Vector cross product. |

*Number theoretic functions*

| | |
|---|---|
| `factor` | Prime factors. |
| `isprime` | True for prime numbers. |
| `primes` | Generate list of prime numbers. |
| `gcd` | Greatest common divisor. |
| `lcm` | Least common multiple. |
| `rat` | Rational approximation. |
| `rats` | Rational output. |
| `perms` | All possible permutations. |
| `nchoosek` | All combinations of N elements taken K at a time. |

*Coordinate transforms*

| | |
|---|---|
| `cart2sph` | Transform Cartesian to spherical coordinates. |
| `cart2pol` | Transform Cartesian to polar coordinates. |
| `pol2cart` | Transform polar to Cartesian coordinates. |
| `sph2cart` | Transform spherical to Cartesian coordinates. |
| `hsv2rgb` | Convert hue-saturation-value colors to red-green-blue. |
| `rgb2hsv` | Convert red-green-blue colors to hue-saturation-value. |

*Others*

| | |
|---|---|
| `dot` | Vector dot product. |
| `besschk` | Check arguments to bessel functions. |
| `bessela` | Obsolete Bessel function. |
| `betacore` | Core algorithm for the incomplete beta function. |
| `erfcore` | Core algorithm for error functions. |

**Directory:** `Toolbox/MATLAB/specgraph` (Specialized graphs)

*Specialized 2-D graphs*

| | |
|---|---|
| `area` | Filled area plot. |
| `bar` | Bar graph. |
| `barh` | Horizontal bar graph. |

| | |
|---|---|
| `bar3` | 3-D bar graph. |
| `bar3h` | Horizontal 3-D bar graph. |
| `comet` | Comet-like trajectory. |
| `errorbar` | Error bar plot. |
| `ezplot` | Easy to use function plotter. |
| `feather` | Feather plot. |
| `fill` | Filled 2-D polygons. |
| `fplot` | Plot function. |
| `hist` | Histogram. |
| `pareto` | Pareto chart. |
| `pie` | Pie chart. |
| `pie3` | 3-D pie chart. |
| `plotmatrix` | Scatter plot matrix. |
| `ribbon` | Draw 2-D lines as ribbons in 3-D. |
| `stem` | Discrete sequence or "stem" plot. |
| `stairs` | Stairstep plot. |

*Contour and $2^1/_2$-D graphs*

| | |
|---|---|
| `contour` | Contour plot. |
| `contourf` | Filled contour plot. |
| `contour3` | 3-D Contour plot. |
| `clabel` | Contour plot elevation labels. |
| `pcolor` | Pseudocolor (checkerboard) plot. |
| `quiver` | Quiver plot. |
| `voronoi` | Voronoi diagram. |

*Specialized 3-D graphs*

| | |
|---|---|
| `comet3` | 3-D comet-like trajectories. |
| `meshc` | Combination mesh/contour plot. |
| `meshz` | 3-D mesh with curtain. |
| `stem3` | 3-D stem plot. |
| `quiver3` | 3-D quiver plot. |
| `slice` | Volumetric slice plot. |
| `surfc` | Combination surf/contour plot. |
| `trisurf` | Triangular surface plot. |
| `trimesh` | Triangular mesh plot. |
| `waterfall` | Waterfall plot. |

*Images display and file I/O*

| | |
|---|---|
| `image` | Display image |
| `imagesc` | Scale data and display as image. |

`colormap` Color look-up table.
`gray` Linear gray-scale color map.
`contrast` Gray scale color map to enhance image contrast.
`brighten` Brighten or darken color map.
`colorbar` Display color bar (color scale).
`imread` Read image from graphics file.
`imwrite` Write image to graphics file.
`imfinfo` Information about graphics file.

*Movies and animation*

`capture` Screen capture of current figure.
`moviein` Initialize movie frame memory.
`getframe` Get movie frame.
`movie` Play recorded movie frames.
`qtwrite` Translate movie into QuickTime format (Macintosh only).
`rotate` Rotate object about specified origin and direction.
`frame2im` Convert movie frame to indexed image.
`im2frame` Convert indexed image into movie format.

*Color-related functions*

`spinmap` Spin color map.
`rgbplot` Plot color map.
`colstyle` Parse color and style from string.

*Solid modeling*

`cylinder` Generate cylinder.
`sphere` Generate sphere.

*Utilities*

`makebars` Make data for bar charts.
`contourc` Contour computation (used by `contour`).
`contours` Contouring over non-rectangular surface (used by `contour`).

*Obsolete functions*

`compass` Compass plot.
`rose` Angle histogram plot.

## Directory: `Toolbox/MATLAB/strfun` (Character strings)

### *General*

| | |
|---|---|
| `char` | Create character array (string). |
| `double` | Convert string to numeric character codes. |
| `cellstr` | Create cell array of strings from character array. |
| `blanks` | String of blanks. |
| `deblank` | Remove trailing blanks. |
| `eval` | Execute string containing MATLAB expression. |

### *String tests*

| | |
|---|---|
| `ischar` | True for character array (string). |
| `iscellstr` | True for cell array of strings. |
| `isletter` | True for letters of the alphabet. |
| `isspace` | True for white space characters. |

### *String operations*

| | |
|---|---|
| `strcat` | Concatenate strings. |
| `strvcat` | Vertically concatenate strings. |
| `strcmp` | Compare strings. |
| `strncmp` | Compare first N characters of strings. |
| `findstr` | Find one string within another. |
| `strjust` | Justify character array. |
| `strmatch` | Find possible matches for string. |
| `strrep` | Replace string with another. |
| `strtok` | Find token in string. |
| `upper` | Convert string to uppercase. |
| `lower` | Convert string to lowercase. |

### *String to number conversion*

| | |
|---|---|
| `num2str` | Convert number to string. |
| `int2str` | Convert integer to string. |
| `mat2str` | Convert matrix to `eval`'able string. |
| `str2num` | Convert string to number. |
| `sprintf` | Write formatted data to string. |
| `sscanf` | Read string under format control. |

*Base number conversion*

| | |
|---|---|
| `hex2num` | Convert IEEE hexadecimal to double precision number. |
| `hex2dec` | Convert hexadecimal string to decimal integer. |
| `dec2hex` | Convert decimal integer to hexadecimal string. |
| `bin2dec` | Convert binary string to decimal integer. |
| `dec2bin` | Convert decimal integer to binary string. |
| `base2dec` | Convert base B string to decimal integer. |
| `dec2base` | Convert decimal integer to base B string. |

*Utility functions*

| | |
|---|---|
| `strings` | Help for strings. |

**Directory:** `Toolbox/MATLAB/timefun` (Time and dates)

*Current date and time*

| | |
|---|---|
| `now` | Current date and time as date number. |
| `date` | Current date as date string. |
| `clock` | Current date and time as date vector. |

*Basic functions*

| | |
|---|---|
| `datenum` | Serial date number. |
| `datestr` | String representation of date. |
| `datevec` | Date components. |

*Date functions*

| | |
|---|---|
| `calendar` | Calendar. |
| `weekday` | Day of week. |
| `eomday` | End of month. |
| `datetick` | Date formatted tick labels. |

*Timing functions*

| | |
|---|---|
| `cputime` | CPU time in seconds. |
| `tic, toc` | Stopwatch timer. |
| `etime` | Elapsed time. |
| `pause` | Wait in seconds. |

**Directory:** `Toolbox/MATLAB/uitools` (Graphical user interface tools)

*GUI functions*

| | |
|---|---|
| `uicontrol` | Create user interface control. |
| `uimenu` | Create user interface menu. |
| `ginput` | Graphical input from mouse. |
| `dragrect` | Drag XOR rectangles with mouse. |
| `rbbox` | Rubberband box. |
| `selectmoveresize` | Interactively select, move, resize, or copy objects. |
| `waitforbuttonpress` | Wait for key/buttonpress over figure. |
| `waitfor` | Block execution and wait for event. |
| `uiwait` | Block execution and wait for resume. |
| `uiresume` | Resume execution of blocked M-file. |

*GUI design tools*

| | |
|---|---|
| `guide` | Design GUI. |
| `align` | Align uicontrols and axes. |
| `cbedit` | Edit callback. |
| `menuedit` | Edit menu. |
| `propedit` | Edit property. |

*Dialog boxes*

| | |
|---|---|
| `dialog` | Create dialog figure. |
| `axlimdlg` | Axes limits dialog box. |
| `errordlg` | Error dialog box. |
| `helpdlg` | Help dialog box. |
| `inputdlg` | Input dialog box. |
| `listdlg` | List selection dialog box. |
| `menu` | Generate menu of choices for user input. |
| `msgbox` | Message box. |
| `questdlg` | Question dialog box. |
| `warndlg` | Warning dialog box. |
| `uigetfile` | Standard open file dialog box. |
| `uiputfile` | Standard save file dialog box. |
| `uisetcolor` | Color selection dialog box. |
| `uisetfont` | Font selection dialog box. |
| `pagedlg` | Page position dialog box. |
| `printdlg` | Print dialog box. |
| `waitbar` | Display wait bar. |

### *Menu utilities*

| | |
|---|---|
| `makemenu` | Create menu structure. |
| `menubar` | Computer dependent default setting for MenuBar property. |
| `umtoggle` | Toggle "checked" status of uimenu object. |
| `winmenu` | Create submenu for "Window" menu item. |

### *Toolbar button group utilities*

| | |
|---|---|
| `btngroup` | Create toolbar button group. |
| `btnstate` | Query state of toolbar button group. |
| `btnpress` | Button press manager for toolbar button group. |
| `btndown` | Depress button in toolbar button group. |
| `btnup` | Raise button in toolbar button group. |

### *User-defined figure/axes property utilities*

| | |
|---|---|
| `clruprop` | Clear user-defined property. |
| `getuprop` | Get value of user-defined property. |
| `setuprop` | Set user-defined property. |

### *Miscellaneous utilities*

| | |
|---|---|
| `allchild` | Get all object children. |
| `hidegui` | Hide/unhide GUI. |
| `edtext` | Interactive editing of axes text objects. |
| `getstatus` | Get status text string in figure. |
| `setstatus` | Set status text string in figure. |
| `popupstr` | Get popup menu selection string. |
| `remapfig` | Transform figure objects' positions. |
| `setptr` | Set figure pointer. |
| `getptr` | Get figure pointer. |
| `overobj` | Get handle of object the pointer is over. |

### *Utilities*

| | |
|---|---|
| `textwrap` | Return wrapped string matrix for given `uicontrol`. |
| `btnicon` | Icon library for `btngroup`. |
| `icondisp` | Display icons in `btnicon`. |
| `ctlpanel` | Initialization of `guide` control panel. |
| `fignamer` | Chooses next available figure name. |
| `tabdlg` | Create and manage tabbed dialog box. |

B

# *Axes* Object Properties

This appendix lists properties and associated values or their description for the *axes* object. Default values are shown in *italics*. Unless stated otherwise, all properties can be set.

`'AmbientLightColor' ColorSpec`

Specifies the color of the ambient light that shines uniformly on all objects in the *axes*. One or more visible *light* objects must be present. `ColorSpec` is a three-element RGB vector or the string name of a standard color.

`'Box' 'on' | 'off'`

Specifies whether the *axes* is enclosed in a box for 2-D views or a cube for 3-D views.

`'BusyAction' 'cancel' | 'queue'`

Determines how to deal with interruptions by other object callbacks; that is, what action to take if a callback to this object is busy. If the `'Interruptible'` property of the *axes* object is `'off'`, `'cancel'` discards callbacks that interrupt the *axes* object, and `'queue'` places the interrupting callback in the event queue to be executed after the current callback terminates. If the `'Interruptible'` property of the object is `'on'`, `'BusyAction'` has no affect.

`'ButtonDownFcn' string`

Specifies the character string to be evaluated whenever you press a mouse button with the pointer over the *axes* object, but not over any descendent or parent objects.

`'CameraPosition' [x, y, z]`

Specifies the position from which the camera views the *axes* scene in *axes* coordinates.

`'CameraPositionMode' 'auto' | 'manual'`

Specifies whether MATLAB automatically adjusts the camera position such that the camera lies a fixed distance from the camera target along the azimuth and elevation specified by the `'View'` property. Set to `'manual'` if `'CameraPosition'` is set.

`'CameraTarget' [x, y, z]`

Specifies where in *axes* coordinates the camera points to.

`'CameraTargetMode' 'auto' | 'manual'`

Specifies whether MATLAB automatically positions the camera target at the centroid of the axes. Set to `'manual'` if `'CameraTarget'` is set.

`'CameraUpVector' [x, y, z]`

Specifies the rotation of the camera around the viewing axis defined by the `'CameraTarget'` and the `'CameraPosition'` properties. `'CameraUpVector'` is a three-element vector (e.g., `[0 1 0]` specifies the positive *y*-axis as the up direction). The default is `[0 0 1]`, which defines the positive *z*-axis as the up direction.

`'CameraUpVectorMode' 'auto' | 'manual'`

Specifies whether the `'CameraUpVector'` is default or user specified. If `'CameraUpVector'` is set, this property is set to `'manual'`.

`'CameraViewAngle' Ø ≤ angle ≤ 18Ø degrees`

Specifies the camera field of view.

`'CameraViewAngleMode' 'auto' | 'manual'`

Specifies whether `'CameraViewAngle'` is set automatically by MATLAB to the minimum angle that captures the entire *axes* scene. If `'CameraviewAngle'` is set, this property is set to `'manual'`.

`'Children' vector`

Contains a vector of handles to all child objects in the *axes*. The order the handles appear is the stacking order with the topmost object appearing first. Rearranging the vector rearranges the stacking order. The *text* objects used to title the plot and to label the *x*-, *y*-, and *z*-axes are default children of *axes* that have their `'HandleVisibility'` property set to `'callback'`.

`'CLim' [cmin, cmax]`

Specifies a two-element vector that determines color axis limits.

`'CLimMode' 'auto' | 'manual'`

Specifies whether `'Clim'` is automatically adjusted by MATLAB. Setting the `'Clim'` property sets this property to `'manual'`.

`'Color' 'none' | ColorSpec`

Specifies the color of the *axes* background. `'none'` specifies a transparent axes. `Colorspec` is a three-element RGB vector or the string name of a standard color.

`'ColorOrder' m-by-3 array`

Specifies a sequence of `m` RGB values to use for line plots.

`'CreateFcn' string`

Specifies the callback string evaluated just after the object is created. Must be set as a default property; for example, `set(Ø,'DefaultAxesCreateFcn', 'grid off')`.

```
'CurrentPoint' [xback,yback,zback;...
                xfront,yfront,zfront]
```

Contains the location of the last mouse button click in the axes in axes data coordinates. The 2-by-3 array contains the coordinates of two data points that lie on a line perpendicular to the plane of the screen passing through the pointer.

```
'DataAspectRatio' [dx dy dz]
```

Specifies the relative scaling of *axes* data units. For example, `[1 2 3]` causes the length of one unit of data in the *x*-direction to be the same length as two units in the *y*-direction and three units in the *z*-direction.

```
'DataAspectRatioMode' 'auto' | 'manual'
```

Specifies whether MATLAB automatically adjusts the `'DataAspectRatio'`. If `'DataAspectRatio'` is set, this property is set to `'manual'`.

```
'DeleteFcn' string
```

Specifies the callback string evaluated just prior to deleting the object.

```
'DrawMode' 'normal' | 'fast'
```

Specifies the tradeoff between drawing speed and accuracy for the painters rendering method.

```
'FontAngle' 'normal' | 'italic' | 'oblique'
```

Specifies the character slant to use for *axes* text.

```
'FontName' string
```

Specifies the font family to use for *axes* text. Helvetica is default.

```
'FontSize' number
```

Specifies the font size in units specified by the `'FontUnits'` property. Ten points is default.

```
'FontUnits' 'points' | 'normalized' | 'inches' | ...
            'centimeters' | 'pixels'
```

Specifies the units for font size. `'normalized'` is with respect to the *axes* height.

`'FontWeight' 'light' | 'normal' | 'demi' | 'bold'`

Specifies font weight.

`'GridLineStyle' '-' | '--' | ':' | '-.' | 'none'`

Specifies the line style used to draw grid lines.

`'HandleVisibility' 'on' | 'callback' | 'off'`

Specifies whether the object's handle can be seen from the *Command* window or callbacks. When set to `'callback'`, the object's handle is visible from callbacks but not the *Command* window. When a handle is invisible it does not show up in a list of children, nor can it be found by the `findobj` function. However, if the handle is known, it can still be accessed by `get` and `set`.

`'Interruptible' 'on' | 'off'`

Specifies whether *axes* object callbacks are interruptible by other callbacks.

`'Layer' 'bottom' | 'top'`

Specifies the placement of tick marks and grid lines with respect to *axes* children.

`'LineStyleOrder' LineSpec`

A character string with linestyles and markers separated by vertical bars |, which are used when creating multiple line plots. MATLAB cycles through these styles and markers ***after*** using all colors in the `'ColorOrder'` property. The default is `'-'`.

`'LineWidth' scalar`

Specifies the line width of axis lines in points (1 point = 1/72 inch). The default is 0.5 points.

`'NextPlot' 'add' | 'replace' | 'replacechildren'`

Specifies where to draw the next plot using high-level plotting functions such as `plot`:

> `'add'`—add plot to existing axes; that is, `hold on`.
> `'replace'`—issue `cla reset`, then plot.
> `'replacechildren'`—issue `cla`, then plot.

`'Parent'` `figure handle`

Parent *figure* object handle.

`'PlotBoxAspectRatio'` `[px py pz]`

Specifies the relative *x*-,*y*-, and *z*-direction scaling of *axes* plotbox (the total rectangular area within the *figure* used by the axes).

`'PlotBoxAspectRatioMode'` `'`*`auto`*`'` | `'manual'`

Specifies whether MATLAB automatically adjusts axis scaling. Setting `'PlotBoxAspectRatio'` sets this property to `'manual'`.

`'Position'` `[left,bottom,width,height]`

Specifies the *axes* position in standard position rectangle format with respect to the lower left-hand corner of the *figure* window. Units are defined by the `'Units'` property.

`'Projection'` `'`*`orthographic`*`'` | `'perspective'`

Specifies the type of graphic projection used.

`'Selected'` `'on'` | `'off'`

Specifies whether object is selected. Selection handles are shown if the `'SelectionHighlight'` property is `'on'`.

`'SelectionHighlight'` `'`*`on`*`'` | `'off'`

Specifies whether selection handles are shown when `'Selected'` is `'on'`.

`'Tag'` `string`

Character string specified by the user to tag the object so that it is easily found using `findobj`.

`'TickDir'` `'in'` | `'out'`

Specifies the direction of axis tick marks. `'in'` is default for 2-D views. `'out'` is default for 3-D views.

`'TickDirMode'` *`'auto'`* `| 'manual'`

Specifies whether MATLAB automatically adjusts the tick direction as the view changes.

`'TickLength' [2Dlength,3Dlength]`

A two-element vector that specifies the tick mark length for 2-D and 3-D views, respectively. The lengths are normalized lengths with respect to the longest visible *x*-, *y*-, or *z*-axis lines.

`'Title' handle`

Contains the handle to the *text* object that is used for the *axes* title.

`'Type' string`

Read-only property that returns the string `'axes'`.

```
'Units' 'pixels' | 'normalized' | 'inches' |...
        'centimeters' | 'points'
```

Specifies the units of measurement for *axes* object `'Position'` property.

`'UserData' variable`

Storage location for the contents of a variable of any class. Only accessible by using `get` and `set`.

`'View' [Azimuth, Elevation]`

Specifies the *axes* viewpoint. Considered obsolete during MATLAB version 5 development, but kept in part because it is easier to manipulate than the camera properties.

`'Visible'` *`'on'`* `| 'off'`

Determines whether the *axes* object is visible. `get` and `set` still work on an invisible object.

`'XAxisLocation' 'top' |` *`'bottom'`*

Specifies the location of *x*-axis tick marks and labels.

`'YAxisLocation' 'right' |` *`'left'`*

Specifies the location of *y*-axis tick marks and labels.

**Properties that affect the *X*-, *Y*-, or *Z*-axis:**

`'XColor', 'YColor', 'ZColor' ColorSpec`

Specifies the color of axis lines, tick marks, tick mark labels, and axis grid lines on the associated axis. `ColorSpec` is a three-element RGB vector or the string name of a standard color.

`'XDir', 'YDir', 'ZDir'` *`'normal'`* `| 'reverse'`

Specifies the direction of increasing values with respect to a right-hand coordinate system on the associated axis.

`'XGrid', 'YGrid', 'ZGrid' 'on' |` *`'off'`*

Determines the presence of axis grid lines on the associated axis.

`'XLabel', 'YLabel', 'ZLabel' handle`

Contains handle of respective axis *text* object labels.

`'XLim', 'YLim', 'ZLim' [min, max]`

Specifies a two-element vector containing the minimum and maximum values of the associated axis.

`'XLimMode', 'YLimMode', 'ZLimMode'` *`'auto'`* `| 'manual'`

Specifies whether the respective axis limits are chosen automatically by MATLAB. If `'Xlim'`, `'Ylim'`, or `'Zlim'` are set, the respective `'XLimMode'`, `'YLimMode'`, or `'ZLimMode'` properties are set to `'manual'`.

`'XScale', 'YScale', 'ZScale'` *`'linear'`* `| 'log'`

Specifies the axis scaling for the associated coordinate axis.

`'XTick', 'YTick', 'ZTick' vector`

Specifies a vector of monotonically increasing data values that identify the tick mark locations on the associated coordinate axis. For no tick marks, set equal to an empty array [ ].

`'XTickMode', 'YTickMode', 'TickMode'` *`'auto'`* `| 'manual'`

Specifies whether the respective axis tick vectors are generated automatically by MATLAB. If `'XTick'`, `'YTick'`, or `'ZTick'` are set, the respective axis tick mode is set to `'manual'`.

`'XTickLabel', 'YTickLabel', 'ZTickLabel' string`

Specifies the character string labels to be placed at the tick mark locations on the associated axis. `string` can be specified as a cell array of strings, a string matrix, a character string with tick labels separated by vertical slash characters, or a numerical vector that is internally converted to a string matrix.

`'XTickLabelMode', 'YTickLabelMode',...`
`'ZTickLabelMode'` *`'auto'`* `| 'manual'`

Specifies whether the respective axis tick labels are generated automatically by MATLAB. If `'XTickLabel'`, `'YTickLabel'`, or `'ZTickLabel'` are set, the respective axis tick label mode is set to `'manual'`.

C

# *Figure* Object Properties

This appendix lists properties and associated values or their description for the *figure* object. Default values are shown in *italics*. Unless stated otherwise, all properties can be set.

'BackingStore' '*on*' | 'off'

Specifies whether a copy of the *figure* is stored in an off-screen pixel buffer. An 'off' setting speeds animations, but bringing the *figure* to the front requires more time since the *figure* must be regenerated rather than being copied from the off-screen buffer.

'BusyAction' 'cancel' | '*queue*'

Determines how to deal with interruptions by other object callbacks; that is, what action to take if a callback to this object is busy. If the 'Interruptible' property of the *figure* object is 'off', 'cancel' discards callbacks that interrupt the *figure* object, and 'queue' places the interrupting callback in the event queue to be executed after the current callback terminates. If the 'Interruptible' property of the object is 'on', 'BusyAction' has no affect.

`'ButtonDownFcn' string`

Specifies the character string to be evaluated whenever you press a mouse button with the pointer over the *figure* object, but not over a descendent object.

`'Children' vector`

Contains a vector of handles to all *axes, uicontrol,* and *uimenu* objects in the *figure.* The order the handles appear is the stacking order with the topmost object appearing first. Rearranging the vector shuffles the stacking order.

`'CloseRequestFCn' string`

Specifies a callback routine that is evaluated whenever you initiate the process of closing the *figure* in a way other than using the `delete` command. The default callback is `closereq`, which uses `delete(get(0,'CurrentFigure'))`.

`'Color' Colorspec`

Specifies the color of the *figure* background color. `Colorspec` is a three-element RGB vector or the string name of a standard color.

`'Colormap' m-by-3 array`

Specifies an `m`-by-3 array of RGB values that define `m` colormap colors for the rendering of *surface, image,* and *patch* objects.

`'CreateFcn' string`

Specifies the callback string evaluated just after the object is created. Must be set as a default property; for example, `set(0,'DefaultFigureCreateFcn','grid off')`.

`'CurrentAxes' handle`

Specifies the handle of the current *axes.* The current *axes* need not be the topmost *axes.* If there are no *axes* objects in the *figure,* `get(gcf,'CurrentAxes')` is an empty array.

`'CurrentCharacter' character`

Read-only property containing the last key pressed in the *figure* window.

`'CurrentObject' handle`

Contains the handle of the object that is under the current point as defined by the `'CurrentPoint'` property. The handle returned in the topmost object in the stacking order. The function `gco` returns this handle.

`'CurrentPoint' [x, y]`

Specifies the *x*- and *y*-coordinates of the last mouse button click in the *figure* in units defined by the `'Units'` property. Measurements are with respect to the lower left-hand corner of the *figure*.

`'DeleteFcn' string`

Specifies the callback string evaluated just prior to deleting the object.

`'Dithermap' m-by-3 array`

Specifies the colormap used for true-color data on pseudocolor displays.

`'DithermapMode' 'auto' | 'manual'`

Specifies how dithered displays are generated:

> `'manual'`—use of the 'Dithermap' array.
> `'auto'`—generate dithermap based on the colors currently displayed.

`'FixedColors' m-by-3 array`

Read-only RGB array of fixed colors appearing in a *figure* that are not obtained from the *figure* colormap.

`'HandleVisibility' 'on' | 'callback' | off'`

Specifies whether the object's handle can be seen from the *Command* window or callbacks. When set to `'callback'`, the object's handle is visible from callbacks but not the *Command* window. When a handle is invisible it does not show up in a list of children, nor can it be found by the `findobj` function. However, if the handle is known, it can still be accessed by `get` and `set`.

`'IntegerHandle' 'on' | 'off'`

Specifies whether the *figure* handle is an integer or floating-point number.

`'Interruptible'` *`'on'`* `| 'off'`

Specifies whether *figure* object callbacks are interruptible by other callbacks.

`'InvertHardcopy'` *`'on'`* `| 'off'`

Specifies whether printed *figures* appear as colored objects on a white background or print as shown in the *figure.*

`'KeyPressFcn' string`

Specifies a callback that is evaluated when a key is pressed while the *figure* is active.

`'Menubar' 'none' |` *`'figure'`*

Specifies the presence of the menubar in the *figure.* Since the Macintosh menubar appears at the top of the screen, it is not common to hide it. However, it is common to hide it on other platforms when the *figure* contains a GUI or dialog box. The function `menubar` handles this automatically; that is, `set(gcf,'Menubar',menubar)` hides the menubar on all platforms except the Macintosh.

`'MinColormap' scalar`

Specifies the minimum number of system color table entries used by MATLAB to store the *figure* colormap. The default is 64.

`'Name' string`

Specifies the title displayed in the *figure* window title bar. For example, `set(gcf,'Name','Hi Jack')` generates a *figure* window title bar `'Figure No. 1: Hi Jack'`.

`'NextPlot'` *`'add'`* `| 'replace' | 'replacechildren'`

Specifies which *figure* to use for the next graphics function:

> `'add'`—use the current figure.
> `'replace'`—issue `clf reset` then use the current *figure.*
> `'replacechildren'`—issue `clf` then use the current *figure.*

See the `newplot` function for additional information.

`'NumberTitle' 'on' | 'off'`

Specifies whether the *figure* window title bar contains the string `'Figure No. N'`, where `N` is the *figure* handle.

`'PaperOrientation' 'portrait' | 'landscape'`

Specifies horizontal (landscape) or vertical (portrait) paper orientation for printed *figures.*

`'PaperPosition' [left, bottom, width, height]`

Specifies the location of the *figure* on the printed page in standard position rectangle format when `'PaperPositionMode'` is set to `'manual'`. Units are defined by the `'PaperUnits'` property.

`'PaperPositionMode' 'auto' | 'manual'`

Specifies whether figures are printed using the `'PaperPosition'` property information or are `'auto'` printed WYSIWYG, centered on the page.

`'PaperSize' [width, height]`

Read only property that returns the size of the current paper type in units defined by the `'PaperUnits'` property.

```
'Paper Type' 'usletter' | 'uslegal' | 'a3' |...
             'a4letter' | 'a5' | 'b4' | 'tabloid'
```

Specifies the current paper type.

```
'PaperUnits' 'normalized' | 'inches' |...
             'centimeters' | 'points'
```

Specifies the units used for paper properties.

`'Parent'`

Parent object handle to *figure* object. Always returns 0.

```
'Pointer' 'crosshair' | 'arrow' | 'watch' |...
          'topl' | 'topr' | 'botl' | 'botr' |...
          'circle' | 'cross' | 'fleur' |...
          'left' | 'right' | 'top' | 'bottom' |...
          'fullcrosshair' | 'Ibeam' | 'custom'
```

Specifies the mouse pointer used in the *figure.*

`'PointerShapeCData' 16-by16 array`

Defines the mouse pointer that is used when the `'Pointer'` property is set to `'custom'`. Each element specifies a pixel, with element (1,1) being the upper left corner, using the following values:

- `1`—color pixel black.
- `2`—color pixel white.
- `NaN`—make pixel transparent.

`'PointerShapeHotSpot' [i, j]`

Specifies the row and column indices in `'PointerShapeCData'` indicating the pointer's selection spot. The default location is (1,1).

`'Position' [left,bottom,width,height]`

Specifies the *figure* position in standard position rectangle format with respect to the lower left-hand corner of the screen. Units are defined by the `'Units'` property.

`'Renderer'` *`'painter'`* `| 'zbuffer'`

Specifies the rendering method used for both the screen and printing.

`'RendererMode'` *`'auto'`* `| 'manual'`

Specifies whether MATLAB selects the best rendering mode automatically and individually for the screen and printer.

`'Resize'` *`'on'`* `| 'off'`

Specifies whether the *figure* is resizable by the user.

`'ResizeFcn' string`

Specifies a string that is evaluated immediately after the *figure* window has been resized by the user.

`'Selected' 'on' | 'off'`

Specifies whether object is selected. Selection handles are not shown on *figure* window.

`'SelectionHighlight' 'on' | 'off'`

Specifies whether selection handles are shown when `'Selected'` is `'on'`.

`'SelectionType' 'normal' | 'extended' | 'alt' | 'open'`

Specifies the type of mouse selection made:

> `'normal'`—Single click using leftmost mouse button.
>
> `'extended'`—**Shift**-click using leftmost mouse button.
>
> `'alt'`—**Ctrl**-click left mouse button on PCs and X-windows platforms, **Option**-click on the Macintosh.
>
> `'open'`—Double click any mouse button.
>
> `'ListBox'` style *uicontrol* objects set `'SelectionType' to 'normal'` to indicate a single mouse click, and they set it to `'open'` to indicate a double mouse click.

`'ShareColors' 'on' | 'off'`

Specifies whether the *figure* shares slots in the system color table. An `'on'` setting is more efficient, while an `'off'` setting may speed *figure* rerendering.

`'Tag' string`

Character string specified by the user to tag the object so that it is easily found using `findobj`.

`'Type' string`

Read-only property that returns the string `'figure'`.

```
'Units' 'pixels' | 'normalized' | 'inches' |...
        'centimeters' | 'points'
```

Specifies the units of measurement for *figure* object properties.

`'UserData' variable`

Storage location for the contents of a variable of any class. Only accessible by using `get` and `set`.

`'Visible' 'on' | 'off'`

Determines whether the *figure* object is visible. `get` and `set` still work on an invisible object.

`'WindowButtonDownFcn' string`

Specifies a callback string that is evaluated whenever you press a mouse button down while the mouse pointer is in the *figure* window.

`'WindowButtonMotionFcn' string`

Specifies a callback string that is evaluated whenever you move the mouse within the *figure* window.

`'WindowButtonUpFcn' string`

Specifies a callback string that is evaluated whenever you release a mouse button after a `'WindowButtonDownFcn'` event occurred in the *figure* window.

`'WindowStyle'` *`'normal'`* `| 'modal'`

Specifies *figure* window style. A `'modal'` window is primarily for dialog boxes and GUIs where the user must respond before being able to bring another window forward.

D

# *Image* Object Properties

This appendix lists properties and associated values or their description for the *image* object. Default values are shown in *italics*. Unless stated otherwise, all properties can be set.

`'BusyAction' 'cancel' | '`*`queue`*`'`

Determines how to deal with interruptions by other object callbacks; that is, what action to take if a callback to this object is busy. If the `'Interruptible'` property of the *image* object is `'off'`, `'cancel'` discards callbacks that interrupt the *image* object, and `'queue'` places the interrupting callback in the event queue to be executed after the current callback terminates. If the `'Interruptible'` property of the object is `'on'`, `'BusyAction'` has no affect.

`'ButtonDownFcn' string`

Specifies the character string to be evaluated whenever you press a mouse button with the pointer over the *image* object.

`'CData' matrix or 3-D array`

The image data as specified by any one of the three *image* interpretation methods.

`'CDataMapping' 'scaled' | 'direct'`

Specifies whether colors are direct or scaled indexed. This property has no effect when true color is used.

`'Clipping' 'on' | 'off'`

Specifies whether images are to be clipped to the *axes* rectangle.

`'CreateFcn' string`

Specifies the callback string evaluated just after the object is created. Must be set as a default property; for example, `set(0,'DefaultImageCreateFcn','axis image')`.

`'DeleteFcn' string`

Specifies the callback string evaluated just prior to deleting the object.

`'EraseMode' 'normal' | 'none' | 'xor' | 'background'`

Specifies what procedure MATLAB uses to draw the object.:

- `'normal'`—redraw the display, performing all tasks required to ensure that all axes objects are rendered correctly.
- `'none'`—do not redraw anything.
- `'xor'`—draw and erase the object by exclusive-oring (XOR) it with the color underneath it. Other covered objects are not destroyed.
- `'background'`—erase the object by redrawing it in the axes background color. Other covered objects are destroyed.

`'none'`, `'xor'`, and `'background'` are faster but less accurate than `'normal'` and are therefore useful when animation or object movement with the mouse is desired.

`'HandleVisibility' 'on' | 'callback' | 'off'`

Specifies whether the object's handle can be seen from the *Command* window or callbacks. When set to `'callback'`, the object's handle is visible from callbacks but not the *Command* window. When a handle is invisible it does not show up in a list of children, nor can it be found by the `findobj` function. However, if the handle is known, it can still be accessed by `get` and `set`.

`'Interruptible' 'on' | 'off'`

Specifies whether *image* object callbacks are interruptible by other callbacks. Only `'ButtonDownFcn'` callback routines can be interrupted. MATLAB checks for interruptible events only when it encounters a `drawnow`, `figure`, `getframe`, or `pause` command.

`'Parent' Axes handle`

Parent *axes* object handle to *image* object.

`'Selected' 'on' | 'off'`

Specifies whether object is selected. Selection handles are shown if the `'SelectionHighlight'` property is `'on'`.

`'SelectionHighlight' 'on' | 'off'`

Specifies whether selection handles are shown when `'Selected'` is `'on'`.

`'Tag' string`

Character string specified by the user to tag the object so that it is easily found using `findobj`.

`'Type' string`

Read-only property that returns the string `'image'`.

`'UserData' variable`

Storage location for the contents of a variable of any class. Only accessible by using `get` and `set`.

`'Visible' 'on' | 'off'`

Determines whether the *image* object is visible. `get` and `set` still work on an invisible object.

`'XData' [1 size(Cdata,2)]`

A two-element vector specifying the *x*-coordinates spanned by the *image.*

`'YData' [1 size(Cdata,1)]`

A two-element vector specifying the *y*-coordinate spanned by the *image.*

E

# *Light* Object Properties

This appendix lists properties and associated values or their description for the *light* object. Default values are shown in *italics*. Unless stated otherwise, all properties can be set.

`'BusyAction'  'cancel' | 'queue'`

Determines how to deal with interruptions by other object callbacks; that is, what action to take if a callback to this object is busy. If the `'Interruptible'` property of the *light* object is `'off'`, `'cancel'` discards callbacks that interrupt the *light* object, and `'queue'` places the interrupting callback in the event queue to be executed after the current callback terminates. If the `'Interruptible'` property of the object is `'on'`, `'BusyAction'` has no affect.

`'Color'  ColorSpec`

Specifies the color of the *light* object. `Colorspec` is a three-element RGB vector or the string name of a standard color.

'CreateFcn' string

Specifies the callback string evaluated just after the object is created. Must be set as a default property; for example, set(Ø,'DefaultLightCreateFcn','set(gcf,'Colormap jet').

'DeleteFcn' string

Specifies the callback string evaluated just prior to deleting the object.

'HandleVisibility' '*on*' | 'callback' | 'off'

Specifies whether the object's handle can be seen from the *Command* window or callbacks. When set to 'callback', the object's handle is visible from callbacks but not the *Command* window. When a handle is invisible it does not show up in a list of children, nor can it be found by the findobj function. However, if the handle is known, it can still be accessed by get and set.

'Interruptible' '*on*' | 'off'

Specifies whether *light* object callbacks are interruptible by other callbacks. 'DeleteFcn' callbacks are inherently not interruptible.

'Style' '*infinite*' | 'local'

Specifies the light source style as being parallel (infinitely far away) or divergent (local).

'Parent' Axes handle

Parent *axes* object handle to *light* object.

'Position' [x,y,z]

Vector containing the *x*-, *y*-, and *z*-coordinates of the *light* object in axes data units.

'Tag' string

Character string specified by the user to tag the object so that it is easily found using findobj.

'Type' string

Read-only property that returns the string 'light'.

`'UserData' variable`

Storage location for the contents of a variable of any class. Only accessible by using `get` and `set`.

`'Visible'` *`'on'`* `| 'off'`

Determines whether the effects of the *light* object are visible. `get` and `set` still work on an invisible object.

F

# *Line* Object Properties

This appendix lists properties and associated values or their description for the *line* object. Default values are shown in *italics*. Unless stated otherwise, all properties can be set.

`'BusyAction'  'cancel'  |  'queue'` (default: *'queue'*)

Determines how to deal with interruptions by other object callbacks; that is, what action to take if a callback to this object is busy. If the `'Interruptible'` property of the *line* object is `'off'`, `'cancel'` discards callbacks that interrupt the *line* object, and `'queue'` places the interrupting callback in the event queue to be executed after the current callback terminates. If the `'Interruptible'` property of the object is `'on'`, `'BusyAction'` has no affect.

`'ButtonDownFcn'  string`

Specifies the character string to be evaluated whenever you press a mouse button with the pointer over the *line* object.

`'Clipping' 'on' | 'off'`

Specifies whether the *line* is to be clipped to the *axes* rectangle.

`'Color' ColorSpec`

Specifies the color of the *line* object. `ColorSpec` is a three-element RGB vector or the string name of a standard color.

`'CreateFcn' string`

Specifies the callback string evaluated just after the object is created. Must be set as a default property; for example, `set(0,'DefaultLineCreateFcn','grid off')`.

`'DeleteFcn' string`

Specifies the callback string evaluated just prior to deleting the object.

`'EraseMode' 'normal' | 'none' | 'xor' | 'background'`

Specifies what procedure MATLAB uses to draw the object:

- `'normal'`—redraw the display, performing all tasks required to ensure that all *axes* objects are rendered correctly.
- `'none'`—do not redraw anything.
- `'xor'`—draw and erase the object by exclusive-oring (XOR) it with the color underneath it. Other covered objects are not destroyed.
- `'background'`—erase the object by redrawing it in the *axes* background color. Other covered objects are destroyed.

`'none'`, `'xor'`, and `'background'` are faster but less accurate than `'normal'` and are therefore useful when animation or object movement with the mouse is desired.

`'HandleVisibility' 'on' | 'callback' | 'off'`

Specifies whether the object's handle can be seen from the *Command* window or callbacks. When set to `'callback'`, the object's handle is visible from callbacks but not the *Command* window. When a handle is invisible it does not show up in a list of children, nor can it be found by the `findobj` function. However, if the handle is known, it can still be accessed by `get` and `set`.

`'Interruptible' 'on' | 'off'`

Specifies whether *line* object callbacks are interruptible by other callbacks.

`'LineStyle' '-' |' --' | ':' | '-.' | 'none'`

Specifies the linestyle used for the *line*.

`'LineWidth' scalar`

Specifies the line width of the *line* in points (1 point = 1⁄72 inch). The default is 0.5 points.

`'Marker' (any standard marker character)`

Specifies the marker symbol placed at the *line* data points.

`'MarkerEdgeColor' ColorSpec | 'none' | 'auto'`

Specifies the color of the marker or edge color for filled markers:

> `Colorspec`—defines the color to use.
> `'none'`—specifies no color, which makes nonfilled markers invisible.
> `'auto'`—sets `'MarkerEdgeColor'` to the same color as the `'Color'` property.

`'MarkerFaceColor' ColorSpec | 'none' | 'auto'`

Specifies the fill color for markers that are closed shapes:

> `Colorspec`—defines the color to use.
> `'none'`—makes the marker interior transparent.
> `'auto'`—sets the fill color to the *axes* color.

`'MarkerSize' scalar`

Specifies the size of the marker in points. The default size is 6 points.

`'Parent' Axes handle`

Parent *axes* object handle to *line* object.

`'Selected' 'on' | 'off'`

Specifies whether object is selected. Selection handles are shown if the `'SelectionHighlight'` property is `'on'`.

`'SelectionHighlight' 'on' | 'off'`

Specifies whether selection handles are shown when `'Selected'` is `'on'`.

`'Tag' string`

Character string specified by the user to tag the object so that it is easily found using `findobj`.

`'Type' string`

Read-only property that returns the string `'line'`.

`'UserData' variable`

Storage location for the contents of a variable of any class. Only accessible by using `get` and `set`.

`'Visible'` *`'on'`* `| 'off'`

Determines whether the *line* object is visible. `get` and `set` still work on an invisible object.

`'XData' array`

Array containing *x*-coordinates defining the *line*.

`'YData' array`

Array containing *y*-coordinates defining the *line*.

`'ZData' array`

Array containing *z*-coordinates defining the *line*.

G

# *Patch* Object Properties

This appendix lists properties and associated values or their description for the *patch* object. Default values are shown in *italics*. Unless stated otherwise, all properties can be set.

`'AmbientStrength'` $0 \leq$ scalar $\leq 1$

Specifies the ambient light strength of one or more visible *light* objects illuminating the *axes*.

`'BackFaceLighting'` `'unlit'` | `'lit'` | *`'reverselit'`*

Specifies how faces are lit when their vertex normals point away from the camera.

`'BusyAction'` `'cancel'` | *`'queue'`*

Determines how to deal with interruptions by other object callbacks; that is, what action to take if a callback to this object is busy. If the `'Interruptible'` property of the *patch* object is `'off'`, `'cancel'` discards callbacks that interrupt the *patch* object, and `'queue'` places the interrupting callback in the event queue to be executed after the current callback terminates. If the `'Interruptible'` property of the object is `'on'`, `'BusyAction'` has no affect.

`'ButtonDownFcn'` `string`

Specifies the character string to be evaluated whenever you press a mouse button with the pointer over the *patch* object.

`'CData'` `scalar, vector, or matrix`

Specifies the color of the *patch* object. You can specify color for each vertex, each face, or a single color for the entire *patch.* The data can be numeric values that are scaled to map linearly into the current colormap, integer values that are used directly as indices into the current colormap, or arrays of RGB values. See MATLAB documentation for a more thorough description.

`'CDataMapping'` *`'scaled'`* `|` `'direct'`

Determines how indexed color data are used to color the *patch* object.

`'Clipping'` *`'on'`* `|` `'off'`

Specifies whether the *patch* is to be clipped to the *axes* rectangle.

`'CreateFcn'` `string`

Specifies the callback string evaluated just after the object is created. Must be set as a default property; for example, `set(0,'DefaultPatchCreateFcn','grid off')`.

`'DeleteFcn'` `string`

Specifies the callback string evaluated just prior to deleting the object.

`'EdgeColor'` *`ColorSpec`* `|` `'none'` `|` `'flat'` `|` `'interp'`

Specifies how color is applied to the *patch* edges:

`Colorspec`—a three-element RGB vector or the string name of a standard color.
`'none'`—the edges are not drawn.
`'flat'`—the color of each vertex controls the color of the edge that follows it.
`'interp'`—interpolated coloring is used.

`'EdgeLighting'` *`'none'`* `| 'flat' | 'gouraud' | 'phong'`

Specifies the algorithm used to calculate the effect of light objects on *patch* edges.

`'EraseMode'` *`'normal'`* `| 'none' | 'xor' | 'background'`

Specifies what procedure MATLAB uses to draw the object:

- `'normal'`—redraw the display, performing all tasks required to ensure that all *axes* objects are rendered correctly.
- `'none'`—do not redraw anything.
- `'xor'`—draw and erase the object by exclusive-oring (XOR) it with the color underneath it. Other covered objects are not destroyed.
- `'background'`—erase the object by redrawing it in the *axes* background color. Other covered objects are destroyed.

`'none'`, `'xor'`, and `'background'` are faster but less accurate than `'normal'` and are therefore useful when animation or object movement with the mouse is desired.

`'FaceColor'` *`ColorSpec`* `| 'none' | 'flat' | 'interp'`

Specifies how color is applied to the *patch* face:

- `Colorspec`—a three-element RGB vector or the string name of a standard color.
- `'none'`—the faces are not drawn.
- `'flat'`—the color of the first vertex controls the color of the face.
- `'interp'`—interpolated coloring is used.

`'FaceLighting' 'none' |` *`'flat'`* `| 'gouraud' | 'phong'`

Specifies the algorithm used to calculate the effect of light objects on *patch* faces.

`'Faces' m-by-n array`

Specifies the connection array that identifies which vertices in the `'Vertices'` property are connected. The faces array defines `m` faces with up to `n` vertices each. Each row designates the connections for a single face and the number of elements in that row that are not `NaN` defines the number of vertices for that face. The `'Faces'` and `'Vertices'` properties provide an alternative and often more efficient way to specify a *patch*.

`'FaceVertexCData' scalar, vector, or matrix`

Specifies the color of *patches* defined by the `'Faces'` and `'Vertices'` properties when `'FaceColor'`, `'EdgeColor'`, `'MarkerFaceColor'`, or `'MarkerEdgeColor'` are set properly. The data can be numeric values that are scaled to map linearly into the current colormap, integer values that are used directly as indices into the current colormap, or arrays of RGB values. See MATLAB documentation for a more thorough description.

`'HandleVisibility' 'on' | 'callback' | 'off'`

Specifies whether the object's handle can be seen from the *Command* window or callbacks. When set to `'callback'`, the object's handle is visible from callbacks but not the *Command* window. When a handle is invisible it does not show up in a list of children, nor can it be found by the `findobj` function. However, if the handle is known, it can still be accessed by `get` and `set`.

`'Interruptible' 'on' | 'off'`

Specifies whether *patch* object callbacks are interruptible by other callbacks.

`'LineStyle' '-' | '--' | ':' | '-.' | 'none'`

Specifies the linestyle used for the *patch* edges.

`'LineWidth' scalar`

Specifies the line width of *patch* edges in points (1 point = 1/72 inch). The default is 0.5 points.

`'Marker' (any standard marker character)`

Specifies the marker symbol placed at the *patch* vertices.

`'MarkerEdgeColor' ColorSpec | 'none' | 'auto' | 'flat'`

Specifies the color of the marker or edge color for filled markers:

- `Colorspec`—defines the color to use.
- `'none'`—specifies no color, which makes nonfilled markers invisible.
- `'auto'`—sets `'MarkerEdgeColor'` to the same color and the `'EdgeColor'` property.

`'MarkerFaceColor' ColorSpec | 'none' | 'auto' | 'flat'`

Specifies the fill color for markers that are closed shapes:

`Colorspec`—defines the color to use.
`'none'`—makes the marker interior transparent.
`'auto'`—sets the fill color to the *axes* color.

`'MarkerSize' scalar`

Specifies the size of the marker in points. The default size is 6 points.

`'NormalMode' 'auto' | 'manual'`

Specifies whether MATLAB or user-provided vertex normal vectors are used.

`'Parent' Axes handle`

Parent *axes* object handle to *patch* object.

`'Selected' 'on' | 'off'`

Specifies whether object is selected. Selection handles or a bounding box is shown if the `'SelectionHighlight'` property is `'on'`.

`'SelectionHighlight' 'on' | 'off'`

Specifies whether selection handles or a bounding box are shown when `'Selected'` is `'on'`.

`'SpecularColorReflectance' Ø ≤ scalar ≤ 1`

Specifies the color of light reflected from the object. If set to 0, the reflected light is the color of the object and light source. If set to 1, the reflected light is that of the light source only.

`'SpecularExponent' scalar ≥ 1`

Specifies the size of the specular spot where light reflects. Most materials have exponents in the range of 5 to 20.

`'SpecularStrength' Ø ≤ scalar ≤ 1`

Specifies the intensity of the specular component of the light falling on the *patch*.

`'Tag' string`

Character string specified by the user to tag the object so that it is easily found using `findobj`.

`'Type' string`

Read-only property that returns the string `'patch'`.

`'UserData' variable`

Storage location for the contents of a variable of any class. Only accessible by using `get` and `set`.

`'VertexNormals' matrix`

If the `'NormalMode'` property is set to `'auto'`, `'VertexNormals'` contains the MATLAB generated normals used to perform lighting calculations. If `'NormalMode'` is set to `'manual'`, the user must supply this matrix.

`'Vertices' matrix`

Contains the *x*-, *y*-, and *z*-coordinates for each vertex. See the `'Faces'` property for more information.

`'Visible' 'on' | 'off'`

Determines whether the *patch* is visible. `get` and `set` still work on an invisible object.

`'XData' vector or matrix`

Contains the *x*-coordinates of the points at the vertices of the *patch.* If `'XData'` is a matrix, each column represents the *x*-coordinates of a single face of the *patch.*

`'YData' vector or matrix`

Contains the *y*-coordinates of the points at the vertices of the *patch.* If `'YData'` is a matrix, each column represents the *y*-coordinates of a single face of the *patch.*

`'ZData' vector or matrix`

Contains the *z*-coordinates of the points at the vertices of the *patch.* If `'ZData'` is a matrix, each column represents the *z*-coordinates of a single face of the *patch.*

# *Root* Object Properties

This appendix lists properties and associated values or their description for the *root* object. Default values are shown in *italics*. Unless stated otherwise, all properties can be set.

`'AutomaticFileUpdate'` *`'on'`* | `'off'`

Specifies whether file modification dates on M-files outside the Toolbox directory are compared against the date of the same file when it is compiled into memory. If `'on'`, the newer file version is compiled into memory. Not documented in MATLAB 5.0.

`'CallbackObject'`

Read-only property that returns the handle of the object whose callback routine is currently executing.

`'Children' vector of handles`

Returns handles of all nonhidden *figure* objects. The topmost *figure* appears first in the list. Changing the order of the handles and setting this property rearranges the stacking order.

`'CurrentFigure' figure handle`

Returns the handle of the current *figure* if it exists; otherwise returns an empty matrix. `set(0,'CurrentFigure',Hf)` makes the *figure* having handle `Hf` the current figure.

`'Diary' 'on' | 'off'`

Used by `diary` command to enable creation of diary file.

`'DiaryFile' string`

File name of diary file.

`'Echo' 'on' | 'off'`

Used by `echo` command to enable echoing of script files into *Command* window.

`'ErrorMessage' string`

Returns text of the last error message generated. Can be set to any string.

```
'Format' 'short' | 'shortE' | 'shortG' | 'long' |...
         'longE' | 'longG' | 'bank' | 'hex' | '+' | 'rat'
```

Used by `format` command to set *Command* window output format.

`'FormatSpacing' 'compact' | 'loose'`

Used by `format` command to set *Command* window output spacing.

`'PointerLocation' [x,y]`

Returns instantaneous *x*- and *y*-coordinates of mouse printer measured from the lower left-hand corner of the screen, in units specified by the `'Units'` property. Can be used to place the mouse pointer at a specific location.

`'PointerWindow'`

Read-only property that returns the handle of window containing the mouse pointer at any instant. If the mouse is not over a MATLAB window, 0 is returned.

`'Profile' 'on' | 'off'`

Enables or disables the MATLAB profiler. Used by the `profile` command.

`'ProfileFile' M-file name`

Full path name of M-file to profile.

`'ProfileCount' vector`

Returns an `n`-by-1 array, where `n` is the number of lines in the M-file being profiled. Each *i*th element contains the number of times MATLAB executed the *i*th line of the file.

`'ProfileInterval' scalar`

The sampling time interval at which MATLAB monitors M-file execution during profiling.

`'ScreenDepth'`

Read-only property that returns the numerical bit depth of the computer screen.

`'ScreenSize' [left,bottom,width,height]`

Read-only property that returns the screen size in standard position rectangle format. Units are defined by the `'Units'` property.

`'ShowHiddenHandles' 'on' | 'off'`

Globally enables or disables visibility of hidden object handles (i.e., objects whose `'HandleVisibility'` property has been set `'off'` or `'callback'`).

`'TerminalHideGraphCommand' string (X-Windows only)`

Specifies the escape sequence issued to hide the graph window on a terminal.

`'TerminalOneWindow' 'on' | 'off' (X-Windows only)`

Specifies whether there is only one window on your terminal.

`'TerminalDimensions' pixels (X-Windows only)`

Specifies the size of the default terminal.

```
'TerminalProtocol' 'none' | 'x' | 'tek401x' | 'tek410x'
                   (X-Windows only)
```

Specifies the type of terminal that MATLAB is communicating with.

`'TerminalShowGraphCommand' string (X-Windows only)`

Specifies the escape sequence used to bring the graphics window forward and hide the *Command* window.

`'Type'`

Read-only property that returns the string `'root'`.

```
'Units' 'pixels' | 'normalized' | 'inches' |...
        'centimeters' | 'points'
```

Specifies the unit of measurement for other *root* object properties.

`'UserData' variable`

Storage location for the contents of a variable of any class. Only accessible by using `get` and `set`.

I

# *Surface* Object Properties

This appendix lists properties and associated values or their description for the *surface* object. Default values are shown in *italics*. Unless stated otherwise, all properties can be set.

`'AmbientStrength'  0 ≤ scalar ≤ 1`

Specifies the ambient light strength of one or more visible *light* objects illuminating the *axes*.

`'BackFaceLighting'  'unlit' | 'lit' | 'reverselit'`

Specifies how faces are lit when their vertex normals point away from the camera.

`'BusyAction'  'cancel' | 'queue'`

Determines how to deal with interruptions by other object callbacks; that is, what action to take if a callback to this object is busy. If the `'Interruptible'` property of the *surface* object is `'off'`, `'cancel'` discards callbacks that interrupt the *surface* object, and `'queue'` places the interrupting callback in the event queue to be executed after the current callback terminates. If the `'Interruptible'` property of the object is `'on'`, `'BusyAction'` has no affect.

`'ButtonDownFcn'  string`

Specifies the character string to be evaluated whenever you press a mouse button with the pointer over the *surface* object.

`'CData'  matrix`

Specifies a matrix of colors at every point given in the `'ZData'` property. If the `'FaceColor'` property is set to `'texturemap'`, `'CData'` is mapped to conform to the *surface*. Both indexed and true color are supported.

`'CDataMapping'`  *`'scaled'`* `| 'direct'`

Determines how indexed color data are used to color the *surface* object.

`'Clipping'`  *`'on'`* `| 'off'`

Specifies whether the *surface* is to be clipped to the *axes* rectangle.

`'CreateFcn'  string`

Specifies the callback string evaluated just after the object is created. Must be set as a default property; for example, `set(0,'DefaultSurfaceCreateFcn','grid off')`.

`'DeleteFcn'  string`

Specifies the callback string evaluated just prior to deleting the object.

`'DiffuseStrength'  0 ≤ scalar ≤ 1`

Specifies the intensity of the diffuse light component falling on the *surface*.

`'EdgeColor'`  *`ColorSpec`* `| 'none' | 'flat' | 'interp'`

Specifies how color is applied to the edges of individual faces that make up the *surface:*

- `Colorspec`—a three-element RGB vector or the string name of a standard color. The default is black.
- `'none'`—the edges are not drawn.
- `'flat'`—the color of the first vertex for a face controls the color of each edge.
- `'interp'`—interpolated coloring is used.

```
'EdgeLighting' 'none' | 'flat' | 'gouraud' | 'phong'
```

Specifies the algorithm used to calculate the effect of light objects on *surface* face edges.

```
'EraseMode' 'normal' | 'none' | 'xor' | 'background'
```

Specifies what procedure MATLAB uses to draw the object:

- `'normal'`—redraw the display, performing all tasks required to ensure that all *axes* objects are rendered correctly.
- `'none'`—do not redraw anything.
- `'xor'`—draw and erase the object by exclusive-oring (XOR) it with the color underneath it. Other covered objects are not destroyed.
- `'background'`—erase the object by redrawing it in the *axes* background color. Other covered objects are destroyed.

`'none'`, `'xor'`, and `'background'` are faster but less accurate than `'normal'` and are therefore useful when animation or object movement with the mouse is desired.

```
'FaceColor' ColorSpec | 'none' | 'flat' |...
            'interp' | 'texturemap'
```

Specifies how color is applied to the *surface:*

- `Colorspec`—a three-element RGB vector or the string name of a standard color.
- `'none'`—the faces are not drawn.
- `'flat'`—the color of the first vertex controls the color of each face.
- `'interp'`—bilinear interpolation is used.
- `'texturemap'`—texture map the `'CData'` to the surface.

```
'FaceLighting' 'none' | 'flat' | 'gouraud' | 'phong'
```

Specifies the algorithm used to calculate the effect of light objects on *surface* faces.

```
'HandleVisibility' 'on' | 'callback' | 'off'
```

Specifies whether the object's handle can be seen from the *Command* window or callbacks. When set to `'callback'`, the object's handle is visible from callbacks but not the *Command* window. When a handle is invisible it does not show up in a list of children, nor can it be found by the `findobj` function. However, if the handle is known, it can still be accessed by `get` and `set`.

'Interruptible' *'on'* | 'off'

Specifies whether *surface* object callbacks are interruptible by other callbacks.

'LineStyle' '-' |' --' | ':' | '-.' | 'none'

Specifies the linestyle used for the *surface* face edges.

'LineWidth' scalar

Specifies the line width of *surface* edges in points (1 point = 1⁄72 inch). The default is 0.5 points.

'Marker' (any standard marker character)

Specifies the marker symbol placed at the *surface* vertices.

'MarkerEdgeColor' ColorSpec | 'none' | *'auto'* | 'flat'

Specifies the color of the marker or edge color for filled markers:

Colorspec—defines the color to use.
'none'—specifies no color, which makes nonfilled markers invisible.
'auto'—sets 'MarkerEdgeColor' to the same color and the 'EdgeColor' property.

'MarkerFaceColor' ColorSpec | *'none'* | 'auto' | 'flat'

Specifies the fill color for markers that are closed shapes:

Colorspec—defines the color to use.
'none'—makes the marker interior transparent.
'auto'—sets the fill color to the *axes* color.

'MarkerSize' scalar

Specifies the size of the marker in points. The default size is 6 points.

'MeshStyle' *'both'* | 'row' | 'column'

Specifies whether to draw row, column, or both row and column edge lines.

`'NormalMode'` `'auto'` | `'manual'`

Specifies whether MATLAB or user-provided vertex normal vectors are used.

`'Parent'` Axes handle

Parent *axes* object handle to *surface* object.

`'Selected'` `'on'` | `'off'`

Specifies whether object is selected. Selection bounding box is shown if the `'SelectionHighlight'` property is `'on'`.

`'SelectionHighlight'` `'on'` | `'off'`

Specifies whether selection bounding box is shown when `'Selected'` is `'on'`.

`'SpecularColorReflectance'` 0 ≤ scalar ≤ 1

Specifies the color of light reflected from the object. If set to 0, the reflected light is the color of the object and light source. If set to 1, the reflected light is that of the light source only.

`'SpecularExponent'` scalar ≥ 1

Specifies the size of the specular spot where light reflects. Most materials have exponents in the range of 5 to 20.

`'SpecularStrength'` 0 ≤ scalar ≤ 1

Specifies the intensity of the specular component of the light falling on the *surface.*

`'Tag'` string

Character string specified by the user to tag the object so that it is easily found using `findobj`.

`'Type'` string

Read-only property that returns the string `'surface'`.

`'UserData'` variable

Storage location for the contents of a variable of any class. Only accessible by using `get` and `set`.

`'VertexNormals' matrix`

If the `'NormalMode'` property is set to `'auto'`, `'VertexNormals'` contains the MATLAB generated normals used to perform lighting calculations. If `'NormalMode'` is set to `'manual'`, the user must supply this matrix.

`'Visible'` *`'on'`* `| 'off'`

Determines whether the *surface* object is visible. `get` and `set` still work on an invisible object.

`'XData' vector or matrix`

Specifies *x*-coordinates of the *surface* points. If `'XData'` is a vector, it is converted to a column and is replicated until it has the same number of columns as the data in the `'ZData'` property.

`'YData' vector or matrix`

Specifies the *y*-coordinates of the surface points. If `'YData'` is a vector, it is converted to a row and replicated until it has the same number of rows as the data in the `'ZData'` property.

`'ZData' matrix`

Specifies the *z*-coordinates of the surface points.

J

# *Text* Object Properties

This appendix lists properties and associated values or their description for the *text* object. Default values are shown in *italics*. Unless stated otherwise, all properties can be set.

`'BusyAction' 'cancel' | 'queue'`

Determines how to deal with interruptions by other object callbacks; that is, what action to take if a callback to this object is busy. If the `'Interruptible'` property of the *text* object is `'off'`, `'cancel'` discards callbacks that interrupt the *text* object, and `'queue'` places the interrupting callback in the event queue to be executed after the current callback terminates. If the `'Interruptible'` property of the object is `'on'`, `'BusyAction'` has no affect.

`'ButtonDownFcn' string`

Specifies the character string to be evaluated whenever you press a mouse button with the pointer over the *text* object.

`'Clipping' 'on' | 'off'`

Specifies whether text is to be clipped to the *axes* rectangle.

`'Color' ColorSpec`

Specifies the color of the *text* object. `Colorspec` is a three-element RGB vector or the string name of a standard color.

`'CreateFcn' string`

Specifies the callback string evaluated just after the object is created. Must be set as a default property; for example, `set(0,'DefaultTextCreateFcn','grid off')`.

`'DeleteFcn' string`

Specifies the callback string evaluated just prior to deleting the object.

`'EraseMode' 'normal' | 'none' | 'xor' | 'background'`

Specifies what procedure MATLAB uses to draw the object:

`'normal'`—redraw the display, performing all tasks required to ensure that all *axes* objects are rendered correctly.

`'none'`—do not redraw anything.

`'xor'`—draw and erase the object by exclusive-oring (XOR) it with the color underneath it. Other covered objects are not destroyed.

`'background'`—erase the object by redrawing it in the *axes* background color. Other covered objects are destroyed.

`'none'`, `'xor'`, and `'background'` are faster but less accurate than `'normal'` and are therefore useful when animation or object movement with the mouse is desired.

`'Editing' 'on' | 'off'`

Specifies whether the *text* object can be edited in place. Can be applied to a single *text* object at a time. When set to `'on'`, the object appears in an editable box with an I-beam cursor. Clicking in the *figure* window outside the object terminates editing and sets `'Editing'` to `'off'`. Alternatively, editing can be set `'off'`. Undocumented in MATLAB version 5.0.

`'Extent' [left,bottom,width,height]`

Read-only property that returns the size and position of the text string in standard position rectangle form using units specified by the `'Units'` property.

`'FontAngle'` *`'normal'`* `| 'italic' | 'oblique'`

Specifies the character slant to use.

`'FontName' string`

Specifies the font family to use. Helvetica is default.

`'FontSize' number`

Specifies the font size in units specified by the `'FontUnits'` property. 10 points is default.

```
'FontUnits' 'points' | 'normalized' | 'inches' |...
            'centimeters' | 'pixels'
```

Specifies the units for font size.

`'FontWeight' 'light' |` *`'normal'`* `| 'demi' | 'bold'`

Specifies the weight of the text.

`'HandleVisibility'` *`'on'`* `| 'callback' | 'off'`

Specifies whether the object's handle can be seen from the *Command* window or callbacks. When set to `'callback'`, the object's handle is visible from callbacks but not the *Command* window. When a handle is invisible it does not show up in a list of children, nor can it be found by the `findobj` function. However, if the handle is known, it can still be accessed by `get` and `set`.

`'HorizontalAlignment'` *`'left'`* `| 'center' | 'right'`

Specifies the horizontal justification of the text with respect to its `'Position'` property.

`'Interpreter'` *`'tex'`* `| 'none'`

Specifies whether the text should be parsed for LaTeX instructions.

`'Interruptible' 'on' | 'off'`

Specifies whether *text* object callbacks are interruptible by other callbacks.

`'Parent' Axes handle`

Parent *axes* object handle to *text* object.

`'Position' [x,y] or [x,y,z]`

Vector containing the *x*-, *y*-, and optional *z*-coordinates of the *text* object in units specified by the `'Units'` property. The default *z*-coordinate is zero.

`'Rotation' scalar`

Specifies text angular orientation in degrees. Compass East is 0 degrees. Compass North is 90 degrees.

`'Selected' 'on' | 'off'`

Specifies whether object is selected. Selection handles are shown if the `'SelectionHighlight'` property is `'on'`.

`'SelectionHighlight' 'on' | 'off'`

Specifies whether selection handles are shown when `'Selected'` is `'on'`.

`'String' string`

Specifies the text string to be displayed. `string` can be a single string, a string matrix, or a cell array of strings.

`'Tag' string`

Character string specified by the user to tag the object so that it is easily found using `findobj`.

`'Type' string`

Read-only property that returns the string `'text'`.

`'Units'` `'pixels'` | `'normalized'` | `'inches'` |...
`'centimeters'` | `'points'` | *`'data'`*

Specifies the units of measurement for *text* `'Extent'` and `'Position'` properties. All units except `'data'` are with respect to the lower left-hand corner of the *axes* plotbox. `'data'` implies use of the data coordinates of the underlying *axes.*

`'UserData'` `variable`

Storage location for the contents of a variable of any class. Only accessible by using `get` and `set`.

`'VerticalAlignment'` `'top'` | `'cap'` | *`'middle'`* | ...
`'baseline'` | `'bottom'`

Specifies the vertical justification of the text with respect to its `'Position'` property.

`'Visible'` *`'on'`* | `'off'`

Determines whether the *text* object is visible. `get` and `set` still work on an invisible object.

K

# *Uicontrol* Object Properties

This appendix lists properties and associated values or their description for the *uicontrol* object. Default values are shown in *italics*. Unless stated otherwise, all properties can be set.

`'BackgroundColor' ColorSpec`

Specifies the background color used to fill the object rectangle. `Colorspec` is a three-element RGB vector or the string name of a standard color. The default color is light gray.

`'BusyAction' 'cancel' | `*`'queue'`*

Determines how to deal with interruptions by other object callbacks; that is, what action to take if a callback to this object is busy. If the `'Interruptible'` property of the *uicontrol* object is `'off'`, `'cancel'` discards callbacks that interrupt the *uicontrol* object, and `'queue'` places the interrupting callback in the event queue to be executed after the current callback terminates. If the `'Interruptible'` property of the object is `'on'`, `'BusyAction'` has no affect.

`'ButtonDownFcn' string`

Specifies the character string to be evaluated whenever you press a mouse button while the pointer is within a 5-pixel-wide border around the object, but not within the object as long as `'Enable'` is set to `'on'`. The callback is meant to be evaluated when a mouse button is pressed while the pointer is within the *uicontrol* object and the object's `'Enable'` property is set to `'inactive'` or `'off'`. The *uicontrol* object's `'Callback'` property specifies the string to be evaluated when `'Enable'` is `'on'`.

`'Callback' string`

Specifies the character string to be evaluated whenever you activate the *uicontrol* object; for example, push a `'pushbutton'` or drag a `'slider'`, etc. `'frame'` and static `'text'` *uicontrol* styles do not invoke callbacks.

`'CreateFcn' string`

Specifies the callback string evaluated just after the object is created. Must be set as a default property; for example, `set(Ø,'DefaultUicontrolCreateFcn', 'disp(''Testing, 1, 2, 3.'')')`.

`'DeleteFcn' string`

Specifies the callback string evaluated just prior to deleting the object.

`'Enable' 'on' | 'inactive' | 'off'`

Specifies whether the *uicontrol* object is enabled. When `'off'`, the *uicontrol* label string is dimmed and is not functional. When `'inactive'`, the *uicontrol* is not functional and is not dimmed, but the `'ButtonDownFcn'` property is functional.

`'Extent' [Ø,Ø,width,height]`

Read-only property that returns the size of the text string used to label the *uicontrol* in standard position rectangle form using units specified by the `'Units'` property. Useful to determine the *uicontrol* size required to hold the desired label string.

`'FontAngle' 'normal' | 'italic' | 'oblique'`

Specifies the character slant to use.

`'FontName' string`

Specifies the font family to use. The default font is platform dependent.

`'FontSize' number`

Specifies the font size in units specified by the `'FontUnits'` property. The default size is platform dependent.

```
'FontUnits' 'points' | 'normalized' | 'inches' | '...
            'centimeters' | 'pixels'
```

Specifies the units for font size.

`'FontWeight' 'light' | 'normal' | 'demi' | 'bold'`

Specifies the weight of the text string.

`'ForegroundColor' ColorSpec`

Specifies the color of the text labeling the *uicontrol.* `Colorspec` is a three-element RGB vector or the string name of a standard color. The default color is black.

`'HandleVisibility' 'on' | 'callback' | 'off'`

Specifies whether the object's handle can be seen from the *Command* window or callbacks. When set to `'callback'`, the object's handle is visible from callbacks but not the *Command* window. When handle is invisible it does not show up in a list of children, nor can it be found by the `findobj` function. However, if the handle is known, it can still be accessed by `get` and `set`.

`'HorizontalAlignment' 'left' | 'center' | 'right'`

Specifies the horizontal justification of the *uicontrol* label string with respect to its `'Position'` property. On PCs and the Macintosh, this property affects only `'edit'` and `'text'` style *uicontrols.*

`'Interruptible' 'on' | 'off'`

Specifies whether *uicontrol* object callbacks are interruptible by other callbacks.

`'ListboxTop' scalar`

For `'listbox'` style *uicontrols,* specifies the index of topmost string displayed in the list box.

`'Max' scalar`

For `'radiobutton'` and `'checkbox'` style *uicontrols,* `'Max'` is the contents of the `'Value'` when the *uicontrols* is `'on'`. For `'slider'` style *uicontrols,* `'Max'` is the largest value you can select and it must be greater than the value specified by the `'Min'` property. The default value is 1. For `'edit'` style *uicontrols,* if `'Max'-'Min'` > 1, the `'edit'` box accepts multiline input strings. If `'Max'-'Min'` ≤ 1, only a single line is accepted. For `'listbox'` style *uicontrols,* if `'Max'-'Min'` > 1, multiline selection is allowed. If `'Max'-'Min'` ≤ 1, only a single line selection is allowed. For `'popupmenu'` *uicontrols,* on X-Window systems only, `'Max'` is set to the maximum index value. `'frame'` and `'text'` style *uicontrols* do not use this property.

`'Min' scalar`

For `'radiobutton'` and `'checkbox'` style *uicontrols,* `'Min'` is the contents of the `'Value'` when the *uicontrol* is `'off'`. For `'slider'` style *uicontrols,* `'Min'` is the smallest value you can select and must be less than the value specified by the `'Max'` property. The default value is 0. For `'edit'` style *uicontrols,* if `'Max'-'Min'` > 1, the `'edit'` box accepts multiline input strings. If `'Max'-'Min'` ≤ 1, only a single line is accepted. For `'listbox'` style *uicontrols,* if `'Max'-'Min'` > 1, multiline selection is allowed. If `'Max'-'Min'` ≤ 1, only a single line selection is allowed. For `'popupmenu'` *uicontrols,* on X-Window systems only, `'Min'` is set to the minimum index value (1). `'frame'` and `'text'` style *uicontrols* do not use this property.

`'Parent' figure handle`

Parent *figure* object handle to *uicontrol* object.

`'Position' [left,bottom,width,height]`

Specifies the *uicontrol* position in standard position rectangle format with respect to the lower left-hand corner of the *figure* window. Units are defined by the `'Units'` property.

`'Selected' 'on' | *'off'*`

Specifies whether object is selected. Selection handles are shown if the `'SelectionHighlight'` property is `'on'`.

`'SelectionHighlight' *'on'* | 'off'`

Specifies whether selection handles are shown when `'Selected'` is `'on'`.

`'SliderStep' [arrow_step trough_step]`

Specifies the normalized amount, 0 to 1, of the `'Max'`-`'Min'` distance used as a slider step size. `arrow_step` is the slider movement made for clicks on the arrows. `trough_step` is the slider movement made for clicks in the trough.

`'String' string`

Specifies *uicontrol* label string displayed on push buttons, radio buttons, check boxes, static text, editable text, listboxes, and popup menus. Multiple items in a popup menu or listbox can be specified as a cell array of strings, a string matrix, or within a string vector separated by vertical slash | characters. Line breaks in multiple-line editable text or static text controls occur between each row of a string matrix, each cell of a cell array of strings, and after any `\n` characters embedded in the string. Vertical slash characters are not interpreted as line breaks.

```
'Style' 'pushbutton' | 'radiobutton' | 'checkbox' |...
        'edit' | 'text' | 'slider' |...
        'frame' | 'listbox' | 'popupmenu'
```

Specifies the *uicontrol* object style to create.

`'Tag' string`

Character string specified by the user to tag the object so that it is easily found using `findobj`.

`'Type' string`

Read only property that returns the string `'uicontrol'`.

```
'Units' 'pixels' | 'normalized' | 'inches' |...
        'centimeters' | 'points' | 'data'
```

Specifies the units of measurement for *uicontrol* object properties.

`'UserData' variable`

Storage location for the contents of a variable of any class. Only accessible by using `get` and `set`.

`'Value' scalar or vector`

Contains the current value of the *uicontrol:*

`'Radiobutton'` and `'Checkbox'`—`'Max'` when `'on'`, `'Min'` when `'off'`.
`'Slider'`—the number representing slider position.
`'PopUpMenu'`—index of item selected.
`'ListBox'`—vector of indices of items selected.

Other *uicontrol* styles do not set this property.

`'Visible'` *`'on'`* `| 'off'`

Determines whether the *uicontrol* object is visible. `get` and `set` still work on an invisible object.

L

# *Uimenu* Object Properties

This appendix lists properties and associated values or their description for the *uimenu* object. Default values are shown in *italics*. Unless stated otherwise, all properties can be set.

`'Accelerator' character`

Specifies keyboard equivalent for the menu item. On X-Windows and PC platforms the sequence for issuing a keyboard equivalent is **Control**—`character`. On the Macintosh the keyboard equivalent sequence is **Command**—`character`. Accelerators work only for menu items that directly execute a callback routine, not items that bring up other menus.

`'BusyAction' 'cancel' | 'queue'`

Determines how to deal with interruptions by other object callbacks; that is, what action to take if a callback to this object is busy. If the `'Interruptible'` property of the *uimenu* object is `'off'`, `'cancel'` discards callbacks that interrupt the *uimenu* object, and `'queue'` places the interrupting callback in the event queue to be executed after the current callback terminates. If the `'Interruptible'` property of the object is `'on'`, `'BusyAction'` has no affect.

`'Callback' string`

Specifies the character string to be evaluated whenever you select the *uimenu* object. A menu with children (submenus) executes its callback routine before displaying the submenus. A menu without children executes its callback routine when you release the mouse button.

`'Checked' 'on' | 'off'`

Specifies whether a check mark appears next to the menu item. This property is not set automatically by selecting a menu item.

`'Children' vector of handles`

Vector of handles to nonhidden submenus, with the highest menu item appearing first. Menu order can be rearranged by rearranging this vector and resetting it.

`'CreateFcn' string`

Specifies the callback string evaluated just after the object is created. Must be set as a default property; for example, `set(0,'DefaultUimenuCreateFcn', 'disp(''Testing, 1, 2, 3.'')')`.

`'DeleteFcn' string`

Specifies the callback string evaluated just prior to deleting the object.

`'Enable' 'on' | 'off'`

Specified whether the *uimenu* object is enabled. When `'off'` the *uimenu* label is dimmed and it cannot be selected.

`'ForegroundColor' Colorspec (X-Windows only)`

Specifies the color of the *uimenu* label string. `Colorspec` is a three-element RGB vector or the string name of a standard color.

`'HandleVisibility' 'on' | 'callback' | 'off'`

Specifies whether the object's handle can be seen from the *Command* window or callbacks. When set to `'callback'`, the object's handle is visible from callbacks but not the *Command* window. When a handle is invisible it does not show up in a list of children, nor can it be found by the `findobj` function. However, if the handle is known, it can still be accessed by `get` and `set`.

`'Interruptible' 'on' | 'off'`

Specifies whether the *uimenu* object `'Callback'` property is interruptible by other callbacks.

`'Label' string`

Specifies the character string to place on the menu item. On X-Windows and PC platforms a & character forces the next character to appear underlined on the menu and allows it to be selected by pressing the underlined character. On the Macintosh, the & character is ignored.

`'Parent' handle`

Returns the *uimenu*'s parent, which is either a *figure* object or another *uimenu* object.

`'Position' scalar`

Specifies relative menu position on the menu bar or within a menu. Top-level menus are placed from left to right on the menu bar according to the value of their `'Position'` property, with 1 representing the leftmost position. Individual items within a given menu are placed from top to bottom according to the value of their `'Position'` property, with 1 representing the topmost position.

`'Separator' 'on' | 'off'`

Specifies the presence of a menu separator above the menu item.

`'Tag' string`

Character string specified by the user to tag the object so that it is found easily using `findobj`.

`'Type' string`

Read-only property that returns the string `'uimenu'`.

`'UserData' variable`

Storage location for the contents of a variable of any class. Only accessible by using `get` and `set`.

`'Visible' 'on' | 'off'`

Determines whether the effects of the *uimenu* object are visible. `get` and `set` still work on an invisible object.

# Index


+ (addition) 7
& (AND logical operator) 84
= (assignment operator) 82
@ (class directory prefix) 278
: (colon notation) 41, 59
, (comma) 10–11, 46, 124
% (comments) 10
... (continuation statements)11
.\ or ./ (division, element-by-element or dot-division) 49
\ or / (division, scalar or matrix) 7, 167
.' (dot-transpose operator) 45
{} (empty cell array) 125
[] (empty numeric array) 88, 171
== (equality operator) 82
.^ (exponentiation, element-by-element) 49–50
^ (exponentiation, scalar or matrix) 7, 169
> (greater-than operator) 82
>= (greater-than or equal-to operator) 82
< (less-than operator) 82
<= (less-than or equal-to operator) 82
.* (multiplication, element-by-element or dot multiplication) 48
* (multiplication, scalar or array) 7, 48, 166
~ (NOT logical operator) 84
~= (not-equal operator) 82
| (OR logical operator) 84
; (semicolon) 6, 10–11, 27, 44, 46, 124, 166
- (subtraction) 7
' (transpose operator) 45

abs 13, 16, 231, 245
acos 14
acosh 14
acot 14
acoth 14
acsc 14
acsch 15
addition (+) 7
addpath 37
airy 17
align 497
all 85, 186
and 291
AND logical operator (&) 84
angle 13, 16, 231
Anonymous FTP 522
ans 6, 8
any 85
area 317
argnames 156–157
Arithmetic Operations [table] 7
Array Addressing [table] 63
Array Construction Features [table] 44
array elements 40
array indexing 40
Array Operations, basic element-by-element [table] 51
Array Searching [table] 65
Array Size [table] 69
array subscripts 41
asec 15
asech 15
asin 15
asinh 15
assignin 150, 285
assignment operator (=) 82
atan 15
atan2 15
atanh 15
auread 397
autumn 367
auwrite 397
axes 412, 415–417, 443
axes box 303
*Axes* Object Properties (Appendix B) 571
axis 305–308, 332, 388
axis Features [table] 305–306
axlimdlg 509
azimuth 348

balance 169
bar 321–322
bar3 321–322
bar3h 321
barh 321–322
base workspace 7, 19, 150, 285
base2dec 96
besselh 17
besseli 17
besselj 17
besselk 17
bessely 17
beta 17
betainc 17
betaln 17
bic 174
bicstab 174
bin2dec 95
bitand 94
bitcmp 94–95
bitget 94–95
bitmax 94
bitset 94–95
bitshift 94
bitor 94
bitxor 94–95
bone 367
box 303, 332
break 140–141
breakpoints 161–162
brighten 371
btndown 496
btngroup 496
btnicon 496
btnpress 496
btnstate 496
btnup 496
bugs 159
builtin 283
Button Click 481
Button Press 481
Button Press Location and Callback Properties [table] 481
Button Release 482

caching the search path 152
calendar 117
Callback Functions 472
Callback Processing 483
Callback Properties, Selection Regions, and Stacking Order 480
callbacks 433, 439–442, 455–459
caller workspace 150, 285
capture 396, 407
cart2pol 16

cart2sph 16
case 141
cat 73, 389
caxis 371–374
cbedit 497
cd 36, 283, 487, 514
cdf2rdf 169
ceil 14, 16
cell 122, 277
cell arrays 121
cell arrays of strings 109–112
cell indexing 122
cell2struct 133
celldisp 123
cellplot 123
cellstr 111, 133, 163
cgs 174
char 98, 111, 133, 277, 294
Check Box Controls 461
child class 294–295
chol 169
cholinc 169, 174
cla 434, 495
clabel 355
class 95, 126, 133, 278–280, 282, 295
class directory prefix (@) 278
class, defined (OOP) 277
Classes, Built-in [table] 277–278
clc 23
clear 10, 20–21, 152
clf 311, 434, 495
clock 113
close 311, 439
closereq 439
clruprop 496
colmmd 174
colon 291
colon notation (:) 41, 59
colorbar 369
colorcube 367
colordef 38, 301–302, 366
colormap 345, 368, 371
Colormap Functions [table] 367
colperm 174
comet 316
comet3 360
comma (,) 10–11, 46, 124
command line editing 8
*Command* window 5, 8, 19, 23
Command Examples [table] 154
comments (%) 10
Common Object Properties [table]432–433
comp.soft-sys.matlab (newsgroup) 524
compan 171
compass 325
compiled functions 151
Complex Functions [table] 16
complex numbers 12, 325
computer 23
cond 169, 231
condest 169, 174
condeig 169
conj 16, 233
constructor 278–280
content addressing 122
Contents.m 153
continuation statements (...) 11
contour 257, 351, 355, 368
contour3 351, 355
contourf 353, 355
contrast 388
Control Placement and Font Considerations 469
conv 190
conv2 231
converter function 293
convhull 211
cool 367
Coordinate Transformation Functions [table]16–17
copper 367
Copy and Paste 407–408
copyobj 417, 442
corrcoef 183, 187
cos 15
cosh 15
cot 15
coth 15
cov 183, 187
cplxpair 16, 187, 231
cputime 118
Creating Different Control Types 460
Creating Menus and Submenus 449
cross 18
csc 15
csch 15
ctranspose 291
cumprod 187
cumsum 187
cumtrapz 187, 249
current figure 310
cursor keys 8

Data Analysis Functions [table]187–188
data visualization 365
date 114
datestr 114
datetick 118
datevec 116
dbclear 161
dbcont 161
dbdown 161
dblquad 247, 251–252
dbquit 161
dbstack 161
dbstatus 161
dbstep 161
dbstop 161
dbtype 161
dbup 161
deal 110, 127–128
deblank 101, 112, 163
Debugging Commands [table] 161
Debugging GUI M-files 478
dec2base 96
dec2bin 95
dec2hex 95
deconv 192
del2 187, 258, 379
delaunay 209, 346
delayed copy 61
delete (file) 33, 36
delete (object) 417, 439, 442
demo 446
det 169
device drivers 400–404
diag 53, 66
dialog 500
dialog Default Properties [table] 500
diary 31
diff 183, 186–187, 254–255
diffuse 385
dir 36, 514
Directory String Creation Functions [table] 36
Disabling Menu Items 454
Disk File Manipulation Commands [table] 36
disp 27, 100
display 281, 291
division, element-by-element or dot division (.\ or ./) 49
division, scalar or matrix (\ or /) 7
dlmread 34
dlmwrite 34
dmperm 174
doc 519
dot 18
dot division (.\ or ./) 49
dot multiplication (.*) 48
dot transpose (.') 45
double 98, 277, 389
dragrect 495
drawnow 315, 482–483
dsearch 211

echo 26–27, 29
Editable Text Box Controls 462
editing M-files 25–26

editpath 37
edtext 495
eig 169
eigs 169, 174
elevation 348
ellipj 18
ellipke 18
else 139–140
elseif 140
empty cell array ({}) 125
empty matrix ([]) 88, 171
empty numeric array ([]) 88, 171
end (as control flow block delimiter)136, 138–142
end (as array index) 61, 233
eomday 117
eps 8, 83
eq 291
equality operator (==) 82
erf 18, 323
erfc 18
erfcx 18
erfinv 18
error 145
errorbar 323
errordlg 502
etime 118
etree 174
etreeplot 174
Euler identity 13
eval 106, 155–156, 433, 439, 450, 471, 480
evalin 150, 284–285, 287
Event Queue 482
exist 33, 36, 515
exit (see quit)
exp 15
expint 18
expm 169
expm1 169
expm2 169
expm3 169
Exponential Functions [table] 15–16
exponentiation, element-by-element (.^) 49–50
exponentiation, scalar or matrix (^) 7
eye 52, 171
ezplot 327–328

factor 17
False (defined) 81
fclose 35
fcnchk 157
feather 325
feof 35
ferror 35
feval 155
fft 227–231, 233
fft2 227, 231
fftn 231
fftshift 227, 231
fgetl 35
fgets 35
fieldnames 131
fields 130
figure 310, 412, 443, 482–483
*Figure* Object Properties (Appendix C) 581
figure Paper Properties [table] 429
file identifier 35
file permission 35
File Utility Functions 514
filesep 36
fill 317–318, 343, 368
fill3 357, 359
filter 184, 231
filter2 231
find 64, 88, 160, 174, 389
findobj 425, 442
findstr 107–108
fix 14, 16
flag 367
flipdim 75
fliplr 65
flipud 65, 143–144
float 390
floor 14, 16
flops 9
fmin 240–241, 245
fmins 240, 242, 245
fopen 35
for 136
format 22
Format Conversion (for sprintf) [table] 104–105
formula 156
forward references 478
fplot 325–327
fprintf 35, 102
Frame Controls 466
frame matrix 395
frame2im 396
fread 35
Frequently Asked Questions (FAQ) 522
frewind 35
fscanf 35
fseek 35
ftell 35
full 174
fullfile 36
function 145, 147
function declaration line 145
function functions 245
functional form 33
funm 169
fwrite 35
fzero 238, 247

gallery 171
gamma 18
gammainc 18
gammaln 18
gca 412, 420–421, 423, 442
gcbf 495
gcbo 495
gcd 17
gcf 412, 420–422, 442
gco 420–421, 423–425, 438–439, 442, 457, 493
ge 291
get 413–420, 442
getfield 132
getframe 315, 395, 482–483
getptr 494
getuprop 496
ginput 314
global 149–150
global variables 475
gmres 174
gplot 174
gradient 187, 257, 379
graphic displays 447
gray 367
greater-than operator (>) 82
greater-than or equal-to operator (>=) 82
greymon 38
grid 302, 332, 337
griddata 211
gt 291
gtext 303
guide 497
GUIDE Tool Functions [table] 497

H1 line 145, 513
hadamard 171
Handle Graphics Functions [table] 442–443
hankel 171
help 511–513
help text 512–513
help topics [list] 512–513
helpdesk 519
helpdlg 501–502
helpwin 515–518
hess 169
hex2dec 95–96
hidden 338
hidegui 495
High-level Dialog Box Functions 501
hilb 171
hist 322, 325